Recording Secrets for the Small Studio

Recording Secrets for the Small Studio

Mike Senior

LONDON AND NEW YORK

First published 2015 by Focal Press

Published 2017 by Routledge
2 Park Square, Milton Park, Abingdon, Oxon OX14 4RN
711 Third Avenue, New York, NY 10017, USA

First issued in hardback 2017

Routledge is an imprint of the Taylor & Francis Group, an informa business

Library of Congress Cataloging in Publication Data
Application submitted

ISBN 13: 978-1-138-40645-2 (hbk)
ISBN 13: 978-0-415-71670-3 (pbk)
ISBN 13: 978-1-315-87950-5 (ebk)

Typeset in ITC Giovanni Std
By MPS Limited, Chennai, India, www.adi-mps.com

Printed and bound in Great Britain by
TJ International Ltd, Padstow, Cornwall

To my parents.

Bound to Create

You are a creator.

Whatever your form of expression — photography, filmmaking, animation, games, audio, media communication, web design, or theatre — you simply want to create without limitation. Bound by nothing except your own creativity and determination.

Focal Press can help.

For over 75 years Focal has published books that support your creative goals. Our founder, Andor Kraszna-Krausz, established Focal in 1938 so you could have access to leading-edge expert knowledge, techniques, and tools that allow you to create without constraint. We strive to create exceptional, engaging, and practical content that helps you master your passion.

Focal Press and you.

Bound to create.

We'd love to hear how we've helped you create. Share your experience:
www.focalpress.com/boundtocreate

Contents

Acknowledgments

This book hasn't been the easiest of births, and would not have been possible at all without the many people who have helped it on its way. First of all, I'd like to thank everyone who has emailed me about my writing, and those who have peppered me with questions at the various seminars, workshops, and conferences I've been involved with—thereby sowing the seeds for this book! I'm also very thankful for the forbearance of all the musicians who have helped me carry out the numerous practical tests which were vital in road-testing the advice offered here. Mike Zufall has also been instrumental in making this project possible by developing a content-management system for the Cambridge Music Technology web resource pages, which has saved me an inordinate amount of time fiddling with HTML. I must also express my gratitude to Bruce Lee, Dr. Elisabeth Wadge, and Rovio's *Angry Birds* franchise for much-needed motivation during the lengthy writing process!

In addition, I'd like to applaud all the interviewers who have done such an immense service to us all by shedding light on top-level studio practice: John Baccigaluppi, Matt Bell, Howard Bilerman, Bruce Borgerson, Richard Buskin, Bill Bruce, Rick Clark, Larry Crane, Eli Crews, Mark Cunningham, Dan Daley, Tom Doyle, Maureen Droney, Chris Eckman, Hans Dietrich Faulhaber, Eddi Fiegel, Tom Flint, Jimmy Foot, Matt Frost, David Greeves, Keith Hatschek, Jason Hiller, Nigel Humberstone, Sam Inglis, Blair Jackson, Michael Jackson, Mark Lewisohn, Dave Lockwood, Howard Massey, Chris Mayes-Wright, Alex McKenzie, Stephen Murray, Henry Owings, Bobby Owsinski, Joey Ramone, J. Robbins, Andrea Robinson, Anthony Savona, Jonathan Saxon, Sue Sillitoe, Dave Simons, Craig Smith, Roman Sokal, Philip Stevenson, Paul Tingen, Marsha Vdovin, Pete Weiss, and Peter Wetherbee.

The following engineers and producers have also generously taken it upon themselves to share their insight into the recording process, and I also offer them my sincere thanks: Chuck Ainlay, Steve Albini, Tom Dowd, Tony Faulkner, Oz Fritz, Jimmy Jam, Kevin Killen, Gerry Kitchingham, Daniel Lanois, Roger Nichols, John Merchant, Shep Pettibone, Jack Joseph Puig, Phil Ramone, LA Reid, Eric "Mixerman" Sarafin, Al Schmitt, Mike Stavrou, Ron Streicher, Bruce Swedien, Butch Vig, and Michael Wagener. Furthermore, many thanks are due to the numerous music-technology authors who have helped inform my own opinions on the subject of recording: Bruce Bartlett, Trevor De Clercq, Karl Coryat, Tim Crich, Joe Dochtermann, Wes Dooley, John Eargle, F. Alton Everest, Carlos Lellis Ferreira, Lynn Fuston, Nikolay Georgiev, Mike Gray, David Greeves, Christian Hugonnet, Sam Inglis, Jürg Jecklin, Mike Major, Jürgen Meyer, Mallory Nichols, Alec Nisbett, Harry F. Olsen, Bobby Owsinski,

Adrian Revill, Hugh Robjohns, Mike Ross-Trevor, Francis Rumsey, Eberhard Sengpiel, Günther Theile, Pierre Walder, Paul White, Michael Williams, Helmut Wittek, Chris Woolf, Wieslaw Woszczyk, and Jörg Wuttke.

The unenviable task of reading through early drafts of this text fell to Timo Carlier, Daniel Plappert, and Sam Inglis, and I am indebted to them all for their in-depth feedback, insightful suggestions, and continual encouragement. I would also like to express my appreciation to the Editorial Department of *Sound On Sound* magazine in general, who have been extremely supportive during this book's rather extended gestation. In preparing suitable images, I have been assisted enormously by Kim Campbell of Linn Records; Stefan Gienger at Munich's Mastermix studios; Matt Houghton at *Sound On Sound* magazine; Blake Lewis of Stewis Media; Neil Rogers of Cambridge's Half Ton Studios; Matthias Schaaff, Rainer Schwarz, and all at SAE Munich; and Chris Woolf of Microphone Data. I also extend my sincere thanks to all the staff at Focal Press whose patience and expertise have played such a large role in bringing this project to fruition, in particular Carlin Bowers, Emma Elder, Mary LaMacchia, and Anaïs Wheeler.

Above all, however, I'd like to thank my wonderful wife, Ute, for her unwavering love and support, as well as for taking on the worst of the bibliography and referencing tasks so graciously. I'd be lost without you, my love! And, finally, I'd like to thank my trainee limb models Lotte and Lara—I can scarcely believe what pros you're turning out to be!

Introduction

WHAT YOU'LL LEARN FROM THIS BOOK

How to achieve release-quality recordings on a budget within a typical small-studio environment, by applying power-user techniques from the world's most successful producers. In my work for *Sound On Sound* magazine I notice the same recording mistakes cropping up time and again, and while it's sometimes possible to salvage a usable mix from poor recordings, that way of working is incredibly long winded and laborious. This book is about helping you make better use of your time and energy, by leading you step-by-step through a series of in-depth practical exercises that demonstrate how to record sensibly in the first place.

WHAT YOU WON'T LEARN

If you want a "How To Record My Band In Three Easy Steps!" quick-fix guide, you'll have to look elsewhere. This book has a much broader remit: To teach you how to be a confident small-studio recording engineer. That means having the skills to record a full range of instruments and ensembles; understanding how different recording methodologies suit different artists and genres; and being able to adapt your approach to get the best out of low-budget gear and untreated acoustics. While I make every effort to fast-track the learning process within these pages, the art of recording is by nature tremendously complicated and full of nuanced decision-making, so this book is ultimately intended to reflect that reality rather than brushing all the subtleties under the carpet.

I'm also not going to show you how to operate any specific brand of studio gear—that's what equipment manuals are for! In essence, this book is primarily about recording technique, rather than recording equipment. The information here is deliberately "platform-neutral," so that you can use it with whatever studio hardware or software you happen to have access to.

BEFORE YOU START

Although I've done my best to make this book accessible to studio newbies, there is nonetheless some basic background knowledge that you'll need in order to get the best out of what I'll be writing about. In particular, I'm assuming that the reader:

- already understands something about the fundamental physics, measurement, and perception of sound;
- has some idea of the main stages involved in the multitrack production process;

- can identify the main functional components of hardware and software recording studios;
- knows how to set up a small stereo monitoring system.

Many modern musicians will already have absorbed this stuff without realizing it, just by coming into contact with other like-minded people and following the activities of their favorite commercial artists. However, if you feel you might benefit from a quick refresher on any of that, or you'd like to clarify my usage of some of the essential technical terms involved, then check out Appendix 1, where I've provided a super-condensed overview of this material.

Most people reading this book will, I imagine, have some recording system available to them, but if you don't have any studio gear at all, then the web resources page at the end of Appendix 1 suggests a few cost-effective entry-level systems to get you off the starting blocks.

HOW TO USE THIS BOOK

Because this book has been specifically designed as a step-by-step primer, you'll get best results if you work through it from beginning to end. Later sections rely on material covered in earlier chapters, so some aspects of the discussion may not make the greatest sense if you just dip in and out. The complexity of the subject matter also increases progressively throughout the book, so later chapters may suddenly feel rather daunting if you skip ahead.

At the end of each chapter there is a "Cut to the Chase" box, which allows you to review a summary of each chapter's main "secrets" before proceeding further. Underneath it are one or two "Assignment" boxes, which suggest practical activities to help consolidate your understanding of each chapter, and these could also serve as coursework tasks within a more formal educational framework. Each of these assignments is deliberately limited in scope, so that it doesn't call on topics before they've been properly discussed, but these limitations naturally reduce as the chapters progress, and are removed completely in Chapter 11. Finally, the "Web Resources" box leads to a separate website containing an extensive selection of related links and audio files, all of which may be freely used for educational purposes.

This book is based on my own wide-ranging research into the studio practices of more than 200 world-famous engineers, drawing on more than 5 million words of first-hand interviews. The text therefore includes hundreds of quotes from these high-fliers. If you don't recognize someone's name, then look it up in Appendix 2 to get an idea of the most high-profile records they've worked on—you'll almost certainly have heard a few of those! If you'd like to read any quote in its original context (which I'd heartily recommend), then follow the little superscript number alongside it to Appendix 3, where there's full reference information for each one. Finally, if you have any further questions or feedback, feel free to email me on ms@cambridge-mt.com.

PART 1
One Source, No Mics

CHAPTER 1

Recording a Machine

Although much of the magic in many productions stems from a marriage of live performers and microphones, I'd like to start this primer by eliminating both these variables from the proceedings, so that we can first concentrate on the bedrock studio techniques you'll need for pretty much every recording job you attempt. As such, our first goal will be learning to record samples from playback machines that directly output an electrical signal—things like radios, TVs, CD players, and mobile devices. On the face of things, this might seem a rather unedifying prospect, but bear with me, because it's actually the quickest way to fast-track your basic technical skills. If you can learn to do this task right every time, then you'll avoid the embarrassment of elementary goofs and the frustration of unnecessary delays once musicians are in the room. In addition, this activity should iron out the most common small-studio configuration problems, thereby preparing your core recording system for the rigors of serious music sessions.

1.1 HOOKING THINGS UP

So what do you need to know to connect the outputs of such equipment to your recording system? First off, you should realize that plugging anything in can generate powerful signal spikes, which could easily damage your monitors and/or ears. Seeing as deafening yourself probably isn't the best way to begin a recording course, do ensure that you mute your speakers or headphones at the outset. The simplest way to do this is by silencing your system's monitoring outputs in some way: you could mute your mixer's master output; you could hit your monitor controller's Mute button or disengage its output selector switches; or you could just turn down the volume control on the amplifier feeding your speakers or headphones.

1.1.1 Connector Basics

Most modern playback devices and sound modules output analog signals at "line level," which is an alternating voltage roughly in the 1V ballpark. Getting

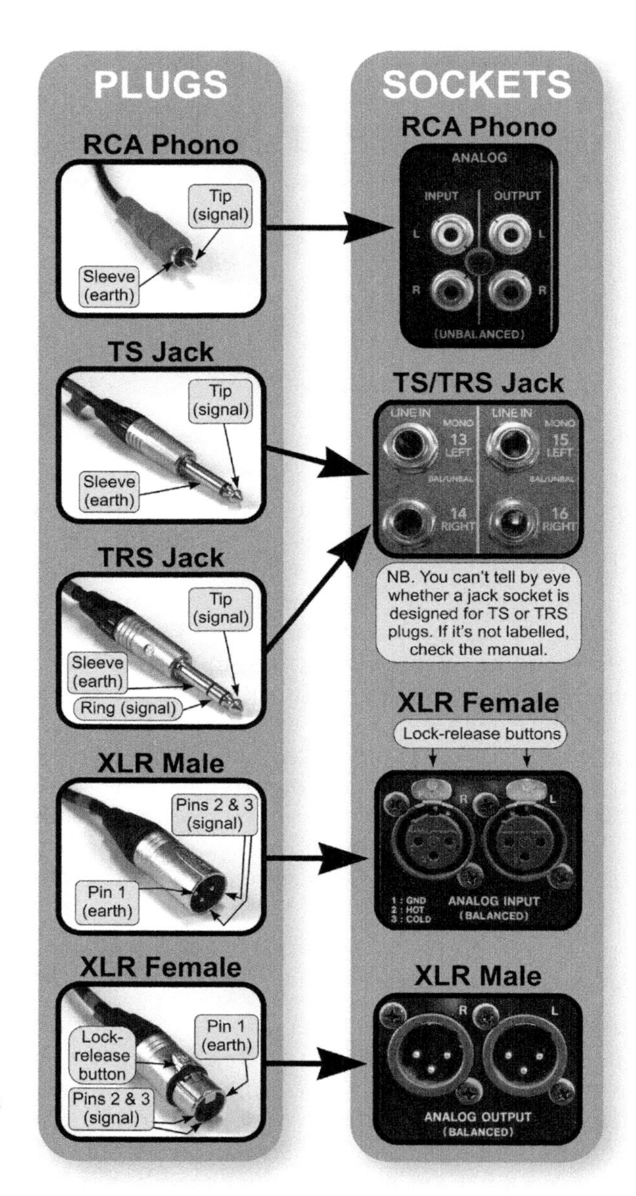

FIGURE 1.1
The most common plugs and sockets for making line-level analog audio connections.

this signal to the inputs of your recording system usually involves one of the hardware connectors illustrated in Figure 1.1.

Each of these provides one earth (or ground) conductor and at least one conductor for carrying audio signals. Where two audio conductors are provided (for example in XLRs and TRS jacks), both still typically carry only one audio signal, but in a so-called "balanced" configuration that better protects it against

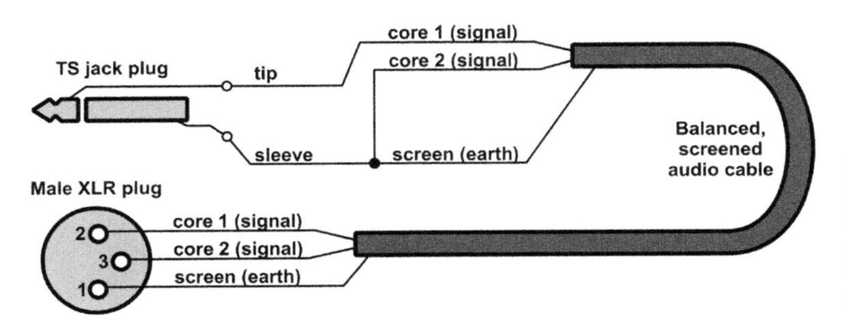

FIGURE 1.2
Schematic diagram for creating a lead to connect an unbalanced TS jack output to a balanced XLR input.

electrical interference *en route* to its destination. Connectors such as RCA phonos and TS jacks don't offer this additional defense, and their "unbalanced" signals are often transmitted at slightly weaker levels too, so you'll usually get a cleaner recording if you use balanced connections wherever possible. Most playback devices have stereo outputs, which means you'll need to make connections for both the left-channel and right-channel audio signals.

There are a few extra things to be aware of with XLRs:

- Because male and female XLR plugs interlock, it's possible to chain several cables into a single longer cable run.
- Many female XLR sockets/plugs have a locking design, so you won't be able to unplug them until you press the special lock-release buttons shown in Figure 1.1.
- Little alignment dimples on XLR plugs and sockets mean that the plugs will only insert fully when the internal pins are correctly oriented. This means you don't actually need to see an XLR socket to plug into it correctly—you can just twiddle the XLR plug between your fingers while holding it gently against the socket until you feel the dimple engage, whereupon the plug can be inserted fully.
- Some XLR inputs are able to feed a 48V "phantom" power supply back along the signal cable, which may damage line-level outputs, so make sure the phantom power is switched off.
- Occasionally you may come across a so-called "combi-jack/XLR" socket, which looks a lot like a normal XLR socket, but which can also accept a jack plug instead if you wish. There's nothing extra you have to know about using these inputs, though—just insert the plug as you'd normally expect, and away you go!

Recording an unbalanced signal through a balanced TRS jack input should be fine as long as you use an unbalanced TS jack cable for the connection. This effectively earths one of the two signal conductors inside the TRS jack socket so that it can accept the unbalanced signal. Balanced XLR inputs can also record unbalanced signals directly if you use a special cable such as the one shown in Figure 1.2.

FIGURE 1.3
Not all balanced TRS jack outputs will operate correctly with a TS jack plug inserted. If in doubt, it's safest to unbalance the signal via a transformer isolator box such as ART's T8 or Ebtech's Hum Eliminator.

Connecting balanced outputs to unbalanced inputs is a bit of a minefield, because the outcome depends on the design specifics of the output circuitry. Although modern equipment frequently has TRS jack output sockets that'll happily permit unbalanced operation if you insert a TS jack plug, you shouldn't assume you can do this unless it's clearly condoned on the unit's rear panel or in its user manual—some output circuits might go up in smoke! Rather than take any risks in this respect, I prefer to use a dedicated transformer-isolation box to unbalance the signal if in any doubt, especially now that companies such as ART and Ebtech offer respectable-quality transformer isolation for only around $30 (£20) per channel. A balanced output can be cabled directly to the transformer isolator's TRS/XLR input without disconnecting or short-circuiting any signal conductors in the process, and the gizmo's outputs will then safely accept TS jack and/or RCA phono plugs to feed your unbalanced recording inputs. Although transformers will inevitably alter the recorded tone a fraction, I wouldn't lose any sleep over this if you find your rig needs them—unbalanced inputs usually only appear on the cheapest recording hardware nowadays, in which case any minuscule transformer-related side-effects should be a long way down your list of quality-control concerns!

While normal TRS jacks mostly carry balanced signals in the studio, a half-size ("minijack") version of the same connector is frequently employed on consumer computer hardware for unbalanced stereo line outputs. In this scenario the two audio-signal conductors carry the left and right channels respectively, while the earth conductor is shared. Headphone outputs use TRS jack/minijack connections in a very similar way, so can be treated as line-level outputs for recording purposes. Functionally speaking, TRS jacks and TRS minijacks are identical, so it's easy to convert between them with cheap adaptor plugs, and there is also a wide variety of adaptors that will take a stereo signal on a TRS jack/minijack connector and split it out onto a pair of TS jacks or RCA phonos.

Occasionally it's possible to make a direct digital connection from a playback unit to a Digital Audio Workstation (DAW) system, bypassing the analog domain entirely. In that event you'll probably have to deal with one of the three digital transmission formats shown in Figure 1.5: coaxial S/PDIF (usually on unbalanced RCA phonos), optical S/PDIF (usually on fiber-optic Toslink connectors), or AES-EBU (usually on balanced XLRs). All of these formats transfer stereo audio, but you'll need a dedicated format-conversion gadget to convert between them if you're faced with incompatible socketry.

FIGURE 1.4
Where an unbalanced stereo output is provided on a TRS jack or TRS minijack connection, cheap adaptors are widely available to convert between these plug sizes (bottom right), or indeed to make the left-channel and right-channel signals available on separate mono connectors such as RCA phonos (bottom left and top).

1.1.2 Choosing Cables

Once you've muted the monitors, decided which audio plugs/sockets you're going to use, and placed your sound source within easy reach of your recording system, you're ready to hook things up. Even cheap-as-chips consumer hi-fi cables will carry line-level signals OK, but I'd definitely recommend using the more robust professional-style cables designed specifically for onstage/studio use. You shouldn't have to spend more than about $20 (£15) each for workmanlike specimens of up to 10 m in length, but do try to avoid anything with molded-on connectors, because they're almost impossible to repair if one of the internal solder joints fails—separate connectors can be resoldered. (Speaking of which, if you're a soldering Jedi then you can save yourself a packet by constructing your own cables in the first place.) Plastic-bodied connectors are also a false economy in my opinion, as the casings have a tendency to crack even at the best of times, and only get more brittle with age.

As long as a line-level audio cable has the appropriate plugs, it'll usually do the job, but there are a few exceptions that you need to be careful about if you're rooting through an unknown box of leads. Firstly, some TS jack cables are designed for connecting the high-level output of an amplifier to a speaker, and don't offer as much protection against external electrical interference. The best way to identify these speaker-level cables is to unscrew one of the TS jack plugs and have a look inside (see Figure 1.6): a line-level cable will usually have a single plastic-insulated core (for the audio signal) surrounded by an earthed wire-mesh "screen," whereas a speaker cable typically has two insulated cores of a thicker gauge without any mesh screening.

Cables designed for digital audio signals are also constructed differently from normal analog leads, but because they're usually clearly labeled by

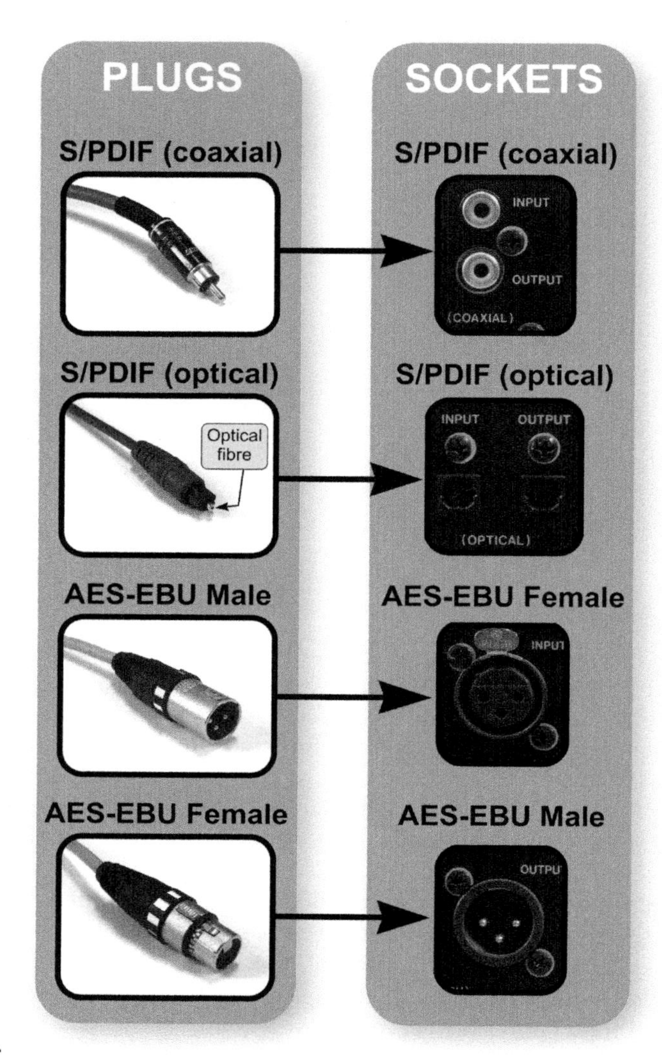

FIGURE 1.5
The most common plugs and sockets for stereo digital audio connections.

the manufacturer (and often noticeably less flexible when handled), you're unlikely to confuse them in practice. As it happens, you can sometimes get away with using analog RCA phono or XLR leads to transmit coaxial S/PDIF or AES-EBU digital signals over short distances without problems, but if you hate the thought of sacrificing session time to data glitches as much as I do, then spending $45 (£30) on a dedicated digital cable is a no-brainer. Another thing to realize when recording a digital source is that your recorder's sample rate must be synchronized to that of the incoming data, so you'll only get the audio coming through correctly once that's been done. The simplest means of doing this is to "slave" the recorder's sample rate to that embedded in the incoming digital data stream, but more complex systems are also possible (see "Digital Clocking & Jitter" in Section 1.4.4 for more details). On computer systems the synchronization options are usually located somewhere in the audio interface's driver settings—although somehow they always seem to be playing "hide and seek" whenever I need to find them!

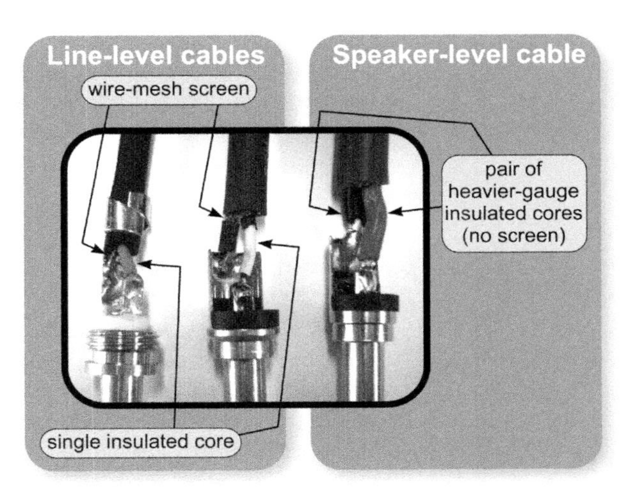

FIGURE 1.6
Here are the TS jack plugs from three different audio cables, with their casings unscrewed. Examining the internal conductors can help to identify which are suitable for line-level signals (left and center), and which for speaker-level signals (right).

1.1.3 Cable Layout

Physically arranging cables in the studio should be mostly common sense, but I've seen enough small-studio catastrophes over the years that I'm still inclined to proffer a few basic tips. The most frequent mistake is using leads that are too short, thereby suspending the cables in the air as first-class tripwires and stressing the internal solder joints and signal contacts of your sockets—not a good idea with budget studio equipment in particular, where the socketry may be pretty fragile to begin with.

It's also preferable to keep audio cables (particularly unbalanced ones) as far away as possible from your mains electrical wiring or any mains transformers, in order to minimize the amount of unwanted interference these may add to the signal you're trying to record. Mains transformers can be pernicious in this regard, because they're so often tucked away out of sight inside everyday electrical appliances and domestic lighting systems. Networking devices such as modems, routers, and hubs are also best given a wide berth if possible.

TWISTED CABLES

A classic newbie mistake is coiling up cables for storage in such a way that they get twisted. A twisted cable is a complete pain in the ass, because the twist's slight torque causes the cable to wrap around itself, so that it refuses to lie flat on your studio floor and generally creates an almighty tangle. Fortunately, it's easy to avoid this, either by using one of several special coiling techniques which neutralize the normal twisting action, or by storing your cables in loose folds or on a cable drum. (For more information on these methods, check out this chapter's web resources.)

Spare a thought for wear and tear, because although studio cables are pretty hardy, there's no sense in throwing money away on unnecessary repairs/replacements. Obviously, do your best not to tread on them, but also try to anticipate anything that might roll over them (wheeled flightcases, office chair casters, trolleys), crush them (doors, table legs, high heels), or melt them (lamps, radiators), and take evasive action. Special mention needs to be made of optical digital cables, however, because these are quite fragile, and won't even tolerate being bent beyond a certain angle without damaging the optic fiber, so keep the kid gloves on when handling those.

1.2 THE LINE-CHECK

Returning to our stated task (namely recording the output of a stereo playback device), in an ideal world you should now be able to start playback, make any necessary internal signal-routing assignments within your recording system, and see the appropriate meters on your recorder lighting up. Back on Planet Earth, however, it's actually very common for source signals to go AWOL, which is why "where's the signal?" is probably the most common setup problem in any studio. As such, one of the recording engineer's primary tasks before every session is to "line-check" the rig, ensuring that all signals are reaching their intended destinations. "There's nothing worse on a session, and nothing worse on an engineer than when things are just not working right," comments Al Schmitt,[1] echoing similar comments from Simon Climie[2] and Stephen Hague.[3] "It's like they say in the boy scouts," continues Schmitt: "Be prepared. Be on top of your game. Make sure everything's working."

1.2.1 Unity Gain

To make the line-checking process as straightforward as possible, you should first try to make sure nothing in the recording path is changing the level of your audio signal at all, which means setting every gain control that precedes the recorder track to its 0 dB or "unity gain" position. There can be hundreds of gain controls even in entry-level recording systems, manifesting themselves as knobs, switches, or faders labeled Gain, Volume, Level, Sensitivity, Pad, or Trim—or sometimes just calibrated in decibels. Digital gain controls usually default to their 0 dB setting, while analog controls often have their unity-gain position marked or detented, so the trickiest part of zeroing them is usually just tracking them all down! Computer audio interface drivers and DAW software mixers in particular often have little gain controls tucked away in nooks and crannies all over the place, and if just one of those has been pulled down inadvertently it can throw a spanner in the works. Some gain controls (such as +4 dBu/−10 dBV sensitivity switches) may offer no unity-gain position at all, while others (such as headphone volume knobs) may have no calibration—in either case start off with those at their lowest gain settings.

Don't forget to check that any Pan or Balance controls are centered too, because both of those are essentially stereo-ganged gain controls. Also, confirm that there are no Mute/Solo buttons engaged, and that any signal-processing facilities have been reset or removed from the circuit—particularly the channel EQ if you're recording via an analog mixer. Once you've set your whole recording chain to unity gain, do the same for the monitoring chain.

RECORDING FROM TURNTABLES

Despite being superseded in the mass market by digital playback formats, the vinyl record shows little sign of dying out, especially now that scratch DJs have elevated it to the status of musical instrument. Capturing the output of a turntable isn't any trickier than dealing with regular line-level sources, as long as you realize you can't just record the cartridge's raw signal—it'll require both amplification and heavy RIAA-standard equalization first. Fortunately, most DJ mixers and turntable-equipped hi-fis have built-in RIAA preamplification to handle all that, so the trick is to take your recording feed from the DJ mixer's master outputs or from the hi-fi amplifier's dedicated line-level recording outputs, rather than using the RCA phono sockets on the turntable itself. Also, be sure that the turntable itself is correctly earthed, otherwise the cartridge won't be properly shielded from unwanted electromagnetic interference. This may require a separate earth wire to be connected between dedicated binding posts on the turntable and preamplifier.

Now you can start the line-check. Level meters should be adequate for the purpose, so your monitors can remain muted for now. Set the playback machine going, turn up any output-level or headphone-volume control it has, and examine your recorder track's level meter. Don't worry too much about what the level actually is for the moment—just check whether there's something there! If you can see something on the readout, then stop the player and check that this reading also dies—a simple step which confirms that what you're metering is really the playback device, and not some other unwanted signal. Now restart the player, pull down the recorder track's monitor-channel fader, unmute your studio monitoring system, and carefully fade up the recorder track to confirm that what you're hearing tallies with what you're seeing. All of which should take no more than 20 seconds if all's well, after which you're ready to set recording levels. If nothing comes through, though, then it's time to troubleshoot.

1.2.2 Divide and Conquer!

The secret to line-check troubleshooting is to make like Daft Punk: Keep a cool head! Unless you work methodically, you'll waste masses of time going down blind alleys. When a signal isn't reaching the recorder track, your first call should be the recording system's input metering: in other words the very first meters the input signal hits after exiting the connection cable. These might be single LEDs next to the input sockets or something much more sophisticated within a computer soundcard's software control utility, but unfortunately they're not always that easy to find—the input-channel metering in most DAW software rarely fits the bill for line-checking, for example, because a signal can so easily be misdirected by the audio drivers prior to that. Whatever form the system's real input meters take, it's paramount that you find out where they are, because they allow you to speedily eliminate a whole section of the studio setup from your inquiries: if those input meters light up but you're still not hearing anything, then something's awry in your recording system; if they don't, then you need to scrutinize your sound source or its connections.

Armed with this information, try to slim down the variables further, as illustrated in Figure 1.7. For example, assuming a no-show on your recording system's input meters, try to find any readout on the player itself, or try plugging headphones directly into its headphone socket. If something's actually coming out of the player, then the finger of blame points at your cabling; whereas if the player's silent, then it's time to dust off its instruction manual or ring the repair shop. Alternatively, let's say audio is indeed arriving at your computer system's recording inputs, but not showing up at the recorder track. Try inspecting your DAW's input-channel metering: if you get a reading, then you can concentrate on tinkering with the routing and channel-assignment options

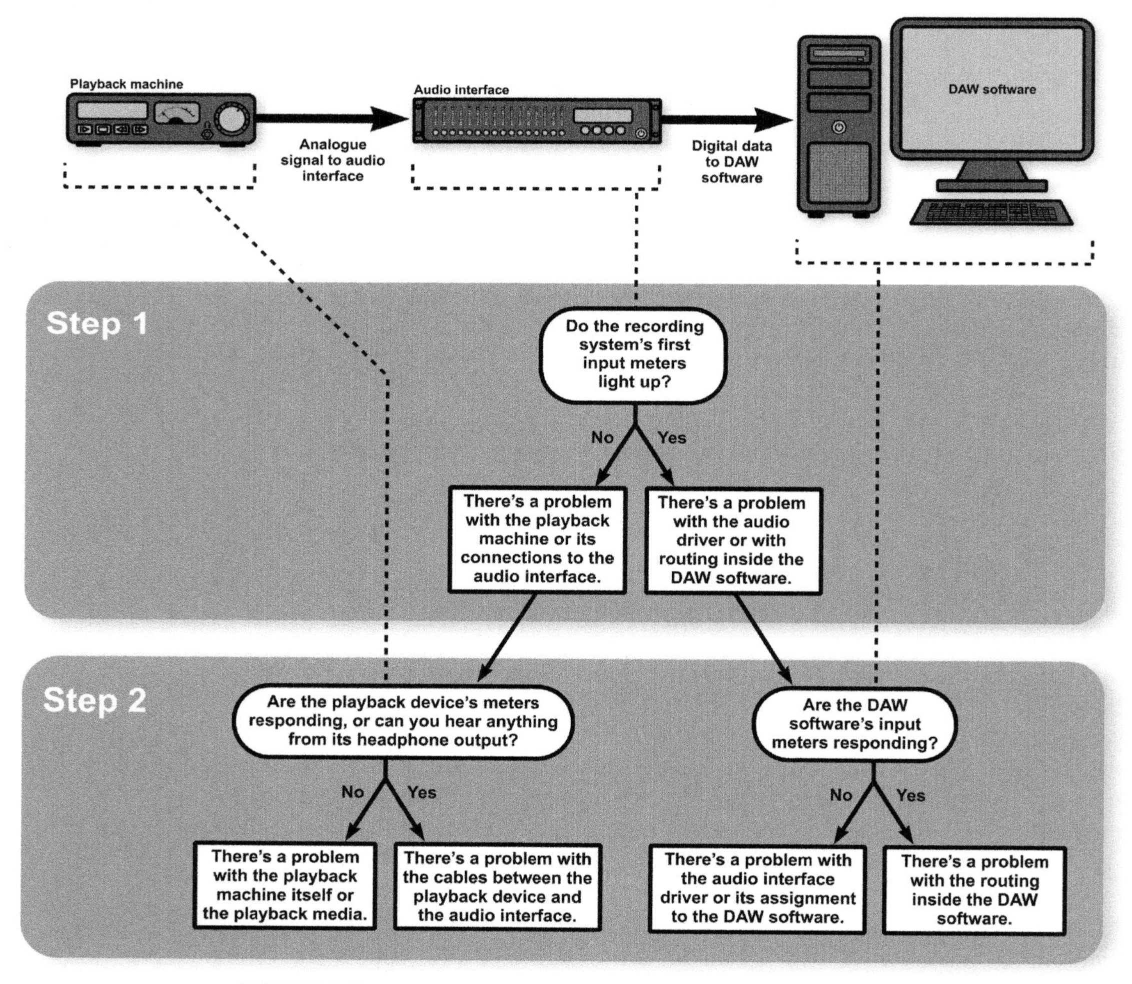

FIGURE 1.7
The "divide and conquer" troubleshooting approach in action. Step 1 divides the field of inquiry in half by using the recording system's very first input meters (those on the computer audio interface itself). Whichever section of the recording chain Step 1 implicates is then divided again in Step 2, narrowing the options further.

in your software mixer; if not, then the input signal has probably been misdirected somehow by your audio interface's configuration settings—or perhaps the DAW is just using the wrong driver.

This kind of "divide and conquer" approach is the cornerstone of effective troubleshooting, irrespective of the gear you're using, so try to get into the habit of thinking this way right from the outset. It might seem a bit pedantic when you're just recording your radio, but unless it becomes second nature you'll quickly flounder on more complicated sessions, especially when you've got frustrated musicians/clients breathing down your neck!

Another way to apply the divide-and-conquer principle is if you find that only one side of a stereo signal is coming through properly, as in Figure 1.8—for

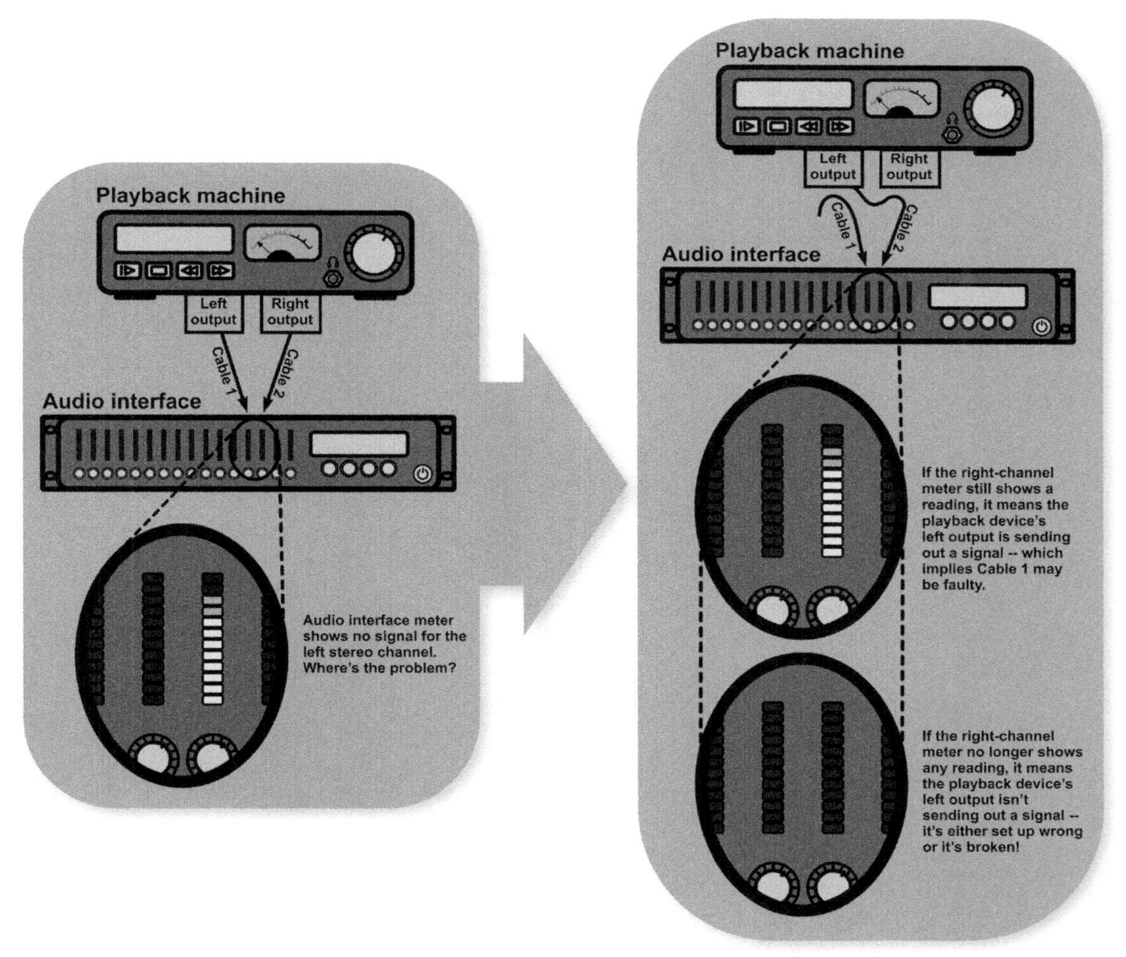

FIGURE 1.8
If one side of a stereo signal's coming through, but not the other, swapping over the cables can help diagnose the source of the problem.

the sake of discussion, let's say the right channel's arriving safely at your system's input meters, but the left channel's missing in action. First, confirm that the right-channel cable is correctly connected to the player by briefly removing the plug in the player's right-channel output socket—if all's well, the right-channel meter reading should disappear as you do this. Then plug the right-channel cable into the player's left-channel output socket instead, looking at your meters to see whether the signal reappears: If it does, you know that both of the player's outputs are working, so you should probably regard the left-channel cabling with suspicion; if it doesn't, then either the player's left-channel output is bust, or there's a stereo balance control set funny somewhere under the hood. Where both of the player's outputs are actually functioning correctly, your next move might be to restore the player's output-socket plugs to their original positions, and then swap the plugs feeding the relevant pair of recording-system inputs instead. If you still only get the right-channel input meter lighting up in that instance, then you've removed the cables from the equation—in other words, there's probably something wrong with the left-channel input itself.

Exactly the same troubleshooting techniques can be applied if you encounter problems getting signals out of the recording system for monitoring purposes. Again, you should take pains to identify which metering stage most reliably reflects what's appearing at the physical output sockets of your recording system, because this gives you concrete information about whether or not your monitoring hardware is at fault. The channel-swapping technique is also very handy if you're only hearing one side of the stereo picture, but with one important safety caveat: You should switch off all amplifiers (including headphone amps and powered speakers) before replugging any monitoring connections, otherwise you risk damage to your listening equipment and/or hearing.

1.3 SETTING LEVELS

You don't want just any signal reaching your recorder track, though—you want something that actually sounds good! The fundamental prerequisite for achieving this is that you set appropriate signal levels throughout your recording chain. For one thing, low-level noise will inevitably be added to your recording by any recording equipment you use, so you want to keep your audio signal level higher than this "noise floor"—the higher the better, in fact, in order to maximize the recording's signal-to-noise ratio. On the other hand, overloading or "clipping" your recording gear by feeding it levels that are too high will produce unwanted distortion, and the further you push the level beyond any unit's capabilities, the more audible this distortion will become. So in the normal run of things your aim is to keep signal levels high enough to minimize noise, but not so high that you trigger undesirable clipping distortion.

SAMPLE RATES AND BIT DEPTHS

The limit of CD-quality sound is largely determined by its standardized 44.1 kHz sample rate (which restricts the upper frequency-response limit to around 20 kHz) and 16-bit sampling resolution (which results in a noise floor at roughly –96 dBFS). Given that most commercial music is distributed in this form (or in a data-compressed file format directly derived from it), you should at the very least record at 44.1 kHz/16-bit if you're planning on releasing anything to the general public.

However, to make best use of the CD noise floor, it actually makes sense to record at a higher bit-depth, so that the digital noise floor of your recording won't rise above that of a CD even if you increase the levels of your recordings during mixing and mastering. For this reason I suggest working at 24-bit resolution instead, which drops the digital noise floor well below the noise floor of any other equipment in a typical small studio—at which point you can stop worrying about it! The downside of 24-bit audio is that it takes 50% more storage space than 16-bit, but nowadays this really isn't a big deal given the ridiculously low cost of digital storage. Some DAW platforms give you the option of recording at "32-bit floating point" as well, but I wouldn't waste further disk space on that—frankly, it's overkill for practical recording purposes, unless you're the kind of person who stores their CDs in the fridge to keep them fresh…

The choice of sample rate for recording is a more contentious issue. In addition to the CD-quality rate of 44.1 kHz, a 48 kHz rate has long been standard in the broadcast and film industries on account of its ease of synchronization with video equipment. To be honest, it matters very little which you choose for music work, although I marginally favor 44.1 kHz so that I don't have to convert the sample rate for CD mastering—a process that can have audible side-effects. However, in recent years manufacturers have begun offering higher rates as well, based on doubling and redoubling the 44.1 kHz and 48 kHz standards to 88.2 kHz/96 kHz, 176.4 kHz/192 kHz, and beyond, extending the upper limit of the captured frequency range well beyond the 20 Hz–20 kHz zone commonly regarded as the range of human hearing. The extent to which frequencies above 20 kHz do actually influence our listening experience is very much a moot point, but there are also well-understood technical reasons why elevated sample rates can actually sound better even below 20 kHz, and many professionals have already voted with their ears and wallets by moving to 88.2 kHz/96 kHz in particular.

On the face of it, this should be a strong incentive for ambitious small-studio owners to follow suit, but there are two big downsides to factor in. The first is that working at a doubled sample rate not only doubles the storage space you need, but it also doubles the strain on every data buss and digital signal processor in your entire studio, which frequently translates into fewer simultaneous record/playback tracks, fewer effects plug-ins at mixdown, and more complicated digital cabling. The second thing is that the difference in resolution brought about by the increase in sample rate demands equal resolution of your recording and monitoring hardware, and many lower-cost devices simply haven't been designed with frequencies above 20 kHz in mind.

As a result, although it's difficult to dispute that elevated sample rates make an audible difference to the audio quality, I wouldn't recommend for anyone on a budget to bother with them. In my opinion, the 44.1 kHz/48 kHz rates are more than a match for practically every small studio I've ever been into, so upgrading your entire studio rig to handle elevated sample rates will rarely ever be an efficient use of resources—especially when seen within the wider context of a music market increasingly reliant on MP3/AAC files and media-streaming technology, which fall well short of even CD fidelity.

This seems quite simple on the face of it, but there are a couple of complications in practice: Firstly, different stages in your recording chain may require different optimum signal levels; and, secondly, there are often so many meters and gain controls on hand that it's easy to get confused about which ones to use. Neither does it make things any easier that there's no "standard" recording rig these days, and that many small-studio users are frequently working with shared or borrowed gear that's unfamiliar. So in response to all this I'd like to explain a step-by-step procedure that I've found to be pretty foolproof for setting decent levels, no matter what studio setup you happen to be faced with.

1.3.1 Find the First Gain Stage

Firstly, try to locate the very first gain stage in your signal chain. To give a simple example, imagine that I'm using an unbalanced splitter cable to record the TRS minijack headphone output of a portable MP3 player into the TS jack inputs of one of those little all-in-one sampling workstations (perhaps one of Akai's iconic MPC range) as illustrated in Figure 1.9. My first gain stage there will probably be the MP3 player's headphone level control.

A more complicated setup is shown in Figure 1.10: a vinyl turntable being recorded via a DJ mixer and a studio mixing console to a stand-alone stereo analog-to-digital converter, which in turn feeds the digital inputs of a software DAW system. In this case the first gain stage will likely be the Input Gain knobs on the DJ mixer's turntable channels.

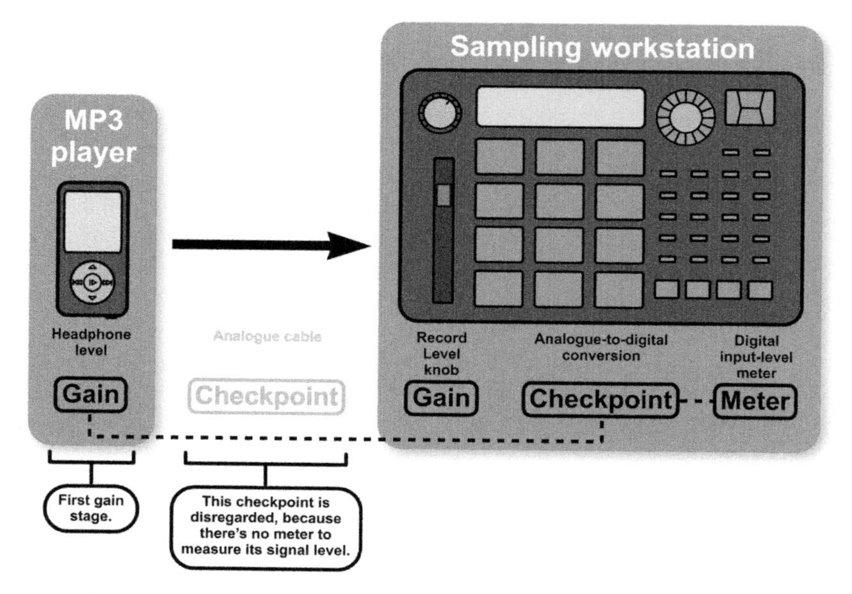

FIGURE 1.9
A very simple recording chain, from a portable MP3 player to a sampling workstation, showing the position of the gain controls, level checkpoints, and meters. The first checkpoint is disregarded because it has no meter to measure the signal level there. Setting a preliminary recording level using this setup would therefore involve adjusting the MP3 player's headphone level control to achieve a sensible reading on the sampling workstation's digital input meter (see dashed lines).

1.3.2 Identify the Important Meters

Next, narrow down which meters you need to concentrate on while setting your levels. The way to do this is to work your way through the signal chain, all the way from the first gain stage to the destination recorder track, looking for "checkpoints" where the signal:

- passes through an analog cable (I call this a "cable" checkpoint);
- is converted between the analog and digital domains (a "conversion" checkpoint).

Optimizing the signal level at a given checkpoint should pretty much guarantee that you'll steer clear of noise and distortion problems between that checkpoint and the previous one—provided that you were conscientious in setting unity gain throughout the signal path back in Section 1.2.1. The most suitable meter for assessing the level in each case will be:

- the closest meter before a cable checkpoint;
- the digital meter closest to a conversion checkpoint.

If you can't find a meter for a given checkpoint that satisfies these conditions before reaching another checkpoint (or the first gain stage), then it means that it can't reliably be measured, so strike it from your list and continue working through the remaining checkpoints as if it didn't exist. (Theoretically speaking, disregarding the signal level at any checkpoint raises the possibility of noise/distortion concerns, but in practice problems very rarely arise, because manufacturers of studio equipment tend to provide metering where it's required.) Bear in mind that some meters may be able to measure signals in a variety of signal-chain positions—for example, the main meters of a small analog mixing console will often display the control-room monitoring signal, allowing you to measure the level of different channels, groups, and returns just by hitting their Solo buttons. (Solo buttons may sometimes be labeled PFL (Pre-Fader Listen) or AFL (After-Fader Listen) to indicate where in the channel path the signal is being measured.)

Let's look at how these principles apply to our two setup examples. In Figure 1.9 there'd be two level checkpoints: the headphone splitter cable and the sampler's analog-to-digital converters. However, assuming there was no metering within the MP3 player (which is quite common), then there'd be no way of measuring the level prior to the first checkpoint, so I'd disregard that one and focus all my attention on the second checkpoint instead, relying on the sampling workstation's digital input meters. (In theory, by ignoring the first checkpoint I'd risk overloading the MP3 player's analog output circuitry if I turned the headphone level up too high. In practice, though, this would be very unlikely given that most MP3 players are designed to play back even the loudest digital files at maximum headphone volume without significant distortion—which is why they seldom require output metering!)

Turning to the more complicated setup in Figure 1.10, the checkpoints would be: the cables from the DJ mixer's main outputs, measured from the DJ mixer's

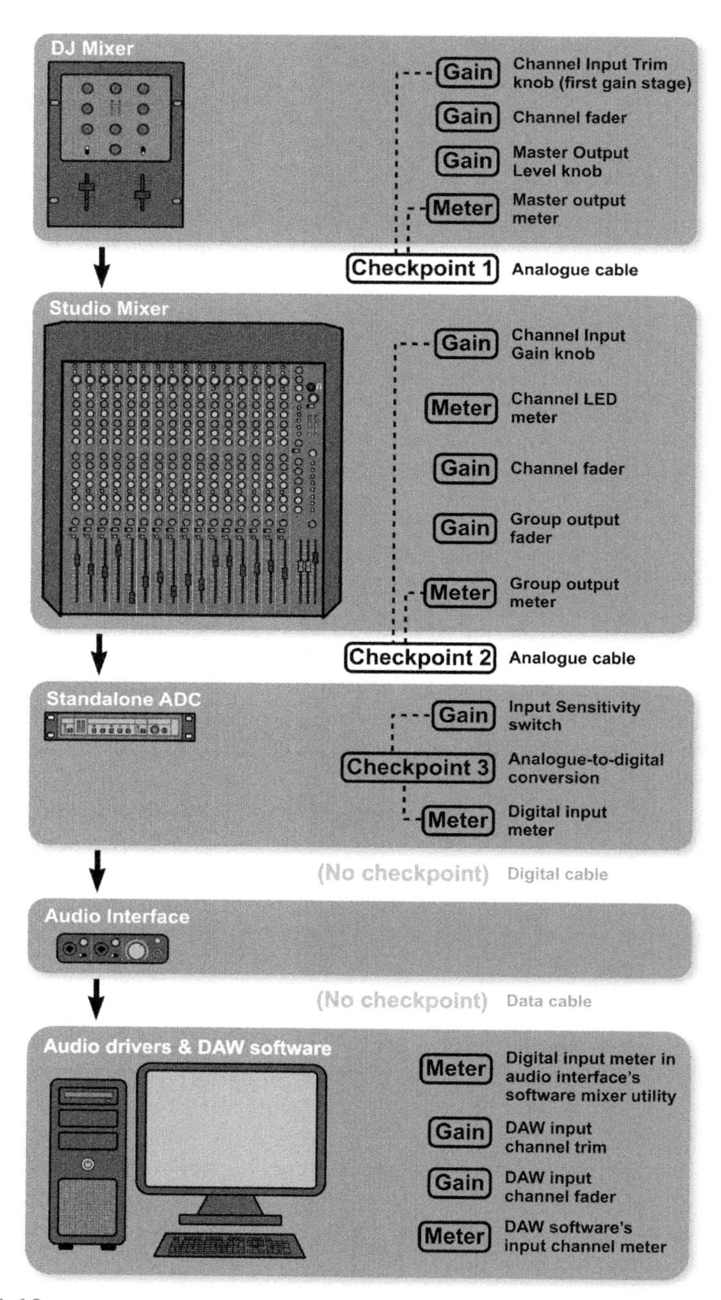

FIGURE 1.10

A more complicated recording chain, from a turntable, through a DJ mixer, a studio mixer, and a standalone ADC, and then into a computer recording system's audio interface. Again, the dashed lines indicate which gain control and which meter would be used for setting each checkpoint's level in the first instance—all other gain stages would remain at their unity setting unless the first-choice control's gain range was insufficient.

master output meters; the cables from the studio mixer's group outputs, measured from its group output meters; and the analog-to-digital conversion stage, measured from the digital meters most closely following it, i.e., those within the standalone converter unit itself. (Note that the cable between the standalone converter unit and the audio interface wouldn't qualify as a level checkpoint because it's not an analog connection.) I'd have to keep my eyes on the meters for all three checkpoints when setting levels, to be sure of capturing the cleanest signal.

DATA-COMPRESSED FILE FORMATS

Most affordable digital studio equipment now offers at least 44.1 kHz/24-bit recording resolution, but older gear and some more modern mobile devices may offer you the option of maximizing the recording time available by using a data-compressed file format such as MP3 or AAC. Despite the fact that the sound quality captured by such compression schemes can sometimes be almost indistinguishable from uncompressed ("linear" or "PCM") digital audio, I would personally advise against them, and here's why. The way most of these algorithms squeeze the data bandwidth is by reducing the recording resolution whenever the human hearing system wouldn't normally notice—contrary to what you might expect, we don't actually perceive everything in a musical signal all the time, and background details in particular are often rendered inaudible by strong foreground sounds. This drop in resolution would be fine if all you were planning to do with your recording was play it back (through one side of a pair of earbuds, natch!), but the moment you apply any mixdown processing to the audio, you'll find yourself unveiling previously inaudible (and unmusical) distortions introduced during the data-compression process.

1.3.3 Adjust Gain Through the System

Now make sure your monitors are muted and restart your playback device—it'll save time if you can shuttle to the loudest section for level-setting purposes. Adjust your first gain stage while watching the meter for the following level checkpoint. The reading you're aiming for will vary depending on the type of meter you're using, so here are some guidelines:

- **Digital Peak Meter.** This shows the instantaneous level of a digital signal, sample by sample. At the top of a digital meter is 0 dBFS, which is the highest level the system can capture before overloading. The basic level-setting tactic here is to make sure the signal is as hot as possible, but without ever hitting the top of the scale. (Digital distortion is one of the nastier-sounding varieties.) There's no need to be too finicky about things, though—if the loudest signal from the playback device registers within the top 6 dB of the meter's scale you're fine.
- **Volume Unit (VU) Meter.** This is normally used for analog signals, and doesn't respond nearly as quickly to fast-moving waveforms as a digital peak meter does. As such, it tends to favor lower frequencies in general, as well as seriously under-reading short-term level spikes (often called transients) that are responsible for both the percussive attack of drums and the note-onset definition of many other instruments such as acoustic guitar, piano, and tuned percussion. For this reason, the 0 dBVU "reference level" marking on such meters is usually designed to correspond to an electrical

level about 20 dB below the actual overload point of the surrounding analog circuitry. Therefore you can set your gain control to give a reading at the meter's reference level and you'll still have around 20 dB of "headroom" to accommodate the unmetered transient peaks without distortion. (If your VU meter eschews the dBVU scale in favor of dBu markings, then you can usually treat the +4 dBu level as 0 dBVU.)

- **Peak Program Meters (PPMs).** These meters are pretty uncommon in small studios, so you're only likely to come across them if you've scavenged some broadcast gear for your recording rig. Like VU meters, they're primarily intended for use with analog equipment, but they respond much better to transients. For this reason, you can set a recording level only about 10 dB below the analog overload point on a PPM without any real danger of distorting unmetered signal peaks. Irritatingly, though, there are several different labeling standards commonly used for PPMs, some of which use decibel markings, while others use an arbitrary numeric scale, so you'll need to work out roughly where the "10 dB below overload" point appears in your specific instance if you're going to use PPMs to judge your gain settings.
- **Uncalibrated Meters.** Metering is one area where manufacturers of budget equipment tend to cut corners, most commonly by replacing fully-featured metering with a single LED. The simplest design just lights up when it senses any signal significantly stronger than the unit's own noise floor, and is therefore mostly just a line-checking tool. More useful for level-setting is the overload/clip LED, which warns you of impending distortion on signal peaks, so your primary concern while setting recording levels is to turn up the gain as far as you can without triggering that. These two single-LED meters complement each other quite well, so they're often twinned in practice, or else have their functions combined via a single variable-color LED.

When you've achieved the meter reading you're after for your first checkpoint, grab whichever gain control is immediately after the first checkpoint and adjust it (if necessary) while looking at the next checkpoint's meter. Continue in a similar manner until you've got appropriate meter readings for all the checkpoints.

Returning to our examples, in Figure 1.9 there's only one measurable checkpoint, so I'd just turn up the headphone volume control until the loudest playback signal registered just under 0 dBFS on the sampling workstation's digital input meters. In Figure 1.10 I'd first grab the Input Trim knob on the DJ mixer's turntable channel, and set that for a 0 dBVU reading on the unit's master output VU meters. Then I'd shift my focus to the studio mixer's group-output VU meters and check whether those were reading around 0 dBVU—if not, I'd adjust the Input Gain knob on the mixer channel(s) receiving the DJ mixer's stereo output. Finally, I'd check the digital meter on the standalone converter unit and if necessary adjust the input sensitivity controls on the converter to achieve the most sensible reading.

WHAT ABOUT ANALOG TAPE?

Because analog tape machines feature in very few small studios these days, I've not complicated my main level-setting discussion by including them. However, if you're faced with recording to tape, then here's how to incorporate it into the scheme.

Firstly, unless you have the luxury of a studio technician, you'll need to acquaint yourself with the overbiasing and alignment procedure of your specific tape machine—if the tape deck's manual isn't enough help here, then find someone who can talk you through it. As well as generally minimizing undesirable signal-loss or tonal coloration for the recorded sound, this also does two things that are absolutely essential for the level-setting process:

- It matches the unit's input and output gain so that you get roughly the same level coming off the tape as you put onto it. Therefore the tape machine's input metering can be used to optimize the signal level in its input and output circuitry simultaneously.
- It determines how hard the tape is driven magnetically for a given input level. You'd need a dedicated fluxmeter to measure this directly, something which would be uneconomical on studio tape machines. Instead, the idea is to establish a known relationship between electrical input level and magnetic flux level using a special test tape during alignment, so that the tape machine's input metering can guide you in minimizing noise and distortion incurred by the tape itself.

What this all means is that you can treat an analog tape machine as a third type of level checkpoint in Section 1.3—let's call it a "tape" checkpoint—and you should use the machine's own input metering to measure the signal level. Although the noise/distortion response of the tape will depend on the specifics of the alignment process, the suggested meter readings in Section 1.3.3 still provide a reasonable starting point. However, it's important to realize that refining your level settings in Section 1.5.2 may involve not only revising your target meter reading, but also adjusting your initial alignment parameters.

One potential fly in the ointment, though, is if a specific gain control doesn't have enough juice to achieve your target meter reading. Here's what to do in that scenario:

- If you need more boost, use the next gain control in line—as long it still precedes the metering point.
- If you need more cut, use the previous gain control in line.

So if, for example, the DJ mixer's Input Trim in Figure 1.10 didn't give me an adequate level on its master output meters, then I might apply further gain by pushing the turntable channel's fader above its 0 dB mark, and even the DJ mixer's master fader too. Alternatively, if setting the optimum level on the studio mixer's group output meters resulted in overloading the standalone converter unit's digital meters (even at the lowest input-sensitivity setting), I might turn down the studio mixer's group output fader to reduce the level.

1.3.4 Listen

With a bit of luck, you should now be able to fade up your studio monitors and hear what you're about to record, clean and clear. With the best will in the world, however, there will still be times when the signal is unacceptably degraded in some way, either by a quirk of the equipment or by good old-fashioned user error. So what can be done to fix such problems?

1.4 TROUBLESHOOTING SIGNAL QUALITY

1.4.1 Pinpointing the Problem

In order to solve any audio-quality problem, you need to find out what's causing it, and whereabouts in the signal chain. Here are some sleuthing tips:

- **Check the Source.** It may seem obvious, but you'll feel like a prize chump if you dismantle your whole recording rig in search of setup gremlins, only to discover that the signal degradation you're hearing is actually part of the playback medium itself! Listen to that LP/tape/disc/file on another playback system if you've not already done so, or plug headphones directly into the player as a confidence check.

"It may seem obvious, but you'll feel like a prize chump if you dismantle your whole recording rig in search of setup gremlins, only to discover that the signal degradation you're hearing is actually part of the playback medium itself!"

- **Check Your Monitoring.** In a similar vein, briefly mute the destination recorder track to confirm that any unwanted noise you're hearing through the monitors is actually coming through your recording chain, and not from some other unnoticed source. Another simple thing to check is that the signal isn't inadvertently clipping on its way from the recorder track through to your monitoring hardware—the big giveaway is when reducing the track's monitor-channel fader level cleans up the distortion, in which case you may need to apply the gain-management techniques of Section 1.3 to your monitoring chain as well.
- **Listen to the Unwanted Signal.** One of the quickest ways to track specific signal-degradation problems is if you've learnt to recognize their tell-tale sound character. For example, analog clipping is generally smoother-sounding than digital clipping, and both are usually more related to changing signal levels than the kinds of distortion generated by faulty connections/gear. Likewise, the generic "hiss" of broadband circuit noise is usually dealt with quite differently to things like hums/buzzes or digital clicks/pops. If your ear can distinguish between these sonic signatures, you've immediately got a head start.
- **Divide and Conquer—Again!** If none of the above provides a definitive answer, then you'll need to adopt a "divide and conquer" approach similar to that described in Section 1.2.2. Occasionally you may discover a mismatch between the noise/distortion levels of the left and right channels of your stereo signal, in which case repatching cables as before can help tracing the signal-path position of a problem. In most situations, though, both channels will be equally afflicted, so rather than repatching cables, the trick is to grab a gain control somewhere in the middle of the signal chain (preferably one of those you used while setting the metered levels) and try gradually backing it off while you listen. Where distortion is the concern, ask yourself whether the distortion remains consistent as you turn down the gain. If it does, then clipping is happening *before* that gain

control in the signal path; if the degree of distortion reduces, then clipping is happening *after* that gain control. Where noise is the concern, ask yourself whether the signal-to-noise ratio remains consistent as you turn down the gain—in other words, whether the noise level reduces roughly in proportion with that of the desired signal. If it does, then unwanted noise is being added *before* that gain control in the signal path; if the signal-to-noise ratio gets worse (i.e., the gain change doesn't affect the noise as much as the desired signal), then the unwanted noise is being added *after* that gain control. These questions answered, you can now reach for another gain control within the portion of the recording chain still under suspicion, and keep repeating the process to home in on the primary culprit.

SOME USEFUL ALARM BELLS

One of the things engineers gain with experience is a kind of sixth sense which provides advanced warning that something, somewhere, is wrong with their setup. Since this instinct is usually hard won, a reward for the litany of embarrassing mistakes everyone makes as they learn, hopefully I can spare you a few blushes by sharing a few little things that always set off my own mental alarm bells:

- **If any of your left-channel gain settings don't roughly match the corresponding right-channel gain settings.** There's probably a stereo balance control set wrong somewhere.
- **If the left and right meters of a stereo recording appear to be responding identically.** This usually means you're inadvertently recording the same mono signal twice.
- **If you have to really crank any gain control when recording a line-level signal.** Most of the time line-level signals shouldn't need much gain adjustment.
- **If turning up a gain control doesn't change your meter readings as much as you'd expect.** There may be an undetected overload (or some unnoticed signal processing) going on between the gain control and the meter.
- **If you spend several minutes trying to decide on the merits of any fine adjustment.** It's not that it's wrong to do this, it's just that it's really easy to imagine differences that aren't actually there—something that has made every engineer in the history of the universe feel like a total idiot at least once in their life. So do yourself a favor: First give the control a good wrench to check it's actually doing what you think it is before you focus on minutiae!

To demonstrate these concepts in practice, here's a quick case study. Imagine I'm hearing a distorted signal coming through the rig in Figure 1.10. First I plug headphones into the DJ mixer and solo the turntable channel, confirming to my satisfaction that the source signal's clean. Then I briefly mute the DAW's recording track to check that a distorted signal isn't inadvertently arriving via another route, and try waggling the recording track's software fader to see whether the degree of distortion changes—this doesn't remedy the problem, so I know I'm not being misled by a fault in my listening system. The smoother sound of the distortion suggests analog clipping to me, so I try fading down the studio mixer's channel input gain control momentarily, then its group output fader: The former reduces the distortion, the latter doesn't, so I'm now pretty sure there's something amiss in the mixer itself. Perhaps the channel fader's been knocked accidentally, some processor has been left in circuit by mistake, or there's some flaky electronic component under the hood—whatever species of technical gremlin it is, I'm a whole lot closer to exterminating it now that I know where it's skulking.

1.4.2 Managing Earth Loops

So far I've mostly focused on tactics for minimizing broadband hiss and clipping distortion, because this is something that's fundamental for any session. However, there are a few other signal-quality concerns that arise from time to time which can easily render a recording unusable: earth loops, digital errors, and feedback loops. You never quite know when any of these will crop up, but they're a lot more common on low-budget sessions, so I think it's important for small-studio users to learn to recognize them and respond appropriately right from the outset.

An earth loop is an unbroken ring of electrical earth conductors, which can arise whenever you cable audio devices together. If two pieces of studio gear have power cables that access the shared earth conductor of the studio's mains electrical supply, then the earth conductor in a signal cable running between them can complete the circuit, as shown in Figure 1.11. The earth loop then effectively operates as a kind of aerial, typically picking up a continuous pitched humming or buzzing from the alternating current of the mains supply itself, sometimes accompanied by various clicks, chirps, and swizzles induced by other electrical equipment in the vicinity, particularly mobile phones and refrigerators—I've even been treated to bursts of police radio on a couple of sessions!

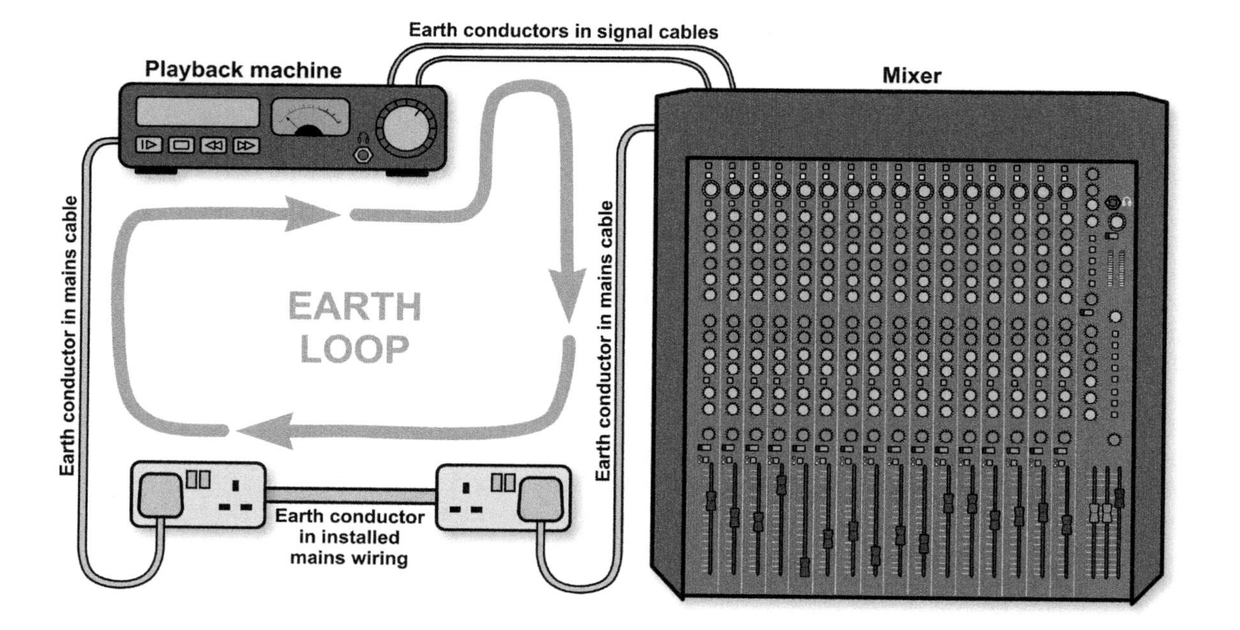

FIGURE 1.11
A simple earth loop between two pieces of audio equipment, in this case a playback machine and a mixer.

One of the reasons to address any earth loops in your system at the earliest opportunity is that they can be a pain in the posterior to remedy, so you *really* don't want to be faffing around with them while there are musicians waiting to play. The biggest obstacle small studios face with earth-loop problems is that they usually only become worryingly audible once there's a whole tangle of earth loops interacting with one another, at which point you simply can't troubleshoot them in the same straightforward way you can a duff cable or a bit of distortion—there are too many variables. As such, these two preventative measures are a better recourse:

- **Use "Star" Mains Wiring.** One way to reduce the impact of any earth loops within your system is to eliminate as much of the building's mains wiring from them. The way to do this is to connect your whole recording and monitoring system to a single power socket. Whether you do this via a dedicated multi-output power distribution unit or just by chaining several multi-output plugboards together is a question of budget as much as anything, but either approach should substantially reduce the audible effects of any earth loops.
- **Use Balanced Connections Wherever Possible.** While this won't break earth loops, it can decrease the level of the unwanted noises enough to sidestep the problem.

"Don't *ever* try to break an earth loop by disconnecting the earth conductor of a mains cable—that's there to stop people getting barbecued by electrical faults. In the grand scheme of things, a little hum's preferable to the silence of the grave!"

If these measures don't give you a workable system (or are impossible to implement in your specific studio scenario), then the only reliable way to deal with earth loops is by first unplugging all the audio cables in your system, with the exception of the cables connecting your studio monitors to their amplifier. In this state, the monitors shouldn't hum or buzz unless they're broken or a wasp's flown into the vent! You can then start rebuilding your studio setup one unit at a time, listening after each addition for earth-loop problems. The moment you hear something untoward, see if you can find a way to break the earth connection in the audio cables you most recently connected.

Where an unbalanced cable is responsible for creating an earth loop, then the easiest way to tackle it is with the same kind of transformer isolator box I've already mentioned in Section 1.1.1, since this interrupts the earth connection. (As we'll see in Section 3.2.1, a DI box can be pressed into service to do the same job, but a simpler transformer-isolator is usually more cost-effective with line-level signals.) Alternatively, if you're a whizz with a soldering iron, you can reduce the level of the hum by inserting a 100 Ohm resistor and a 100 nanoFarad capacitor in parallel between the earth connector and the earth conductor at one end of the cable, and a similar scheme can also be applied to the unbalanced-to-balanced adaptor cable previously shown in Figure 1.2—both these configurations are illustrated in Figure 1.12.

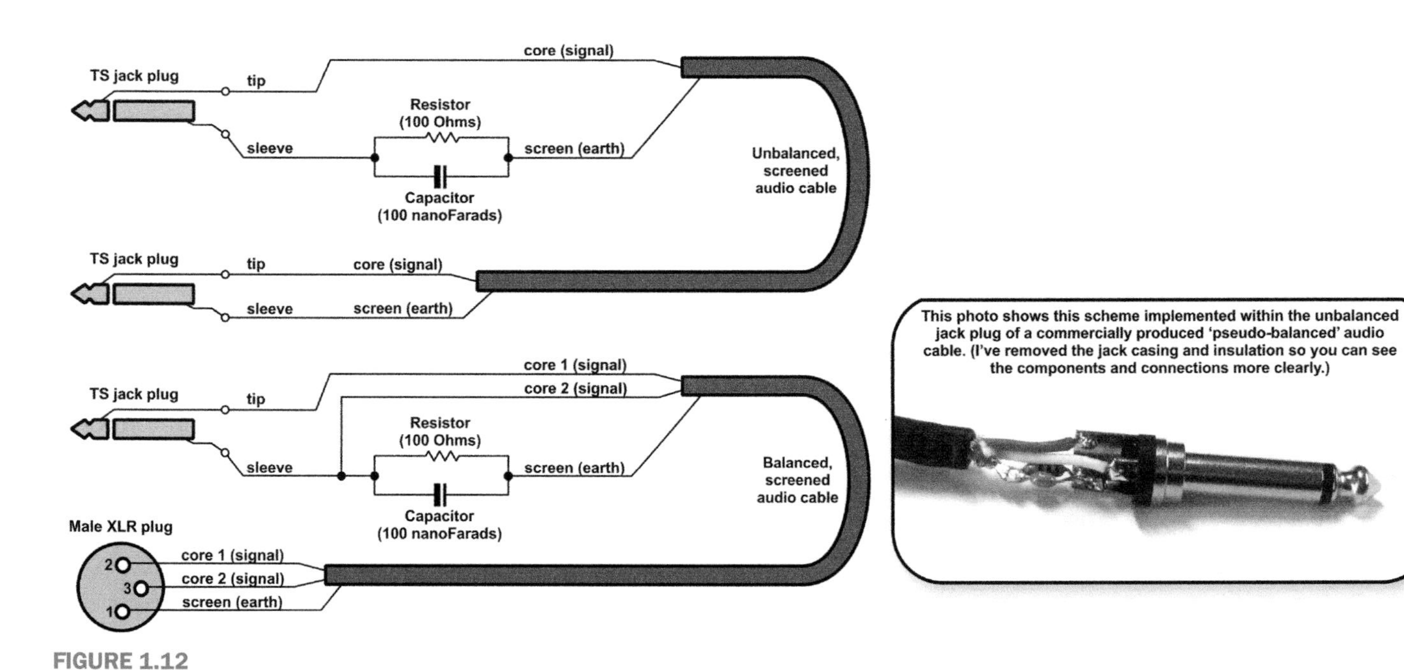

FIGURE 1.12
Two cables which can reduce earth-loop hum when connecting the outputs of unbalanced gear. The upper schematic shows a lead for connecting an unbalanced TS jack input to an unbalanced TS jack output; the lower schematic and the photograph show a pseudo-balanced lead for connecting an unbalanced TS jack output to a balanced XLR input.

Transformer isolation will also work with balanced connections, but in a fully balanced scenario there's also a cheaper option: You can actually break the loop by snipping the earth conductor away from the connector pin at one end of the audio cable without compromising its operation. (Cutting the earth conductor on unbalanced connectors is inadvisable because it makes the signal more vulnerable to unwanted interference.) However, in my experience this dodge is hardly ever necessary in smaller systems, so I wouldn't recommend it unless you've already tried all other possible solutions, because it makes your customized cable ineffective for interfacing with unbalanced gear and it'll need soldering to repair! Whatever you do, though, don't *ever* try to break an earth loop by disconnecting the earth conductor of a mains cable—unlike the earth conductor in an audio cable, a mains power cable's earth conductor is there to stop people getting barbecued by electrical faults. In the grand scheme of things, a little hum's preferable to the silence of the grave!

EARTHLESS SYSTEMS

Having too many earth connections can cause earth loops, but you can also encounter similar-sounding problems if your studio has no mains earth connection at all. Fortunately, an earthless "floating" system is pretty rare, because there's usually at least one piece of equipment in any setup that connects to the mains electrical earth via its power cable. However, low-budget home recordists often rely heavily on consumer-grade electronic equipment powered via external "wall wart" power adaptors or figure-eight mains connectors (Figure 1.13), neither of which make an earth connection. So, for example, if you were recording a battery-powered MP3 player via a USB-powered audio interface to a laptop with a wall-wart power adaptor, and you were monitoring on headphones, there'd be nothing electrically connecting the whole system to earth.

In practice, for line-level signals this probably won't impinge drastically on the signal quality as long as you get your gain settings right. If it does, though, you can usually scotch the unwanted noises by improvising an earth connection. Try making contact between a bit of the recording system's exposed metalwork and some other earthed item, such as an unpainted plumbing pipe or the metal casing of an earthed domestic appliance. If that's awkward to do in practice, you can make it easier by plugging a spare cable into an unused audio socket of the recording system, and using that as an earthing extension, gaffer-taping the free end so that its earth conductor touches some earthed metalwork. Alternatively, if you have any earthed audio gear you're not using for your recording, a redundant connection between that and your recording system might also do the job.

FIGURE 1.13
Audio equipment powered via a "wall wart" mains adaptor (left) or a figure-eight mains plug (right) typically makes no connection to the mains earth.

1.4.3 Avoiding Digital Errors

Errors in the data stream of a digital audio signal are usually identifiable as little click/pop sounds that don't bear any obvious relation to the rest of the audio signal—they don't get worse for louder signals, and you can't usually predict exactly when they'll occur. There are several common causes for them in the small studio, each of which calls for different remedial action:

- **Damaged Optical Media.** If the surface of a CD or Minidisc is damaged, and that sufficiently interferes with the player's laser optics to defeat the medium's built-in error-correction mechanism, then you may get audio glitches and drop-outs. Unlike most types of digital error, these often occur in a regular rhythm, directly related to the disc's revolution speed—a useful troubleshooting clue, especially if you can actually see the disc spinning in the player. There's not a tremendous amount you can do to fix a damaged CD or Minidisc, but at least you can save yourself some troubleshooting time by spotting the nature of the fault swiftly.
- **Cable Issues.** When audible glitches are combined with sporadic error messages from your digital recording device along the lines of "Digital Clock Unlocked!" or "Digital Synchronization Error!", this usually means you've got a problem with a digital cable or your digital clocking/sychronization scheme isn't set up properly. It's most likely to occur if you're using normal XLR or RCA phono cables instead of correctly specified AES-EBU or S/PDIF cables, but can also happen if an optical S/PDIF cable's internal filament has been damaged. Even bending an optical cable too sharply *en route* between input and output may cause the data to become unreliable if it causes too much light to refract out of the optical fiber, rather than being reflected successfully along it. However, if you use purpose-designed digital cables, and treat them carefully, this kind of data corruption shouldn't ever make it onto your radar.

MULTI-CHANNEL DIGITAL CONNECTIONS

Although we're only focusing on recording stereo in this chapter, it's not unusual for even quite unassuming studio setups to have a multi-channel digital connection somewhere in the recording chain, most frequently between the output of a multi-channel analog-to-digital converter box and the corresponding digital input on a digital recorder or computer audio interface. There's no need to panic about multi-channel digital connections, though, because they're usually no more complicated to deal with than stereo connections. The eight-channel ADAT standard uses the same Toslink connections as optical S/PDIF, so budget gear frequently has Toslink sockets which are software-switchable to use either data format. Eight-channel AES-EBU works electrically, using thick multi-core cables with D-Sub connectors (9-pin or 25-pin) equipped with locking screws that can be fastened by hand. Unbalanced audio lines are best kept away from any D-Sub cable if possible, to avoid collecting modem-like interference on your recordings. It's important to treat D-Sub plugs with care, because the pins are thin enough to be bent if handled roughly, and you should be aware that the allocation of signals to pins varies between different manufacturers, so be prepared to consult your equipment manuals when hooking things up for the first time.

- **Sample-Rate Mismatch.** Whenever you send a digital signal down a cable, it's important that the sampling frequencies of the source and destination units are properly synchronized (even if they're nominally operating at the same sample rate), otherwise the receiving unit will periodically misread the incoming data, causing a series of digital pops and clicks. Most budget-friendly equipment will usually prevent you getting yourself into difficulties here by refusing to unmute a digital input unless the unit's sample rate has been set to synchronize with the incoming audio data stream's embedded clocking information. Equipment aimed more directly at the professional market may not be as forgiving, however, and the remedy will usually involve a bit of menu/manual surfing in order to track down the necessary digital clocking and synchronization settings.
- **Software Driver Problems.** If your recording system is based around a computer, by far the most common cause of unwanted digital detritus on the audio signal is the computer's software configuration. The permutations and pitfalls here are innumerable, so rather than getting too bogged down in specifics, let me suggest some general strategies that usually lead me to a suitable fix. First, invest in a dedicated computer recording interface, instead of trying to use the built-in audio hardware—you can get something for under $100 (£70) that'll work more reliably and record at a much higher quality. Then check that you have the latest version of your audio interface's software driver, and that you've correctly selected it for recording purposes in your DAW's user preferences. Close down any unnecessary applications running alongside your DAW software to minimize the load on your computer's CPU resources—bear in mind that some programs may be running as background processes, so they may not immediately be visible. For similar reasons, remove any unnecessary plug-ins from your DAW's recording project. If you're using an external audio interface connected via a data connection such as USB or FireWire, try to connect it directly (rather than via any kind of multi-port hub), and try different sockets on the computer in case some are handled by its motherboard in different ways. Finally, find the audio-interface driver's configuration settings and try increasing the buffer size parameter, which reduces CPU drain, albeit at the expense of increasing the processing delay incurred between the audio inputs and outputs (which shouldn't cause any practical problems with this chapter's recording task). If none of that helps, then try the soundcard manufacturer's technical support channels for ideas, and search the Internet for similar reports from any other users of your particular interface—there's usually someone out there who's already got the solution!

1.4.4 Feedback Loops and Comb-Filtering

While setting up your recording rig, it's not hard to unwittingly send a signal from some point in your recording chain back to an early point, creating a circular signal path called a "feedback loop." If the overall gain around the loop is above unity, then the signal level in the loop will rapidly accumulate and overload, typically resulting in an unpleasantly distorted high-pitched squealing called "howlround" or "howlback." Howlround isn't easy to ignore, so you'll know pretty quickly if you've got a positive-gain feedback loop in your recording chain, but if the gain around the feedback loop is less than unity, the

DIGITAL CLOCKING AND JITTER

If all you wish to do is digitally connect a stereo playback machine to your recording system, then you shouldn't have to tangle with anything more complicated than a single S/PDIF or AES-EBU cable. However, some larger-scale DAW systems may incorporate several independent digital converters, at which point things get a bit more complicated: Either you have to find some way of synchronizing all their sample rates, because the destination device can't slave itself to several sources at once; or you have to have additional circuitry at the inputs to continuously convert the incoming sample rates to match the destination sample rate. The former solution is cheaper to implement (so it's the one you're most likely to meet in small studios), but if you're not careful it may lead to digital glitches and can also introduce subtle sample-rate fluctuations (called "jitter"), which make analog-to-digital conversion less accurate. Because such problems may afflict any of a recording system's inputs, even this chapter's simple stereo recording task may require you to tangle with them, so here are some tips that should hopefully keep you out of trouble.

The most common scheme for synchronizing digital equipment together is to distribute a dedicated digital "word clock" signal from a master unit, to which a number of slave units are synchronized. In small systems the usual approach is for one of the units to "daisy-chain" its word-clock output to each of the slaves in turn using special digital clock cables, whereas more complex multimedia rigs may use a dedicated multi-output "master clock" unit to feed each slave directly instead. (See the first two setups in Figure 1.15.) In either scenario, every slave unit needs to be switched into its designated word-clock synchronization mode. When using the daisy-chaining approach, it's best for the master unit to be something that's doing analog-to-digital conversion, because slaving any digital converter to an external word-clock signal will usually increase its jitter. If there's more than one ADC in the system, then choose the higher-quality unit to act as the slave, because it'll normally have superior clock-reading mechanisms to minimize the added jitter.

Word-clock cables are very similar to digital audio cables in design, but use BNC connectors featuring a bayonet-style locking system to hold them in place—a quarter-turn of the plug's rotating collar secures it in place. One complication with BNC word-clock cabling, however, is that every chain of BNC cables needs a 75 Ohm impedance at the end, to stop the clock signal reflecting back and forth along the chain. In many cases the clock input's circuitry will already provide this, in which case all you need to do is connect the cable, but there are some units that can switch to a higher input impedance as well. The higher-impedance termination is provided as a means of avoiding the processing delays incurred by each unit's word-clock circuitry, should these begin to compromise synchronization between the first and last units in a clocking daisy chain. The way this works is that the final unit in the chain is connected to the master with a 75 Ohm impedance as usual, but all the intervening units maintain a higher input impedance which allows them to tap into this signal using special BNC "T-piece" splitters without rendering it unreadable. (See the third setup in Figure 1.15.)

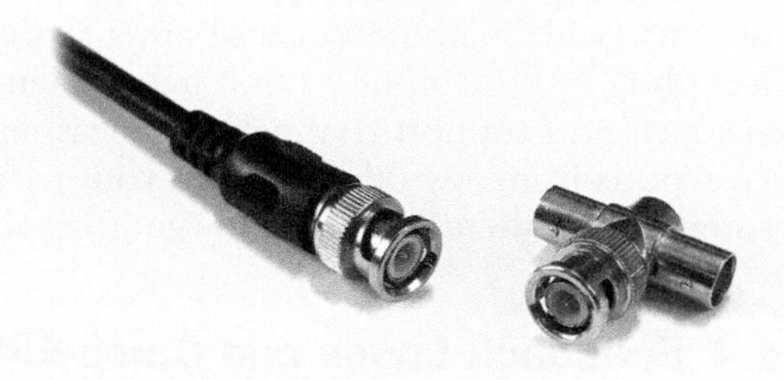

FIGURE 1.14
A BNC cable and a "T-Piece" splitter, as used for digital word-clock connections.

problem becomes more insidious. Where the feedback level is reasonably high or the feedback loop's gain is near unity, you may hear a pronounced pitched ringing, or a strangely metallic "in a drainpipe" timbral quality, but if the feedback is slight, there may only be a subtle (though usually undesirable) tonal change to show for it.

You also need to be on guard against situations where a portion of your input signal is inadvertently split off, delayed by some means, and then mixed back in with the undelayed signal further down the signal chain. The time-difference between the undelayed and delayed signals causes a tonal effect called comb-filtering (or "phase cancellation"), which often sounds quite similar to a subtle feedback loop. You might think this kind of comb-filtering would be an unlikely eventuality, but it's actually quite easy to fall foul of, given that every piece of digital audio equipment inevitably causes a small (but significant) processing delay between its inputs and outputs.

The most challenging thing about both these problems is recognizing them in the first place, so it pays to tune your ear to how they sound under controlled conditions, because you'll then spot them more quickly in the wild. (I've provided some pertinent audio examples in this chapter's web resources.) Even then, you may sometimes question whether you're hearing subtle feedback-loop or comb-filtering artifacts, or whether your ears are just playing tricks on you—in which case the best sanity-saver is to use headphones to compare the sound of the player's raw output with your main monitoring signal.

Once you're aware what's wrong, it's not usually too difficult to diagnose the causes. In my experience, both these issues are most likely to arise where you're recording through the same small mixer that's also routing the recorder's monitoring signal to your loudspeakers/headphones. One wrong button-press can easily route the monitor channels back to the recording inputs (creating a feedback loop through the recording system) or the input channels to the mixer's monitor outputs (mixing the original undelayed signal with the delayed signal returning from a digital recorder). Likewise, many computer audio interfaces now incorporate complex DSP-powered mixers that operate independently of the DAW software, and they seem to take a perverse glee in encouraging very similar routing gaffes. It may only be your monitoring signal that's suffering, not the recording itself, but it's still important to banish this malaise before hitting Record, because it can disguise various sonic subtleties you may want to assess—especially when we start working with microphones later on.

1.5 BEYOND REASONABLE: REFINING YOUR RESULTS

By this point you should be able to hit Record on pretty much any comparatively modern studio system and capture a passably clean-sounding version of whatever your playback machine's putting out. But if you're serious about the art of recording, then that simply isn't good enough! You want to get the

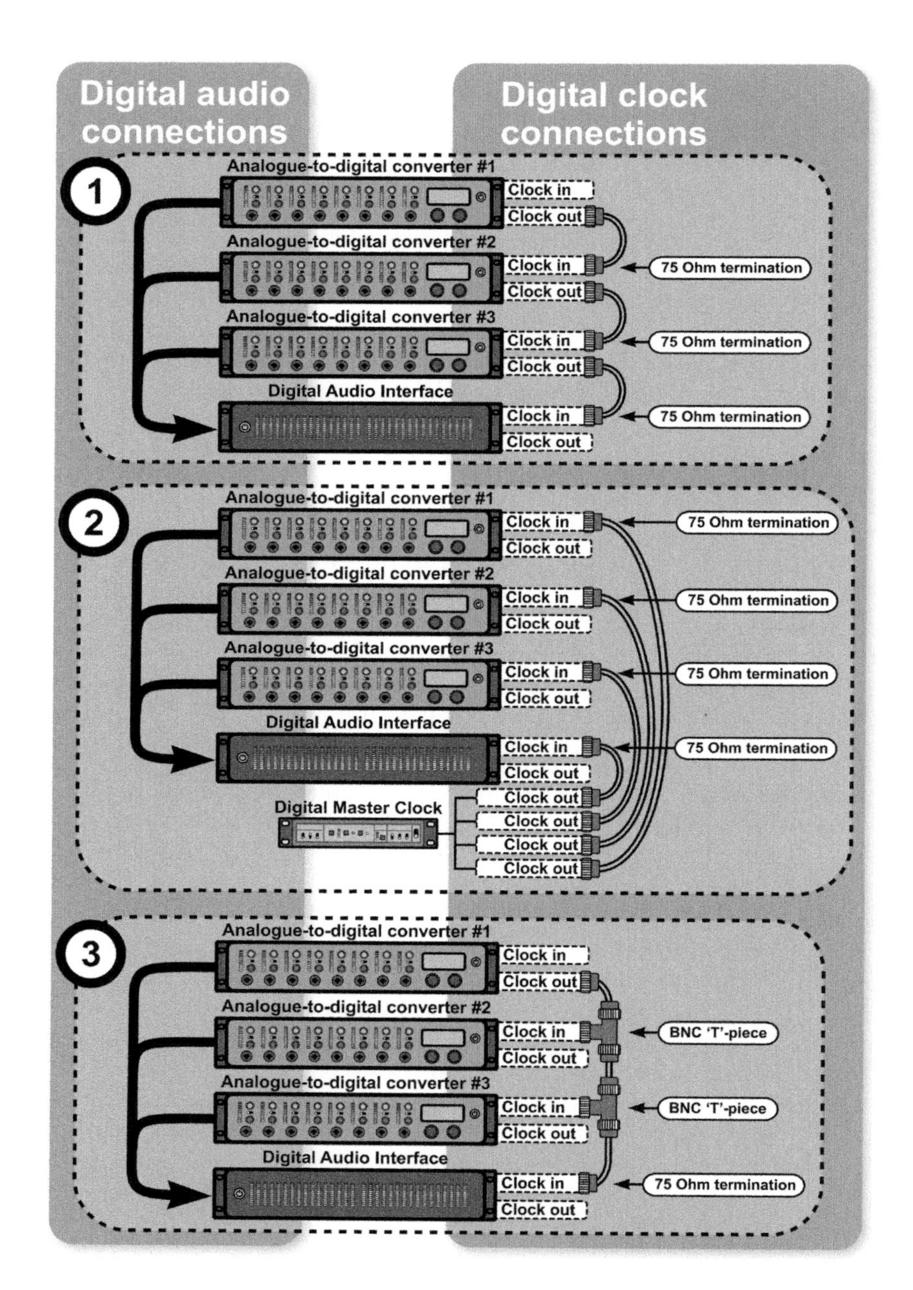

FIGURE 1.15
Three different digital clocking arrangements for interfacing several external analog-to-digital converters with a single computer audio interface: (1) a simple daisy-chain system; (2) a system synchronized from an independent digital master clock unit; and (3) a daisy-chain system using BNC "T"-piece adaptors.

maximum quality out of a given system, especially if you're at the mercy of outdated and/or low-budget gear which may struggle to deliver truly usable results even at the best of times. So let's look at how we can refine the basic approach so far in order to boost quality.

1.5.1 Improving the Source

One of the essential truths about any recording job is this: The better your source sounds, the better your recording will sound. Furthermore, the sonic benefits of adjusting a sound source directly will usually dwarf any improvements attained via your studio gear, and often cost less time and money into the bargain. Admittedly, there may not be anything you can do to improve the sound of the simplest playback devices, but if you're recording from a cassette deck, say, you'll almost certainly get a bigger step-up in quality by cleaning and aligning its playback heads than you'd get by doubling the cost of your recording system or switching to a 96 kHz sample rate.

Given the impact that the source quality has on the recording, one of the big learning curves of studio engineering is discovering how to get the best out of each specific sound source you're presented with, so if you're not happy with the recorded sound of any playback device (DAT, LP, cassette, wax cylinder...), either do a bit of background reading to harvest general-purpose tips or find someone who can clue you in on the best specialist tweaks.

1.5.2 The Aesthetics of Noise and Distortion

When I talked about setting initial levels using your meters back in Section 1.3, I deliberately erred on the side of caution with my suggested meter readings, on the basis that excess distortion is usually trickier to remedy at mixdown than excess noise. In doing this, however, I've allowed you to dodge an important judgment call that every engineer must make at each level checkpoint in their own unique rig: What do *you* think is the best balance between noise and distortion?

Answering this question is partly a question of learning the nuances of your own equipment, so if you're starting out I'd suggest fast-tracking your development in that respect—by which I mean experimenting with alternative gain settings at this point, while there aren't any musicians hanging around getting bored! Work through each of the level checkpoints you identified in Section 1.3.2 and try hitting them with a higher signal level, bearing in mind that you may need to turn down a gain control later in the recording/monitoring chain to avoid unintentionally overloading later checkpoints. Discovering how each of your pieces of gear responds to overloads is an important factor in getting the best from your studio, not least because it tells you in practical terms how much headroom your specific VU meters and PPMs provide, and thus what target meter readings are sensible. And once you accept the inevitability of a trade-off between noise and distortion, it becomes perfectly defensible on technical grounds to allow some overloads if their audible consequences turn out to be a reasonable price to pay for an improved signal-to-noise ratio. While you can often afford to leave lots of headroom for signal peaks when recording at 24-bit resolution with decent project-studio equipment, there's also nothing to say you shouldn't light up a few red LEDs with your signal peaks on a cheap and noisy mixer, or flatten the odd transient against the 0 dBFS digital ceiling of an old 12-bit sampler, if your goal is substantially lower noise.

NOISE FINGERPRINTS

Despite the best efforts of your gain controls, you may occasionally find yourself with no choice but to record a signal which is noisier than you'd like—when sampling an old cassette tape, for example. There's specialist noise-reduction software for such eventualities which can improve matters after the session, but its effectiveness will often depend on whether there's a noise-only section of the recording (called a "noise fingerprint") which can be analyzed to help tailor the processing precisely. So if you can't escape background noise while you're recording, make sure you record a few seconds of it on its own to improve your chances of removing it later.

However, it's equally important to acknowledge that overloading any studio equipment has an aesthetic dimension too—I'd wager that every studio box ever invented has been purposely overdriven on at least one hit record! Part of the point of trying out your gear with hotter signal levels is to hear what it actually sounds like, so that you can draw on that knowledge for creative ends if necessary—whatever the meters are telling your eyes, it's your ears that should always have the ultimate authority. Pushing any bit of analog equipment out of its comfort zone will add a unique blend of distortion products, and the appealing tonal color of these additions is a big part of the appeal of the most sought-after studio hardware. Analog tape machines are particularly revered by many engineers for their ability to smoothe and fatten high-level drum transients, for instance, but even the abrupt, harsh onset of digital clipping is put to artistic use on occasion, particularly in more electronic styles. As an engineer, you're better equipped to know when the time is right for each of these if you're able to conjure their contrasting characteristics to your mind's ear at will. One word of caution, though: It's a lot easier to add distortion while mixing than to take it away, so if you're in any doubt about how much distortion to add while tracking, I'd always suggest playing it a little safe.

1.5.3 Simplify the Signal Path

Another way you can improve signal quality is by simplifying the signal path. Fundamentally, the less circuitry you have between your signal source and your recorder track, the fewer opportunities there'll be for any type of audio degradation. Clearly some of you will face restrictions here—perhaps you're working on a college system that you're not allowed to repatch—but don't let that stop you improving what you can.

In the first instance, you should ask yourself whether there's any more equipment in the recording chain than you need. So in Figure 1.10 you'd get a cleaner signal by connecting the DJ mixer directly to the standalone digital converter—indeed, bypassing the studio's main mixer like this is extremely common amongst professionals. Another thing to avoid is unnecessary digital conversion stages. So if the DJ mixer in Figure 1.10 were actually a digital mixer rather than an analog one, you might be able to take a digital feed directly from that into the computer audio interface, completely removing the mixer and standalone digital converter from the equation—as well as speeding up the gain-setting process by eliminating all but one level checkpoint. In a similar vein, try to turn off (or "bypass") any unused signal processing facilities within your equipment as this will reduce the likelihood that it tampers with what you're recording, and will often reduce the amount of circuitry in the signal path too.

Cabling should be kept as simple as possible, so if you can use a single cable between two pieces of equipment, rather than relying on chains of cables and adaptor plugs, you'll likely improve the recorded sound a little, as well as reducing the number of variables to troubleshoot if you hit a snag during the line-check. Choosing your audio connections wisely may also help streamline your cabling, particularly if you can obviate the need for transformer isolators. Finally, notwithstanding my warnings in Section 1.1.3 against using cables that are too short, overlong cables can also impact on recording quality, so aim for a happy medium in that department.

DEALING WITH PATCHBAYS

A patchbay (or jackfield) is basically a means of extending the socketry of many different studio units so that it's all in one place, at which point you can conveniently connect or "patch" things together using short cables called "patch cords." Although the general trend toward software-based recording rigs has now rendered patchbays redundant in many small studios, there are a few things that you should know if you're working in a system which uses them.

Firstly, the jack sockets provided on some patchbays may not be designed to accept standard quarter-inch plugs—bantam jack plugs are smaller, for instance, and older "B-type" jacks have a significantly different shape despite sharing the same diameter. Trying to use any type of jack plug with the wrong type of jack socket will usually result in an unreliable connection, and can easily damage the socket's internal contacts too, so consider yourself warned!

Beyond that, though, the best way to navigate any specific patchbay is to ask the person who wired it up, because small-studio patchbays are far from standardized in my experience: Earthing schemes may be quirky to say the least; labeling will frequently be missing or misleading, without adequate indication of normaling (whereby certain sockets may automatically be connected if unpatched); and a random selection of sockets and patch cords will usually need cleaning and/or repairing. Indeed, trying to line-check or troubleshoot earth-loop hum through various unknown patchbays has so often caused me to chew my own face off in frustration that I now routinely avoid patchbays while recording wherever possible. Besides, even if a patchbay works as it should, it'll still make the wiring between any two pieces of connected equipment more complicated than it need be, so direct connections may also sound better, even if they don't look as neat.

1.6 AFTER YOU HIT RECORD

You're not off the hook, though, even if you've got the world's most high-fidelity signal arriving at your recorder. While the recording's happening, it's still your job to keep eyes and ears open. If you see an unexpected overload on any of the crucial meters you identified in Section 1.3.2, especially the one that's measuring the actual recorded level, then at the very least you should check that section of the recording for distortion later on. (The "peak hold" function of many digital meters can be very handy in this regard, so I'd recommend switching it on if you have it.) But whatever the meters say, if you hear unwanted distortion while you're recording, then stop where you are, adjust your gain controls accordingly, and retake.

And once the recording's completed you should always listen back to it, if only to make sure you *did* actually record something—no recording device is foolproof and, besides, every audio engineer presses the wrong button at least once in their career! Much better to discover a mistake immediately, while the gear's still all set up, rather than waiting until you've packed everything away. Playing back what you've recorded isn't just about saving you from silly slip-ups, though: it's the final stage in a quality-assurance process that reaches back all the way to your line-check. When you've taken such trouble to get a great signal to your recorder, it'd be daft not to guard against potential failures in the recording medium itself.

1.7 KEEPING UP WITH THE MUSICIANS

If you've grasped everything so far in this chapter, then you should now be equipped to record line-level signals. However, until you can confidently hit Record within a couple of minutes of switching on your source playback device, you aren't really ready to record anything other than a machine. So given that Chapter 2 involves working with real performers, you'll still need to consider how you'll go about reaching that kind of session speed.

"Once the recording's completed you should always listen back to it, if only to make sure you *did* actually record something—no recording device is foolproof and, besides, every audio engineer presses the wrong button at least once in their career!"

The main weapon at your disposal is preparation. Even if you can't get your hands on the source playback device until the session itself, it shouldn't be too difficult to find out beforehand what socketry it has. Armed with this information, you can deal with almost everything up to Section 1.3.3. Doing a preliminary dry run of the session, with the help of a substitute playback machine if necessary, will take you even further, highlighting whether any of the signal-quality issues described in Section 1.4 are inherent in the recording system itself, so that you can address them before the clock's ticking. With all of that in the bag, come session time the job should be a breeze: Plug in the new playback machine, tweak a couple of gain controls, and burn rubber!

But your gear isn't the only thing that needs preparing: You also have to prime your ears and your mind. In other words you have to learn how far you can push your meter readings to get the best noise/distortion trade-off; expand your ability to recognize different signal-quality problems when you hear them; and build your troubleshooting powers to the stage where no level/routing snag delays you longer than a minute or so. All of these things improve with experience, and the best way to gain experience is through practice, practice, and more practice.

In light of that, I'm going to draw this chapter to a close with a series of assignments designed to consolidate your understanding of the principles we've covered so far. In the process you'll also be preparing your studio gear for the fray, as well as fast-tracking your own engineering skills so that you can operate at a more musician-friendly pace. But just before we get to those assignments, let's quickly recap on the main concepts in this chapter.

CUT TO THE CHASE

- The basic procedure for recording a line-level playback device is: Mute your monitors, select your connections, and hook everything up; set unity gain through the system and carry out a line-check; set preliminary gain based on meter readings; unmute your monitors to troubleshoot signal degradations and refine the signal quality as much as you can; record the signal and then listen back to check it over.
- Use balanced connections wherever possible: It'll reduce unwanted noise on your recordings and make earth loops easier to manage. If you're using XLR connections for line-level signals, make sure phantom power is switched off. Balanced inputs can usually be connected to unbalanced outputs with a specially wired cable, but if you want to feed a balanced output to an unbalanced input, then it's safest to go through a dedicated transformer isolator unless the output is specifically designed to cope with unbalanced operation.
- Dedicated studio/stage cables without molded connectors are well within most budgets, so don't settle for less. Dedicated digital cables are also worth the investment, even though analog cables can sometimes substitute for them in a pinch.
- When laying out cables for a session: Keep audio cables away from sources of interference such as mains leads, power-supply transformers, and data cabling; try not to create obvious tripping hazards; and think ahead to prevent needless damage to cables or sockets.
- The "divide and conquer" approach is the most reliable troubleshooting technique at your disposal, but it saves time in the long run if you also learn to recognize how different types of audio degradation sound.
- When setting levels, you need to achieve the best balance between noise and distortion whenever your signal passes through an analog cable, undergoes digital conversion, or is committed to analog tape. This involves finding the best meter for each of these level checkpoints, and then selecting the most appropriate gain control(s) for any adjustments.
- The best way to improve your recording is to improve the quality of the sound source. Simplify the signal path to keep your recording clean and minimize troubleshooting time—bypassing the main studio mixer or any patchbays if possible. Where unacceptable background noise levels prove unavoidable, record a noise fingerprint to assist with later restoration processing.
- Working at digital resolutions above 24-bit/48 kHz is rarely the best use of a small-studio budget, but recording to data-compressed digital formats is almost always a false economy. If you're going to use analog tape, make sure it's properly aligned, and familiarize yourself with how different alignment variables affect its general sound and response to overloads.

- Reduce earth-loop problems by using "star" mains wiring and balanced cables. Any troublesome individual earth loops can be broken by cutting a balanced audio cable's earth connector or using a transformer isolator. Earth loops in unbalanced lines can be tackled with transformer isolation too, or by using special pseudo-balanced cables. Never disconnect the earth conductors in any mains plug or cable. If your whole recording system has no direct mains earth connection, you may need to improvise one.
- Digital errors are usually caused by: damaged recording media; inappropriate cabling; unmatched sample rates; or software audio-driver problems. Inattention to digital clocking can increase the unwanted effects of jitter upon your digital converters.
- Feedback loops and comb-filtering can easily arise in small-studio systems, and may subtly undermine the tone of your recordings in a way that's easily missed.
- Preparation is vital if you want your recording sessions to go smoothly and swiftly.

Assignment 1: The Dry Run

The purpose of this assignment is to identify the controls and meters you need for recording; to line-check and troubleshoot the recording system itself; and to give you a clear idea of the meter readings you should be aiming for at each point in your system.

- Find yourself a stereo line-level playback device, and connect it to your recording system as described in Section 1.1.
- Work your way through the entire recording and monitoring signal path, drawing yourself a flowchart that indicates every single gain control (don't forget switches!), mute button, and stereo pan/balance control. While you're at it, set all of them to unity gain and reset/bypass any audio processing facilities you come across.
- Line-check your recording and monitoring system as discussed in Section 1.2.
- With reference to Section 1.3, indicate on your flowchart the position of every level checkpoint between the first gain stage and the recording track, and mark the location of the best meter for measuring each checkpoint's signal level. (You should also make a note of any level checkpoints you can't measure, as those may be under suspicion if you hear distortion later on.) Now set your gain controls according to the guidelines in Section 1.3.3.
- Listen carefully to the signal you're about to record, keeping your ears peeled for any of the audio-quality problems described in Section 1.4. Troubleshoot as necessary.
- Record your signal, maintaining a watchful eye on the relevant meters and listening carefully for anything unexpected that might warrant further fixes. Play back the recording to check you've captured it intact, and then save/store it somewhere safe.
- Now follow the guidelines in Section 1.5 to improve the signal quality: investigate whether you can adjust the source playback machine for better results; see how hard you can drive each level checkpoint in turn before distortion side-effects become unappealing; and simplify the signal chain in any way you can.
- Record the tweaked signal, again watching and listening carefully for problems, and play it back to check it over.
- Compare the two recordings side-by-side within your recording system, evening out any loudness differences between them by adjusting their relative playback levels. Give special attention to the audibility of the recorded noise floor, the impact of distortion on the timbre, and how any transients in the source signal have been

affected. Whether or not the second recording turns out to be an improvement on the first, try to isolate which adjustments to the recording chain were primarily responsible for the most audible differences—if you're in any doubt, perform additional test recordings to probe the variables further.

Assignment 2: Line-Checking & Troubleshooting

This assignment is designed to speed up the development of your line-checking and trouble-shooting skills.

- Get hold of a stereo line-level playback device, if possible a different one to that used in Assignment 1. Connect it up (Section 1.1), do a line-check (Section 1.2), set levels (Section 1.3), troubleshoot any signal problems (Section 1.4), and refine the recording quality if necessary (Section 1.5). If you've already worked your way through Assignment 1, this shouldn't take you more than about five minutes.
- Turn down your monitors, and then give a mildly sadistic friend (preferably a fellow student of the recording arts with a passing knowledge of your studio system) precisely two minutes to repatch audio connections and tweak equipment settings while you turn your back, their express intention being to try to fool you. For safety's sake you should make sure they don't touch any mains connections or phantom power switches, or indeed do anything else that's likely to actually damage the kit, but otherwise try to give them a free hand so that they can get things into a proper tangle.
- When their two minutes are up, hand over the stopwatch to your persecutor so they can time how long it takes you to get the system back up and running again. And try to resist the urge to throw things at their smugly amused face…
- It shouldn't require too many repetitions of that exercise to dramatically accelerate your troubleshooting skills. If you can repay your fellow student by performing the same controlled sabotage on their system, all the better—and if you're really serious about boosting your engineering mojo, then try swapping over and debugging each other's systems too!

WEB RESOURCES

On this book's companion website you'll find a selection of resources to support this chapter, including:

- audio demonstrations of different types of signal degradation for ear-training purposes, as well as example files showing the scale of side-effects from some affordable transformer isolators;
- specimen flowcharts for Assignment 1, based on the example setups shown in Figures 1.9 and 1.10;
- further reading on topics such as analog metering standards, tape-machine alignment, digital interfacing, studio wiring, cable coiling, and patchbay design;
- links to some manufacturers of affordable transformer isolators and pseudo-balanced cables.

http://www.cambridge-mt.com/rs-ch1.htm

CHAPTER 2

The Human Element

Now that some essential technical skills are under your belt, you should be in a position to start grappling with the most important and complex part of any music recording session: the performer. We'll stick with line-level and digital signals for the moment, and make it our task this chapter to learn about capturing musical instruments which generate no acoustic sound of their own: synths, sampling workstations, laptops with live performance software, and scratch DJ turntables. And because these kinds of instruments are often recorded as overdubs, we're also going to talk about the practicalities of monitoring the instrument against whatever backing track you have. (This chapter's web resources include a variety of prerecorded stereo backing tracks in different musical styles which you can download if you don't already have a production of your own to overdub on.)

I realize that most people reading this will probably be performers themselves, and may therefore be multitasking as both artist and engineer during recording sessions. However, working with other people is a far more challenging prospect, so I'm going to focus primarily on that in this chapter. Fundamentally, though, the same engineering principles apply in either situation, the main difference being that you'll cut yourself more slack on the engineering side of things than anyone else is likely to!

2.1 PREPARING FOR THE SESSION

If you've learnt anything from the assignments in Chapter 1, then I'm assuming you've already line-checked your recording and monitoring chains in advance of the session, you've made sure that all the relevant gain controls and meters are at your fingertips, and you've got the right cables for the instrument's output socketry. But that's just the tip of the iceberg...

2.1.1 Studio Layout

Because the instruments we're recording in this chapter don't make any acoustic noise, the best place to set them up is in your control room—whether that's the glistening nerve-center of a high-grade project studio, or just the bedroom containing your PC. When you're recording your own performance, this keeps the studio controls within easy reach while you're playing, but even if you're recording someone else, the advantage of working in the control room is that everyone can use your main studio loudspeakers for monitoring purposes. From an engineering perspective, it can be tricky to judge the timbre of recorded sounds over headphones (or "cans"), and there are also many aspects of headphone listening that can interfere with a musician's ability to perform:

- the characteristic sensation that the music is originating inside your cranium, rather than happening in front of you;
- the slightly unsettling psychological effect of being isolated from the environment you're performing in, because the headphone earcups stop you hearing the room's natural acoustic reflections;
- the physical restrictions imposed by the tether between your headset and the headphone outlet;
- the communication difficulties that arise from the fact that it's tough to listen and talk at the same time, especially during highly collaborative sessions where several bystanders are brainstorming freely with the performer;
- the tendency for pitch-perception to become less reliable, particularly at higher listening volumes, which can cast doubt over tuning decisions.

HEADPHONE EXTENSION CABLES

If you're looking for a headphone extension cable, don't go for the flimsy variety designed for domestic multimedia use, because the skinny wires and cheap molded connectors rarely survive long under studio conditions, in my experience. My recommendation is to combine a studio-grade TRS-to-XLR cable with one of those chunky barrel adaptors which have an XLR socket at one end and a TRS jack socket at the other. Not only is this a whole lot more robust, but you can also further lengthen the cable using spare XLR mic cables if you need.

Clearly, though, many of you may have no speakers in your recording system at all—or you may be working too late at night to fire them up—so you have to make the best of it. If you have any open-back headphones available, these are a good choice for this kind of overdubbing, because they normally give a more natural sound and also make it easier to hear conversation and other incidental noises while you're working. Open-back headphones aren't as advisable when working with microphones, because they let the monitoring signal escape into the room and spill onto your recording, but that doesn't mean they're not fair game in mic-less situations.

When someone you're recording has headphones of their own that they like, then suggest they bring them along to the session, because the timbre of different models can vary enormously and some people are very partial to a certain balance. (Just make sure you have a spare pair of headphones in the wings as a backup in case they're actually rubbish!) The volume level at which a performer normally monitors may also have a bearing on which headphones are most suitable for them, because the human ear becomes more sensitive to the high and low extremes of the audible frequency spectrum as listening levels increase—headphones with a brighter tone can prove quite fatiguing on the ear at higher listening levels, for instance, whereas models with hyped bass response may improve the subjective appeal of quieter playback.

If you need more headphone outputs from your recording system than you actually have, the cheapest workaround is one of those little adaptor plugs that can split a single headphone output to a pair of sockets. I hate those things, though, and not just because they typically stress the socket contacts. The main problem is that the adaptor plug ties both listeners to the same headphone volume control. If you like listening a lot louder than the performer does, say, either you end up deafening the talent, or you're forced to make sonic judgments at a listening level that presents the audible spectrum with an unfamiliar frequency balance. Even if both of you like listening at the same volume, some sets of headphones work more efficiently than others, so it's not uncommon to notice a significant level mismatch between two pairs of cans fed from the same socket. Moreover, splitting a headphone output like this will reduce the maximum volume the headphone amplifier can deliver for each listener. Overall, then, if you don't have enough headphone outputs, I'd thoroughly recommend supplementing your rig with a dedicated multi-output headphone amplifier that has a separate volume control for each output. A four-output box like this needn't cost you more than about $75 (£50), although it is worth checking out models around the $150 (£100) mark if you want to guarantee bone-splintering volume levels even with less sensitive headphones. Whatever headphone amplification method you choose, though, do get some stereo

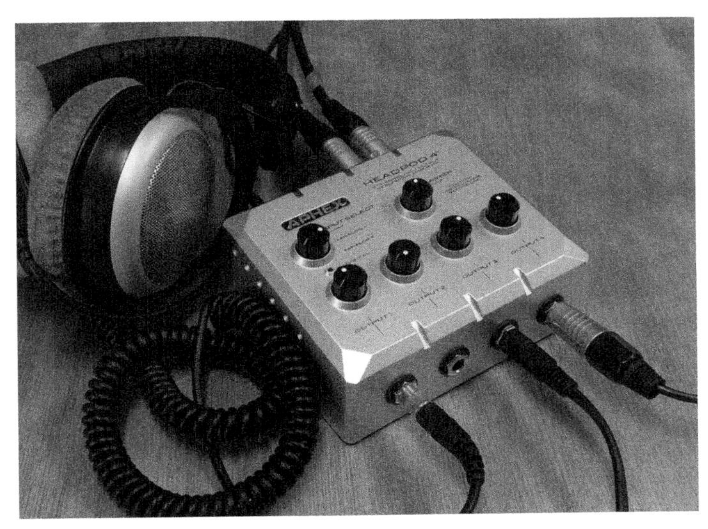

FIGURE 2.1
A dedicated multi-output headphone amplifier, such as the Aphex Headpod 4 shown here, can be very useful for small-studio work, because it gives several listeners independent control over their own listening level.

headphone extension cables, so that the musician will always have enough freedom to move while performing, and any other people in the control room can be positioned well out of harm's way!

Many of the instruments we're considering here need to rest on something like a tabletop or keyboard stand, so set that up in advance, along with a chair or music stand if you think either might be of assistance. While you're at it, consider where your cable runs would be best located (with reference to Section 1.1.3), and make sure that mains outlets and headphone sockets are placed sensibly—headphone and power cables are often rather short. Because communication is so vital, you should go out of your way to plan decent sight lines between you and your performer, and indeed anyone else directing the session. This is especially true if you're unable to use speaker monitoring, because you can transmit all sorts of information visually even when the headphone monitoring militates against verbal communication: You can cue a musician when to start/stop playing; indicate a specific note that's pitched a bit flat; express your emotional response to a killer take; or just warn someone they're in danger of spilling coffee over themselves! It's not a bad idea to have a small lamp around too, because the controls of some electronic instruments can be notoriously difficult to navigate in low-light conditions.

2.1.2 Written Materials

Although some musicians are content to overdub parts entirely by ear, almost any session will proceed a lot more smoothly with the help of some pre-prepared written materials. At the very least, it makes sense for the engineer to know what each of the major song sections is called, where it starts and ends in the timeline, its tempo and time-signature, and how many musical bars it contains. I normally put all that info onto a single piece of paper (see Figure 2.2) called a "cue sheet," but there's nothing to stop you using the metric grid and timeline markers of your DAW software in lieu. Not only does a cue sheet save time whenever the artist says, "Let's start four bars before the third chorus," but it also standardizes everyone's terminology when discussing takes. Some structural terms, such as "bridge" and "breakdown," can mean very different things to different people, so if you're not careful it's surprisingly easy to end up recording parts over the wrong song sections, or losing/obliterating some unrepeatable moment of performance heroics by mistake.

Useful as a cue sheet is, though, it won't give you much help if you anticipate working with the performer to learn/refine a new musical part in detail. In that case, jotting down the music's chord structure and main melody/bass lines prior to the session can speed up progress tremendously. Daniel Lanois[1] is a big fan of this idea, creating elaborate production charts using large charcoal-drawing books so that he's never at a loss during fast-paced musical discussions: "You have to be able to say, 'I know *exactly* what you mean.' Not, 'Oh, where's that?' and waste people's time." Finally, make sure you have plenty of spare paper and functional writing implements handy, if only for your own note-taking purposes, as we'll come to in Section 2.3.1.

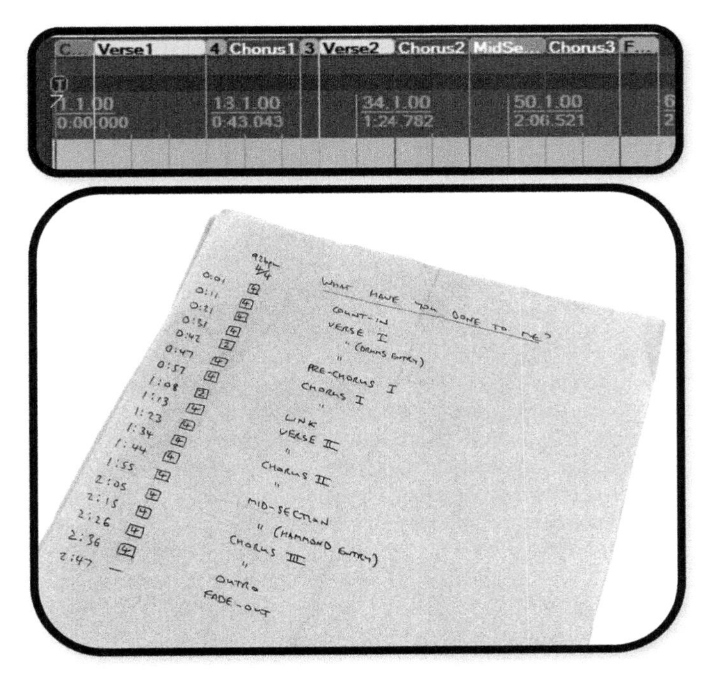

FIGURE 2.2
To keep up with the creative flow when overdubbing, it's essential that you're able to locate to any point in the project quickly. A DAW's internal locator markings (top) can help you here, but alternatively a paper cue sheet (bottom) will work with any recording system, whether hardware or software.

2.1.3 Creature Comforts

Although technicalities are important in the studio, you can also improve results by thinking more holistically about the performer's needs and the frame of mind you'd prefer them to be in while recording. Most musicians give of their best when they're comfortable, both physically and emotionally. "The most important thing an engineer has to remember," says Al Schmitt,[4] "is that he's got to keep the musicians happy."

It's always a good idea to offer appropriate refreshments (whether that means water, coffee, beer, meths…), and snacks are a welcome addition on any session lasting longer than a couple of hours. This isn't just about courtesy either (endearing though that is), because it's also a time-worn method of increasing productivity—you'll almost always get better work done if everyone's well hydrated, energy levels are high, and people's tastes in chemical mood-enhancement are catered for.

In fact, where you're on sessions that span a day or more, I strongly suggest proposing some kind of schedule for communal mealtimes, and not just for nutritional reasons. Although you never want to stick too rigidly to the agreed

BUILDING PRODUCTIONS FROM SCRATCH

Until Part 4 of this book, I'll for the most part be considering the task of overdubbing new parts alongside existing recorded backing tracks. However, if you'd rather create your own productions from scratch while carrying out the earlier chapter assignments, then don't let me stand in your way! The first challenge you may encounter, though, will be keeping the first performer sufficiently in time when there's nothing for them to play against. Sure, it's easy enough to use a metronome to prevent a musician's tempo drifting too much between takes, and that carries little danger of stamping out natural performance variations that are the heart and soul of more natural-sounding styles. However, if you're aiming for the stricter timing regimentation heard in many mainstream chart styles, then it makes sense to prepare some kind of guide backing track for the performer to listen to while recording.

This might be as simple as a metronome "click track" to keep them in time with your DAW's tempo grid, although it's rarely the best solution in practice because metronomes often sound very unmusical, and can interfere with the rhythmic groove if they're too foursquare. For this reason I usually prefer to use some kind of drum/percussion guide track that fits musically with the style being performed—free percussion loops are dime-a-dozen on the Internet these days, and frequently bundled with DAW software, so lack of availability is really no excuse. If the rhythm loop you like is at the wrong tempo, then there's plenty of freeware that can time-stretch it to suit. When choosing a loop for guide-track purposes, I usually like to make sure there's something useful going on between the main beats of the bar, because that helps keep timing on track even when powerful on-beat notes (such stomping chords or percussive backbeats) obscure the guide track's main beats. If the first performer's tuning also needs help, then a simple chordal synth pad running along with the guide loop will usually suffice from a technical standpoint, but again you may get a better performance if you use a particularly evocative sound or some kind of rhythmic figuration.

Another issue is that simple loop-based guide tracks won't allow any ebb and flow in the tempo, of the kind that good musicians provide instinctively when left to their own devices. If this concerns you, then you should perhaps investigate the in-depth tempo options available within many software DAW systems these days—these will let you create a tempo "map" which varies throughout the musical time-line, and in some cases you can even extract this map from a freely recorded guide part rather than having to input tempo values directly. Whatever you choose to do, just remember to include at least two bars of count-in before the start of the music.

There's no end to how far you can go with the guide-track concept if you're handy with MIDI sequencing, but mocking up the whole arrangement beforehand can be a mixed blessing. On the one hand it may give everyone a better idea of the final context within which their part will exist, which can make choosing timbres and performances easier. But, on the other hand, there's a risk that presenting the performer with an apparently finalized arrangement may prevent them developing some new riff, countermelody, or arrangement idea that knocks spots off the ones you initially planned.

One final consideration when constructing a project from scratch is that recording initial parts without any backing track makes it very easy to misjudge the recorded tone. Us humans are much better at comparing timbres than at evaluating them in isolation, so Toby Wright[2] suggests having some familiar commercial releases (preferably similar in musical style to your production) on hand to calibrate your ears from time to time during the earliest tracking stages. If you happen to be working with unfamiliar control-room monitoring which makes those commercial tracks sound tonally imbalanced, you might also consider putting an equalizer into your main mix buss to compensate for this, as this can also make finding appropriate tones a little more instinctive. Once you've got a couple of parts recorded, they should themselves provide enough of a tonal reference point that you can forget about comparing with the commercial releases from then on. "If you have basic tracks on tape and you're going to another place to do vocals and overdubs," confirms John Agnello,[3] "they're your frequency barometer."

times (because creative and musical considerations should always take priority), just the presence of a notional "stop for lunch" point can do a surprising amount to focus and motivate everyone's work in the run-up to that. And if you're anything like as absent-minded as me, it'll remind you to eat something before you start feeling faint...

2.1.4 Framing the Session

"Every performance is affected by the environment it's in," says Oz Fritz.[5] "Many studios have a neutral, institution-like feeling about them... the exact opposite of what you want in a creative environment." This is why there's a long tradition in professional circles of scattering drapes, posters, and music memorabilia around the studio, as well as setting up moody/novelty lighting. You don't need a black belt in Feng Shui to benefit from "framing the session" like this, though, as long as you keep a few general concepts in mind.

- **Shrinking the Space.** For many players, performing on their own in a large space makes them feel rather exposed and defensive, inhibiting their performance. This is where things like drapes and "gobos" (free-standing acoustic screens) can help a lot, creating a much more intimate and protective niche. Strong overhead lighting can also be quite intimidating, especially if you've got those starkly utilitarian fluorescent strips that blight so many classrooms, factories, and offices. Switching off overhead lights in favor of more localized lighting—even if that's just a well-targeted desk lamp and the glow from your computer screen—will therefore make most recording environments feel cozier and more emotionally secure.

"While a 'Spartan chic' approach to control-room decor may impress some clients, featureless minimalism may cause confidence problems for performers who'd rather not be the center of everyone's attention, so a modicum of attention-diverting clutter in the studio is frequently a positive thing."

- **Choosing the Visual Focus.** Another thing that'll sap most people's self-confidence is being too closely scrutinized. Jimi Hendrix was a case in point when recording vocals, remembers Eddie Kramer:[6] "He'd always face the other way. He hated to be looked at." While a "Spartan chic" approach to control-room decor may impress some clients, featureless minimalism may cause confidence problems for performers who'd rather not be the center of everyone's attention, so a modicum of attention-diverting clutter in the studio is frequently a positive thing. Angling your lighting a little away from the talent can help too. Sometimes there's a good argument for adjusting the studio setup so that you're not directly facing the performer while working your equipment, because that means you can avoid eyeballing them the whole time under the entirely defensible pretext of minding the controls. That said, there are some musicians who'll only come out of their shell when playing to the gallery, in which case shifting more focus onto them by turning the session into a mini-gig crammed full of band members and hangers-on may be just the ticket. Another important issue for computer users is that your computer screen will exert a strong draw on everyone's eyes, and can easily distract far too much attention away from what people are actually hearing. "[A lot of artists] look at music, instead of listen to it," asserts Steve Lillywhite.[7] "I don't look at the screen, ever." For this reason, I personally prefer a "for my eyes only" screen position while tracking, wherever possible, and sometimes switch it off entirely for critical auditioning.

- **Setting the Tone.** The concept of surrounding artists with imagery that resonates with the style of their music should be pretty instinctive to most people. It isn't hard to work out that a nostalgic display of pastel-hued merchandising from the heyday of Swing is unlikely to jibe well with your average Nu Metal scratch DJ. However, your studio decor also has the potential to help or hinder the atmosphere and productivity of your session in other ways. For example, if your studio looks like something from an interior design catalog, it can be tough to get performers to relax because they're self-conscious about spreading out all their necessary musical paraphernalia, or anxious that they might scuff your immaculate leather sofa. Happily, this is one problem most real-world budget studios don't have to contend with very often—indeed, one of their foremost advantages is that they frequently occupy domestic-style rooms and familiar rehearsal spaces that naturally encourage you to make yourself at home. Given how useful humor is for boosting morale on many sessions, any little badges or curios you leave lying around that'll reliably elicit a chuckle can do a lot to diffuse tension if nerves start fraying under pressure.

SHOULD I SEND THE PLAYER A MIX IN ADVANCE?

Some performers like to come to an overdubbing session well prepared, particularly classically trained musicians, who are less comfortable with improvisation. For other players, however, their most inspired response may only spring forth during their first flush of emotion upon hearing the track fresh. So when it comes to deciding whether to send them a work-in-progress mix prior to overdubbing, there can be no definitive answer. My feeling is that people who favor preparation will usually request a mix without prompting, in which case there'll rarely be grounds for denying them that. Where a mix isn't specifically requested, the only situation I might send one anyway would be if there were some specific guide parts that had to be replaced under time pressure. However, I'd be reluctant to do this under less constrained circumstances, because many players will instinctively gravitate toward parts that work best on their instrument if you don't immediately shove them toward some preconceived line.

- **Sources of Inspiration.** Circumventing creative blocks is one of the hardest things to do in the studio, particularly when deadlines are looming, so most producers keep a few things around their room that can help reinvigorate a stalled production. Probably the most celebrated example of this is Brian Eno and Peter Schmidt's "Oblique Strategies" (a deck of cards offering cryptic artistic suggestions), but there are plenty of quirky junk-shop items that can inspire equally fresh lines of enquiry: an illustrated history of the ancient Mayans, a collection of bizarre old LPs, a dog-eared orchestration textbook, a suggestively shaped ethnic instrument… in short, anything that has the potential to send thoughts in unexpected directions. A brief word of warning, though: Don't be tempted to bring mainstream entertainment vehicles like TV sets, games consoles, and Internet terminals into the studio on these grounds, because it'll probably backfire, derailing your workflow with incessant distractions even when the going's good.

Now I'll be the first person to concede that stunning performances have occasionally been achieved by placing musicians in unpalatable situations, pushing their creativity outside the comfort zone. It seems every producer has at least one faintly implausible anecdote about half-starving the keyboard player and then forcing him to record inside a washing machine during the spin cycle—or something like that! However, combative and confrontational approaches tend to be most appropriate and effective when dealing with confident and experienced players with a strongly defined performance style that just doesn't happen to fit the context of the production in hand. Working with the less experienced and commanding musicians that most small-studio engineers encounter on a regular basis, you'll almost always get better results if your first impulse is to frame the session from their perspective, trying your hardest to make them feel most comfortable. Sure, if that tactic draws a total blank you might consider playing hardball with alternative session-framing tactics, but I'd think twice even then, because deliberately antagonizing the talent can seriously damage your working relationship with them, especially if the musical results you're gambling on fail to materialize.

2.1.5 Setting up the Recorder

When your performer nails their part straight away, it can be one of the greatest thrills you'll get in the studio with your clothes on. The other 95% of the time, though, you'll need to record more than one take to get the performance you're looking for. "Almost all good recordings are built from the best parts of many performances," says Bob Rock.[8] "Getting better is simply a matter of hard work." Different engineers build their master takes in different ways, but for now we're going to learn the simplest: retaking on a single recorder track as many times as necessary to achieve the required result. On this basis, you can go ahead and prepare your recorder for this, creating/selecting a free track and routing your recording chain to it. If you're working in a software system, then naming and coloring the new track sensibly will help you keep tabs on it easily during the session.

> "When your performer nails their part straight away, it can be one of the greatest thrills you'll get in the studio with your clothes on. The other 95% of the time, though, you'll need to record more than one take to get the performance you're looking for."

Of course, if the part you're recording is of any length, it's also doubtful you'll get a flawless performance of it all in one take, so you'll almost certainly need to rerecord some specific sections, a process referred to as "punching in" (or "dropping in"). So let's say the player slam dunks a full take, with the exception of a couple of isolated phrases. While you might try recording the whole thing again on the off-chance you get perfection next time, it'll be less risky just to do a couple of punch-ins to replace only those dodgy phrases.

In order to use this facility, though, you need to familiarize yourself with the input-monitoring facilities of your specific recorder. Hardware multitrack machines normally provide two monitoring modes for any track that's record-armed (see Figure 2.3):

- **Input (or Source).** This routes the track's input directly to the track's monitor output, whatever you do with the recorder's transport controls (i.e., the Record, Play, Stop, Pause, Fast Forward, and Rewind buttons, as well as any shuttle/seek/locator functions). This mode is well suited to developing and rehearsing ideas with your instrumentalist, because you can listen to the player's live input alongside playback from the rest of the multitrack backing without having to hit Record.
- **Auto (Auto Input).** Here, if the multitrack is recording, or if it's stopped, the track's input will go directly to the track's monitor output, as in Input mode. However, whenever the multitrack is playing back, the material

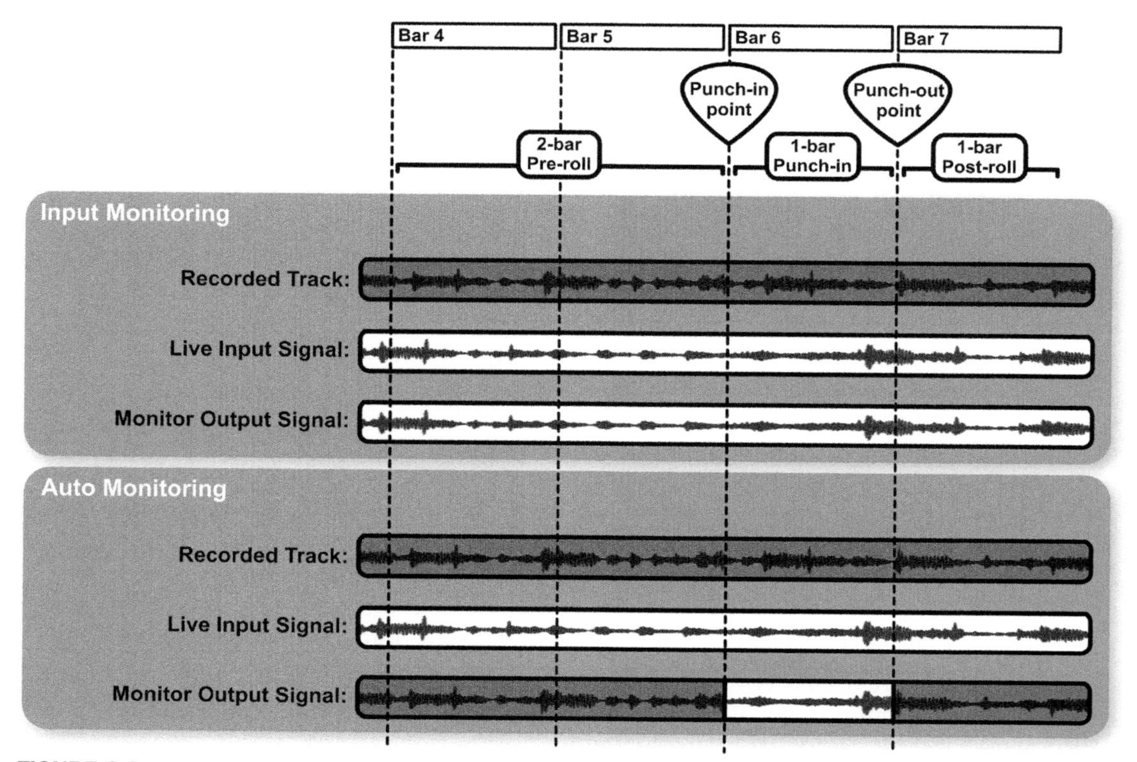

FIGURE 2.3
This diagram demonstrates how Input and Auto monitoring modes operate during a typical punch-in.

recorded on the track is sent to the track's monitor output instead. This mode is specifically designed for punching in, and to explain why this is, let's imagine I wanted to repair a mistake in bar six of a take I'd just recorded (as in Figure 2.3). I'd maybe start playback at bar four, giving the player a two-bar "pre-roll" during which we'd hear the pre-recorded take for orientation purposes. During the pre-roll the performer would play along to get into the swing of things, but wouldn't hear their live playing until the punch-in began at bar six—at which point the prerecorded track's signal would be muted and the player's input signal would be heard instead. After punching out at the beginning of bar seven, we'd once again hear the prerecorded track until I stopped the machine (the "post-roll").

In software, things can get a bit more complicated. Every DAW platform implements its monitoring functions slightly differently, so you'll have to do a bit of detective work to discover which combination of track switches and project preferences allows you to recreate these two different input-monitoring functions. In some cases, you may find that Auto mode is called "tape-machine style," for instance, while in others the Input/Auto terminology isn't used, so you have to work things out from first principles. Whatever the technicalities in your DAW, the crucial thing is that you work out how to implement both these fundamental input-monitoring modes, and also learn how to switch between them in the blink of an eye.

Another issue you may have to contend with is if your DAW's latency (i.e., the unavoidable processing delay incurred by a live input signal as it passes through the computer on its way to your monitoring hardware) is sizable enough to impinge on the musician's ability to perform. Although latencies below 15 ms or so are barely detectable in rhythmic terms, above this the instrument will begin to feel like it's responding a little sluggishly. Up to around 30 ms latency most players will probably be able to take this in their stride, compensating for the delay by marginally anticipating the beat, but once you get beyond 50 ms the performance timing will certainly suffer unless the part has very little rhythmic content by nature.

ALTERNATIVES TO SOFTWARE MONITORING

If you're a computer user, then for the purposes of this chapter you should stick with your DAW's standard "software monitoring" facilities—in other words, you should monitor the input signal through the DAW, using the software's internal mixer to adjust the balance between the input signal and the backing track in your loudspeakers/headphones. However, it's as well to be aware at this point in our progress that there's also a popular alternative to software monitoring, which is usually called "low/zero-latency monitoring" or "direct monitoring."

This involves splitting the input signal so that it simultaneously feeds both the software's recorder track and the post-software monitoring chain. For this to work properly, the DAW's software monitoring must be disabled (otherwise it'll cause comb-filtering, as mentioned in Section 1.4.4), and the mix you hear on your loudspeakers/headphones has to be created outside the DAW software. The necessary splitting and mixing duties can be executed in the analog domain, banishing latency altogether, or in the digital domain via a dedicated DSP chip within the computer's audio interface, thereby avoiding any delays caused by the computer's software (usually the most punitive). I'll fully explore the ramifications of these schemes in the next two chapters, but the main thing to realize for now is that your computer audio interface may already have such latency-reduction facilities built into it, and you should ensure these are switched off at the moment so that they don't subvert the operation of the software monitoring.

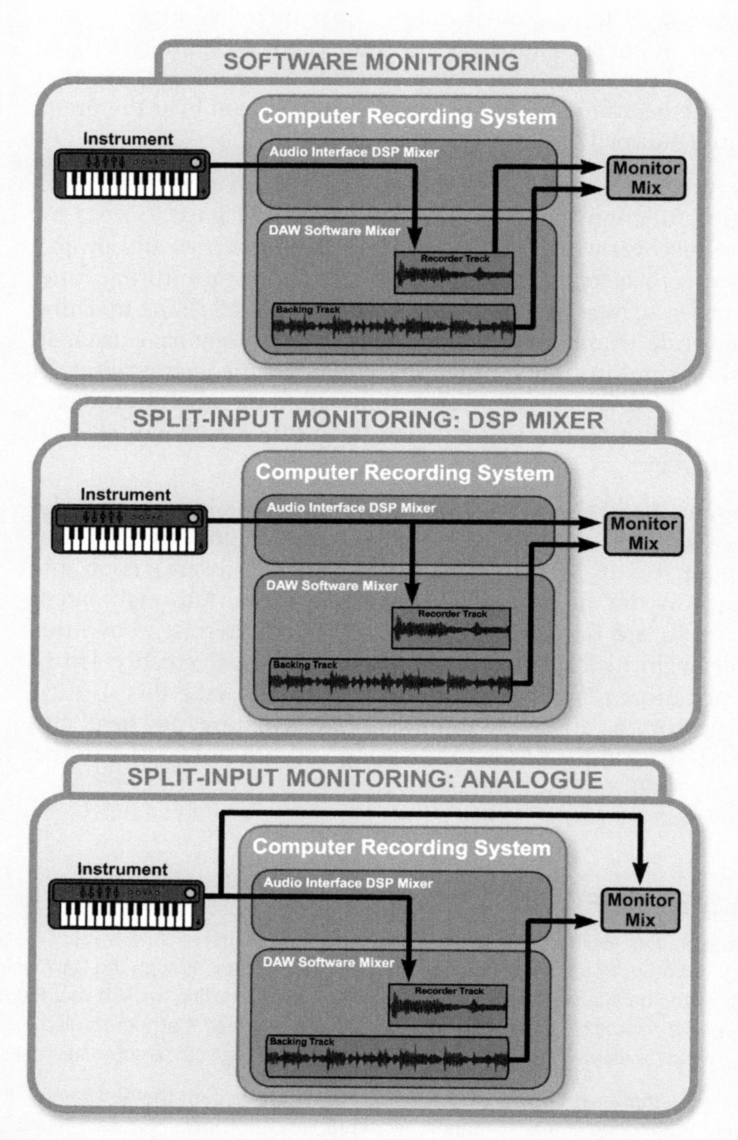

FIGURE 2.4
These three block diagrams show how a software monitoring setup (top diagram) compares with two different types of split-input monitoring, the first using a DSP-based digital mixer (middle diagram) and the second using analog hardware (bottom diagram).

The easiest way to reduce the latency of your system is to go into the audio-interface driver's configuration settings and decrease the audio buffer size—at normal 44.1 kHz/48 kHz sample rates you'll need a value below 300 samples if you're aiming for sub-15 ms latency. However, this will also increase the strain on your computer's CPU, which may mean your audio starts glitching before your performer's happy with the timing, even if you already did your best to lighten the machine's workload in Section 1.4.3.

In this case, the first thing to try is removing all processing plug-ins and virtual instruments from your project. If this usefully reduces the CPU load and/or delay, then return the project to its original state and see if you can slim down the number-crunching requirements more selectively. You may be able to substitute troublesome plug-ins or instruments for more well-behaved alternatives, for instance, or else bounce down (or "freeze") their output to a new audio file so you can disable them—you can always reverse these changes at mixdown, when higher latency is rarely a concern. In general, third-party effects are likely to be less efficient of CPU resources than those bundled with the DAW, especially those oriented toward mastering activities. It's also worth pointing out that plug-ins powered by external DSP hardware (such as those designed for Universal Audio's UAD cards) will incur some transmission delays as data is sent to and fro, so I prefer to avoid those at the tracking stage.

If combining all these measures still doesn't deliver manageable system latency, then either you have the kinds of underlying computer problems that require specialist attention, or you need to change your audio interface. Occasionally, circumstances may dictate that you use a PC's built-in audio sockets, rather than dedicated audio interfacing, in which case a generic ASIO driver such as Michael Tippach's freeware ASIO4ALL may allow you to work at usably low latency—but it won't make the audio quality any less ropey!

One final point where monitoring delays are concerned: It takes roughly a millisecond for sound to travel a foot (30 cm) through the air, so moving 10 feet (3 m) closer to the monitors will effectively reduce your system's apparent latency by 10 ms. If a more distant playing position can't be helped, then this may be a strong argument for giving the performer headphones, even if everyone else in the room continues listening via loudspeakers.

2.1.6 Checking the Monitoring

For the purposes of this chapter, we're going to constrain ourselves to using a single monitor mix, which will be shared by everyone in the control room. As monitoring arrangements go, this is the apex of simplicity, but you still want to minimize the amount of session time spent adjusting it before you start recording. So hook up some other line-level source (perhaps the playback machine from Chapter 1) to your recording chain and check that both input signal and backing mix are coming through clearly, and at roughly the right relative levels. If you're using headphones, make a point of listening to every pair to check for malfunctions or unwanted distortion, and give the plugs

MEASURING MONITORING LATENCY

The only reliable way of finding out the latency of a studio DAW system is to measure it yourself. Here's a fairly straightforward method:

- Find a stereo line-level sound source and use it to play a mono sound, which will result in identical signals appearing at its left and right outputs. The sound should also contain well-defined transients, so that any subsequent timing discrepancies are easy to measure. A synth or sampler should fit the bill here, but you could also use a simple playback device if you've got a suitable mono recording for it to play. (You can download an audio test file from this chapter's web resources if you need one.)
- Get hold of a separate audio recording device that can capture stereo. This doesn't need to be complicated or high quality, as long as you can get line-level signals into it and there's some way to export its recordings back into your DAW.
- Connect the source's left-channel output directly to the left-channel input of the separate stereo recorder. Connect the source's right-channel output to your DAW system, routing it and setting the gain as if you were going to record it, and then connect one of your studio system's main monitoring outputs to the right-channel input of the separate stereo recorder. (This setup can be seen in Figure 2.5.)
- Adjust the input gain of the separate stereo recorder if necessary to avoid clipping, and then record a few seconds of the source signal. (There's no need to record anything on the DAW system itself.)
- Create a fresh DAW project file, into which you should import the recording from the separate stereo recorder. Zooming in on the waveform display, you'll be able to see how much the DAW system's latency has delayed the right-channel waveform, and you can then measure it using your software's time displays.

If you're unable to lay hands on a separate stereo recorder, an alternative is to simultaneously record both the sound source and the studio's control-room monitoring signal to separate tracks in your DAW. Your system's latency will then be revealed as the time offset between the two recorded waveforms. However, if the track which receives the "loopback" signal from the control-room monitoring outputs itself feeds the monitor mix, you'll get instant feedback howlround, so it's important to guard against that in some way—for example by pulling the fader on that track's monitor channel all the way down. If you're on a PC, there's freeware such as Centrance's Latency Test Utility and Oblique Audio's RTL Utility that will perform this kind of loopback latency test without any danger of feedback. Neat as these are, though, they won't pick up any latency incurred by plug-ins within your DAW project, so they won't give you as complete a picture of real monitoring predicaments.

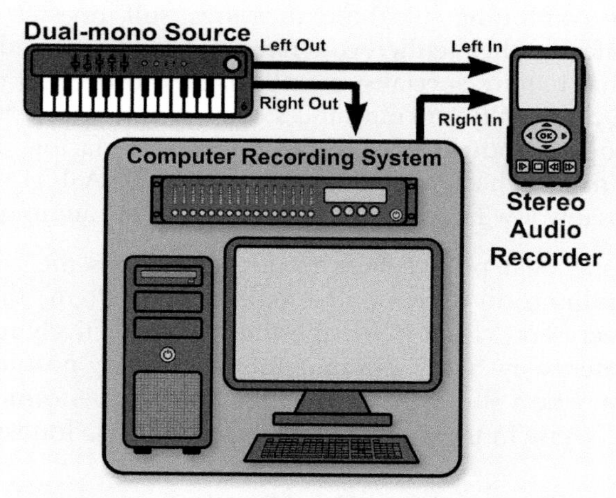

FIGURE 2.5
A simple setup for measuring a DAW system's audio latency.

a gentle wiggle to check for intermittent solder joints or socket contacts. The goal here is that, once the player arrives, you should only have to nudge the recording track's monitor fader a bit, and maybe the loudspeaker/headphone volume levels, before the monitoring is good to go. One of the most frustrating things for a performer is when they're forced to sit on their hands for ten minutes while the engineer gets their headphones working—almost as if it were a surprise that musicians might need to hear themselves while recording!

While you're checking all that, briefly reassess any default balance of your backing parts in relation to the type of instrument and part you're about to record. If the rhythmic accuracy or groove of the overdub is of primary importance, then you might try pulling out any parts that muddy the water there, particularly any guide parts designed only to showcase the arrangement, and which won't be used in the final mix. However, it will normally make sense to include everything you've recorded so far, so that you can more easily design the overdub's musical part and timbre to fit within that context—after all, the better it fits, the more straightforward the mixing process will be. By the same token, any of the existing parts that you envision having obviously audible special effects (atmospheric syncopated echoes, say, or super-heavy flanging) are worth presenting in that way for monitoring purposes, so that the musician can react to them appropriately.

LISTENING LEVELS WHILE OVERDUBBING

A lot of musicians like to listen loud while performing, often to emulate the visceral experience of performing live through a fusion-powered PA, and there's no arguing against this in terms of production priorities if it actually gets the performances you're seeking. However, high listening levels can cause irreversible damage to the human hearing system, especially if maintained for hours on end during extended studio sessions, so I consider it your moral duty as an engineer to alert the talent if you think they're endangering their long-term career prospects in this manner. From time to time, though, you'll find yourself stuck in a studio with people who will cheerfully ignore you, adopting a defiantly cavalier attitude not only to their aural health, but also to yours. In the company of such personalities, the engineer's natural impulse to serve the music results in a strong temptation to just go with the flow and hope for the best, so as not to sour the mood. If the nebulous threat of premature hearing loss isn't enough to help you resist that urge, let me provide some other reasons that are more closely related to the job in hand.

The biggest practical difficulty with trying to engineer while listening at deafening volumes is that you immediately lose perspective on the quality of what you're recording. It's programmed into our psychological make-up that louder playback will sound more appealing (so you won't work as hard on refining the timbre) and more exciting (so you won't work as hard on the performance). "I like to listen loud," comments Jeff Powell,[9] "but I also know you need to turn it down and see what you can hear. If you can't hear your cool guitar sound on a medium to low volume, you probably don't have a very good, or exciting, guitar sound." In addition, it's easier to hear all the details in a mix when you turn it up, which means you're less likely to leave enough space in your arrangement for those details to be heard at more moderate listening levels. And, of course, it'll probably only be a few minutes before hearing fatigue renders your decisions pretty worthless at low levels too, and probably for the rest of the day at least. So if you're really interested in serving the music, do take precautions to protect your ears if you feel obliged to work at uncomfortably high monitoring levels. Overall I reckon you'll do a better engineering job if you wear earplugs while the monitors are blaring, and then evaluate the results at more reasonable levels during the inevitable cigarette breaks.

So why not go the whole hog and polish things further with detailed channel processing, high-quality mixdown effects, computer-controlled parameter automation, and even mastering-style processing? You could argue that this would make engineering judgments even clearer, and also ease the load at mixdown if you're handling that yourself. However, personally I'd advise against getting too carried away, particularly on smaller-scale projects. First and foremost, you're more likely to fall foul of latency or general reliability problems with more processing in circuit. But apart from this, I'm not convinced that mixing as you record is actually such a good use of time for small-studio productions, because it encourages questionable mixing practices. For example, the order you mix your tracks can significantly alter the way the mix turns out, and in some cases the most sensible mixing order may even be the opposite of the tracking order—many producers record pop vocals last, for instance, whereas it's often smartest to fade them up first of all at mixdown.

Furthermore, I think over-sweetened monitor mixes can actually have a negative impact on the recording quality. It's easy, for instance, to add too many effects to an arrangement that isn't yet complete, which means that there's less room in the mix for any new overdub. The result is that you won't try quite as hard to hunt down characterful and engaging raw materials. Compound that effect over several stages of overdubbing and you have a recipe for blandness, unless you already have a great deal of experience at avoiding this pitfall.

One final thought on the matter. During the process of building an arrangement via overdubbing, everyone involved with the process may end up hearing the production hundreds of times over, and it's a natural byproduct of such overexposure that things can begin to feel slightly stale, no matter how exciting the music may appear to fresh ears. Leaving your monitor mixes comparatively unadorned while tracking, leaves you more scope to reinvigorate the artist's enthusiasm at the mixdown stage.

PERFORMANCE HEADROOM

I introduced the concept of headroom back in Section 1.3.3, but the meter readings at that point were based on the assumption that you were recording a machine. When humans are involved, things aren't nearly as predictable: Not only are no two human performances ever identical, but a lot of people will naturally play louder while actually recording than they do when you ask them to play along with the backing track for the purposes of level-setting. Because of the slow response and built-in headroom of VU meters, allowing an additional safety margin of 2–3 dB on their meter reading should be plenty to account for this, but with PPMs you may want to bring the peaks down 5–6 dB for peace of mind, and I'd allow an extra 12 dB where digital peak meters are concerned (i.e., peak readings between −18 dBFS and −12 dBFS), given that digital systems usually have extremely low noise and distinctly ugly-sounding distortion characteristics. Even so, you should still watch your meters like a hawk during the first few takes particularly, keeping your gain controls poised to react swiftly if any peaks start getting scorched.

2.2 THE MUSICIAN STROLLS IN...

You should now be ready for the musician to stroll in. Once the instrument is unpacked and you've plugged it in, the troubleshooting and gain-management skills you developed during Chapter 1 should equip you to capture the performance in high definition. Viewed purely from an engineering perspective, your initial workflow might therefore proceed along the following lines:

1. Generate some arbitrary output from the instrument so you can line-check.
2. Play the backing track so everyone can find a suitable loudspeaker/headphone monitoring volume.
3. With the recording track in its Input monitoring mode, ask the performer to play along with the backing mix while you set levels.
4. While the performer continues to play, balance the instrument against the rest of the monitor mix using the recording track's monitor-channel fader (i.e., the main track fader in most DAWs).
5. Refine the instrument's recorded timbre if necessary, in discussion with the musician and anyone else with a stake in the outcome.
6. Hit Record.

Real-world sessions, however, will frequently play havoc with such neatly laid plans, for the simple reason that the last of the steps listed above is more important than any of the other five.

2.2.1 The Importance of Hitting Record

If the musician plays something magical before you've hit Record, you've flunked the most important challenge of all, which is to serve the music. "Probably the worst thing that could happen at a session," confirms Phil Ramone,[10] "would be missing a take." "Vibe is all about spur of the moment," adds Wyclef Jean.[11] "Being ready when the thing happens. You don't want to miss it when it happens...There's no second takes for vibe." On some occasions that may well prevent you addressing some (or all!) of those other five steps until *after* you've started recording the first take, but you just have to take that in your stride when it's performances you're looking for. "If something goes wrong, work around it," asserts Jacquire King.[12] "You never want to stall. Just hit Record, even if something is not to your satisfaction engineering-wise—if you record something that is inspired, it doesn't really matter."

> "If the musician plays something magical before you've hit Record, you've flunked the most important challenge of all, which is to serve the music."

So imagine that a friend of the band you're recording offers to lay down some keyboard overdubs for them between gigs. She's brought her well-loved stage piano with her, so she parks it on the stand you've so thoughtfully provided and asks to have a listen to the song you've all been slaving away at, seeing as she hasn't heard it yet. Within eight bars, the genre-busting awesomeness of your Hawaiian trance rock has her completely freaking out—dancing round the room, punching the air, the whole bit. By the time the chorus arrives, the urge to join in this new musical revolution proves overpowering, and she flings herself at her keyboard as one possessed—the same keyboard you've barely finished powering up and plugging into your rig. Now is the time to hit that red button. It doesn't matter whether *you* feel ready to record, because the musician certainly is, and may never again be as swept away by the music once the power of its first impression has dissipated.

This is why your preparations are so important. You need to have all your gear set up so that "even without hearing the music, you can hit Record and get a decent sound," says Justin Niebank,[13] for instance. "If you think it's going to take you a day to set up," adds Jacquire King,[14] "well then do that day of setup before somebody is waiting on you." Assuming you've already line-checked your own system, and sensibly prepared the multitrack recorder, then it'd be the height of bad luck in this case if the keyboard player's signal didn't find its way to the recorder—the only thing you've been unable to defend against is a fault in the instrument itself. And by using a playback device to make ballpark gain and monitoring settings in advance, you've also dramatically increased the chances that our guest will be able to hear herself, and at some kind of useful level, from the moment she touches the keys, without a bunch of noise or distortion breaking the mood. At which point you're a hero, because the recording should, at the very least, be usable. Give it another twenty seconds, during which you're frantically twiddling the audio interface's Input Gain knob with one hand and the recording track's monitor-channel fader with the other, and you'll have refined both the signal quality and the monitor balance—all in good time for that blistering middle-eight solo that's about to sear itself onto a new page of musical history. Hero? That'll be *super*hero, thank you very much!

GETTING THE BEST OUT OF THE DIGITAL MEDIUM

There continues to be heated debate in the recording industry about the pros and cons of different recording formats for studio work, but most small-studio users don't have much choice in the matter—they only have the budget for digital recording technology. In this context, though, John Kurlander[15] and Mike Poole[16] both mention a trick that helps you make the best of this situation: Don't ever listen to the input signal directly. Monitoring through the digital system forces you to adapt your recording technique to what you're hearing, which means you'll naturally try to factor out any undesirable characteristics inherent in the recording medium.

2.2.2 Playing for Time

Now while it's absolutely your responsibility to deal with seat-of-the-pants scenarios like this when they arise, that certainly doesn't mean you should work that way by default—quite the opposite, in fact, because it leaves rather too many opportunities for Lady Luck to rain on your parade. Maybe someone dropped the keyboard while they were loading it into the car, or it's powered up with its default (and wholly inappropriate) "Astral Didgeridoo" patch, or you plugged into the wrong output by mistake, or you forgot to arm the recorder track. Although preparation is the bedrock of any session, there's only so much you can do before the fact. Which is why one of the real secrets of successful engineering is learning how to play for time. For example, when our hypothetical keyboardist first dumped her keyboard on the stand, you might have said: "Hey, is that a real Casiotone? I've heard that does a mental Wurlitzer!" Whether or not this is a big fat lie, flattering any musician's taste in hardware never does any harm, and if it naturally triggers a few seconds of showing off at the ivories, you'll have a gift-wrapped opportunity to line-check, adjust your recording-path gain controls, and set a preliminary monitor-channel fader level. Within a minute, you can then pipe up with "Can't wait to hear what you're going to do on this song!" and dive back into Plan A, safe in the knowledge that you're all but guaranteed a decent-sounding result, no matter how soon you're called upon to hit Record.

In larger studios, the assistant often steps into the breach here too, taking beverage orders and dropping a choice one-liner about the battery in the engineer's hearing aid, during which time you've surreptitiously played a few notes to service the dull but important technicalities and given that cheeky rascal a good-natured kick in the shins. It's possible to pull off similar stunts single-handed too if you've got the gift of the gab, but it takes a bit of practice to maintain a diverting patter while troubleshooting missing signals or concentrating on meter readings. This is where your session framing can really help out: If the detailed map of Discworld on your studio door or the enormous inflatable Godzilla behind your mixing console prompts 30 seconds of awestruck perusal or self-propelling discussion, then they've earned you a valuable head start.

Clearly, though, such sleight of hand won't work every time, and also carries its own risk: that you'll inadvertently derail a fragile creative impulse that had already germinated just before the musician crossed the studio threshold. On every single session you've got to weigh up that risk against the risk that you'll hit a deal-busting technical obstacle because you didn't have time to nip it in the bud before hitting Record. There's no "one size fits all" game plan that'll work on every session and, frankly, that's all to the good, because otherwise recording wouldn't be nearly as much fun! Just so long as you keep asking yourself what's in the best interests of the music, you shouldn't drastically misread the situation too often. When (not if) you perpetrate a truly epic fail, however, you can console yourself that there are few things more powerful than a thoroughly cringe-inducing memory at spurring you to further hone your chops—and there isn't one of the world's finest engineers who couldn't regale you with a dozen toe-curling tales from their rookie years, each of which informs their session technique to this day.

"When (not if) you perpetrate a truly epic fail, you can console yourself that there are few things more powerful than a thoroughly cringe-inducing memory at spurring you to further hone your chops."

2.2.3 When to Tackle the Sonics

So much for the essentials of capturing a signal: what about trying to make that signal sound incredible? Again, the extent to which you'll be able to grapple with that without unduly compromising the quality of the performance is very much a judgment call on any session. On the one hand, anything that slows down the natural pace of a musician's creativity may endanger the spontaneity of the following take or, worse still, prevent embryonic ideas from gathering sufficient momentum to be of benefit within the session time available. On the other hand, a gorgeous recorded timbre will frequently fuel a player's imagination, and can save an enormous amount of time and effort at the mixdown stage.

Personally, I'm a great believer in involving the musician in the decision-making process here. In other words, if at any time during the session I think some extra attention to the sonics may help the end result, I'll say something like: "I think we may be able to improve the sound here. Can we just try some things out for a couple of minutes?" Such a request very rarely gets a negative response, because every performer wants to sound their best, as long as you're not daft enough to pop that question while everyone's right in the middle of working through an important idea. (Even then the worst you're likely to get is "Can we quickly finish this first?"—provided that it's not the umpteenth time you've stuck your oar in at the wrong moment!) Explicitly bringing the musician on-side in this manner has two big things going for it:

- You can ask the performer for help. Their understanding of the mechanics of their own instrument will frequently outstrip yours. I've already stressed that changing the sound source itself usually bears fruit more quickly than footling with your recording chain, so that makes the musician the best friend you've got when time is of the essence.
- You'll put less strain on your performer's patience. They understand your reasons for the hiatus, and you gave them the right of veto in case they preferred the sound as it was or were focused on more pressing musical concerns. In addition, they now have a rough idea how long the hold-up's going to last—waiting around's always a whole lot easier when there's a clear end in sight.

The same trick is equally relevant if you're fortunate enough to be working with musicians who appreciate that designing a great tone is as much in their interests as yours, and who are willing to spend some time pursuing that goal before attempting takes. This is certainly preferable from an engineering standpoint, so take full advantage of this concession wherever it's on offer. Just remember to clarify the terms of engagement, as I've already mentioned. "How about we spend the next half hour finding just the right sound for this song, and then plan to blast out some takes around eleven?" This lets the performer know precisely where they stand, and also gives anyone else involved in the session the option to sit out the tone-tweaking activities if they want to keep a clear head for concentrating on the quality of the actual takes. Personally I like to slightly overestimate the time we'll need, not only because it gives me a bit of leeway, but also because it can generate a few positive vibes on those occasions when you apparently complete your tweakery ahead of schedule. However, whenever you're sound-mining for any longer than about five minutes, you should always be vigilant for any signs of waning stamina, and step on the gas well before this jeopardizes the creative atmosphere.

Sometimes, though, catching a mercurial talent's first outpourings effectively railroads you into recording a less-than-stellar sound, because if some parts of the very first performance are the bee's knees, then it's indisputably in the interests of the music to keep them, whatever technical difficulties that may cause at mixdown. Any mix engineer would rather be faced with salvaging a poor sound than a poor performance! Within that context, there's little to be gained by tweaking the recorded tone after the first take's gone down, because you'll just be adding a further mix problem: namely, matching the new sound with that of the first take. That's not to say I'm proposing a defeatist attitude, though, because you should nonetheless remain alert for possibilities to address any sonic deficiencies as the session proceeds. For example, an alteration to a musical part may mean rerecording from scratch, in which case you'd be well advised to pounce on whatever chance that gives you to refine the sound without mixdown penalties. If an instrument plays several distinct roles in different sections of an arrangement, then the mix engineer will often seek to contrast those parts anyway, so any timbral changes you make there shouldn't add significantly to the mixing workload. Even when you're only able to make enhancements between one song and the next, or between one session date and the next, that's still got to be better than nothing.

2.2.4 Refining the Sound

Whatever time you have available to deal with sonics, it'll usually feel too short, so it's important to prioritize. For this reason, try to work from coarse tweaks to finer ones rather than obsessing about little details early on. There are thousands of things you *can* do to change a sound, but what sets more in-demand engineers apart is the knowledge they've accrued about what changes will tend to reward you with the most dramatic improvements for each different type of instrument. So try to develop this critical faculty early on. If there's some instrument setting that makes a huge difference to the sound (say, choosing the sample or synth patch you're going to use), then resist the temptation to skate over that. Even if you think you've found the right sound in two seconds, take a little extra time to sift out a couple of other strong contenders, and then do a swift shootout to eliminate the weaklings on the shortlist—more often than not your first choice of sound will end up on the cutting-room floor. By the same token, when you find yourself tweaking something that's not making an enormous difference, there are two questions you need to ask yourself: "Is there a faster way?" and "Can I do this later?" I'm not trying to suggest that you should deliberately defer the hard work of finding great sounds, but the reality of small-studio work is that you're going to be battling the clock

RECORDING WITH PROCESSING

On a professional level, it's long been common practice for engineers to process their input signals on the way to the recorder. One reason for this is purely technical: Many renowned practitioners learnt their trade in analog studios, where preprocessing can help mitigate some technical shortcomings of the equipment. For example, compressors and limiters can turn down signal spikes, allowing you to keep your signal further above a tape recorder's noise floor before you run into unacceptable peak distortion. Tape recordings also gradually lose high-frequency definition as they're repeatedly played back during the lifecycle of a production, so pre-emphasizing the input signal's upper spectrum with an equalizer can help compensate for this, with the added benefit that redressing any overemphasis at mixdown reduces the relative audibility of the tape's HF-heavy background hiss. And, of course, there's always a limit to how many physical processing units you have available in an analog studio, so you'll get more sonic value out of them if you use them during both the tracking and mixing stages.

Beyond this, though, most working engineers also treat processing pragmatically, as a means of making their workflow more efficient—if you can manage to record every track sounding the way it should at mixdown, it's easier to judge the sonics of each overdub in the proper context, and the full arrangement should then pretty much mix itself. From that perspective, any processing that can be "printed in" as part of the initial recordings is one less thing to worry about before deadline.

Despite all that, however, it's my firm belief that most users of small studios are much better off recording without any processing at all. For a start, decent 24-bit digital recording technology is now so affordable that it's child's play to record cleanly without any recourse to compression or limiting. Also, the fidelity of a digital recording doesn't deteriorate over time in the way an analog recording does, so there's no call for pre-emptive processing on that account either. The virtually unlimited plug-in count of modern software DAW systems means most people *can* leave all their processing to mixdown too, an added incentive being that lower CPU drain while tracking decreases the likelihood of digital glitches or general system instability.

However, my biggest reasons for urging you to forget about processing during tracking aren't technical. It's as well to remember that most of the big-name engineers whose recording techniques you read about are working on projects as part of a team—which may include DAW operators, studio assistants, drum/guitar technicians, voice coaches, roadies, caterers... Within that context, they usually have more time available to spend on pure engineering matters. By contrast, on a low-budget session you'll usually be transporting, setting up, and operating all the equipment single-handed within an extremely limited timeframe, whereupon the simple technicalities of piloting the gear at speed become overwhelming unless you adopt a streamlined approach—especially if you're faced with donning the arranger's or producer's hat as well, or if you happen to be the band's drummer! Performers will rarely look kindly on you holding up their session to twiddle compressor knobs, even if you manage to avoid the danger of overcooking your settings in the heat of the moment.

But on top of that you have to ask yourself what your priorities should really be while recording. Over the years I've mixed dozens of home-brew multitracks for *Sound On Sound* magazine's "Mix Rescue" column, and the worst recordings usually come from people who've chickened out of tackling sonic problems at source, and have then compounded the error by trying to fix the unfixable with a zillion plug-ins. If you have any time left over between sorting out your levels and pressing Record, then it's always best ploughed into making the source itself sound as good as possible. And an important ancillary benefit of treating signal processing as a last resort is that it disciplines you to work harder on your actual recording chops, which means you need it less anyway!

a lot of the time, in which case it's better to prioritize those things that will advance the timbre furthest before time runs out, and those things that you can't address after you've hit Record.

It would be utter foolishness for me to try to define what a "good" sound is, because one man's meat is frequently another's poison in this respect. If you're unsure what sounds work in your genres of choice, pretty much the only way you can develop your own instincts is by listening widely to relevant commercial productions. That doesn't mean you have to be an avid consumer of every kind of music you record, though, so long as you're working with musicians who specialize in their chosen style—in which case you'll be able to rely heavily on their tastes to guide you whenever you're at a loss. One of the classic recipes for disaster on self-produced projects is when a band wants to push itself in an unfamiliar stylistic direction, but with the help of its usual engineer, who (it transpires) actually knows even less about the demands of that style than the band does!

So while there's very little point in my lecturing you on sonic aesthetics, there are some general suggestions I'd make for skirting around some of the most common practical pitfalls:

- **Enlist the Musician's Help.** There's always something new to learn about any instrument, which means your player may know a quicker way to reach a target sound. So don't just knit your brow and struggle on taciturnly when results aren't immediately forthcoming; discuss your concerns with the instrumentalist. I find it's best if you speak in terms of more abstract sound and arrangement issues ("Can we take some of the edge off the front of each note?" or "Perhaps something more nasal would contrast better with those other piano chords"), rather than suggesting concrete changes ("Can you increase the envelope attack time?" or "Perhaps we should add more 2 kHz"), because you'll often discover that the performer has a better idea of what'll help ("That spike's only one of the sample layers in the patch. Let's just try switching it off first," or "How about this death-metal ukulele patch instead?").
- **Listen in Context.** Your aim as an engineer should be to make the whole record sound fabulous, but that isn't the same as saying that each part should sound fabulous in isolation. Far from it, in fact, because there's only ever a limited amount of space in any mix. Giving your most important voices and instruments enough elbow room to sound glorious may well result in background parts which sound pretty hideous when soloed—but of course the audience never hears them that way. "You're going for a sound that's appropriate for the song," underlines Roy Thomas Baker,[17] "not necessarily what's a good sound, and what's appropriate can vary greatly." So never let any musician sign off on their sound before everyone's heard how it works when they're playing something representative over the backing track. This is one of the biggest mistakes small-studio engineers make, so I can't stress its importance enough!

- **Flex that Monitor Fader.** One of the most reliable signs that an overdub's timbre is well suited to your production is when you can hear all its most desirable tonal qualities without it feeling too loud overall. In this happy event, you're also more likely to find that it's easy to set the recording track's monitor-channel fader for a respectable rough mix balance, and that you can leave it alone most of the time. Whenever you feel that you can't hear the personality of the overdub unless it's way too loud, or that you're forever fiddling with its monitor-channel fader throughout the track, that probably means the balance is trying to tell you some home truths about your sonics. From an arrangement perspective, however, a part which can be pushed high in the balance without getting in the way of other more important parts is usually serving the production much more efficiently than one which can only be tolerated at low levels.

RECORDING WITH EFFECTS

Many of the instruments we're considering in this chapter commonly have effects processing built into them, but it's usually a good idea for this to be switched off while tracking, otherwise it can interfere with the mix engineer's work later on. That said, don't reject the idea of printing in effects on reflex, because many high-profile engineers (Rhett Davies,[18] Kevin Killen,[19] and Jay Messina[20] amongst them) have no qualms about making this kind of sonic commitment well before mixdown. Some keyboards and samplers can implement incredibly complicated multi-effect chains that would be extremely time-consuming to recreate at the mix, while the vintage analog effects in classic hardware synths frequently prove inimitable by digital means, so if everyone in the room loves the sound they're hearing and you've no clue how you'd get that sound yourself, why not print them and sidestep some unnecessary mixdown aggro? As long as you don't record every instrument with heavy effects, it really shouldn't tie the mix engineer's hands unduly—although do try to be as conservative as you can with delay-based effects and artificial reverberation, which are more likely to be problematic at the mixing stage.

On occasion, though, you may be able to have your cake and eat it too by recording an additional track of "dry" signal (without any of the effects) concurrently. On the off-chance that the effects prove OTT at mixdown, you'll then have the option of mixing in some dry signal to reduce their relative level, or of just replacing them entirely. Studio-oriented synths and samplers often have more than one pair of outputs for precisely this kind of reason, and it'll usually be possible to extract a separate dry signal from any laptop-based performance system that uses a dedicated computer audio interface. Once your musician has worked out how to squirt the right signals out of the instrument, all you have to do is connect them all up to separate inputs of your recording system and arm a larger number of tracks for recording, which hopefully shouldn't cause you to break a sweat by now. Where an instrument is equipped with MIDI sockets, an alternative strategy is to record a MIDI track alongside, so that you can retrigger the same performance at a later date with different effect settings—or indeed a completely different MIDI instrument entirely.

- **Go Easy on the Distortion.** Although distortion can be the very soul of a lot of instrument sounds, a frequent shortcoming of small-studio multitracks is too much drive. The problem is that the noisy elements of a distorted sound always cut through a mix better than the instrument's pitched components and attack character, so anything you overgrill will tend to sound spongy and distant. It also means that players are often misled into thinking that the sounds on commercial records have been driven harder than they actually were, so they end up unwittingly exaggerating their own settings. My own rule of thumb is that if any musician dials up a distorted sound while listening to it on its own, I'll usually want to reduce that drive setting by 25% or so once we're hearing it in its proper context. It's a doddle to add more distortion at mixdown if you really need it, but excess printed-in fuzz is one of those things that's virtually impossible for a mix engineer to fix later on.

- **Look out for the Low End.** Back in Section 1.5, I advised against processing your input signal while recording, but there's one processor that I make an exception for: the high-pass filter. This is a basic type of equalizer that cuts away low end below a certain "cutoff" frequency, and the further down the spectrum you go, the more it cuts. The most useful design has a control to

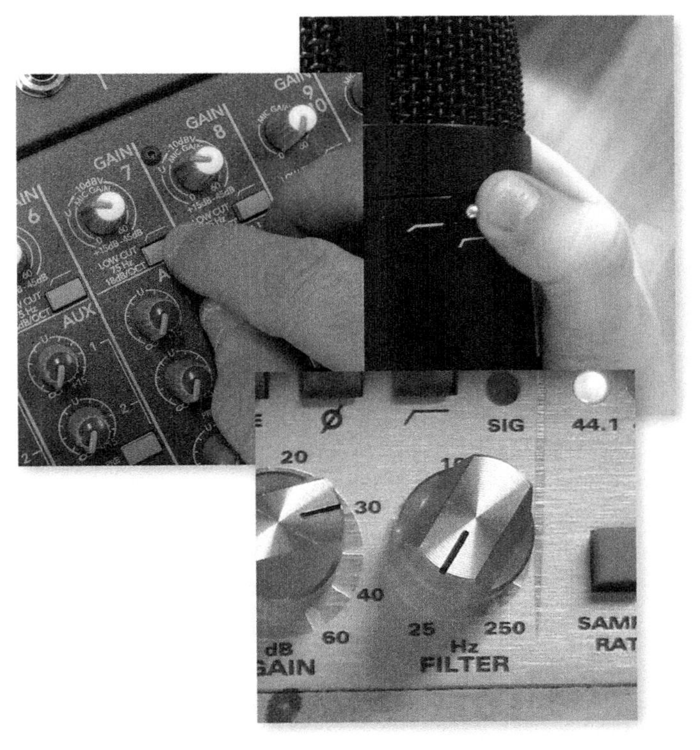

FIGURE 2.6
Although fixed-frequency high-pass filters are available on many mixers (top left) and microphones (top right), mic preamps with a variable-frequency design (bottom) are significantly more useful for small-studio recording.

adjust the cutoff frequency, but on more budget equipment you may just get a rudimentary HPF on/off switch with preset cutoff. One of the reasons high-pass filtering is useful is that many sound sources (especially scratch DJ turntables) output high levels of unwanted low frequencies, below the frequency range of the sound you're ostensibly recording, or even below the 20 Hz low-frequency limit of the audible spectrum. If you don't remove such signals, they may force you to leave more level headroom than is actually necessary for the desirable frequency content—to the detriment of your recorded signal-to-noise ratio. So if you see your speaker's woofer cones flapping about wildly for no real reason, or you're getting massive sporadic level surges that seem out of character with your input signal's longer-term meter readings, check whether high-pass filtering can clear up the problem without undesirably thinning the musician's tone. Where you have a choice of cutoff frequencies, err on the side of setting the filter too low rather than too high.

- **Check it in Mono.** Whenever you record anything stereo, it's a good idea to check what happens to the sound when you combine the left-channel and right-channel signals—in other words, when you listen in mono. (If you don't have a Mono button in your monitoring setup, then check out one of the mono-switch DAW plug-ins in this chapter's web resources.) You should be particularly wary of any keyboard/sampler preset with a super-wide stereo image, because this is often the result of over-egged stereo-widening effects (slathered on by the manufacturer to impress potential customers auditioning on headphones) which can cause untold damage to your carefully chosen timbre when it's summed to mono. Synth bass parts, above all, should be carefully mono-checked, paying particular attention to frequencies below 80 Hz. If you notice a level drop in this spectral region when switching to mono, or there are unmusical sub-80 Hz level inconsistencies from note to note, then your low end isn't going to fare well on quite a lot of end-user playback systems—which means it won't really cut the mustard in most mainstream commercial genres. (In other words, you'll either need to change the patch, or you'll have to high-pass filter it around 80 Hz and fill the blank more powerfully and predictably by recording an additional purpose-designed "sub-bass" synth line.) Keyboard patches based upon naturally miked stereo samples may suffer tonal changes in mono too, or the balance of the instrument's different pitch registers may alter considerably. Some people scoff at the idea of checking mono-compatibility these days ("Er... haven't we had stereo for, like, fifty years?"), but in truth it remains as relevant for mass-market music consumption as it ever was, because many real-world listening and public-address systems are still mono. Leaving that aside, though, your mix engineer won't thank you, either, if reducing the image width of any of your stereo recordings (in other words, partially summing them to mono) guts the tone quality.

RECORDING MIDI PERFORMANCES

Although this isn't really a book about MIDI sequencing, recording a live performer's MIDI data involves a lot of the same considerations as recording their audio, so much of this chapter's content will be of relevance. On a practical level, computer DAW users may encounter latency problems if they want to use CPU-hungry virtual instruments in this scenario, but the beauty of MIDI recording is that you can often use a more lightweight instrument for recording purposes, only substituting the proper sound source at mixdown when latency is less of a concern. Increasing the amount of sample data a sample-based software synth loads into the computer's RAM (rather than streaming it from a hard disk) can also improve your system's overall latency performance. If neither of those workarounds help, though, remember that even if recording at a suitably low latency introduces some audio glitching in your monitor mix, that should vanish once you reinstate a higher audio buffer-size setting post-recording.

2.3 IN-SESSION MULTITASKING

In the small studio, almost everyone on a session is likely to be multitasking, which means the engineer is increasingly being relied upon to act as a producer and/or arranger to some extent, simply because a suitable specialist is rarely on hand to fill those roles. When a recording session is progressing at lightning speed, it may be all you can handle just keeping up with all the technical stuff, and there's no shame in that; but the moment you find yourself with any headspace to spare, there are lots of other things you can do to help achieve an end result that everyone can be proud of.

2.3.1 Take Notes

Of these, taking notes is probably the most important, because it can save so much time and frustration. Here are some things that are frequently worth committing to paper:

- **A "To Record" List.** This should be a list of every part you need to record during the session, and where it belongs in the production. (I usually try to get hold of this before the session, but somehow on low-budget projects it almost never materializes in time.) If any new parts are developed during the tracking itself, note those down too. The main reason for this list is so you won't have to schedule another whole session just to record those four bars that slipped everyone's mind at the end of a long, boozy night's work. However, putting this list somewhere everyone can see it (make a wall chart if you like) also turns it into something of a motivational tool, because everyone knows how far they've got. If you're making good progress, it adds to the positive vibe; if not, the seriousness of the situation will be plain for all to see, and conscientious musicians will automatically respond by shifting up a gear without a cross word from anyone.
- **A Track List.** The main impetus here is that you never want to lose anything you've recorded. It's not only a waste of time, but it's also a massive

morale-killer for any musician trying to recapture a moment of magic they thought was already in the bag—especially if either the sound or performance prove to be unrepeatable. For each project you work on, the track list simply needs to tell you where to find everything you've recorded. On analog recorders you might use a traditional-style pre-printed "track sheet," with blank boxes representing each physical track, but in digital systems where trackcount is more negotiable I tend to just list all the parts I've recorded along with some short-form indication of where on the recorder (and in the timeline) it resides. In DAW systems, a separate track list may not be necessary, provided that you're disciplined about naming your audio tracks sensibly, although if your particular software platform offers alternative "playlists" or "virtual tracks" for each visible recording track, then paper may still have a useful role to play.

- **Recall Notes.** Anything you do on a session that you'd rather not work out from scratch twice should be documented first time around. In the case of Section 2.2.1's hypothetical keyboardist, for example, it would do no harm at all to make a note of your gain settings, so you'll be even more likely to capture her volcanic muse faithfully next time she graces your session.

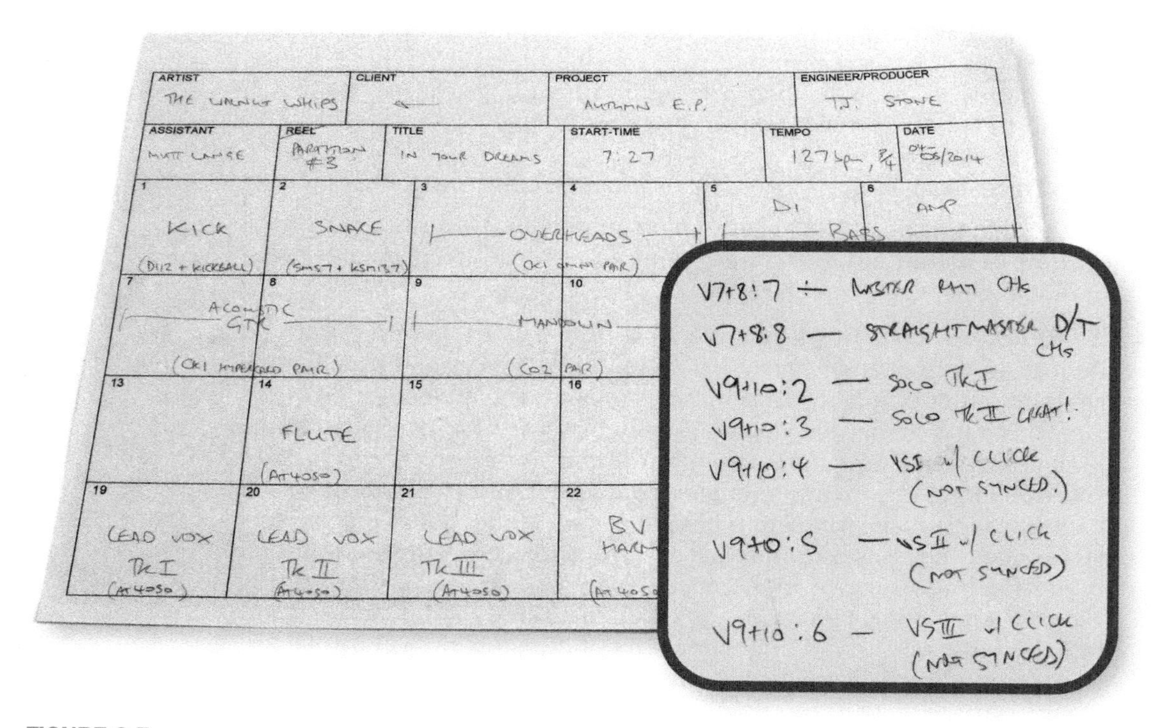

FIGURE 2.7
If you're working on a recorder with a limited track count, a traditional track sheet (main picture) is vital so that you avoid recording over something unintentionally. However, even where track numbers are more negotiable, it still makes sense to keep notes about where each take is within your recorder's file system (inset), so important parts don't slip down the back of the virtual sofa.

Similarly, it's the work of an instant to jot down the keyboard patch she used, but it might take ages to find it again amongst thousands of other presets, should you ever need to change the part or introduce that instrumental texture into another of the band's songs. If you've used any kind of processing in the recording chain, its settings should certainly be recorded somewhere too, for similar reasons. Where settings are very involved, a screenshot or photo may be easier than paper notes, especially now that so many people have cameras built into their mobile phones. Some DAW software platforms also give you the facility for typing in track/project notes, so you can cut down on the wood pulp by using those instead.

- **Session Minutes.** This is just a fancy name for anything important that's said or agreed during the recording process. That might be an opinion from the synth player about which of two countermelody alternatives he prefers; a suggestion from the turntablist that the final seven seconds of his scratch solo should be faded gradually in the mix; or a warning from your laptop whizzkid that he's dropped an uncleared sample of the *Thunderbirds* theme tune into his dubstep breakdown. Write it all down, and you'll look like the most professional person in the room when you're the only one who can answer those inevitable "What did we say we were going to do here?" questions.

2.3.2 Evaluate the Performance

If you pay attention to nothing else about a performance, you should try to make sure the instrument is in tune. It shouldn't be a big deal to do this now that every self-respecting DAW has a bundled tuner plug-in, but unfortunately poor tuning continues to blight home-brew recordings with depressing regularity. Even your grandma will have cottoned on to the fact that pitch-correction software exists these days, but you should be assertive in combating any laziness this may engender on the part of performers. This is partly because fixing tuning in software is almost always a very poor use of time compared with just tuning properly in the first place. Minimizing the unmusical side-effects of pitch-correction algorithms can take ages, even with cutting-edge technology, and even then the results will frequently sound rather artificial. Moreover, some types of sound respond very badly to pitch-processing, which means that there are still a lot of tuning problems that may prove beyond redemption in this way—particularly anything that's polyphonic with strong transients and/or dense high frequencies. Seeing as Grandma's on our minds, I'm sure she'd agree that a stitch in time saves nine.

What a lot of young engineers don't realize is that tuning isn't just a pitch issue, because it also affects how well instruments blend with each other, and consequently how easily they'll fit into the final mixdown. As such, a nicely tuned instrument will give you more leeway when trying to find a sound that'll work in the mix, so reminding the musician to tune regularly (or tuning the instrument yourself if you turn out to be more capable at it!) is worth the energy if only to make your own life easier. With electronic instruments you

FIGURE 2.8
Poor tuning is one of the most common failings of low-budget recordings, and also one of the least defensible, given that the tools to deal with it are so cheap.

should only need to tune once for each arrangement you're overdubbing on, although the tuning of older analog synths may drift over time, so rechecking those every couple of takes is more advisable. You certainly shouldn't assume that any synth, even a virtual instrument, will automatically be in tune with the backing, because not only might the backing itself be slightly flat/sharp, but pitch perception is also highly dependent on musical context, which means that a tuning that fits one section of your song may not sound nearly as pure in another.

Beyond the tuning and timbre issues which impact directly on your engineering role, though, you can also benefit the production as a whole if you keep your ears open to the quality of the performance itself, if only so you don't look like a complete numbskull every time anyone asks "So what did *you* think?" As with choosing sounds, judging musical parts and performances isn't some kind of exact science, and the only way you'll get better at it is through experience—listening to commercial records, watching musicians at work, comparing different takes over many different sessions, and re-evaluating tracking decisions with the benefit of hindsight once projects reach their final form. Again, though, let me suggest a few general things to listen for, which should help start you on the right road:

- **Emotion.** The way you feel about a performance should always be the biggest reason for keeping it, regardless of any more tangible and easily expressed concerns. Clearly your own feelings must yield to those of the artist if it's not your music, but that doesn't mean your own responses have no value at all.

- **Balance.** In other words, listen for whether all the elements of the performance are at an appropriate subjective level relative to each other from a musical perspective. Does the keyboardist's left hand feel like it's overpowering the right, for instance, or is a heavy right thumb imbalancing its main rhythmic riff? Are there melodic fragments within the musical part that could be emphasized to prevent them being obscured by more accompanying elements?

"The way you feel about a performance should always be the biggest reason for keeping it, regardless of any more tangible and easily expressed concerns."

- **Phrasing.** A performance with good phrasing propels the listener through time by indicating where the music's heading. That might mean something as elementary as making the on-beat note of some riff a little louder, longer, or more legato than the following off-beat note, underlining the musical hierarchy between them; but it might equally be enshrined in the long-term build-up of an extended solo, taking you all the way from a stripped-back breakdown through to the powerful full-band entry of your final chorus. So how can you gauge the effectiveness of a musical part's phrasing? Well, my favorite method is to try singing/beatboxing it for myself—yes, even if the overdub's some kind of glitchcore mayhem, anyone can summon some kind of half-recognizable impression! The reason I think this works is that conscious thought is usually the enemy of good phrasing, so if the player's concentrating on getting all their fingers in the right positions, the engineer is listening for the tuning and signal quality, and the producer is thinking about the way the part works with the harmonies and arrangement, then clunky phrasing can easily go unnoticed. Because using your voice is so instinctive, it usually seems to make appropriate phrasing choices spring from your unconscious much more readily. (And besides, regular glitchcore beatboxing is surprisingly beneficial for the sinuses…)
- **Articulation.** Does the nature of each note/sample's attack support its role in the music? Some keyboard patches may soften the attack of notes that overlap others, while sampled musical phrases will make a very different impression depending on whether their onset is abrupt or fades in more softly. In addition, how do the lengths of notes/samples relate to the other parts in the arrangement? Rarely is this more important than with bass lines, where the positioning of each note's end point will frequently be as critical to the arrangement as where each note begins.
- **Pace.** Although the tempo of the production will already have been determined by the time overdubbing takes place, the subjective impression of how fast or slow that tempo *feels* can change when you add an overdub. If the whole song suddenly seems to be dragging, for instance, this can be a clue that your player isn't quite locking into the rhythm properly, or they're clogging up the groove with too many syncopations or busybody figurations. Another common ailment is where a new instrument swings its eighth or sixteenth notes in a way that doesn't match the swing in the existing parts.

In addition, because so many of the decisions people make about musical parts and performances are emotionally charged (and rightly so), they are highly vulnerable to bias from non-musical factors. You'll experience a take very differently if you're watching the performer's movements and facial expressions, for instance, whereas consumers won't have the benefit of these visual cues. Here's Don Was[21] talking about recording Mick Jagger, for instance: "He'd give it the full stage bit and, in fact, it was so awesome I had to keep looking down at the console… I didn't want to be biased in my assessment of his vocal. So I had to look down and just listen to make sure it was working, because otherwise I could have lost my objectivity." Given that personal interaction with the musician is often essential to session dynamics, averting/closing your eyes during recording may not be a viable option, so there's a good argument for ratifying subjective judgments primarily during playback. (This also allows the musicians themselves to participate in this process with a clearer head, of course, which is especially important where the player happens to be you!) The graphical waveform displays on a DAW screen can also lead you astray when you're evaluating performance timing, so I'll often turn them off (or zoom out far enough that they can't be seen) where rhythmic concerns are a high priority.

"There's a fine line between responding innately to the real qualities of your groove and trying to validate dysfunctional rhythms by wishfully bopping along."

Even miming along or tapping your foot while playing back a take can potentially overlay fictitious emotional and rhythmic cues, so I'll occasionally force myself to close my eyes and sit still where an important decision hangs in the balance. (It makes you look more like you're concentrating too, which can't hurt!) While many styles want a beat that incites people to jump up and shake their thang, when you're trying to create that kind of music in the studio there's a fine line between responding innately to the real qualities of your groove and trying to validate dysfunctional rhythms by wishfully bopping along.

2.3.3 Support the Performer

Lending a hand in appraising the quality of the performances is already a valuable service, but there's also a lot you can do to support players in their quest for the perfect take. The biggest way you can do this is by conducting yourself in a manner that allows the talent to feel at ease in your company, hopefully building on the careful studio preparations you already did back in Section 2.1. If you need someone to teach you social skills or a sense of humor, then I'm afraid you're always likely to be fighting an uphill struggle in any music recording situation (unless you happen to work entirely on your own music)—honestly, quit now and plumb your efforts into something that doesn't rely so fundamentally on personal interaction! Fortunately, however, musicians are a pretty sociable lot on the whole, and typical small-studio recordists are musicians themselves, so this should hopefully come naturally. That said, I do have a few recommendations for developing the most productive studio manner.

In my opinion, far too many engineers forget how intimidating a studio can seem to non-techies. "Sometimes, people are scared when they walk into a control room," says Dennis Sands.[22] "It's not their world, and it's got to be frightening when you know you're dependent on this, but it looks so unbelievably complex." John Porter[23] concurs: "If you've never been in a recording studio before, it's such a weird environment, especially if you're used to playing in front of hundreds of people in clubs and there's terrific feedback between you and the audience. Suddenly, you're in this room with headphones on, hearing something that probably doesn't sound at all like what you're used to hearing. It can be very disconcerting." As such, a few words of explanation from you can work wonders: what you're doing at each stage of the process, why you're doing it, and what you need the player to do. It'll soon transpire if your guest is actually a seasoned session veteran, in which case you can tone down the running commentary, but most of the time you'll immediately disperse a small cloud of anxiety from around your player. Not only that, but your openness about the recording process should immediately begin building a sense of mutual trust; your inclusive attitude to them (a non-specialist) will imply that you consider them a colleague/peer, improving the chances that they'll return that respect; and the mere dialogue encourages a positive ethos of free communication and general goodwill.

Leaving musicians guessing is rarely a good idea outside the technical realm either. Whenever they produce something on behalf of the session (whether it's a choice of sound, a recorded take, or a suggested change to the musical part), about the worst response they can get is none at all—it's incredibly unnerving as a performer if you're left in limbo without any indication of how your input has been received. Even if everybody on the session hates that specific contribution, it's vital that somebody says something about it as quickly as possible after it's been offered, so that the player isn't just left hanging, imagining the worst. If there's no one else to fulfill this important function, then it's your responsibility to step up to the plate.

I'd suggest not taking yourself too seriously in the studio either, no matter how good an engineer you feel you are or how fancy-looking your rig. "People are often very surprised working with me quite how idiotic I am in the studio," comments David Kosten,[24] for instance. "I'm a total clown and I like people to have a good time and laugh and not realize that they've done so much work." Coming across too strongly as Mr Big Shot Engineer is one of the most reliable ways to put musicians on edge: Either you'll put the frighteners on them, or they'll suspect you of being a shyster and go on the defensive. Plus, any awe your charade may inspire will be short lived, because there's no engineer on the planet who's immune from stupid errors, and when your comeuppance arrives, everyone's much more likely to start questioning and/or resenting your inflated opinion of yourself. So take my advice and leave your ego at the door, concentrating instead on serving the talent and, by extension, the music itself. Within this context, try to keep any mistakes you make out in the open, being the first to poke fun at them if they're trivial, and the first to accept responsibility for them when they're more serious. There's no point trying to sweep problems under the carpet, either, because fixing things at the earliest opportunity is usually the best means of cutting your losses, and also avoids squandering the trust you've worked so hard to build up with your musicians/clients. "If you realize you fucked up a track or you made a bad decision, fix it," asserts Eric "Mixerman" Sarafin[25]. "If that requires starting from scratch on the track, do it… Everyone fucks up. It's how you deal with those fuck-ups that sets you apart."

TIPS FOR PUNCHING IN

The way you handle the punch-ins during tracking can have a significant bearing on the quality of your eventual master take. First and foremost, whenever you're planning to punch into an existing recording, make sure the performer is already playing along with the backing track before you reach the punch-in point, and continues to play right through the drop-out point too. This helps to ensure that sustain tails and between-note transitions sound natural on either side of the newly recorded snippet. Even taking that precaution won't protect you from the occasional "bump" at one of the punch-in boundaries, though, so these must be chosen with care. Punching in during clear gaps in the performance is obviously the minimum-risk option, but where this isn't possible the best spots to go for will usually be fractionally before well-defined note onsets (especially ones with strong transients) or else at locations where any lumpiness will be veiled by another event in the arrangement, typically a loud drum hit.

In Section 2.1.5, I already mentioned that switching between Input and Auto monitoring modes should be a reflex, and the reason for this is that you may need to switch between them on the fly. Although Auto is usually the default mode for normal patch-up work, some players find it too disconcerting trying to play along silently to their pre-recorded performance prior to the punch-in point, with the result that their rhythmic feel suffers, or they instinctively play too forcefully because they can't hear their own live signal. Jay Graydon[26] comments on this in relation to singers, for instance: "It's most important to put the singer's mic in Input mode on the recorder at least one line before the punch-in. If not, the singer will sing louder at the punch point due to not hearing him/herself in the headphones before the punch point." Performers without much studio experience may simply assume there's nothing that can be done about this, so won't bring the matter up, which means you have to keep your ears open for tell-tale signs: namely, repeated instances of dodgy groove or overenergetic performance within the first couple of beats of the punch-in zone. Don't switch over to Input mode without telling the talent you're doing it, though, and why you think it might help.

Most multitrack recorders now provide some facility for punching in and out automatically at user-definable time locations. There's no denying the usefulness of this when you're recording yourself, but I don't like using it when recording other musicians because it always seems to slow down the work rate—by the time I've set up the necessary boundary points with sufficient accuracy, I could have done the same punch-in manually three times over! While those working on analog tape will naturally approach any punch-in with a certain amount of trepidation, for fear of spoiling an existing recording by fumbling the transport controls, the ubiquitous Undo button banishes any such anxiety for users of digital systems. The worst thing that can happen is you have to redo an occasional botched punch-in, but is that such a big deal? Think of it as just another opportunity to have a good laugh at yourself. A much bigger deal to my mind is inserting an unproductive 30-second time-lag between thinking "let's replace that bar" and actually doing it, especially in any situation where numerous different punch-ins are required.

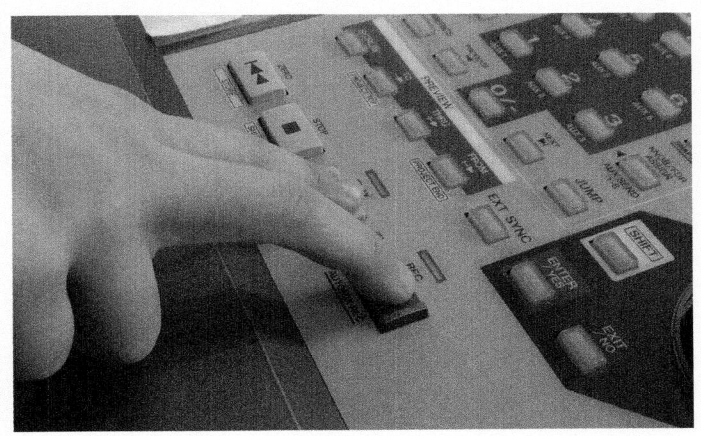

FIGURE 2.9
Although you can easily set up automatic punch-in/out points with most digital recorders, punching in manually on the fly is usually much more straightforward, and carries very little risk of upset because of the ubiquitous Undo function.

Another big no-no is giving the talent the impression that your mind's not fully engaged with the matter in hand. "If you're excited about what you're doing, that will translate to the musicians," says Rafa Sardina,[27] "You need to be there in the moment… connected continually, not detaching yourself from the session." Don't check your Twitter feed while they're tuning up; don't doodle elaborately on the tracksheet; don't whip out your nail file to catch up on your manicure regimen—stay focused on serving the music, because there's always something more you can do either to support the goals of the current session activity or to anticipate and plan better for the next. If you treat the musicians with respect in this way, I find they'll often go out of their way to warrant that respect when the Record light's shining, at which point everyone's a winner.

Small-studio engineers have to deal with performers of all skill levels, and each will have their own unique blend of strengths and weaknesses. Getting the best from a given player therefore means playing to their strengths, and one way you can do this is by adjusting the way you build up the final recording. Although I've currently tied your hands a little by stipulating that you build up your master take on a single track using punch-ins, there are still plenty of other variables you can juggle on the fly. For a start, you can choose how much of the musical part in question you record in one go. Highly skilled and assured instrumentalists will typically deliver better consistency of tone and more fluid phrasing if you record the whole track top to tail, or at least work on extended sections. This is especially pertinent when recording improvised lead solos, since it's very difficult to get these sounding natural by constructing them piecemeal. With less proficient performers, however, the sheer number of retakes required to achieve a respectable performance without flubs will most likely knock their confidence, as well as unnecessarily draining their more limited stamina. In that case, punching in for a couple of bars (or even beats!) at a time may be a more productive approach. That said, even the ropiest performer will often benefit from the warm-up of a full-length take to kick off proceedings, so unless you're already familiar with the talent's capabilities, that's probably the best way to start off in any case, during which you should have ample chance to assess what caliber of player you're dealing with.

An initial run-through take will also highlight those sections the performer finds most challenging, information that you can take advantage of later on. The window of opportunity for successfully capturing such sections may be quite small, because the player not only needs to be well warmed up, but also fresh enough that their concentration is still sharp, so I will frequently structure the workflow of the session accordingly, getting the musician into the swing of things with a couple of simpler segments before tackling the trickiest bits. Also, keep your ears pricked for any musical passages the instrumentalist can sail through with ease, because rattling through one of those can be just the thing to lift spirits if morale's flagging, or if everyone needs a break from grappling with some thornier spot. "With each take you peel away some of your excitement," comments Josh Homme.[28] "If you don't have a good take by take 10, you should move on and come back to it later."

And, speaking of breaks, this is another area where there's scope for positive intervention. When signs of performance frustration begin to show, or when everyone's beginning to find it hard to evaluate the emotional quality of the recording, there's no underestimating the value of five minutes in the company of tea and biscuits, preferably contemplating some non-studio vista. Creative blocks also have a habit of solving themselves subconsciously during moments of respite—which is why every famous studio's "smallest room" has usually witnessed the birth of at least one hit-making hook! Bear in mind, though, that the primary purpose of breaks, from a production standpoint, is to reinvigorate everybody's ears, creativity, and morale, not to give the engineer a breather, so don't you waste that opportunity to head off any technical hitches or prepare for foreseeable upcoming session requirements.

"Creative blocks have a habit of solving themselves subconsciously during moments of respite—which is why every famous studio's "smallest room" has usually witnessed the birth of at least one hit-making hook!"

The monitor mix also allows you to exert some influence over proceedings, because most performers will naturally play more aggressively if they're low in the monitor balance, whereas they'll show more restraint and nuance if they're a touch too loud. And this isn't just a performance issue, either, because altering the nature of a player's "touch" on their instrument will often impact the sonics too. A heavily accented note isn't usually just louder than a softly played one—in many sample-based keyboard patches it'll also have a harder attack and a brighter timbre, for example. Because of this, there's often an argument for encouraging the musician to play more softly during recording where a given part needs to loom large in the final mix, because such a recording can sound fuller and closer to the listener when faded up to an appropriate level relative to the rest of the arrangement. That said, the level of any overdub in the mix will also affect the production's groove, so with rhythmic parts you need to be a little wary of recording to an out-of-balance monitor mix, in case it impairs everyone's ability to judge the performance timing correctly. It's also worth stressing that the needs of the performer should trump those of the engineer when adjusting the balance of a single communal monitor mix. Note that less confident performers will often soldier on with an unhelpful headphone mix out of misguided politesse, because they don't want to be seen as wasting session time, so it's pivotal that you lay any such fears to rest as swiftly as possible.

There are plenty of more direct ways you can take an active role in improving a performance as well, but we already have plenty of new ideas to be getting on with for now. So we'll pick up this subject again in the next chapter once we've gained some practical experience of applying what we've covered so far. For now, just try to do what you can for the performer behind the scenes, and give them the time they need to define their part and deal with the practical challenges of performing it. "The most important thing in the studio besides technical ability," says Kevin Killen,[29] "is to have patience. You have to wait for great performances to happen and be ready to capture them."

CUT TO THE CHASE

- A big advantage of overdubbing electronic instruments in the control room is that you can monitor over loudspeakers. Where you have to use headphones, though, choose cans that suit each listener (and their preferred listening level) and consider open-back designs for improved fidelity and communication. If you don't have independent volume controls for each person, an additional multi-output headphone amplifier is helpful. A setup with good sight-lines can mitigate the inherent communication difficulties when using headphones. Make sure all monitoring hardware is functioning properly prior to the session, and that you've got a good backing balance.
- A good cue sheet (or equivalent information programmed into your DAW system) will speed up your workflow, and things like chord charts and notated musical lines may also be worth generating in advance.
- Try to make performers as comfortable as possible. Suitable refreshments help here, as can manipulating the studio layout, shrinking the space around the musician and choosing the visual focus. Studio decor not only influences the general mood, but can also provide inspiration or distraction in times of need.
- When using a DAW system's software monitoring, you should aim to bring its latency below 15 ms. If buffer-size settings can't achieve this on their own, try removing or replacing any active plug-ins to reduce strain on the computer's CPU, and avoid mastering-oriented effects or those powered by external DSP hardware. Distant speaker placements may also add to the problem.

- Your most important job is to hit Record at the right moment, even if that means fundamental technical and sonic decisions must be deferred. Fastidious preparation helps you work around this fact, but you can often play for time to an extent.
- If you'd like to work on tones, ask the performer. This confirms the time is right, and allows you to enlist their help. Always try to give a clear idea of how much time you'll need for this. If you're unable to settle on the perfect timbre before the first take, look for opportunities to address this as you progress from one musical part to the next. When searching for sounds, make the best use of time by adjusting the most audible things first, and don't try to judge the suitability of a sound out of context or while it's overloud in the balance. Be on the alert for excess distortion, mono-incompatibility, and unwanted low end. Where effects processing is integral to an instrument's character, don't be afraid to print it into the recording, but do try to minimize delay/reverb effects if possible to avoid painting the mix engineer into a corner. As a safety precaution, you can simultaneously record the dry sound or the MIDI trigger data to a separate track if you wish.
- Once overdubbing is underway, take notes to document session progress, keep track of takes, preserve settings for later recall, and record any spoken remarks that may have a bearing later. If this leaves any spare thought-capacity, use that to form your own opinions about the performances, but be wary of factors that can bias your subjective decisions.
- Your studio manner can make a big difference to the performer's state of mind. Demystify your own activities for them and don't intimidate or antagonize them by taking yourself too seriously, trying to cover up your mistakes, and generally acting bored.
- With unfamiliar instrumentalists it's usually best to start off with one take of a reasonable length, so you can gauge their strengths and weaknesses. For weaker players, working on smaller sections will often prove to be the most efficient use of time, whereas stronger players will usually deliver more compelling performances if you record longer chunks. Record the most challenging sections of a musical part as soon as the musician has properly warmed up. If energy levels droop, try working on a simpler section of the part as a brief respite, or take a quick break to regroup.
- Adjusting the monitor balance can usefully influence a performer's touch, but overly skewed listening levels may also mislead them rhythmically.
- To make punch-in boundaries as imperceptible as you can: Get the musician to play right across them; place them just before a transient note onset; or conceal them behind a loud event in the backing track. Be prepared to switch between your recorder's input-monitoring modes if you hear the performer struggling with the start of any punch-in.

Assignment 1: The Dry-Run

The purpose of this assignment is to familiarize yourself with the session-preparation and overdubbing procedures described in this chapter without the additional pressures of dealing with a separate performer.

- Select an existing project you can practice overdubbing on. There are some specimen projects in this chapter's web resources if you have nothing suitable of your own.
- Lay hold of some kind of line-level electronic instrument you can perform with for the purpose—you don't need to be an expert with it, because you're free to make the part as simple as you need it to be in this instance.
- Set up the instrument and monitoring arrangements as explained in Section 2.1.1, line-check the input chain, and set recording levels. Put together a cue sheet (or something similar using markers within your DAW system) as in Section 2.1.2.

- From your recording system's instruction manuals, learn how to do punch-ins (both on the fly and automatically) and how to quickly switch between Input and Auto monitoring modes, as discussed in Section 2.1.5. Users of computer systems should confirm that software monitoring is active, and then play something through the system to check for problems with comb-filtering or monitoring latency.
- Play the backing track to adjust the monitor mix balance and overall playback volume, and then play along to find a comfortable monitor-fader setting for your recorder track.
- In the light of Section 2.2.4, think about where the part you're about to perform fits within the arrangement as a whole, and choose/refine the sound as necessary to better suit that role—as well as to cater for your own personal tastes!
- Record a take. With reference to Section 2.3.2, play it back and assess the quality of your own performance. Take notes of those sections that you may be able to improve upon, and retake them, making use of manual or automatic punch-in facilities as required. Make a point of trying out both Input and Auto monitoring to get an insight into how they feel from a performer's perspective. For similar reasons, experiment with some alternate monitor-mix balances to see how they affect your performance. When you're done, play back the master take in its entirety one more time to confirm that all the patch-ups you originally noted have been completed, that the punch points are inaudible, and that no further punch-ins are required.
- Repeat the last two steps for at least two other parts. Try to record at least one of each of the following: a sustained pad or effects-based soundscape; a riff or accompaniment with a strong rhythmic element; an expressive lead solo part of some type.

Assignment 2: Overdubbing in the Wild

Now that you've had some practice with the mechanics of the overdubbing process, and have developed a basic feel for the workflow, you should be ready to start dealing with the more complicated scenario of working with a separate performer.

- Select an existing project you can practice overdubbing on, and find an understanding player of some line-level electronic instrument to help you out. Arrange a session with them, giving them free reign to create one or more original parts. If they ask to hear the track in advance to develop ideas, then by all means give them a rough backing mix to listen to in their own time.
- Prior to the session, find out as much as you can about the instrument you'll be faced with, and do everything in your power to ensure that you'll get usable input and monitoring signals, even if you have to hit Record the moment that instrument's plugged in. Review the equipment setup you used for Assignment 1, and adjust it if necessary to suit the needs of your upcoming session. Prepare the written materials you might need for the session, including at least some indication of the chords and bass line in preparation for any discussions while developing musical parts.
- With reference to the suggestions in Sections 2.1.3 and 2.1.4, ask yourself whether you've prepared everything you can to make the musician comfortable while performing and to nurture a creative environment in your control room.
- Fifteen minutes before the session, double-check that the backing balance is coming through all monitoring devices cleanly and at a reasonable volume. Also confirm that you've got the necessary cables on hand to connect up the instrument when it arrives.

- Once the performer walks through the door, you should always try to prioritize their needs as a musician over your needs as an engineer. As long as the instrument is plugged in and powered up, you should hit Record the moment they start playing along to the track. If you are able to buy yourself some time to deal with engineering matters before that moment comes, then make that time count: Nail the gain-setting and monitor mix first, before moving on to any other sonic concerns; and then focus on the sonic changes that make most difference, particularly any controls on the instrument itself, enlisting the help of the player as much as you can.
- After the first take, explain to your guest that you're planning to build a master take on a single track using punch-ins—feel free to blame me for this arbitrary limitation! Discuss whether you should keep any sections of that first take or retake it all from scratch. Listen back to the first take if that helps, and note down any decisions as necessary. For the purposes of this exercise, try to stay positive in your comments and defer to the performer if any serious conflict of opinions occurs.
- From here on in, you're on your own! I can't predict what exact course your session will take, but as long as you insure you're always serving both the music and the performer, any mistakes you make should be forgivable, because your heart will be in the right place! Just make sure you save any project data regularly, and stay on the lookout for signs that a break and/or some nourishment might be in order.
- Following the session, make notes of any details that may be of use for future sessions—the gain settings you used, for example, or the location of any particularly useful controls on the player's instrument. In addition, make a list of any technical errors that compromised the recording quality or wasted time; any situations where engineering issues interrupted or impeded the performer's creative process; and any interpersonal misjudgments that broke the mood or caused you to cringe! Needless to say, you'll want to work out in your mind how to avoid those mishaps before the next session comes along.
- Try to carry out another couple of sessions like this, preferably with different performers if possible, the idea being to transform some of the simpler engineering tasks into habits, thereby freeing up valuable brainpower for the extra complexities I'll be introducing in Chapter 3.

WEB RESOURCES

On this book's companion website you'll find a selection of resources to support this chapter, including:

- a selection of backing tracks in a variety of different musical styles, complete with tempo/key indications and structure/chord charts for performers;
- audio example files demonstrating some of the performance parameters highlighted in Section 2.3.2;
- links to free loop download libraries, freeware time-stretching algorithms, and some of the products suggested in this chapter;
- further reading about computer monitoring schemes and latency reduction methods, together with an audio file suitable for testing system latency;
- some glossaries of musical terms.

http://www.cambridge-mt.com/rs-ch2.htm

CHAPTER 3

Instruments with Pickups

So far the only things we've talked about recording are line-level and digital signals, so now let's expand our repertoire to include signals from pickups. This significantly increases the range of instruments we can record, taking in electric guitars and basses, electric pianos/clavinets, and any acoustic instruments that have pickup systems fitted (most commonly acoustic guitars and other stringed instruments).

3.1 RECORDING PICKUP SIGNALS

Electric guitars and electromechanical keyboards typically use magnetic pickups to convert the physical vibrations of metal components in the instrument directly into electrical signals. These signals are then rendered audible by a connected amplifier and loudspeakers, and the nature of this amplification system is frequently essential to the desired sonic character. This is why it's common practice in the studio to record such instruments by miking the amplified sound—a much more complex subject we'll explore in later chapters. However, the unamplified "Direct Injection" (DI) signal can also be very useful for recording purposes. Small-studio acoustics rarely allow powerful low frequencies from bass instruments to be captured faithfully through a microphone, for instance; and recording a DI signal has the advantage that it allows you to re-evaluate or experiment with amplifier sounds after recording, either by rerecording the DI signal through a real miked-up amplifier ("reamping") or by applying amplifier-emulation software at mixdown.

Although the outputs of most electric instruments are quarter-inch TS jack sockets, you're unlikely to get a good recording by hooking those straight up to normal unbalanced recording inputs, because most pickups only operate properly when connected to special high-impedance input circuitry. Some TS jack inputs can be switched into a high-impedance mode to cater for such eventualities, usually via a button labeled "Instrument" or "Hi-Z." A more flexible option, however, is to use a dedicated DI box instead, because most of these not only provide a high-impedance TS jack input socket for the instrument and a balanced XLR output socket for recording purposes, but they also feature a TS jack output socket labeled "Thru" or "Link" which can be connected to the player's amplifier, so that they can perform with that as they'd normally do.

Some amplifiers have a dedicated line output which can be recorded in place of a DI signal, but this rarely sounds as clean, so I almost always favor a DI box if one's available. There are also some specialized DI boxes that can safely accept an amplifier's speaker-level output signal, sitting in-line between the amplifier and the speaker—some can even substitute for a speaker, retaining some of that "real amp" character in the recorded timbre without waking the neighbors. However, because the speaker is typically responsible for taming the frequency extremes of an instrument amplifier's raw output, you're unlikely to achieve a usable timbre in that instance without artificial speaker simulation. Analog speaker simulation is usually built into DI boxes that can handle amplified signals, but if you can switch this off for recording purposes you'll often get a better end result, given the range and quality of software speaker-modeling bundled with most DAW systems these days. To be honest, though,

FIGURE 3.1
Some jack inputs are switchable to accept instrument-level signals with the correct impedance matching (left), but if yours doesn't then you'll want to use a DI box instead (right).

there's little point in DI'ing an instrument's amplified signal if you're able to mike it up instead, because it'll hardly ever sound as good.

Softer acoustic instruments such as acoustic guitar, mandolin, and orchestral strings frequently have piezo-electric pickups fitted, primarily to allow onstage amplification during live performances. Again, though, such systems always sound rather unnatural, so reaching for a mic is almost always a better option in the studio. However, if you've no choice in the matter, then consider getting hold of a DI specifically designed for piezo-electric systems, because the input impedance of a normal DI box isn't always high enough to squeeze the best tone out of them.

GOOD OLD-FASHIONED FLATTERY

Flattery is a pretty good way to get anyone on your side, and is at the heart of one little psychological trick that seasoned engineers have been using since time immemorial. The principle is simple: Mention to the musician that some esoteric technical maneuver, reserved only for special occasions, is to be used for their overdub. Whether or not it actually makes any audible difference, it often spurs the player into giving a better performance. So before you write off, say, hand-woven platinum cables as a total waste of money on purely sonic grounds, try telling your guitarist (in suitably hushed tones) that you've had one smuggled out of darkest Mongolia and transported to the studio under armed guard specially for their overdub. You may be surprised how much better the player makes it sound...

Once you've managed to get a DI signal into your recording system, the gain-setting process shouldn't hold any mysteries, but you have to be a bit more careful about potential sources of signal degradation. This is especially the case with magnetic pickups, some of which are very sensitive to electromagnetic interference. This is a good reason to banish mobile phones from the studio (if their distracting influence on the creative process weren't already enough) and you should also pay attention to the physical orientation of the instrument in your room, because rotating its position through 90° can make all the

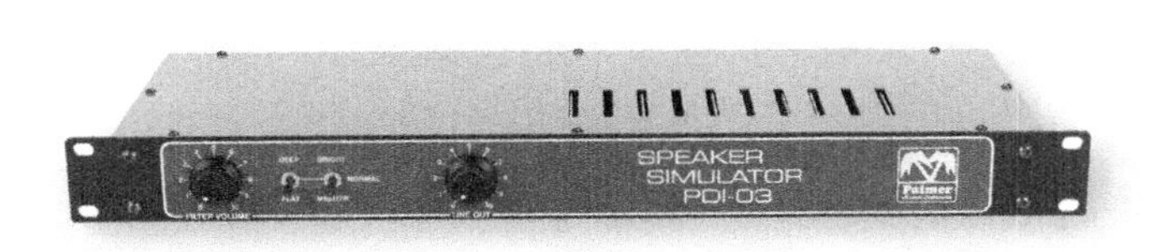

FIGURE 3.2
A couple of more specialized Direct Injection devices: Palmer's PDI03 (left), which can accept the output of an instrument amplifier and includes speaker simulation; and L R Baggs' Para DI (right), a DI box with extra-high-impedance input options for piezo pickups.

difference in the world to the amount of hum, buzz, and general sonic rubbish it hoovers up. Whatever type of pickup you're using, keep instrument cables as short as possible. "With short cables, you get all the bottom end as well as a nice top end," says Mike Fraser.[1] "As soon as you lengthen the cable, the magic of the sound goes away, and you have to add top end." It's also good to lay instrument cables well away from mains and data lines, because of the typically low level of many instrument signals and the unbalanced nature of their TS jack interconnections.

3.2 MONITORING CONSIDERATIONS FOR PICKUP RECORDING

3.2.1 Silent Instruments

Where a pickup-equipped instrument produces no appreciable acoustic sound in the recording room, the simple monitoring setup we discussed last chapter may well continue to fit the bill. It's possible you may encounter latency problems on some DAW systems if your recorded signal is feeding a software amp simulation, and if you can't solve these by substituting a CPU-light stop-gap just for tracking purposes, then an alternative is to monitor through a hardware amp-modeling device instead—you can get something basic but workable for under $75 (£50). The simplest way to set it up is to record both the DI signal and the amp-modeller output simultaneously, onto separate recorder tracks. This means that the performer can record, play back, and do punch-ins listening to the amp-modeled signal at all times, while you retain the flexibility to completely remold the guitar sound at mixdown using the DI track.

You may already have discovered during last chapter's assignments how awkward it can be to make engineering decisions while listening to a monitor balance designed primarily to benefit the performer. A way round this is to give your musician headphones, feeding those with a personalized foldback mix created via your studio system's auxiliary sends, while everyone else in the room continues listening to the main monitor mix. If you're already restricted to using headphones anyway, then arranging this should be pretty simple, just as long as you're sure to use "pre-fader" auxiliary sends (i.e., sends fed from before each monitor channel's fader)—otherwise any changes you make to your own monitoring balance will alter the foldback balance too. The moment the talent has a separate mix, though, you need to be a little more careful about protecting people's ears from any signal spikes generated while you're powering up equipment, plugging cables around, or switching phantom power, because muting your control-room monitors won't mute any foldback mixes. The safest option is usually to mute the recording chain's input channel, but whatever approach you take, you should verify that this does actually mute the foldback feed—the Mute buttons on some mixers don't affect pre-fader sends, for instance.

ACTIVE OR PASSIVE?

Some DI boxes require power to operate correctly, whether this power is supplied directly from the mains electrical supply, from an internal battery, or from phantom powering (via the unit's XLR output). These "active" designs include internal preamplification, so are often preferred for handling the low-level signals of unpowered "passive" pickups (the most common kind). However, this active circuitry can distort if subjected to much higher-level signals, for example from battery-powered active pickup systems, in which case an unpowered passive DI box may be more suitable, because its lack of active circuitry will usually allow it to accept higher levels without problems. With either design, though, you'll regularly find that an input Pad switch is supplied to extend the range of signal levels that it can handle. Another common facility is an earth-lift switch, which interrupts the earth connection between the DI box's input and output sockets, and can therefore help fix some earth-loop problems (see Section 1.4.2). In this capacity a DI box might well be used for interfacing line-level devices with your recording system, in which case the greater headroom typical of passive designs would make most sense.

From a technical standpoint, active pickup systems will usually have the advantage of cleaner and clearer sound quality, and the signal will also suffer less when traveling through long or low-quality guitar cables. The downside, though, is that you need to make sure their batteries have enough juice, so it's safest to replace the battery before any important recording session if you can. (The same also applies, of course, to any other essential studio gear that doesn't have mains powering, such as DI boxes, stompbox-effect pedals, tuners, metronomes, tazers …)

Given the practical benefits of loudspeaker monitoring, there's no point in moving wholesale to headphone listening until it's absolutely necessary. Any moderate rebalancing of the performer's monitoring experience in the control room (such as the common request for "more me") can usually be handled perfectly well by using the performer's headphones merely to supplement the main loudspeaker mix everyone's hearing. Again, open-backed headphones can work very well in this application, but closed-back models can also be pressed into service by using them one-sided, sliding the unused earcup slightly off the performer's ear so they can still hear the loudspeakers unhindered.

3.2.2 Working with a Separate Live Room

Where you're DI'ing an acoustic instrument, or the player can't perform comfortably without hearing their guitar/keyboard through its usual amplifier, the monitoring approach starts to get more complicated, because the direct sound of the instrument in the control room interferes with the engineer's ability to evaluate its recorded timbre via the main monitor mix. The traditional solution to this problem involves moving the player out into an acoustically isolated live room instead, along with their foldback monitoring system. This tracking method might seem completely irrelevant to you if your own studio is crammed into a single room, but bear with me while I explain a little about how the monitoring works in two-room setups, because one-room workarounds are best understood in relation to that.

Instrumentalists who prefer to judge their performance by listening to their recording signal will usually be better served by closed-back headphones, because these will make the acoustic sound in the live room less distracting for them. Loudspeakers might also still be usable in this case too, provided that the instrument is naturally quiet enough to be obscured by its own monitoring

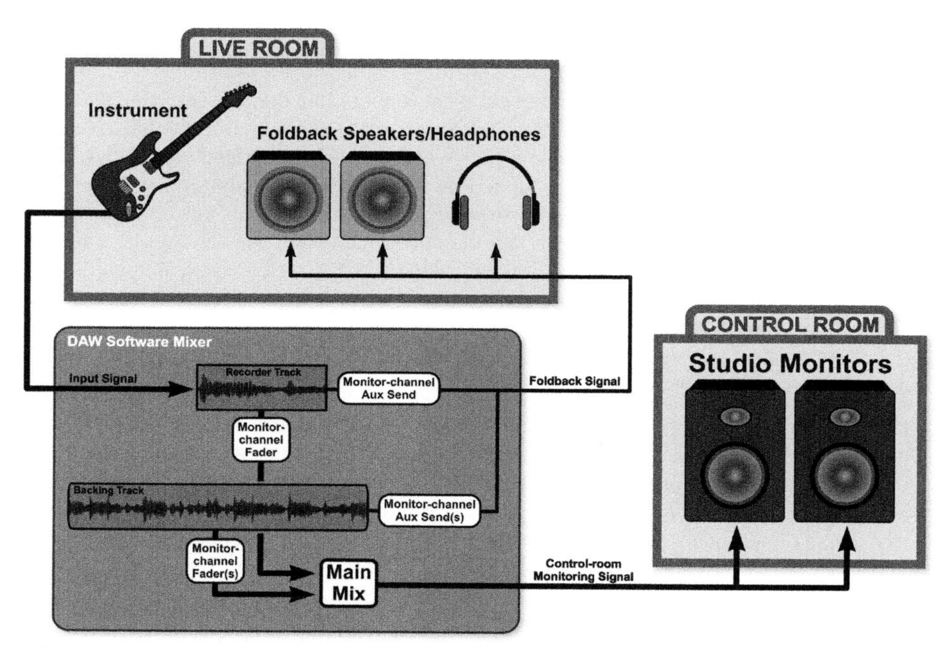

FIGURE 3.3
A two-room monitoring and foldback setup using DAW's software monitoring facilities.

signal when listening to the foldback mix at a comfortable overall volume level—and also provided that the loudspeakers don't trigger so much sympathetic physical vibration in the instrument itself that it sets off a feedback howlround. The same Input and Auto monitoring modes we used in the last chapter should continue to work fine here.

A musician who requires real acoustic interaction with their instrument during a performance might monitor the backing track over speakers, open-backed headphones, or one-sided closed-back headphones, so that they can easily hear their direct sound at the same time. Adjusting the instrument/backing balance then becomes a question of tweaking the volume level of the instrument and/or the playback volume of the monitor loudspeakers/headphones—you could even set up a PA system for monitoring purposes if the instrument's super-loud. Under these circumstances the input signal is best left out of the foldback mix entirely, leaving only the track's playback signal for post-recording evaluation purposes. Different systems require different approaches to implementing this "No Input" monitoring mode:

- On most DAW systems, you can switch off software input monitoring completely. (Where yours can't, a more laborious alternative is to mute the recorder track's monitor channel during recording and unmute it for playback.) However, this approach will also eliminate the input signal from your control-room monitor mix, so it's only really suitable for recording your own playing unless you also feed the input signal direct to your monitor mix by some other means—via the DAW's input channel routing, via your audio

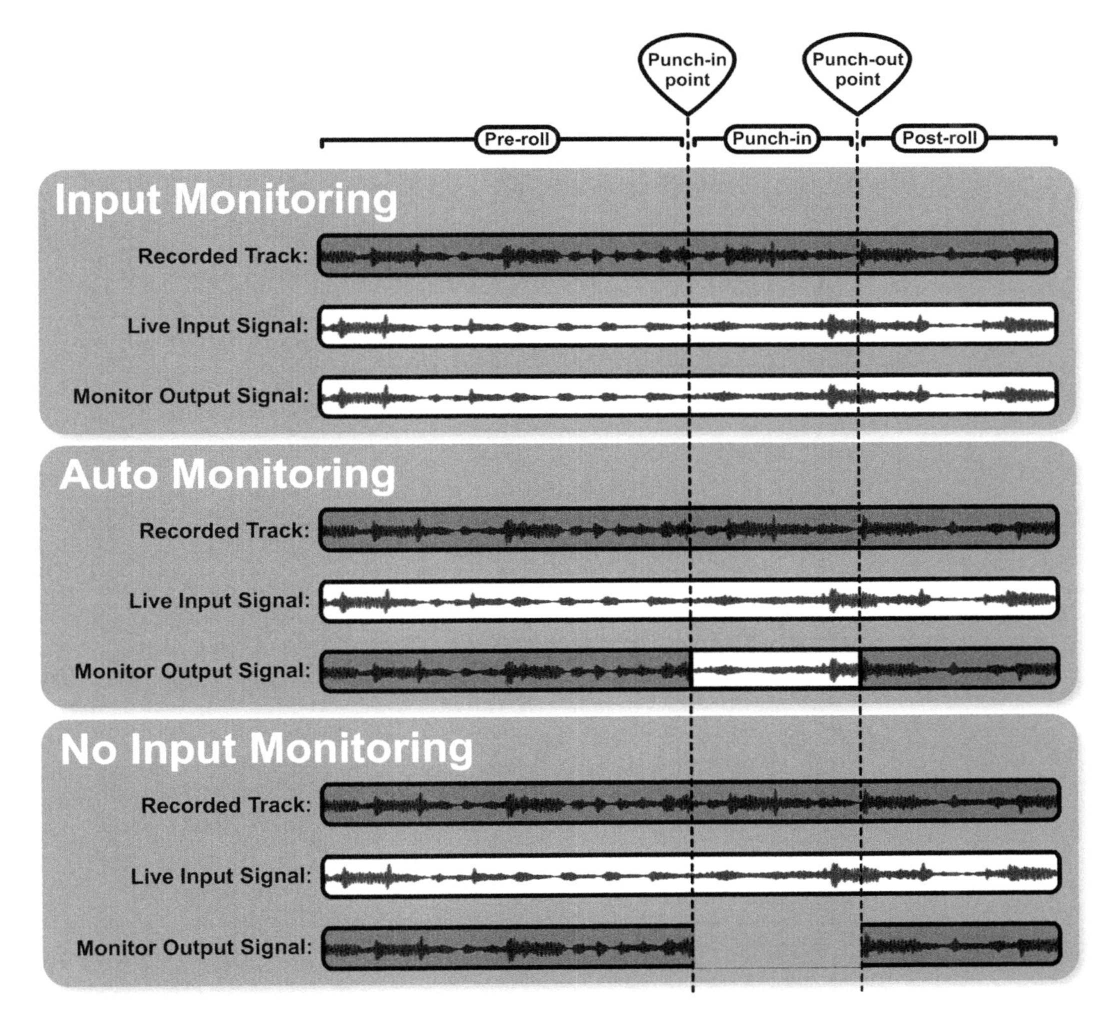

FIGURE 3.4
This diagram compares the No Input monitoring mode with the Input and Auto monitoring modes already introduced in Chapter 2.

interface's onboard DSP mixer, or via some analog hardware working outside the DAW system entirely. Bear in mind, though, that monitoring the input signal before it reaches the digital domain means that you may not detect data glitches on your recording until after you've started laying down takes.

- Leave the recorder track's monitoring mode set to Input or Auto, but turn down the foldback send on that track's monitor channel. This leaves the software monitoring signal intact in the control-room monitor mix, but also prevents playback signals on that track from reaching the foldback mix, so whenever the performer wants to hear what's been recorded you'll need to fade up the send manually, or else invite them into the control room to listen to the main monitor mix with you.

Whatever the monitoring arrangements, insure that it's as straightforward as possible for you to check the foldback mix for yourself during the tracking session. Personally, I like to have a spare pair of headphones in the control room permanently monitoring the foldback feed, so that I can put them on while adjusting the balance the musician's hearing. However, some mixers make it easy to solo the auxiliary send feeding the foldback, allowing you to hear it directly through the main monitors if you prefer (assuming no one else in the control room is relying on those). Whichever method you use, though, don't jump to the conclusion that a performer's just whinging about nothing until you've checked what's coming out of *their* headphones. "Don't just use the control-room headphones," says Clif Norrell,[2] "because his 'phones might be flapping out or doing something strange. You need to hear what he is hearing."

3.2.3 Studio Communications

The inherent acoustic isolation between the live room and the control room can make interacting with the performer a struggle. In most small studios you'll usually get enough sound leakage through the adjoining door that you can just about shout instructions back and forth, and a sight-line between the rooms can also help, especially if you agree a few studio-specific hand signals in advance (things like "tape rolling," "keep going," "play along," and "just listening back") to supplement the timeless comic appeal of mime. However, from the performer's perspective (always the overriding concern), even a combination of these two isn't ideal for discussing parts and takes in any detail, and can also leave musicians feeling rather stranded and "out of the loop" creatively. This is why I invariably prefer to set up a dedicated communications system to enable free conversation between the rooms.

DSP-DRIVEN LOW-LATENCY MONITORING

In Section 2.1.5, I discussed various ways to reduce monitoring latency in DAW systems, but you may still struggle to bring this below 15 ms on some systems—and even if you can, there will be some performers who still find that degree of delay off-putting. For this reason, many audio interfaces now include a built-in DSP-based mixer which allows you to send the performer's input signal to their foldback mix without passing through the computer's operating system, typically bringing the audible latency to well under 5 ms, regardless of how large a buffer size your audio driver needs to deliver glitch-free operation. One downside of working this way is that you have to switch off software input monitoring completely to avoid comb-filtering between that and the DSP mixer's input-signal feed in the foldback mix—sometimes called "double monitoring." As such, you may not be able to use the Auto monitoring mode that's so convenient for punch-in recording. However, some DAW applications provide a workaround scheme (such as Steinberg's ASIO Direct Monitoring) whereby the DAW software is granted control over the DSP mixer, muting the input signal only during playback, thereby reinstating the Auto monitoring facility.

If the concept of DSP-driven low-latency monitoring appeals to you, then my main piece of advice is this: Make sure you get to grips with exactly how your audio interface's DSP mixer works before the session! Some of them can be fiendishly complex, merrily providing plenty of rope to hang yourself with, especially when you're trying to keep an eye on your DAW's mixer at the same time.

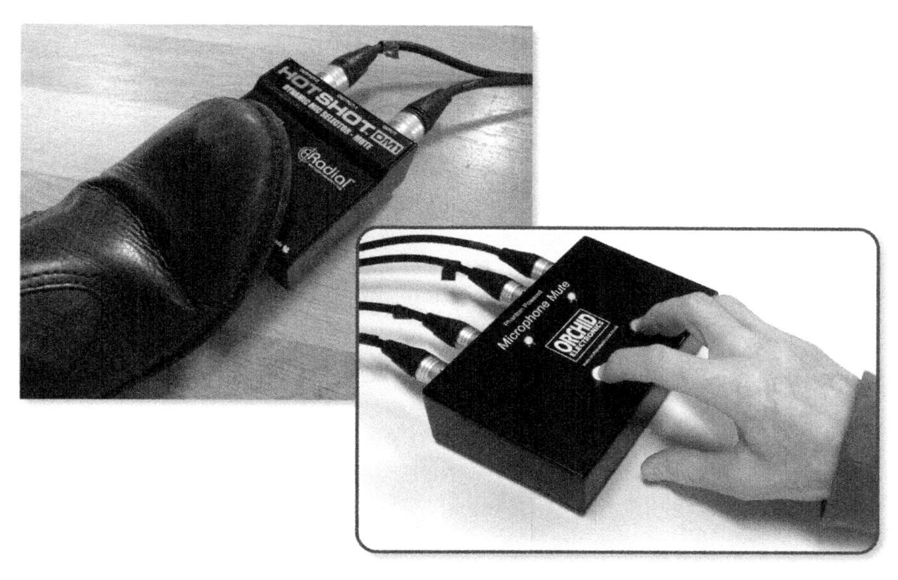

FIGURE 3.5
Two affordable mic switchers: Radial's Hot Shot (above left) footswitch, which can be used for a single dynamic talkback/listen mic; and a phantom-powered Microphone Mute desktop unit from Orchid Electronics (below right), which provides independent switching for two talkback/listen mics and works with both dynamic or condenser models.

The simplest setup involves a "talkback mic" for the engineer, which directly feeds the talent's foldback mix, and a "listen mic" for the musician, which directly feeds the main monitor mix, the result being that a comment in either room will be heard in the other. However, it's normally beneficial for these mics to be switch-activated, so that they're not audible during playback or recording, otherwise the sound of the control-room monitors will compromise the foldback mix, and the acoustic sound of the instrument will break through into the main monitor mix—thereby undermining the very purpose of using a two-room setup in the first place! Where loudspeakers and/or open-back headphones are being used for monitoring in both rooms, using a talkback mic and a listen mic together for communication purposes carries a risk of generating feedback howlround, although you can usually avoid this by positioning the mics close to each person's mouth, angling the mics away from any monitor loudspeakers, and choosing fairly conservative levels for the mic signals in each mix. Another weapon against howlround is if the switch activating the talkback mic also automatically turns down (or "dims") the control-room monitors, a feature frequently found on larger mixers and monitor controllers—an added bonus of this feature being that the engineer can also make comments during recording and playback without spill from the control-room monitors upsetting the foldback mix balance.

There are dozens of ways of implementing studio communications, both in hardware and software, and I'm not going to bore your pants off by trying to catalog them all here. Whichever scheme you choose, the most important thing in my view is the nature of the switching. In the first instance, it's vital to have dedicated hardware buttons to engage each of the mics. Depending on the setup, these might be a mixer's talkback/listen buttons, computer keyboard shortcuts (preferably of the single-key variety that don't require you to hold down modifier keys at the same time), MIDI controller keys, or bespoke mic-switcher gizmos—it doesn't really matter, just so long as anyone can easily press them to instantly open communications. Trying to use faders or foldback-send level knobs to open your talkback/listen mics is extremely cumbersome by comparison, as is trying to click or drag anything in a software DAW with the mouse, and those kinds of kludges are also more difficult for anyone other than the engineer to operate. So don't settle for anything less than real, physical switching!

Some people configure their setups so that the talkback mic switches on automatically whenever the multitrack recorder stops. While this function may initially seem labor-saving, I'm not a huge fan of it myself, because it makes it easier for catastrophically indiscreet control-room banter (who, me?) to reach the ears of the performer inadvertently, and also stops you shouting suggestions during playback or cueing the performer with a count-in during recording.

What kind of mic should you use for talkback and listen mics? Well, if you've not yet had any experience working with mics, then I'd suggest trying a "dynamic" model, simply because it should be fairly robust and foolproof to use—even a cheapy $25 (£20) one will do fine as long as it has an XLR output. Cabling such a mic to a microphone input on your recording system will be straightforward, and all you'll have to do then is make sure you get appropriate signal levels, in exactly the same way you would with line-level sources. You're never going to win any Grammies for the sound of your talkback or listen mics, so don't stress about that—you just need to be able to decipher what people are saying through them!

"While automatic talkback switching may initially seem labor-saving, I'm not a huge fan of it myself, because it makes it easier for catastrophically indiscreet control-room banter (who, me?) to reach the ears of the performer inadvertently, and also stops you shouting suggestions during playback or cueing the performer with a count-in during recording."

I already established in the previous chapter that checking the monitoring arrangements should be part of your regular session-prep routine, but the importance of this increases now that you've got separate control-room and foldback mixes going on, as well as a couple of communications mics to factor in. So make sure that your talkback mic is actually working, and at a comfortable level relative to the backing-track mix, because you'll be slower off the starting blocks if you can't immediately communicate with the musician. That said, if you hit thorny line-checking or signal-degradation problems at any point, it's

FIGURE 3.6
A "powersoak" box such as this Palmer PDI06 can be used to quiet the output of a high-gain instrument amplifier without drastically affecting its overdriven tone.

usually easier to abandon using talkback while you're sorting those out—just leave the live-room door open so you can converse naturally while you focus all your brain-power on tracking down the technical gremlins.

3.2.4 One-Room Monitoring Workarounds

For those without the luxury of a separate live room (in other words the majority of small-studio owners), a little lateral thinking can often pay dividends. In this respect, the first thing to say is that anyone with a pretty minimal studio setup shouldn't discount transporting it to another venue which *does* provide two rooms, particularly if it's somewhere that can be used for free—for example a friend's house, a rehearsal space, or perhaps a local church hall or college at evenings/weekends. If the mere thought of lugging your main recording rig around induces migraine, there's often a good economic case for investing $350 (£250) or so in an additional portable digital multitracker for on-location overdubbing purposes—at least when you compare that to the cost of adding an extra room to your house or hiring a commercial studio for a day! Who cares if it only holds eight tracks? You only need a stereo backing bouncedown from your main studio system to overdub against, which leaves you plenty of spare tracks to work with. (And it's not just small-studio folks who do this kind of thing—David R. Ferguson[3] used a Roland VS2480 multitracker to record Johnny Cash for the Rick Rubin-produced "American" albums, for instance.)

Assuming you're stuck working in a single-room setup, though, the trick to getting the best out of it is to fake as much of a two-room monitoring experience as you can. From your perspective as the engineer, that means trying to reduce how much direct sound from the performer reaches your ears. If there's any way you can turn down the level of the instrument without affecting the performance, then that's worth investigating—for high-gain amplified sounds; a dedicated "powersoak" box between the amplifier and the speakers can do this, although only the more expensive ones can do so without destroying the overdriven tone as well! Positioning the sound source as far as you can from your monitoring position will usually help, of course, as will angling it away from you—instruments and amplifier cabinets all emit more sound in some directions than others. Studio furniture can also be strategically rearranged to absorb or deflect sound, and it's perfectly possible to build something akin to a temporary isolation booth in many one-room studios with nothing fancier than mattresses, quilts, spare mic stands, bungee cords, and gaffer tape.

Despite such measures, headphone monitoring will usually be unavoidable in the circumstances, although you may be able to get away with the lesser isolation of open-back or semi-open-back cans in order to benefit from their typically greater fidelity. (Just don't then turn them up so loud that you can't make useful sonic judgments, as discussed in Section 2.1.6.) Where closed-back headphones are unavoidable, you should be conscious of the fact that they'll block high frequencies disproportionately, so any of the instrument's in-room sound you can still hear will be dulled. If you try to compensate for this when choosing sounds, your recordings will tend to overaccentuate the highs.

As far as the talent is concerned, nothing separates one-room and two-room situations, from a monitoring perspective: The chief concern will still be whether or not their in-room sound is necessary for performance purposes. Where they'd prefer to listen over your control-room loudspeakers, remember that you yourself can still use headphones at the same time, either for additional acoustic isolation from their instrument, or to evaluate the quality of the input signal when No Input monitoring is being employed. (From a practical standpoint, though, you'll probably find it less hassle to set up your headphone monitor mix via auxiliary sends, as if it were a foldback mix, leaving the main monitor mix for the musician's needs.) A major operational plus-point for one-room setups is that dedicated talkback and listen microphones are rarely required, which makes communication with the performer freer. However, if you've gone crazy enough with your DIY booth-building to require talkback/listen facilities, then just treat the booth as if it were a separate live room for setup purposes.

The reality of one-room studio life, however, is that you may never achieve a complete remedy, so there'll always be an unwelcome element of guesswork involved whenever you try to judge the quality of your input signal while the source instrument's playing. An ingenious workaround for this is to use an effects processor to delay your own monitor mix, so that you hear the instrument's input signal a few seconds after its direct sound in the room. This makes it much easier to focus on the quality of the DI sound independently. However, if this isn't practical in your situation, then you've no other choice but to slow the pace of your session by developing each sound via a series of test recordings. In other words, you make your best stab at getting a suitable timbre while the musician's playing, and then record a section so that you can listen back to it more critically against the backing mix. Based on the information gleaned, you take a more educated guess at how to improve on your first effort. Rinse and repeat a few times, and you'll eventually find the sound you're looking for.

I say "eventually," though, because this process isn't half tedious if the cards don't fall in your favor! For this reason it's extremely important to be upfront with the performers about what you're going to do, and why you have to do it.

FIGURE 3.7
If an electric guitarist wants to monitor without headphones using a real amp, this diagram shows some of the ways you might alleviate the resultant monitoring difficulties within a small one-room studio: The player is distanced from the control-room loudspeakers with his or her amp facing away from the monitoring position; a sofa and an upturned table are drafted in as impromptu half-height gobos, blocking the path between the amplifier cabinet and the engineer, but not between the player and control-room loudspeakers; and various quilts have been used to reduce the amount of guitar amp sound reflected toward the monitoring sweetspot. Additional improvements might be brought about by using supplementary headphones for the engineer or by using a powersoak in conjunction with the guitar amp to reduce its acoustic output.

Try something like: "I really want to get the right sound for you here, but we can't properly hear what the recording sounds like while you're playing. Is it OK if we do some test recordings for 20 minutes before we start doing any takes?" Most players recording in smaller studios will understand the practicalities of the situation, cutting you the slack required to make them sound good. When they're forewarned like this they'll also instinctively "hold fire" a bit on the creative side until you're ready to catch them. Even so, if the sound-hunting stage ends up trespassing on the wrong side of the 20-minute mark, I'd recommend taking a short break to clear everyone's heads and ears before doing any real takes. Furthermore, if a musician's muse suddenly takes flight while you're in the middle of working on tones, it's still incumbent on you to hit Record at that instant—whatever the sonics.

3.3 BUILDING A PERFORMANCE WITH COMPING

In the last chapter, I deliberately restricted you to recording a single track for each overdubbed part, building up a master take on that track with the help of punch-ins. The main alternative to this is recording several takes on different tracks, and then comping the best bits of those together at your leisure to create a final

SLAVE REELS AND SLAVE PROJECTS

One of the difficulties of using tape-based multitrack recorders is their limited track count, which can seriously cramp your style where comping is concerned. One workaround is to bounce down a submix of several backing tracks onto a single new track or stereo pair, freeing up the original tracks. The downside, though, is that you can't then process the individual constituents of each submix separately at mixdown, because you'll end up recording over them. For this reason it's much more common practice to bounce backing submixes to a *separate* tape called a "slave reel" wherever lots of free tracks are required, transferring any completed comps back onto the "master reel' once you've finished constructing them. Not only does this maintain flexibility at the mixing stage, but with analog tape it also saves the master reel from the unavoidable wear and tear incurred by intensive overdubbing and auditioning. "[For] the recordings I made with Michael [Jackson]," elaborates Bruce Swedien,[4] "my big worry was that if those tapes got played repeatedly, the transient response would be minimized. I heard many recordings of the day that were very obviously done that way, and there were no transients left on those tapes. So what I would do would be to record the rhythm section on a 24-track tape, then take that tape and put it away and wouldn't play it again until the final mix. And, holy cow, what a difference that made—it was just incredible!"

However, you can only really work with slave reels if you can synchronize them to the master reel when you want to reimport your comped parts. The most convenient way to do this is with twin multitrack machines and a specialized synchronizer unit: Matching synchronization signals (typically SMPTE timecode) recorded onto the tapes themselves are read by the synchronizer, which then governs the transport controls and playback speed of one tape machine to keep it in step with the other. However, there's also a much cheaper method (Warwick Kemp's Image Shift Monitoring) which can synchronize any two playback devices directly, without the use of timecode, as long as one of them has a manual varispeed control—it's just rather more laborious! (Full details of the ISM technique can be found in this chapter's web resources.)

Although the concept of making a slave reel was originally developed for hardware recording devices, don't necessarily discount the idea in the software domain. Where you're working on a large and complex production, it's not too hard to reach a point where your track count is straining the data-transfer bandwidth of the computer's storage media, and the processing demands of so many tracks can also prevent you from achieving a low enough latency for software monitoring while recording. In either case, bouncing down backing submixes into a fresh "slave project" is a good solution, and it has the added benefit of simplifying your monitor mixing while recording, as well as preventing you from unwittingly upsetting pre-existing parts during the cut and thrust of tracking and comping procedures.

patchwork performance. The big advantage of working this way from a production perspective is that you can mix and match tiny sections of different performances, perhaps even from different moments in the timeline, to generate the most emotionally powerful result—a crucial issue when session time with a star performer is in short supply and you don't want to risk losing the tiniest scrap of material. The multi-take approach also suits more restrictive studio setups, because it's not as reliant on a communication system for the musicians—you can just wind 'em up and watch 'em go, reserving the face-to-face conferences for between takes.

But the picture's not all rosy. For a start, comping can be tremendously time consuming, and can easily involve several hours of painstaking auditioning, decision-making, and audio-editing work on top of the basic tracking time required; whereas a master take constructed using punch-ins is finished as soon as the musician is. This is why there's rarely much to gain from comping backing parts that only fulfill a supporting role—save that extra effort for production hooks, lead/solo lines, and maybe important countermelodies. Piling up masses of alternate takes also risks tiring out less experienced players before you've aced the toughest sections, and tends to lull engineers into an uncritical mindset so that they waste the opportunity to optimize the raw material by influencing the performer directly. You can do a zillion takes, but if they're all mediocre you've still got a snowball's chance in hell of comping together something spine-tingling.

This is why I wanted you to get to grips with the punch-in method first of all, because it naturally promotes a much better attitude. It forces you to listen back to your recordings and evaluate them properly, since that's the only real way to determine which sections to retake. This effectively holds a mirror up to the musician, which tacitly challenges them to try to outdo themselves with each new take—an enormously powerful motivator. Working with punch-ins also lends the workflow a natural efficiency, because it focuses the majority of your tracking time on sections that need most improvement; keeps everyone abreast of how much work still remains, so they can pace themselves accordingly; and allows the fullest contribution from the one person who can affect the outcome more than anyone else, namely the performer. So when you're building tracks via comping, the secret to getting maximum results in minimum time lies in retaining as many of the positive features of the punch-in method as you can. There are a couple of practical ways to encourage this: (1) limiting your track count and (2) using a comp sheet.

3.3.1 Limit Your Track Count

For every part you choose to construct through comping, I recommend deciding in advance how many tracks of raw material you're going to allow yourself. I normally use a minimum of three if I'm going to bother comping at all, but even for exposed lead vocal parts you shouldn't need more than eight—unless you've only got 45 minutes with your "A"-list diva and you'd rather be up all night editing than drop a single note!

LAYERING AND DOUBLE-TRACKING

Layering is where you record the same performance more than once, and then blend several of those performances together at mixdown—where only two performances are involved, it's normally called "double-tracking." A lot of modern music styles trade heavily on layered guitars and vocals in particular, most commonly using the technique to thicken the texture, make the performance appear more consistent, and increase the production's apparent stereo width (by distributing the layers across the panorama). The main parameters you need to think about while overdubbing layers are:

- how closely you match the performances. Looser matching will give more of an impression that several parts are playing at once, and tends to suit rootsier styles, while tighter matching will generate a more obviously "studio-enhanced" sound that's akin to a single part with an added effect. In the latter case, detailed audio editing is frequently used to take up the slack where a player finds it impossible to perform two sufficiently similar takes.
- how many layers you record. The more layers you add, the larger your ensemble appears, and this often seduces small-studio artists into overdoing things. The flipside of adding lots of layers is that the expressive performance nuances of each one become progressively more difficult to hear, leading to a bland and emotionless result—a problem that plagues far too many of the low-budget productions I hear. Fewer layers will usually also sound bigger and closer to the listener in your final mix, for a given playback volume.
- how to balance the layers in the mix. All your layers don't have to be at the same level. Indeed, one of the most common applications of double-tracking is where the second performance is quite low in level, surreptitiously smoothing and thickening the main part without most listeners consciously spotting its presence at all.

One of the advantages of using comping is that the unused alternate takes can be pressed into service as additional layers should the need arise later on. But even if you've only recorded a single take, you may be able to conjure up double tracks at mixdown by copying between similar musical sections, presuming that you didn't generate those by copy/paste editing in the first place—it's the slight performance variations between layers that create the thickness in a layered texture, so combining copies doesn't produce the same effect at all.

When you're deliberately overdubbing parts for layering purposes, a salient question is whether you should let the performer monitor layers they've already recorded as the session progresses. For rhythmic parts I wouldn't normally advise this, because the rhythmic information within the previous layers will invariably mislead the performer, reducing their timing accuracy. However, where you're double-tracking melodic lines, it may prove hard work to match note lengths and performance inflections if the previous layers aren't present to guide the player. In this case, panning previously recorded layers well off-center in the foldback mix may help the instrumentalist hear their current performance better when using Input or Auto monitoring modes.

Finally, if you're planning to layer up multiple distorted guitar parts for rock music, I wouldn't recommend recording all of them through digital amplifier emulation if you can avoid it. Even though this technology is now quite mature, there's still something about it that doesn't sound quite as satisfying to my ears when it's layered. If you really have no choice, at least try to use different manufacturers' algorithms for different parts, because that seems to help narrow the quality gap a little. Whatever approach you end up using, however, giving each new layer a slightly contrasted sound will often make for a beefier-sounding mixed result, and changing the layering slightly for different elements of a riff (the chugs versus the powerchords versus the string-bends, say) can actually make it sound more musically engaging. For a final perspective on guitar double-tracking, however, I'll defer to AC/DC's engineer Mike Fraser:[5] "The thing about AC/DC is that they don't double-track their guitars. I keep saying to young bands: 'If you want a really big guitar sound, just get a really good [single-tracked] guitar sound.' Of course, when you double something it sounds bigger, but in the end result you have less guitar: You have to turn it down because it takes up too much space."

When you've decided on your track count, approach each of the tracks with a punch-in mentality, taking and retaking until you obtain equally compelling performances on each track, such that the eventual comp becomes a "best of the best" version. Personally, I prefer to do the takes in a kind of round-robin system. Let's say I've decided to use five tracks. I'll start off by recording one take after another onto different tracks, only retaking one if it's obviously meritless. Then I'll listen back to the first track with the musician and we'll ask ourselves whether we think we can get something better—usually that answer will come back in the affirmative, simply because we've effectively rehearsed the part five times by this point! (That said, there are also times when something fleeting and precious is captured while the musician is still slightly off-guard, so you should always keep your ears open for those delicate early-take gems.) We'll then retake as much of that track as necessary to bring it up to the new quality level we've established.

Once this is done, we'll continue revisiting the tracks in turn like this until each of them feels equally strong from a performance perspective. That's not to say any of them will be flawless, because one of the strengths of comping is that it gives the performer a safety net, so they can push the envelope in pursuit of the most inspired delivery without worrying about the inherently hit-and-miss nature of such risk-taking—as long as at least one of your tracks feels incredible for any given snippet, it doesn't matter how many other tracks you end up leaving on the cutting-room floor. So the main thing is that every phrase on every track should have at least something that feels good enough to make the master take—if not, punch that phrase in again!

3.3.2 Use a Comp Sheet

The other big thing you should do while tracking multiple takes is to make notes of which phrases in each track are worth keeping. This involves creating a "comp sheet," which is basically a sketch of the part's musical timeline, with space alongside it for you to indicate how well each section has been performed on each track. Some kind of musical notation works well for parts that are mostly preplanned, but if the creative process is more fluid and improvisatory (or you just don't like music notation), a more conceptual map of the bars and/or phrases can suffice too. Whatever form your timeline sketch takes, working on manuscript paper can still be quite convenient, because a spare stave above the timeline provides a ready-made grid to organize your annotations—as you can see in Figure 3.8. I'd strongly suggest writing lightly in pencil, so that it's easy to revise your notes as the round-robin process progresses, and it also helps if your timeline sketch is printed or written in ink, so you don't erase that at the same time as your pencil scribblings!

The main things you're looking to note down on your comp sheet for any given track are:

- **Solidly Usable Sections.** That means anything that you'd be willing to accept for your master take if the dog ate all your other tracks, without

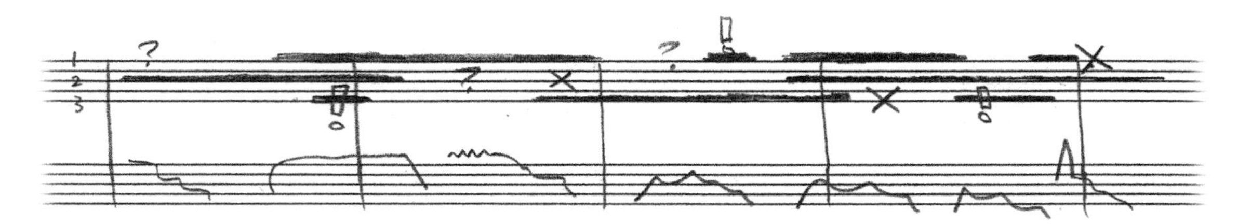

FIGURE 3.8
Whether or not you use traditional musical notation, musical manuscript paper can be very handy for making comp sheets. Here you can see a four-bar guitar solo which has been mapped out graphically on the lower stave, while separate lines of the upper stave are being used to evaluate three different recorded takes for comping purposes.

feeling you had to rerecord. Until you get an unbroken string of these markings across the whole part, the job's not finished. Personally, I don't really feel safe until I have at least two bankable tracks available all the way through the part, however many tracks I happen to be comping across.

- **Moments of Magic.** These may be immediately recognizable "that's the keeper!" licks, or lunatic follies that might none the less have a glint of genius about them given the right context. It's things like these that really make a performance touch the listener, so you want as many of them as possible, and preferably sprinkled liberally throughout the part. If all your tracks are already solidly usable for a given section, but you have no unmissable highlights, then that's a good reason for retaking some of them—and if you make a point of telling the talent you've already got plenty of usable material in the bag, they're more likely to go for broke.

These notes provide valuable reassurance that you're covering all your bases: You're getting enough bread-and-butter material to build a full-length master take that won't embarrass anyone; and you're catching enough pearls of brilliance to justify the effort you're putting into the comping process. Furthermore, by the end of the session you'll already have some clue about how to direct your editing efforts when compiling the master take itself, so I'd advise carrying out the comp fairly soon afterwards, while the meaning of your own scribbled hieroglyphs is still fresh in the memory! However long it takes you to finish the comping, though, don't give a hostage to fortune in the meantime by recording new overdubs against one of the uncomped takes. "Then, when you put the next overdub on, you don't know what to relate to," explains John Simon,[6] "[and] when you try to bring everything up to mix it together, nothing makes sense any more." Your limited track count should prevent the comping process itself being too challenging, as long as you're *au fait* with your recording platform's audio-editing procedures, and bear in mind that the advice I gave in Section 2.3.3 about placing your punch-in boundaries is equally applicable when choosing good locations for edits.

TUNING TIPS

Ropey tuning continues to blight most of the home-brew multitracks I hear, and I think this is partly because people try to judge issues of pitch while listening too loud—pitch perception actually changes subtly at different playback volumes. Loud foldback monitoring may be the only way to get an inspired take from the player in the live room, but that doesn't mean you have to listen that loud in the control room. I also find that pitching issues often become more obvious on a smaller speaker, particularly where bass instruments are concerned, because it emphasises the mid-frequency tuning cues that best survive through small mass-market playback systems. The time to be particularly careful about tuning is when doubling parts at the octave, because that tends to make tuning discrepancies between the parts more obvious. One useful trick to remember is that balancing the upper part a little quieter than the lower one in this situation will usually make the octave appear sweeter.

You may encounter some instruments that can't easily be retuned. Anything with a thousand strings, tines, or sounding rods is probably best left well alone, for instance, unless you've got a specialist professional on speed-dial, while some acoustic instruments which are ostensibly easier to tune may not resonate as effectively when shifted sharp or flat to match a production that's not at concert pitch (which normally puts "A" at 440 Hz). In such cases your recorder's varispeed control may be able to bail you out instead, effectively shifting the tuning of the whole project to match the overdubbing instrument. Once recording is completed and varispeed disengaged, everything you've recorded (including the new overdub) returns to the production's home pitch. Although doing this does have some side-effects for the overdub's sonics, they should be negligible unless you're trying to adjust the pitch by more than about a semitone.

With electric guitars and basses, it's not enough simply to tune their open strings, because you should also check the instrument's "intonation," in other words the relative pitching of the fretted notes. The quickest test here is to tune the open strings and then verify that each string's twelfth-fret octave is also pitching accurately. If not, then someone will need to twiddle the screws in the instrument's bridge assembly to correct this—this is something that many guitarists can do themselves, but otherwise an appointment at the guitar shop can prevent the nightmare of having to retune the instrument for each new chord. Be aware that good intonation may be thrown out if you change the gauge or brand of a guitar's strings, and that it may also suffer if the strings are old.

3.4 IN SEARCH OF THE BEST PERFORMANCE

In Section 2.3.3 I suggested some things you could do behind the scenes to help the talent nail a great take. Otherwise, though, I've so far recommended adopting a fairly deferential stance as regards molding the performance itself, much as you might when working to realize someone else's clearly defined artistic vision—whether that comes directly from an established writer, artist, or band, or from some professional third-party producer or musical director. However, many lower-budget overdubbing sessions are more or less rudderless ships in production terms: The musicians themselves may not have a clear idea of what parts they need, or how to perform those parts most effectively, and there's no veteran hand on the tiller to provide the necessary direction. In that climate, you may be as well qualified as anyone to help chart a new course when the session hits choppy waters on the creative front, especially if you have specific expertise pertinent to the situation—better guitar chops, say, or a training in harmonic voice-leading. It's also quite a natural transition from thinking about sounds to thinking about production. "I basically produce by engineering," says Ken Caillat,[7] for instance. "If I can't get all of the sounds to come up and be a good mix, it's like, "What's wrong?" If the sounds aren't the problem, then what *is* the problem?"

"Let musicians exhaust their own strategies first before chipping in with your own. Apart from anything, gamely following through on their ideas (however daft they may at first appear) provides incontrovertible evidence that you're on their side, so your intent is less likely to be misconstrued if you offer your own contributions."

Clearly, though, there's always a danger of a "when I want your input I'll ask for it" response to any non-technical suggestions coming from the engineer, so you have to tread carefully, especially when working with anyone for the first time. For example, it's best if you let musicians exhaust their own strategies first before chipping in with your own. Apart from anything, gamely following through on their ideas without shooting them down (however daft they may at first appear) provides incontrovertible evidence that you're on their side, so your intent is less likely to be misconstrued if you offer your own contributions.

The last thing you want is for an artist to feel like you're trying to turn them into something they aren't, or for a performer to feel you're undermining them in front of their band mates—who cares what you think if the artist hates it, or if you cause the band to implode? "Mutt [Lange] was always very careful not to intimidate the musicians," recalls Tony Platt,[8] for instance. "I remember on one occasion… we spent a long time trying to get a bass part. Mutt is a bass player himself, so I know he could have gone out and played the part in two minutes flat, but that would have been very intimidating to the bass player and might have destroyed his confidence completely. So Mutt was quite prepared to sit there for the best part of a day and coax the bass part out of this guy." In small studios there will inevitably be occasions where technical limitations or the lateness of the hour preclude much experimentation, but you'll minimize bad vibes if you express misgivings in terms of those practical concerns, rather than arguing on creative grounds. In other words, "Perhaps we should try that once the main part's in the bag?" is likely to lose you fewer allies than "I don't think that'll work, so let's not waste time on it."

One thing that's worth stressing to performers in the studio is that the validity of an idea is frequently impossible to judge unless you persevere with it long enough to hear what it actually sounds like, so trying something out doesn't mean committing to its appearance on the record. The best way to underline this point is by the unflinching ruthlessness with which you bury any idea of yours that the musicians decide against when the results are actually in front of them. "I love being proven wrong," says Joe Chiccarelli,[9] "because then you know what the right way is." Doing this without a backward glance makes musicians feel more relaxed about trying your ideas, knowing that you're not going to react to their veto as if someone just drowned your fluffiest kitten.

WORKING IN TEAMS

In most small-studio situations the engineer will be engaged by the musicians being recorded, and will be directly answerable to them. As such, you'll primarily show your professionalism by being on time, well prepared, courteous, and efficient. (It helps if you don't smell like a curried Stilton too.) However, if you're working as part of a production team, you also have to realize that successful teams need both a clear leadership structure and a culture of mutual support.

In other words, if you're not the producer, then you should know who is, and should be doing your best to make that person's job easier. This usually means being "invisible," in other words performing your tasks so quickly, effectively, and unobtrusively that you never distract the musicians from their job. If a musician or the producer actively notices you, then that usually means you're doing something wrong! Another skill you should cultivate is anticipating the direction of the session and preparing for it in the background, so that everything that's required for the next activity seems to appear magically just as it's needed. "Anything that disrupts the session breaks the vibe, and vibe is what the music is all about," remarks Scott Kieklak.[10] "Technical chops are important, but being able to avoid disruptions to the flow of the session is even more important." Relevant information must be fed up the chain of command in a timely fashion too, and guesswork is to be avoided when answering questions from superiors. While it's bad form to directly oppose or contradict the producer during open discussion, they'll also expect (and respect) a quiet word of warning if you spot them careening toward some embarrassing technical error or social *faux pas* unawares.

When you find yourself in the lead role of a team yourself (even if that team is just you and an assistant), then you should try to have a clear plan for the session so that you keep the team working as a well-organized unit. Leading by example will usually engender most respect amongst underlings, and loyalty can be fostered both by providing people the opportunity to learn on the job, and by apportioning credit where it's due.

3.4.1 Performance Pointers

When working with new people, performance-related suggestions are probably a lower-risk starting point than weighing in about how the part itself should be written. Here are some general-purpose strategies that can help improve results, irrespective of your level of insight into the technicalities of specific instruments:

- **Maintain a Positive Spin.** Presenting your reactions with a positive spin should be a default position—it may not endear you to everyone, but it sure gives you the shortest odds! If the player peels off a blinding lick or inspired fill, acknowledge it, if only with a smile or a quick thumbs-up. If you're feeling the groove, don't just sit there like a gargoyle—let it move you a little. Don't let a take go by without commenting at least in passing about some admirable feature. ("Great lead-up into the middle-section!" "I wish I could play rhythm like that." "That last chord was so evil, you'd be burnt at the stake in certain parts of rural Herefordshire." You get the idea ...) If you think the last take sucked, the player probably thinks so too, so just asking their opinion ("How was that one for you?") will often get them kicking their own ass without any assistance from you. When you have to play Bad Cop, build on something positive that's already been achieved. ("You nailed that first verse, so let's see if we can match that feeling here." "I think we've already got a stronger take on track two, but I reckon we can trump that if we give it another go." "I'm sure I heard a better one from you onstage." "You smoked both those riffs, but the linking part didn't seem to live up to them for me.") The more you accentuate the positive when you can, the fewer flounce-outs you're going to get on those difficult occasions when there's no alternative but to trot out umpteen variations on "It's still flat."

- **Be Informative.** "Do it again" doesn't tell a player anything useful. What wasn't good enough? All of it? The last note? The tuning? The rests?! Even if you can't say anything very concrete, at least try to give some indication of whether or not they're heading in the right direction. ("You were actually closer with the previous take, I think.")
- **Favor the Abstract.** Interfering with the nuts and bolts of an instrumentalist's playing technique is one of the most hazardous things you can do as an engineer. With more spontaneous musicians, it can spook them into thinking too consciously, killing the vibe stone dead; and it can quickly antagonize more cerebral types by implicitly belittling their technical abilities. No matter how much of a virtuoso you happen to be, you'll always be on safer ground politically by keeping your comments fairly abstract, stating the problem in broadly emotive terms and allowing the musician to develop the solution best adapted to their own abilities. So with a violinist I'd give "Can you make that line smoother somehow?" a proper chance before trying anything more specific like "Slur the bowing and stay on the 'A' string." Sure, if a player requests technical assistance, then provide it if you can, but normally I'd treat it as a last resort.
- **Find Some Shared Imagery.** If merely waxing lyrical can't conjure up the right emotional state in the performer, try finding some shared imagery. Visual images are often most potent, such as adverts (Marlboro Man, the sepia-tinged world of Hovis, those ridiculous slow-mo perfume spots), movie scenes/characters ("King of the world!", "Of all the gin joints ... ", "Well, do ya, punk?"), or real/imagined locations (flying over the Grand Canyon, playing to a packed arena, trapped in an ice cube). However, any shared experience is fair game, whether that's memories of high school, historical events, or the personalities of mutual acquaintances.
- **Look to the Lyrics.** Because vocals are frequently added late in the tracking process, it's easy to overlook the value of responding to the lyrical message while overdubbing instrumental parts. Indeed, this is one reason why some producers insist on working to a scratch vocal from the very earliest stages of a project (of which more in Section 11.1.2). Whether you go that far or not, a copy of the lyrics can provide a useful emotional focus for the entire project.
- **Deal in Questions.** Try couching requests in the form of questions. For example, by saying "What about a high 'E'? Is it even possible to get up there?", rather than "I think the high 'E' would be better," you're making it easier for the performer to disagree with your opinion, because you've made

ENGINEER'S QUICKSTART GUIDE: GUITARS

The bewildering variety of guitar and bass-guitar designs means that choosing the right instrument for recording any overdub is pivotal. Indeed, it's in recognition of this fact that some high-profile recordists retain collections of suitable instruments even though they're not players themselves. It's doubly important for rock and pop productions, which rely heavily on guitar textures, because combining different guitars typically gives fuller sonics, and contrasted timbres are terrifically useful for arrangement purposes. So I urge you to ask any guitarist you overdub with to bring some alternative instruments with them, whether they're their own or borrowed from obliging acquaintances.

Old strings won't usually help the sound of any guitar, but if you do sensibly change them for a recording session, try to do it the day before, so that the strings have a chance to finish stretching in—the tuning of nylon strings in particular can be difficult to nail down during the first 24 hours. If the height of the strings above the fretboard (the "action") isn't enough, they're more likely to buzz against the frets, so listen closely for this at the first opportunity, because it's impossible to correct after recording. Raising the action may require adjustment of the instrument's bridge assembly or neck curvature (via a "truss rod" in the neck itself), something that any guitar shop can do if the guitarist isn't comfortable tackling it personally.

If there's a choice of pickups available on any instrument, then it's sensible to ask the player to try some of those different options early on when you're hunting for sounds, because they'll usually make a big difference, and any onboard tone controls will also be extremely important. If you're experiencing balance problems within the sound (maybe one string is overwhelming the others, or the upper strings aren't counterbalancing the lower strings), it may indicate that the pickups would benefit from some height adjustment—typically a job for a specialist. Another critical tonal factor is the type of pick used, if any, so there's no sense in relying on your guitarist here when you can spend a few pennies on a selection of your own. "I generally have a pack of picks with me," says Jack Joseph Puig,[11] "that range from medium to thick, metal, wood, plastic, felt, and smooth and jagged edges. This part of the equation is as important as selecting the proper amp." The picking action warrants attention too. "If the guitarist wants the sound to be a little more bright or aggressive," continues Puig, "you can suggest that he move his picking hand a little further toward the bridge, or maybe turn the pick sideways so the string catches the side of the pick a bit."

FIGURE 3.9
The choice of guitar pick can make a surprisingly big sonic difference, so small-studio engineers are well advised to keep a selection of their own.

it clear that it's negotiable. Plus they can also save face if they're not skilled enough to hit a high 'E', by claiming that it's not physically possible—even if you could play the part much better yourself, there's nothing to be gained by rubbing their nose in it.

FIGURE 3.10
Scenes from classic films like these are fantastic common currency when trying to communicate specific emotional moods you'd like the performer to convey in their playing.

- **Play Through Any Mistakes.** Given that punch-ins and comping can repair any performance mishaps, you should encourage all performers to keep playing even if they make a mistake. Not only does this tend to improve the performer's creative flow overall, but it's also surprising how the effort of playing through a momentary instability very often unearths gold dust on the other side. "It's the mistakes, sometimes, that are the magic," says Jez Coad,[12] "and a major, major part of my job is to say, 'OK, let's go again, but play that—I know it's not what you meant to do, but there's some unbelievable magic there!'"
- **Eyes Closed.** There are many potential advantages of getting an instrumentalist to perform with their eyes closed. Firstly, it blocks out distracting visual cues from the studio, thereby focusing more of the player's attention on the subtle emotional undercurrents of what he or she is hearing. In addition, though, it forces the performer to memorize their part and play by feel (not by sight), both of which can help make a performance less inhibited.
- **Posture and Movement.** The way musicians hold their bodies while performing will alter how they play, so it definitely matters whether your guitarist is lounging back on the sofa, sitting on the edge of a stool, or standing up, because each posture is more likely to suit different parts. Bear in mind that some chairs can impede the playing action too—cellists,

for example, will have difficulties playing in low or overly comfy chairs, or where prominent armrests interfere with their bowing action. You can also influence a performance by getting the musician to move while playing. This is very pertinent when you're grappling with rhythmic issues, because a player who moves with the music will usually play tighter, in my experience. Asking the talent to take a single rhythmic breath on the beat before an important entry is another time-honored means of improving their timing. Physically expressing the emotional content of a phrase while playing it (think of the archetypal classical maestro) is natural to some, but more reticent performers may need some extra encouragement to try this for themselves.

- **Be the Conductor.** While I'm not proposing you go all Herbert von Karajan, performers who rarely perform on their own will find it easier to relax if they're following someone else's lead—even if all you're doing is indicating where the start of each phrase is. Furthermore, the nature of your physical motions and facial expressions may also incite more enthusiasm in the performer. (Or helpless bouts of the giggles, which are rarely a bad thing either!) By way of evidence, here's Leonard Cohen talking about producer Bob Johnston:[13] "The best thing that Bob did was he went up to the glass when people were performing and he put his arms up and he bobbed his head and he smiled like he was hearing the music of the spheres. As a performer, when you see that, you feel like you're doing something right."

ENGINEER'S QUICKSTART GUIDE: STRING INSTRUMENTS

Acoustic instruments almost always sound better miked than DI'd, but if you'd still like to try recording a string soloist for this chapter's assignment, then here are a few tips. The first thing to realize is that players will almost always finger parts with their left hand as close to the tuning pegs as they can ("first position," in the lingo), simply because it's technically easier. However, many of the richest string tones are derived by moving the hand up toward the bridge into higher-numbered positions. Sliding the hand between different positions while playing also has great expressive potential. If you'd like to hear what some of these options sound like, try asking the musician to play the musical line on a lower string (so they have to shift to a higher position) or not to cross strings (which forces them to shift between positions). Using higher positions also allows the instrumentalist to avoid playing their topmost open string, which is usually of a different construction and has a tone that will tend to zing out rather obtrusively—especially on any violin worth less than a pair of Jimmy Choos.

The importance of bowing should not be underestimated. Again, string players will often play everything with separate bow strokes unless you ask them to do otherwise, whereas "slurring" groups of notes into single bow-strokes often improves the phrasing. Asking a player to "use more bow" or "use more vibrato" is also a handy way of getting them to play with more expression—classical players in particular will often favor a more measured delivery instinctively.

On a practical level, bear in mind that cellos and double basses have a metal spike that rests on the floor during playing. This may skid on hard flooring surfaces (and may also damage them), so it's worth being prepared with a small rug or an offcut of carpet to head these problems off.

3.4.2 Arrangement Tips

Once you've built up some trust, the musicians you're working with may well appreciate fresh input on arrangement matters too, although again it's better to give them ample opportunity to solve a problem for themselves first. Whole books have been written on the subject of arrangement, so I'm not going to try to cover all bases here—from an engineer's perspective you can get a long way just by considering some of the following fundamental principles:

- **Contrast.** If your production features different musical sections, then you can draw more attention to each section's arrival by increasing the contrast between them. The enemy of contrast is any part that goes on doing the same thing across a section boundary, so it's a good idea to avoid that. Also, if there are any moments that need to feel really powerful, those can be enhanced by preceding them with sections that have been deliberately softened in some way. Contrast isn't just a concern "horizontally" through time; it's also important "vertically" in terms of differentiating the separate musical elements in what you are recording. Wherever any two instruments play at the same time, it'll be a lot easier to hear them independently (and indeed mix them) if their personalities are distinct—one sustained and the other staccato, perhaps.
- **Variation.** Although repetition is important for groove and memorability, the longer anything repeats in an arrangement, the less you consciously notice it, broadly speaking. So if you want to make your arrangement as arresting as you can, you need to keep refreshing the listener's attention in the more repetitive parts by varying every third or fourth iteration slightly to catch their ear anew. If you just record a loop and then copy and paste that across your whole track, you'll clutter up your mix with a part people won't actually notice most of the time. A variation doesn't have to be anything fancy, though—just altering or missing out a note will often be fine.

"Wherever any two instruments play at the same time, it'll be a lot easier to hear them independently (and indeed mix them) if their personalities are distinct—one sustained and the other staccato, perhaps."

- **Register.** When two important parts happen to reside mostly within the same frequency range, it's going to make life difficult at mixdown: You won't be able to make both of them sound very good, because they'll be fighting for the same frequencies; and they'll also tend to create an unnatural frequency build-up overall. Your choice of timbres can help avoid this, but you should also try to put musical parts in appropriate pitch registers. This is something that many instrumentalists fail to do naturally, because there is usually one region of their instrument that they're most comfortable with, which is why you can bring some valuable objectivity to bear here. Just shifting the octave may be enough, and with guitars, a capo is a simple way of squeezing many more harmonic inversions from an amateur player's limited chord vocabulary. However, effective use of register may also mean restricting the range of chordal instruments, for example dropping a Rhodes part's left-hand notes, or suggesting that a guitarist uses only the lower three strings. Again, a lot of musicians will tend to play any chord in its most thickly textured version by default, even though it's frequently the thinner registrations that are more suitable for fitting in against other tracks. Finally, keep an ear open for which pitch registers (or even which individual notes) actually provide the best sonics for the instrument in question, as this may also inform the exact choice of part.

ENGINEER'S QUICKSTART GUIDE: ELECTROMECHANICAL KEYBOARDS

You don't meet many electromechanical keyboards in small studios nowadays, because so many musicians are content to use modern digital emulations instead—or at least not concerned enough about the sonic differences to put up with the cost, maintenance, and transport problems of owning the real thing! If you do encounter the genuine article, though, the main thing to say is that tuning is no job for the faint hearted, so if this is a concern, it's worth sorting out well in advance, and probably with professional help.

- **Efficiency.** It's an immutable fact that the more parts you put into an arrangement, the less space each one of them has to sound good in the final mix. "People think, 'If I throw a hundred instruments on here, it will sound huge,'" says Frank Liddell,[14] "but it's often the other way around." Neil Dorfsman[15] echoes this sentiment: "I generally try to get by with as few parts as are necessary to get the job done. Even if I hear another part, I'd also rather you hear it in your head when you listen, rather than recording it. That's a hard trick to pull off; only the really masterful records achieve that sense of being literally understated." One way to fight clutter is to make each part more efficient from an arrangement perspective. So you might obviate the need for a new part by adding more rhythmic interest to a simple note-per-bar bass line, by introducing a few broken-chord figurations to add harmonic texture to an oboe solo, or by embellishing a repetitive piano oscillation with snatches of countermelody. By the same token, if a new part includes harmonic support that's already amply provided by an existing part, then why not strip out the harmonic element and focus more on the melodic or rhythmic elements of the new idea?
- **Gaps.** Almost any part will have some psychological "gaps" in it, even when it's sounding all the time, in other words, brief moments where nothing new is happening. If you can aim the main action within any new part to coincide with the gaps in existing parts, and likewise create gaps in the new part to accommodate existing points of interest, then it'll lighten the mix engineer's load when it comes to making everything audible. (For a great recent example of this, check out the way the main chorus riff in Carly Rae Jepsen's enormo-hit "Call Me Maybe" slots neatly around the lead vocal.) Similarly, it can be a mistake to try to fill all the gaps in your arrangement with your earliest overdubs, because that leaves nowhere to go later.

- **Arches.** If you think of a piece of music as a series of resolutions with arches of tension between them, it's usually important that no part undermines an arch by resolving too early. The most common way people do this is by hitting the key's home chord before it's required, or by laboring the tonic note too often in their melodies. If you want to build up harmonic tension, the best way to do it is to withhold the chord/note of resolution. Similarly, if the rhythmic energy of your chorus relies on your adding sixteenth-note percussion, then bringing this in two bars earlier may weaken that section's arrival.

DEALING WITH CLASSICALLY TRAINED MUSICIANS

Compared with folk or popular musicians, players raised in the classical tradition tend to be less comfortable creating their own lines or improvising parts on the hoof. Although this needn't at all stand in the way of getting good results, I've seen numerous sessions falter simply because they failed to acknowledge the fact that classical players usually give of their best when they've prepared, practiced, and (preferably) memorized their parts in advance. So if you're not a classical musician yourself, let me offer a few tips for avoiding this pitfall.

When you've already got a part in mind, do your level best to provide it in sheet-music form. If you're not confident with musical notation yourself, then ask the musicians to notate the part themselves, either before or during the session. (Don't rely on them bringing the necessary pencils and manuscript paper to the studio, though, because they often don't, in my experience! Loose-leaf paper is better than bound pads, because it makes it easier for people to avoid page turns mid-take.) Most players can transcribe a recorded guide part by ear, but screenshots of a MIDI part within a software DAW's piano-roll editor will usually make that process significantly quicker. If your computer sequencer can generate rough-and-ready notated parts from MIDI data semi-automatically, that can speed things up even more, but make sure you check that the part's within your target instrument's pitch-range; that you choose an appropriate clef for the stave; and that you've tidied up (or quantized) the source MIDI data's timing and removed note overlaps as much as possible without actually changing the part's fundamental rhythms, because this will make the printout clearer to read. While the player's writing out their part, take the opportunity to discuss the phrasing, rhythmic stresses, and overall volume contour of the required performance with them, asking them to make whatever markings they need on the page to remind themselves of any decisions. It also improves communications during overdubbing if you write bar numbers in at the start of each line—so much so that I always insist on it myself.

If the onus of creating the part falls primarily on the performer, then leave ample time to develop and rehearse it before you actually need to record. Some classical performers get very self-conscious feeling their way around their own instrument looking for ideas, so this is one situation where looping a section of the backing mix and leaving the musician alone with it for a few minutes can really pay off. But while you're fetching the coffee, do keep an ear open for moments that seem to work particularly well, and compliment those spots to steer the player in the right direction. Getting the musician to note down any particularly juicy bits is also worthwhile, because once you have a handful of these you can often build most of the final part out of them like a jigsaw, after which it becomes much easier to think of smaller sections to fill remaining gaps.

Where inspiration's proving thin on the ground, suggest that the player notates a few of the arrangement's other main lines, which can help on two counts: Firstly there may be aspects of those that can be repackaged into the current part; and, secondly, it may highlight gaps in a competing line which will suggest where your new part should be most active. Writing down the constituent notes of underlying chords is useful for some players too, because it indicates the most obvious resting points for a line, harmonically speaking. You may even find it's instructive for the talent to trundle through a few highlights of their concert repertoire for you, so you can trawl

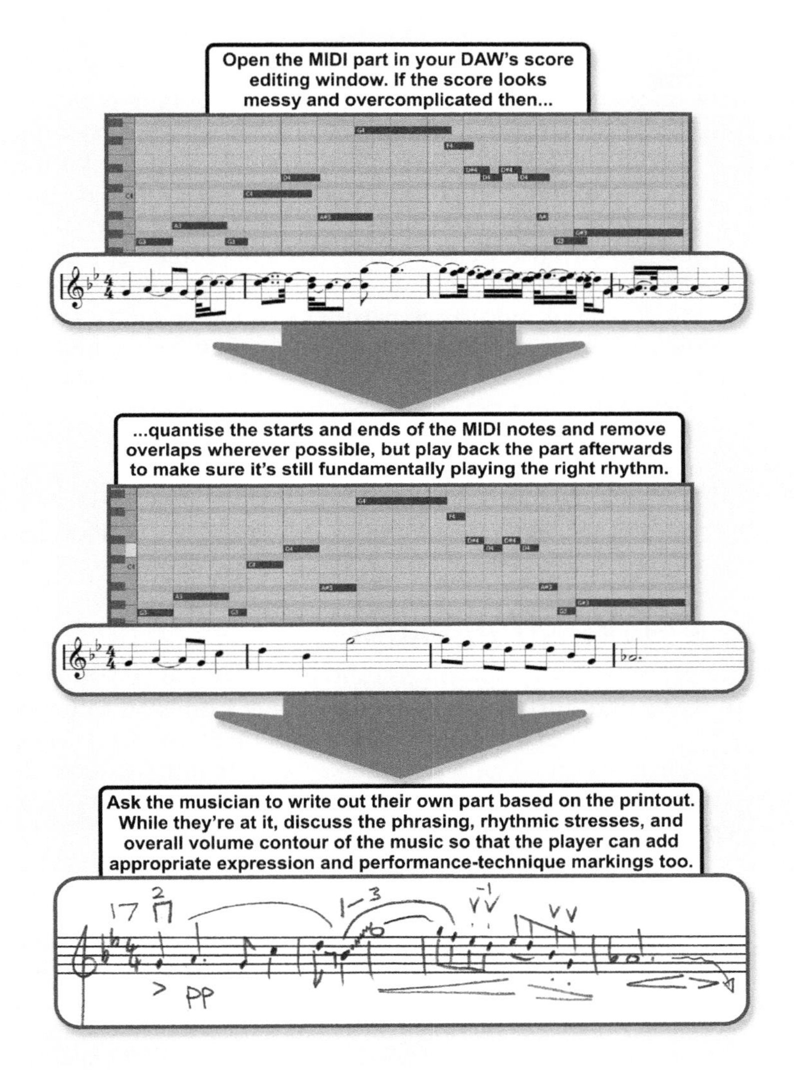

FIGURE 3.11
Even if you aren't confident creating traditional musical notation yourself, you can still help classical performers by providing them with a basic score printout, from which they can handwrite their own parts.

them for evocative-sounding ornaments or specialized performance techniques that are unlikely to spring up naturally during improvisation.

Whatever part you come up with, though, don't let the performer record it cold straight after writing it. Leave them to get used to it on their own while you dust the valances for a few minutes, and only start going for takes when it sounds like they're really inhabiting the part emotionally. One thing classical performers are very good at is practicing a part, so they'll typically improve their performance much more quickly this way (or in a rehearsal situation) than they will just by racking up loads of takes.

- **Countermelody.** Snatches of melody in subsidiary parts can give your music a powerful forward momentum, so try sneaking them into overdubs wherever they don't distract from some more important lead line or hook. Countermelody tends to be most noticeable (and therefore effective) at the high or low registral extremes, which is why the bass line in many commercial hits frequently competes with the lead vocal for melodic supremacy (Brian Malouf:[16] "I try to have the bass and vocals be the counterpoint to each other"), and also why G-Funk's trademark super-high synth leads were such an effective device. When trying to come up with melodic ideas, it's excellent practice to examine the melodic contour of the production's lead vocal or hook riff, to see whether you can reinforce those in the listener's memory by presenting fragments of them in your countermelodies.
- **Dissonance.** Perfectly tuned parts that fit the prevailing harmony will tend to blend better into the arrangement, whereas off-tuned or dissonant notes will stick out more—something you can use to good effect from an arrangement perspective. For example, the snare drum in AC/DC's "Back In Black" blends well with the mix despite its power, because it's tuned to the song's key note, whereas the wayward tuning of the "mob" vocal hook in Usher's "OMG" keeps that clearly audible even when it's at a very low level in the balance.

3.5 NEXT STOP: MICS!

Over the last three chapters you should have accumulated enough information to set up and drive an overdubbing session from start to finish, and we've also explored many of the less tangible issues involved in sharing a studio with living, breathing musicians. If you take the time to hone these skills through this chapter's assignments, you should be well prepared for Part 2's next big challenge: dealing with microphones.

CUT TO THE CHASE

- A DI box or dedicated high-impedance instrument input should be used for pickup signals. Specialized higher-impedance DI boxes are advisable for piezo-electric pickups. If you decide to use a speaker-level DI box, be careful not to damage the amplifier by running it without a load. An active DI box will usually work better for passive pickup electronics, while passive DI boxes are more suitable for higher-level active-pickup and line-level signals.
- Pickup signals are very prone to interference, so listen for noise levels when positioning the player and try to keep cables as short as you can, avoiding mains transformers and digital cabling.
- A separate foldback mix makes your engineering job easier, not least because it allows you to put the musician in a separate room. If the performer and engineer can't hear each other directly, set up some kind of communications system, preferably involving separate hardware-switched talkback and listen mics.
- Where a DI'd instrument makes an acoustic sound, you must decide whether it helps or hinders the musician's performance. If it helps, then the player can balance themselves against the foldback mix acoustically, using loudspeakers or open-back/one-sided headphones in conjunction with No Input monitoring; if it hinders, then closed-back

headphones are safest, using Input/Auto monitoring. Always check that monitoring arrangements are operational immediately before the session, and that talkback/listen mic levels are already set up sensibly.

- Even if you can't use a recording venue with a separate live room, try faking a two-room scheme as far as you can with closed-back headphones, improvised baffling, and power-soaks. If that doesn't work well enough and you don't have the facility to delay your own monitor mix, then you'll have to allocate extra session time for test recordings.
- While comping has many advantages in the small studio, it also has hazards, which can be mitigated by limiting the target track count and by using a comp sheet. If your recorder won't support enough tracks for the comping approach, create a slave reel/project.
- When layering and double-tracking, the main aesthetic choices are how closely you match the performances, how many layers you record, and how you balance the layers in the mix. Whether the performer should monitor previously recorded layers while overdubbing new ones will depend on the nature of the part.
- Make tuning judgments at reasonable volume levels. For guitars and basses, check the instrument's intonation as well as its open-string tuning before recording. A recorder's varispeed function can work around small untreatable tuning offsets with negligible side-effects.
- As an engineer, it's safest to wait until the musicians have exhausted their own ideas before proffering any creative direction of your own, and then to favor performance over arrangement suggestions at first. Don't judge an artist's suggestions in the abstract without trying them out, so that you encourage them to extend you the same courtesy.
- When refining performances: Be positive and informative, and deal in questions; favor abstract imagery and seek inspiration in the lyrics; suggest performing with eyes closed and playing through mistakes; try to use posture and movement in your favor; and consider conducting in some way.
- Important factors for any arrangement include contrast, variation, register, efficiency, gaps, arches, countermelody, and dissonance.
- With guitars and bass guitars, concentrate first on choosing the most suitable instrument, pick, and pickup, as well as tweaking on-body tone controls. Keep an ear open for fret buzz/rattle that might indicate a poorly configured action.
- For string instruments, try to get good use out of left-hand fingerboard positions and bowing possibilities, and plan in advance for cello and double-bass spikes.
- Classically trained musicians usually perform better when parts are notated and practiced in advance. Most musicians can notate their own parts if required, but make sure they include phrasing/expression markings and bar numbers. When developing new parts, try notating existing parts or playing favorite concert repertoire for inspiration, and then allow enough time for practice before doing takes.

Assignment 1: One-Room Dry-Run

In this assignment you'll get some experience of comping and layering, as well as working around the compromises of one-room overdubbing using delayed monitoring and test recordings if necessary.

- Select an existing project you can practice overdubbing on, and find a pickup-equipped instrument you can use for the recording—again, don't worry if you're no great performer, because the simplest of parts will do for now.
- In addition to the session preparations discussed in Chapters 1 and 2, work out how best to get a DI signal into your recording system (Section 3.1) and what

one-room monitoring scheme you're going to use (Section 3.2). If you'd like to try DSP-driven low-latency monitoring, now's the time to get that up and running, working out how this affects your choice of monitoring modes and listening carefully for the comb-filtering side-effects that might indicate double monitoring. For this assignment, you should try building at least one master take via comping, so make sure you can access enough spare tracks for this (comp from just three raw takes for now), migrating to a slave reel/project if necessary.

- Once you've dealt with connection, level-setting, and monitoring technicalities, develop a short lead-solo/countermelody part (with reference to Section 3.4.2) and a sound to go with it (revising the suggestions in Section 2.2.4), using delayed monitoring and/or test recordings to help you. Record three takes of raw material for comping purposes, using a comp sheet to focus both on what's solidly usable and on what's magical. When you're done, stitch together the master take. If you discover any weak moments remaining, mark them on the comp sheet, and go back and rerecord those sections once you've comped as far as you can.
- Build a new rhythm part using a different sound, but this time returning to Chapter 2's punch-in method of creating a master take. Once the first track is done, record two additional layers of the same part, but only monitoring the layer you're recording. Once all three layers are done, experiment with balancing them together for different effects, and ask yourself whether the rhythmic matching is close enough—if not, retake the more wayward track sections. Finally, record a double-track for your comped solo line too, but this time include the comped line in your monitor mix while recording to get a sense of how that affects your ability to perform.

Assignment 2: Two-Room Overdubbing

Assignment 1 should have given you an appreciation of the mechanics of the multi-take recording process from both engineer's and performer's perspectives, which means we can once again get a separate performer involved, and also experiment with a two-room setup.

- Select an existing project you can practice overdubbing on, and find a player with a pickup-equipped instrument who can help you. Arrange a session with them, with the stated aim of generating one or more original parts, including at least one lead solo. Discuss the musician's session requirements, provide them with any guide mixes or written material they need in advance, and find out as much about the nature of the instrument and pickup system as you can (so you can prepare your recording chain). Also ask whether the instrument makes any significant acoustic sound, and whether your player would prefer to hear that or the recorded signal during the performance.
- However minimal your studio space, do your best to arrange some kind of two-room setup for the benefit of this assignment—if that means putting you or the performer in a corridor, a bathroom, or a closet, then so be it! Work out how to create a separate foldback mix for your performer, and set up a communications system with talkback/listen mics, dedicated hardware switching, and some means of listening to the foldback mix in your control room. If you're going to use No Input monitoring, then make sure you can still hear the input signal in the control-room monitor mix, and that you know how to feed playback from the recorder track through to the foldback mix when necessary. Immediately before the session,

check that all monitoring hardware is working, that the levels of the backing track and the communications mics are sensible in both the control-room monitor mix and the foldback mix.

- At the start of the session, explain to the performer that you're going to be working in two rooms, and check that the communications system works for both of you at the earliest opportunity. As in Chapter 2, be ready to hit Record the moment the performer starts playing along with the track. If you're not yet happy with the sound by that point, take any opportunity to refine sonics while the basic parts are still under construction.
- When it comes to recording the solo line, explain to the performer that you'll be creating it using comping techniques, and that you'd like to record three raw tracks for the purpose. As such, they have more freedom to experiment without worrying about making mistakes. Use a comp sheet to keep notes on your progress, noting down any of the performer's preferences as well as your own views. If the performer would like to get involved with the comping process itself, by all means do the comp during the session, but otherwise leave the comping process until afterwards so that it doesn't disturb the player's creative flow.
- For other parts, feel free to use the single-track punch-in procedure for recording a master take, but make a point of layering at least one of the parts, if only for the purpose of gaining more experience of what this technique can offer.
- Use the ideas in Section 3.4 where possible to help the musician achieve the best part and performance, but also follow the musician's lead if they suggest alternative approaches, and don't be tempted to discount an idea before you've both heard what it sounds like.
- As in Chapter 2, make notes after the session about anything you may want to refer back to in future, as well as contemplating any errors or misjudgments you made. After a day or two, listen back to all your overdubs from the session, and experiment with different balances of them against the backing parts to re-evaluate the quality of the sonics, performances, and arrangement decisions.
- Before progressing to Part 2, try to carry out another couple of sessions like this, preferably with different instruments and performers.

WEB RESOURCES

On this book's companion website you'll find a selection of resources to support this chapter, including:

- some concrete examples of workable communications setups in specific software and hardware studio environments, as well as links to handy communications gadgets;
- audio examples demonstrating different DI'ing approaches, the tonality of different string-instrument fingering positions, and the sound of guitar fret buzz;
- further reading on audio editing techniques, slave reels, low-latency monitoring, and the preparation/maintenance of the main instruments mentioned in this chapter;
- links to recommended resources about performance-direction and arrangement techniques.

http://www.cambridge-mt.com/rs-ch3.htm

PART 2
One Source, One Mic

By working your way through Part 1, you should have acquired a bedrock of essential technical skills:

- Connecting line-level sources and instrument pickups to your recording system's inputs.
- Line-checking and troubleshooting your studio rig for recording purposes.
- Managing gain controls appropriately throughout the recording and monitoring chain, and scotching all the most common signal-degradation problems.
- Setting up suitable monitoring and communications equipment, and getting the best out of a DAW's software monitoring.
- Using punch-ins, comping, and layering to prepare for mixdown.
- Documenting all aspects of the session.

Furthermore, you'll have been developing those intangible characteristics of the recording engineer that are just as important to the success of a session: being in the right place at the right time, and prepared for anything; getting the best out of the various personalities you meet in studios; supporting progress invisibly when the going's good, but tactfully lending a hand where creative obstacles threaten the music.

So from now on I'm going to assume you've got that all under your belt, because it's on that foundation that I want to begin building an understanding of microphone technique.

CHAPTER 4

Basic Vocal Recording

Recording vocals is one of the commonest requirements in any studio, but that's not the only reason to start our discussion of microphone technique here, because vocal production is where the quality gap between amateur and big-budget projects tends to be starkest. Professionals know that the singing usually drives record sales more powerfully than all the other parts combined, so it's commonplace for a large chunk of production time (or even the majority) to be spent entirely on vocals. The more experience you gain recording singers, the classier your productions will sound. So that's where we're going to start.

Tutorials with titles like *Studio Vocals On A Budget!* are ten-a-penny these days, most of them based on a very similar formula—what I like to call the "vanilla" vocal-recording setup. There are excellent reasons why this setup is so often considered a safe bet, and indeed why the majority of commercial vocal tracks are recorded this way, but the unfortunate fact is that a lot of people really don't seem to get very good results with it at all! So I'd like to dedicate this chapter to explaining how the vanilla setup works, why it's so popular, and how to get the best out of it.

4.1 THE MICROPHONE

4.1.1 Condenser Design

There are many different ways of building a microphone, but the "condenser" or "capacitor" design is by far the most common choice for vocals, partly because it adds only a comparatively low level of noise to the output signal. The human voice has a greater "dynamic range" (the level difference between its loudest and softest sounds) than almost any other source you're likely to meet in the studio, so a low-noise mic is vital if you want to capture a singer's quietest utterances cleanly.

The other reason a condenser is usually preferred is because of the nature of its "diaphragm," the bit in the mic's business end that vibrates in sympathy

FIGURE 4.1
In this picture of AKG's C414B XLS large-diaphragm condenser mic, you can actually see the circular diaphragm through the mic's protective metal grille.

with the air to transduce an electrical signal. The comparative thinness of this diaphragm means it can move fast enough to provide decent high-frequency response in general, and most large condenser-mic diaphragms also exhibit an inherent 8–12 kHz resonance, which contributes further upper-spectrum emphasis—the combined result being a glossier and more upfront vocal sound at mixdown.

The size of the diaphragm is a critical design parameter in any mic, and amongst condenser designs a relatively large diaphragm (around an inch across) is the overwhelming favorite for vocals. Not only do larger diaphragms typically provide lower noise, but they also have greater inertia, so that they don't respond quite as fast to transients or high-frequency extremes, lending the recording a slightly softer, smoother sound character. In addition, large-diaphragm condensers are almost always designed such that "on-axis" sound sources (i.e., those directly in the mic's firing line) are given a hefty high-frequency boost, a built-in brightness enhancement that's usually welcomed as another means of getting the vocal to cut through the mix.

4.1.2 Cardioid Polar Pattern

An important concern when recording vocals in the studio is the amount of room reverberation picked up by the microphone. Most engineers do their best to minimize it, because:

- reverb tends to pull a vocal backwards into the mix, when the goal is often to have the singer out front;
- a recognizable "medium cupboard" reverb character will make the production sound cheap, as well as contradicting the spatial illusion from any artificially added delay/reverb effects;
- otherwise dynamics processing applied to the vocal will sound less natural because it'll effectively vary the apparent level of the reverb from moment to moment;
- pitch-correction software is more likely to misinterpret a recording with reverb.

If your singer is on-axis to the microphone, then you can reduce the level of the primarily off-axis room reflections by using a "directional" microphone whose sensitivity tails away as you move off-axis, the most common being the "cardioid" type which reaches a "null" of maximum rejection at the mic's rear. Figure 4.2 plots this directional sensitivity on a circular graph called a "polar diagram," and you can clearly see the characteristic inverted heart shape that gives the cardioid pattern its name.

4.1.3 Valves and Transformers

I've researched the mic preferences of hundreds of top-name engineers, and my conclusion is that most of them favor vocal mics which feature both a

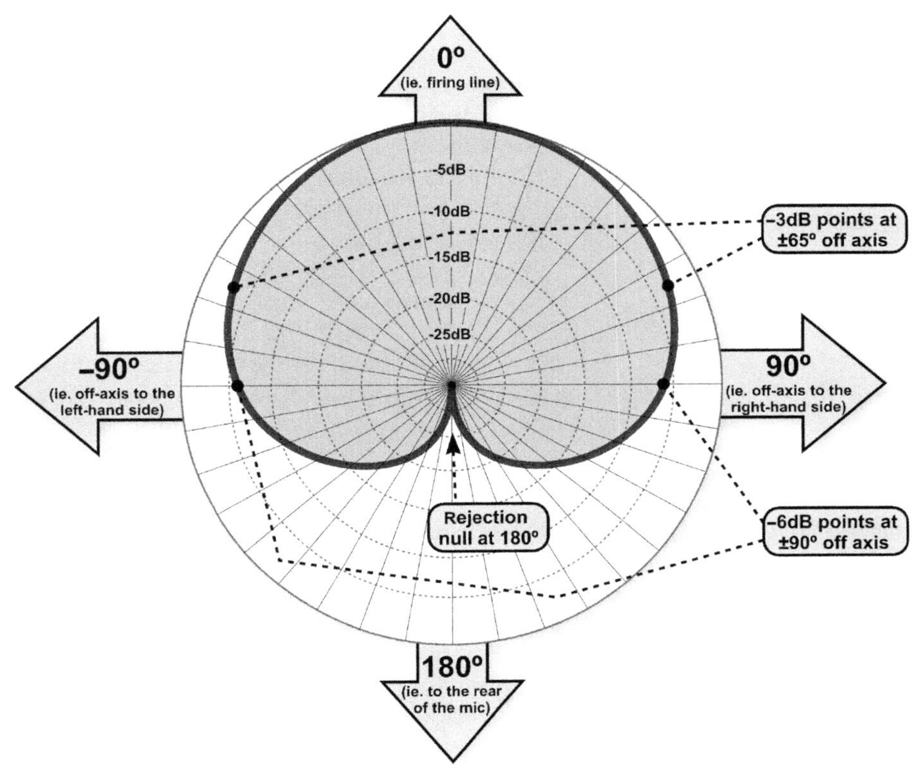

FIGURE 4.2
This polar diagram shows the directional pickup characteristics of an idealized cardioid microphone.

valve (or "tube") stage and an output transformer—both of which are present in classic vocal mics such as AKG's C12, Neumann's M49, U47, and U67, Telefunken's ELA M251, and Sony's C800G. The subtle tonal colorations and distortions incurred by such circuit components often thicken vocal recordings in a flattering way, which is why the appeal of valves and transformers has endured despite technological advances that might arguably have rendered them obsolete. Where you've got several large-diaphragm condensers to choose from, this may help you narrow things down further.

4.1.4 Price and Reputation

In recent years, outsourced manufacturing has flooded the microphone market with ridiculously inexpensive mics, and there's no denying that many great recordings have been made with mass-produced products like these. However, I regularly do shootouts between mics ranging in price from under $80 (£50) to well over $3000 (£2000), and I still maintain that you're much more likely to get a great recording with a more expensive mic. Costlier mics will typically

"The pot-luck flattery of a bargain-basement mic can occasionally outshine the more honest presentation of a pricier model for a certain individual's voice, but that shouldn't mislead you into thinking that more expensive mics are a waste of cash."

stay truer to the source, so rarely give a bad result, whereas I find that a cheap mic's tonal colorations usually make it something of a one-trick pony for vocals: Once in a while the mic will beautifully complement a specific singer, but in the majority of cases it won't, leaving you with a recording that's somewhere between mediocre and appalling. It's hardly surprising, therefore, that the pot-luck flattery of a bargain-basement mic can occasionally outshine the more honest presentation of a pricier model for a certain individual's voice, and that shouldn't mislead you into thinking that more expensive mics are a waste of cash. Once you've done dozens of shootouts, a more revealing trend surfaces: The expensive mics always tend to make the shortlist, whereas the cheaper mics frequently crash and burn.

But this isn't the only reason why I'd instinctively always reach for more expensive vocal mics. It's also because cheaper mics tend to succumb to technical failures more quickly, in my experience—and usually right when you're in the middle of a take! Whether this is because of cost-cutting on components, manufacturing tolerances, and/or quality control I couldn't say, but I've had enough budget mics die on me mid-session that I've learnt to be a bit circumspect about using them for important vocal recordings.

And there's one other thing that you should weigh up in your mind when it comes to mic selection. There are some microphone companies who have decades of experience in their R&D departments, a catalog full of well-respected high-end studio products, and a hard-earned reputation to defend. And there are some companies who don't. Given that both these types of company are now offering ranges of budget-friendly mics aimed at the project-studio market, whose do you think I'd recommend trying first? Hmm…

4.1.5 "Best Guess" Versus "Ideal"

Now don't misunderstand my recommendations in this section. I'm definitely *not* suggesting that anyone can pick the ideal mic for a given singer without using their ears. That'd be idiocy, because mic choices for vocals are so heavily dependent on the personality of the voice you're recording. However, on real vocal sessions, it's imperative that you're prepared to catch lightning in a bottle the moment the artist steps up to the mic, which means taking an educated guess about which mic to put up first. "The very first time they sing might be the best take," warns Stephen Lipson.[1] "That's why having a go-to vocal chain is absolutely essential." Clearly, if you have more than one microphone available, and a cooperative performer, the best way to narrow down the contenders is to compare them—something we'll return to properly in Section 4.6.1.

CONNECTING, POWERING, AND CARING FOR CONDENSER MICS

It's a doddle to connect up most condenser microphones: Just cable between their male XLR output connectors and the female XLR connectors of your recording system's microphone inputs. All condenser mics are active devices, so they'll normally need phantom power from the mic input to work—if phantom power is switchable, it's best to switch it off before plugging or unplugging the cables, although hot-plugging XLRs shouldn't usually cause any damage in practice. Some microphones are able to operate on battery power, but I don't recommend using those most of the time, because their headroom and noise performance are often compromised as a result.

At the lower end of the market a recent trend is microphones that connect directly to a computer via USB, drawing their power from there. As much as it's the job of any engineer to get the best results with whatever equipment's available for a session, I wouldn't seek out such mics personally. Some of my reservations about them are practical: It's not as easy to put them 30 m away from the recording system; they force you to use software monitoring; and you can't use them with non-computer systems. However, they're also not a very future-proof investment. The moment you decide to record anything more than just the USB mic, you lay yourself open to software driver conflicts with other audio-interfacing hardware. In addition, you may be unable to use the mic with successors to the USB standard once it's inevitably superseded.

Valve mics almost always require their own separate power supply, because phantom power isn't capable of providing high enough voltages. A special multi-pin cable will usually connect the microphone to its power supply, and then you cable an XLR output on the power supply to a microphone input on your recording system. For best results, the valve should be properly warmed up before use—at least twenty minutes is ideal here, but you can usually get away with recording after five minutes at a pinch. You won't hurt a valve mic by hitting Record before it's warmed up, but the tone just won't have settled down properly, so you may subsequently find it tricky to comp between early takes and later ones.

Finally, because a condenser mic's diaphragm is so delicate, it can easily be damaged if the mic is dropped or hit. The diaphragm's range of movement is utterly minuscule too, so you don't want to let dust collect on it and degrade its performance: Keep the mic in a sealed protective case when not in use, and put something like a plastic freezer bag over it if you have to leave it set up overnight. Any moisture on the diaphragm will cause it to malfunction electrically (although usually not permanently), so be careful with vocalists who spit a lot while enunciating, and avoid condensation problems by allowing any newly arrived mics time to acclimatize to your studio temperature before getting them out of their cases.

4.2 MOUNTING AND CABLING

Although singers often use handheld mics on stage, in the studio it's usually better to mount the mic on a stand so that you have more control over the positioning and can minimize handling noises. If you have a choice of similar-sized mic stands available (or you're buying your own) then the most reliable one will usually be the one that's heaviest and which uses the least plastic in its construction. In addition, check over the main joints (particularly between the main shaft and the boom arm, and between the upper and lower segments of a telescopic shaft) to confirm that they're in good nick. Now, I'm sure you're not one of those people who abuse stands by repositioning them without

FIGURE 4.3
Out yourself as a noob in one easy step...

loosening the clamps (heh-hem!), but a lot of people do, which means every collection of small-studio mic stands always seems to have one or two which droop involuntarily no matter how hard you tighten them. Sometimes you can revive an ailing stand by cannibalizing bits of its fallen comrades, but cheaper hardware is much less likely to be field-serviceable in this way. This is a good reason to spend a little extra money on the pricier modular stands if you're investing in hardware yourself, especially if the manufacturer has a full catalog of spare parts with which you can resurrect casualties, rather than having to buy whole new stands.

4.2.1 Stability

The quickest way to spot a recording newbie is by how they handle mic stands. The most common mistake is to compromise the stand's stability by allowing the main stand shaft to protrude below the base, so some of the feet don't rest properly on the floor; by placing the stand's base too far from the source, so you overextend the boom arm; or by failing to place one of the stand's feet directly under the boom where possible. Save yourself the bills for mic repair and dental work—don't let the stand topple onto the singer (or anywhere else for that matter)! If your mic is particularly heavy, then find something to counterweight the other end of the boom arm (those Velcro-strapped wrist/ankle weights used in fitness training, for example) or load down the stand's base with some extra mass (five-liter plastic petrol cans filled with sand are a cheap option there).

4.2.2 Shockmounts and Mic Clips

One problem with directional condenser mics is that they don't just respond to vibrations in the air; they're also liable to pick up any physical vibration that reaches their casing, whether that's caused by something coming into direct contact, or it's transmitted via the microphone stand or cable. This is why most large-diaphragm microphones are supplied with a special shockmount which suspends them flexibly, helping to isolate them from vibrations coming through the stand—for example traffic rumble. If your mic doesn't have its own suspension shockmount, then find out whether the manufacturer sells one separately or check out general-purpose replacements such as Rycote's InVision series. To be honest, though, with vocal recordings it's rarely a disaster if the most suitable microphone has no shockmount, because generic low-frequency rumble can easily be removed with high-pass filtering at mixdown, and most other problems can be avoided by stopping the singer touching the mic stand or tapping their foot too vigorously.

3/8-INCH VERSUS 5/8-INCH THREADS

The threaded end of a microphone stand's boom arm, which screws into the mic-mounting clip, will normally have either a 3/8-inch (9.5 mm) or a 5/8-inch (15 mm) gauge, the former being more common in Europe and the latter more common in the U.S.A. If your mic clip and boom-arm thread turn out to be incompatible, you can get cheap little adaptor widgets such as those shown in Figure 4.4 to convert between thread gauges. Where fitting such an adaptor into a 5/8-inch mic clip, make sure the slotted edge is facing outwards so that you can easily remove it again, if necessary, with a coin, a broad-bladed screwdriver, or (my personal favorite) a blunt old cutlery knife.

FIGURE 4.4
Adaptors for converting between 3/8-inch and 5/8-inch mic-stand screw threads.

Whether you have a proper suspension shockmount or not, another classic beginner's mistake is to spin the mic clip/shockmount to screw it onto the stand. Not only does this take ages, but if the mic is already in the clip, then you're taking a big risk of dropping the mic while you're twirling it around. A much better method is to hold the clip/shockmount (already containing the mic, if you like) in one hand, and then, with the other hand, loosen the boom-arm clamp, twiddle the boom arm between your fingers to screw it into place, and retighten the boom-arm clamp.

One thing that many home-studio folk don't seem to understand is the purpose of the little flat disc that screws onto the end of the microphone boom arm right underneath your microphone clip. This is called a "locking ring" or "jam nut," and you can screw it upwards underneath the mic clip to hold it in place more securely. There's often another one on the stand's main telescopic shaft for locking the boom arm in place. These serve a very useful purpose in the cut and thrust of onstage miking, but most of the time I think they're overkill in small-studio situations unless you're particularly concerned about a mic clip or boom arm twisting under the weight of a heavy microphone. Don't remove the locking rings, though, because you never know when they'll eventually come in handy!

4.2.3 Cable Dressing

We've already talked about laying cables sensibly in the studio, but with microphones you also need to consider how you "dress" the cable round the stand. If a cable hangs out of the bottom of a mic unsupported, or spans between

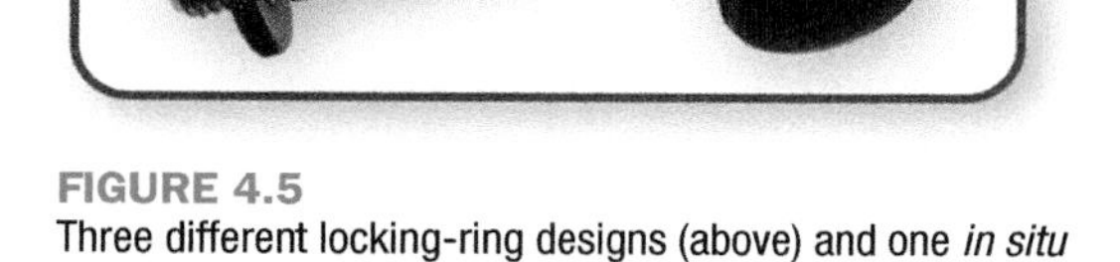

FIGURE 4.5
Three different locking-ring designs (above) and one *in situ* on the end of a mic stand's boom arm (below).

the mic and the stand with any tension, physical vibrations from the floor or stand can defeat your suspension shockmount by traveling up the mic cable instead. To avoid this, I suggest looping the cable around the stand's boom arm so that there's about 12 inches of slack hanging beneath the mic. Alternatively, a plastic cable clip (provided with some mic stands) or some gaffer tape can achieve a similar end. In the latter case, just make sure to leave enough cable slack to allow repositioning of the mic, and it makes sense to fold the end of each gaffer strip back on itself about a half an inch so that it's easier to remove at the end of the session.

Some people also employ plastic clips or tape to train the cable around the rest of the stand, so that if someone trips over the cable it pulls primarily on the lower section of the stand, not the boom arm, reducing the likelihood of the whole shebang keeling over. However, I think life's too short for that myself, so I usually just loop the cable a couple of times around the stand on its way to the floor, which fulfills roughly the same function. (Obviously you don't want to go mad with that either, otherwise disentangling the stand at home time becomes purgatory.) If you're left with any spare cable between the mic stand and your recording system, then it's good practice to tidy that together somewhere close to the mic stand, since the stand's the more likely to need moving during the session.

4.3 MIC POSITIONING

Almost all large-diaphragm mics are "side-fire" or "side-address" (as opposed to "end-fire/address") designs, which means you're meant to sing into them from the side. In the case of a cardioid mic, it's most sensitive side (the "live side") will usually be indicated on the casing by the manufacturer's logo and/or the mic's model name. If it's not clear, though, do take a couple of moments to plug the mic in and check it for yourself—singing into the wrong side of a cardioid mic sounds pretty horrid, but there always seems to be one student in every recording class who does it!

In order to place a vocal mic effectively, there are lots of things you need to know about how a singer projects sounds, and about how the microphone picks them up.

4.3.1 Miking Angle

No sound source ever radiates its spectrum evenly in all directions, and this is why the angle from which you mike a singer can make a big difference to

FIGURE 4.6
The microphone stand on the left displays many of the common mistakes made by novice engineers: The boom arm is overextended, and has been set up horizontally where it's most prone to drooping; none of the stand's feet is directly beneath the boom arm, which destabilizes the whole stand; the cable is pulled tight at the mic end, rendering the suspension shockmount less effective; and there's no spare cable on the floor by the stand in case it needs to be repositioned. The setup shown in the right-hand picture is more advisable, as it avoids these problems.

the recorded tone. The first thing to understand is that high frequencies by their very nature are more directional and easily obstructed/absorbed, whereas low frequencies are better at passing through or around physical objects. The human voice's high frequencies project in a broad downwards "beam," roughly along the line of the nose, which means that you get a brighter tone below nose level than you do above. Above nose level the high end will be more muted, but there'll also be less sonic contribution from the throat and chest, so the midrange tone usually remains a little more consistent at different distances from the head when miking slightly from above.

"The human voice's high frequencies project in a broad downwards "beam," roughly along the line of the nose, which means that you get a brighter tone below nose level than you do above. Certain consonants are also directional."

It's worth mentioning that some performers contort their faces asymmetrically while singing, in which case miking from one side will have a significant bearing on the high-frequency balance. Mike Stavrou[2] makes a point of listening for this while standing in front of the singer: "As they sing a long 'ah,' I sway from side to side listening to the tone of their voice. Lowering your head helps, so you can hear the treble bounce off their palate. Each side is always different… You can't equalize the left side to sound like the right, because the blend of harmonics leaving their mouth is actually different."

Certain consonants are also directional. For example, "s" sounds (often called "sibilants") typically project horizontally from the mouth, so any microphone position on that plane will normally overemphasize them. Plosive sounds (things like "p" and "b") are created by releasing a sudden puff of air from behind the lips, which normally shoots directly forwards, therefore a mic directly on-axis to the mouth will be hit by this, producing a characteristic blast of low frequencies called "popping." Mark Neill[3] and Elliot Scheiner[4] both mention miking vocalists from above lip level for this reason: "[It's] a great way to get the sibilance and 'p' pops under control," remarks Neill.

REDUCING PLOSIVES AND SIBILANCE WITHOUT MOVING THE MIC

Occasionally you may find yourself in a recording situation where it's tricky to move the vocal microphone out of the line of fire of plosives (or indeed spit), in which case a "pop shield/filter/screen" can bail you out. This normally takes the form of two layers of fine nylon mesh supported by a lightweight hoop, and is held in front of the microphone via a flexible gooseneck clamped to the mic stand. You can get purpose-built pop shields for under $30 (£20), but for an even cheaper alternative try one of those circular wire-mesh splatter guards for frying pans, or bend a wire coat hanger into a frame and cover it with a thin nylon stocking. Whatever design you go for, though, do leave at least four inches' distance between the pop shield and the mic, otherwise it won't

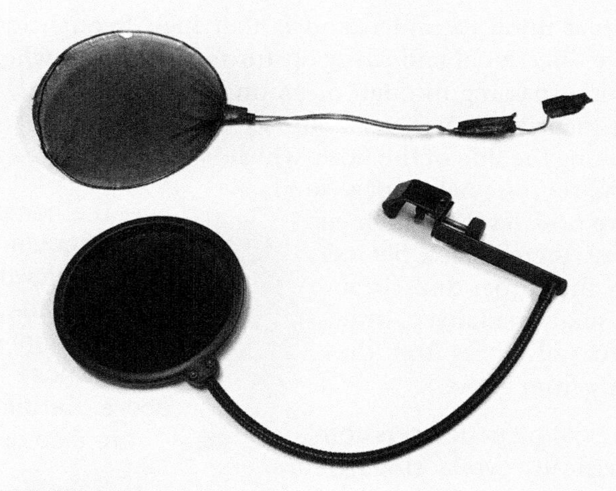

FIGURE 4.7
Although professional-looking commercial pop shields (lower picture) are reasonably affordable, an even cheaper (but still effective) option is to bend an old wire coathanger into a hoop and then cover it with thin nylon stocking material (upper picture).

work properly. Bear in mind, though, that a number of high-profile engineers, amongst them Bruce Botnick, question the acoustic transparency of pop shields, so you shouldn't just set one up on reflex unless it's really necessary. "I still don't like using them," comments Botnick.[5] "I can hear them."

Some people suggest shielding against plosives and/or sibilance by taping a pencil to the body of the mic so that it covers the center of the diaphragm.The effectiveness of the "pencil trick" is a moot point amongst pundits, but whether or not it works for you, it will also affect the entire vocal sound to some extent, and its efficacy will vary if the singer moves around, so I don't favor it myself.

A final tip for reducing the severity of plosives comes from Geoff Emerick:[6] Get the singer to replace every "p" with a "b." You'd be forgiven for thinking that this would make a nonsense of the lyrics, but in practice the semantic context almost always insures that listeners still perceive the correct consonant, and the wind blast on the microphone diaphragm is usefully reduced. (Of course, argument still rages over whether Pink Floyd used this technique for "Another Brick In The Wall"…)

FIGURE 4.8
The pencil trick.

4.3.2 Miking Distance

How far away from the singer's mouth should the microphone be? While there's a broad consensus amongst professionals that it should probably be closer than 18 inches, there's clearly no "industry standard": for instance, Al Schmitt[7] has mentioned 9–12 inches, Steve Churchyard[8] and Elliott Scheiner[9] 6–8 inches, and Eddie Kramer[10] 4 inches. The reason for this, in my view, is that changing the miking distance has a powerful influence on the recorded tone.

When you get within about 3 feet of any directional mic, it starts noticeably boosting the output signal's low frequencies—a phenomenon called the "proximity effect." With a cardioid mic, the difference between 12-inch and 4-inch positions may be as much as 12 dB at 50 Hz, but even up around 300 Hz (well into the vocal range) you'll get a 4–5 dB rise. Think about that for a moment: That's a huge tonal change for only moving the mic 8 inches. This aspect of mic positioning is therefore an important tonal control. "If [the singer thinks] it sounds too boomy, then I'll have them move back," explains Steve Albini,[11] "and if it sounds too thin then I'll have them move forward."

This isn't the only tonal effect, though, because there's also a fair bit of mid-range variation with distance because of the way different bits of the singer's anatomy interact, especially (as mentioned in Section 4.3.1) when you're miking from below the nose line. Another issue is that a vocal's non-resonant attributes (i.e., consonants and other lip noises) are produced by comparatively small bits of physiological apparatus, whereas the more resonant sounds are created by much larger areas of the body. What this means in practice is that the former reduce in level more quickly than the latter as you increase the miking distance. However, the scale of both these changes is dwarfed by that of the proximity effect, which is another argument for having a variable high-pass filter in your recording chain, because raising the cutoff frequency can rein in any excess bass boost for whichever miking distance best suits the rest of the vocal spectrum. (In the absence of a variable filter, a fixed-frequency filter on the microphone or preamp can serve similar ends, but it's a relatively blunt tool by comparison.)

4.3.3 Mic Orientation

Although many small-studio engineers assume that a microphone should always be used on-axis, alternative orientations of the mic can provide further timbral control. For any real-world directional microphone, moving the sound source off-axis doesn't just change the mic's output level, it also affects the recorded frequency response. Large-diaphragm cardioid designs are often the worst offenders in this respect, especially budget models. Broadly speaking, the general trends will be a reduction in high-frequency energy and a weakening of the proximity effect's bass boost as you move off axis (whether to the side of the mic or over the top of it), but every different mic will also contribute other more complex tonal fluctuations as well. For cheaper mics, these spectral foibles mean it's rarely wise to deviate beyond about 30° off-axis, but classier mics may still deliver useful vocal sounds at twice that angle.

You may have seen studio pictures of vocal mics hanging upside down from their suspension shockmounts, and this setup has practical advantages if your ceiling and mic-stand height allows it. The main advantage is that by moving the mic's body, shockmount, and cable above its diaphragm, you get more space below the mic for things like music stands. In addition, there are sonic reasons for mounting valve microphones in this way, because the warm air rising from the valve in the mic's casing can't then generate turbulence noise by wafting past the diaphragm.

4.3.4 Positioning Consistency

Because small changes in microphone positioning can profoundly alter the recorded vocal sound, a singer's natural performance movements can cause serious sonic difficulties. Indeed, one of the toughest challenges any mix engineer can be faced with is when the vocal tone becomes a moving target because of this. One way of mitigating such effects is to use more distant mic positions

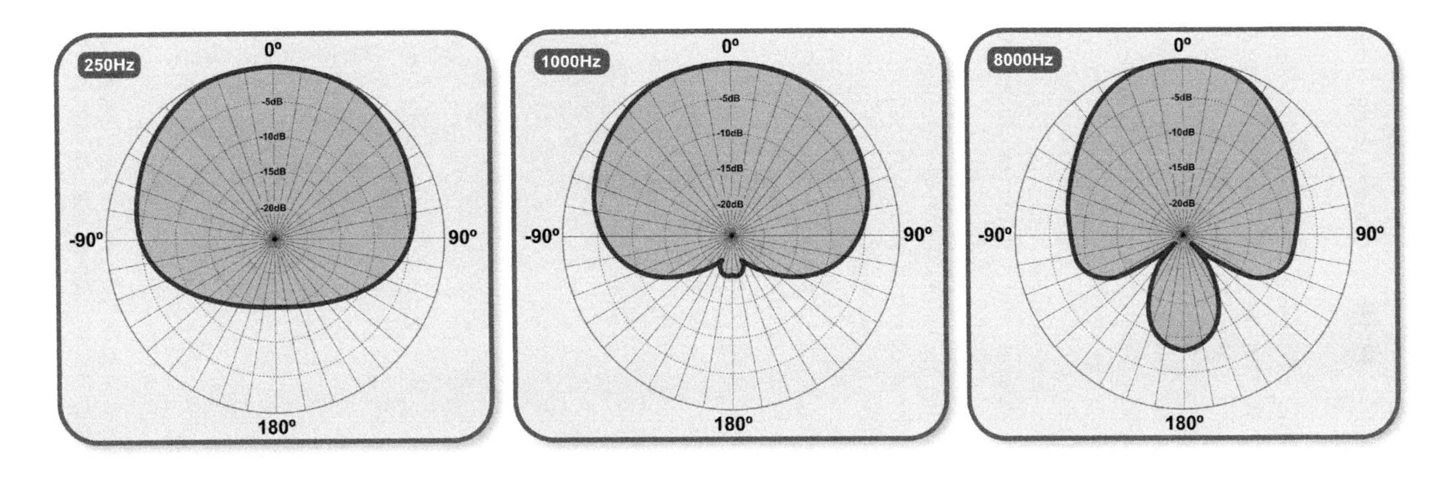

FIGURE 4.9
These polar diagrams show the directional response of a real-world large-diaphragm cardioid condenser mic (in this case the Rode NT2) at three different frequencies. The increasing high-frequency directionality and decreasing low-frequency directionality are typical of large-diaphragm condenser mics.

(around the 12–18 inches mark), because you get less change in proximity-effect bass boost for a given front-back movement when you're further away, and the singer also moves less off-axis for a given range of side-to-side motion, as shown in Figure 4.10.

However, where a closer mic distance is critical to the recorded tone, an alternative approach is to try to stop the performer moving as much. For stage vocalists, a much-prized skill is the ability to "work" the vocal mic while singing, modulating their level and tone by adjusting their position relative to the mic. In a recording situation, though, this usually causes more problems than it solves with the vanilla setup, especially with budget mics. John Hudson:[12] "The way that a cabaret singer moves the mic... whenever he sings a loud note is the worst thing you can possibly have in a studio environment, because the whole sound changes." Fortunately, many musicians can scale back such movements if you ask them to, without any negative ramifications for the quality of the performance.

More instinctive movements are a trickier issue, though, because those are often an intrinsic part of the musician's natural response to the music—try to stop them and you may end up killing the vibe. The best tactic for squaring that circle is to provide the singer with some physical object that they can use to anchor their mouth roughly into position, but which still allows them to move the rest of their body to the music. This is where a pop shield can be

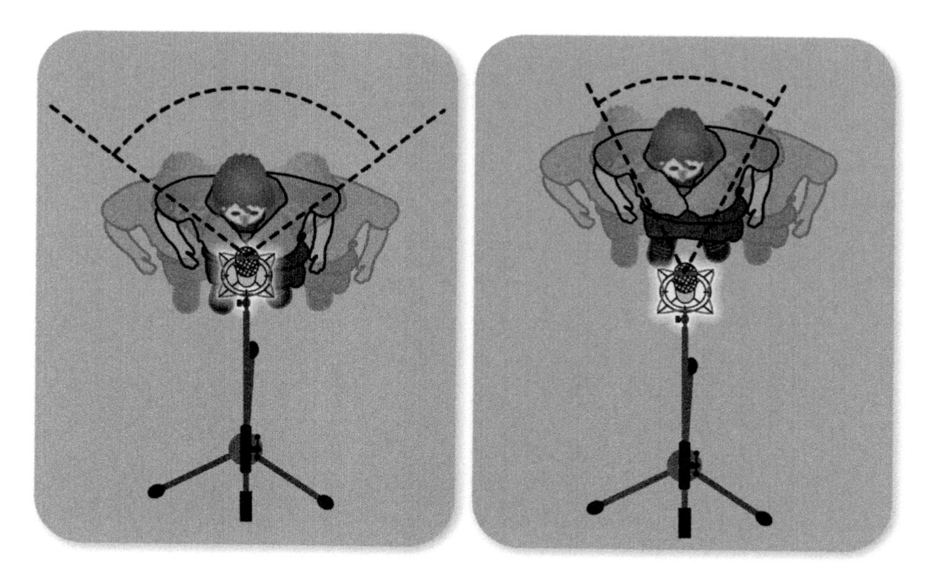

FIGURE 4.10
When the vocal microphone is close to the singer (left-hand diagram), movements during performance can put the sound source well off-axis to the microphone, leading to considerable inconsistencies in both level and tone. Miking from further away (right-hand diagram) reduces this problem, but at the risk of picking up too much reflected sound from the room.

great, because it provides a clear visual distance indicator and will also work with singers who close their eyes while performing—they just have to keep their noses in contact with the mesh. Another alternative is to put up a second mic stand with just a mic clip on it, and get the singer to hold that clip in their hand, singing slightly over the top of it as if into an imaginary mic, so that the high frequencies aren't blocked.

Maintaining a consistent position in relation to the mic can still be a problem even with the least animated of performers, because most vocalists will tend to sing in the direction they're looking. For example, if you've set up your mic with the singer looking straight ahead, but they then perform lyrics from a music stand that's below eye level, they'll tilt their head downwards, upsetting your carefully adjusted sound. Personally I prefer to gaffer-tape the lyric sheet to the end of a mic stand at eye level, not only because this avoids the miking problem, but also because it gives me more flexibility when positioning the lyrics than when using a traditional music stand. If you position your microphone off-axis to the singer's mouth (which is normally sensible), most singers will subconsciously counteract that by trying to sing directly at the microphone. Because of this, I like to explain the situation to the vocalist and find something else in the recording room that they can look at and "sing to" while overdubbing: a mark on the wall, the end of another mic stand, a life-size cutout of Marilyn Monroe…

Finally, bear in mind that all vocalists have some leeway in terms of how they position their head while singing, and you can use this to your advantage. Singing slightly downwards tends to give a less strained and more resonant sound, for instance, while singing upwards makes the sound more raw and cutting—as Lemmy is undoubtedly aware!

4.4 ACOUSTICS CONSIDERATIONS

Recording vocals isn't just a question of positioning the singer and the mic, though, because the studio environment's acoustics are also a major concern.

4.4.1 Room Reflections

As discussed in Section 4.1.3, it's sensible to minimize the amount of general room reverb you record along with your vocal signal. Although the directionality of the microphone helps here, it'll rarely be enough in any normal room, so putting some sort of acoustic padding around the singer is well-nigh essential. If you already have some acoustic treatment in the recording room, then you may be able to get away with just locating the singer with their back to it, but in most cases it's usually better to sling something behind them specially—it's rear-wall reflections that are most problematic, because they bounce sound into the sensitive side of the mic. Blankets/curtains or winter quilts (the thicker and heavier the better) are project-studio staples in this role, and with good reason given that they can be easily mounted on spare mic stands, lightweight clothes rails, or wall/ceiling hooks. Try to bring them as close to the singer as

FIGURE 4.11
When you want to capture a dry vocal sound with a cardioid mic, the most effective place to put absorptive materials is on the sensitive side of microphone—in other words behind and above the singer. Here you can see a typical small-studio vocal-recording setup where bedding quilts have been arranged over a cheap lighting stand to make effective use of this concept.

you can without cramping their style, and preferably treat the whole live side of the mic, all the way round to ±90° off-axis—a cardioid polar pattern only takes about 6 dB off the level of sounds arriving perpendicular to its firing line, remember. Don't forget that the ceiling is often the reflective surface closest to the microphone, so it's very important to treat that too.

Moving closer to the mic will increase the dry level of the vocal, but I'd prioritize tonal considerations (as discussed in Section 4.3.2) over reverb-reduction there. In other words, choose the miking distance for the tone, and then add more padding around the singer if you're getting too much room sound. I often see people setting up a curved acoustic baffle (such as SE Electronics' Reflexion Filter) behind the microphone when recording vocals. While such devices can be extremely useful for reducing spill in ensemble-recording situations, when overdubbing with the vanilla setup the improvement will be negligible unless you're using a (usually dirt-cheap) microphone that sounds horrible off-axis and the room you're working in is less than about eight feet across. Most of the time the cardioid's rear rejection will eliminate the room sound from that direction fine on its own. However, if you do decide to use a baffle, I'd recommend putting it on a different stand to the mic, rather than mounting the mic on a bracket within the screen itself. This is partly because it gives you much more positioning flexibility, and partly because some of those screens are so heavy that it's tricky to stop the stand collapsing or overbalancing under the weight.

General room reverb isn't the only concern, though. Reflected sounds will always arrive at the microphone after the singer's direct sound, because they have to travel further through the air, and (as we mentioned in Section 1.4.4) combining any sound with a delayed version of itself can cause an undesirable comb-filtering effect. If you've already taken precautions to reduce room reverb, then you'll likely have scotched any strong single reflections already, but it's possible to be caught out none the less if the singer has a solid music stand within their padded cocoon. To avoid this, imagine that the music stand's surface is a mirror, and then make sure that the singer can't see the mic in the mirror—this often results in a flatter-than-usual stand angle, but you may also be able to tilt the stand more steeply if you peg/tape the sheets in place to stop them falling over. (Don't forget the line-of-sight issues I raised in Section 4.3.4, though.)

COMB-FILTERING: THE NITTY GRITTY

Now that we're digging into mic technique in earnest, it's helpful to understand a bit more about how mixing any signal with a delayed duplicate of itself results in the frequency-response undulations of comb-filtering. To start with, let's consider a sine-wave signal, the simple audio waveform from which all complex musical sounds can theoretically be built. If I feed the same sine wave to two mixer channels, the peaks and troughs of the waveforms in both channels will be exactly aligned in time ("in phase"), which means that mixing them together will simply produce the same sine wave, only louder. If you gradually delay the audio going through the second channel, however, the peaks and troughs of the two sine waves shift out of time-alignment ("out of phase"). Because of the unique properties of sine waves, the combination of the two channels will now still produce a sine wave of the same frequency, but its level will be lower than if the two channels were in phase, and we say that "partial phase cancellation" has occurred. When the second channel is delayed such that its peaks coincide exactly with the first channel's troughs (and vice versa), the two waveforms will combine to produce silence ("total phase cancellation").

Now let's scale things back up to deal with real-world sounds, made up as they are of heaps of different sine waves at different frequencies, each one fading in and out as pitches and timbres change. If we feed, say, a vocal mic's output signal to our two mixer channels, instead of a single sine wave, any delay in the second channel will have a dramatic effect on the tonality of the combined signal, rather than just altering its level. This is because, for a given length of delay, the phase relationships between sine waves on the first channel and those on the second channel depend on the frequency of each individual sine wave.

So, for example, a 0.5 ms delay in the second channel will put any 1 kHz sine-wave components (the waveforms of which repeat every 1 ms) completely out of phase with those on the first channel, resulting in total phase cancellation. On the other hand, any 2 kHz sine-wave components (the waveforms of which repeat every 0.5 ms) will remain perfectly in phase. As the frequency of the sine-wave components increases from 1 kHz to 2 kHz, the total phase cancellation becomes only partial, and the level increases toward the perfect phase alignment at 2 kHz. Of course, above 2 kHz the sine-wave components begin partially phase canceling again, and if you're quick with your mental arithmetic you'll have spotted that total phase cancellation will also occur at 3 kHz, 5 kHz, 7 kHz, and so on up the frequency spectrum, whereas at 4 kHz, 6 kHz, 8 Hz, and so on the sine-wave components will be exactly in phase. It's this characteristic series of regularly spaced peaks and troughs in the frequency response that we call comb-filtering.

It takes only a minuscule delay to cause audible comb-filtering: Just 0.000025 s (a 40th of a millisecond) between the two channels will cause total phase cancellation at 20 kHz, but you'll also hear partial phase cancellation at frequencies below this. As the delay increases, the comb-filter response marches further down the frequency spectrum, trailing its pattern of peaks and troughs behind it, which themselves get closer and closer together. However, when the delay times reach beyond about 25 ms or so, the comb-filtering weakens as our ears begin to detect the delayed signal as a distinct echo instead.

What connects all this theory back to the subject of vocal mic technique is the fact that sound propagates through the air at about one foot per millisecond. So if your singer addresses a microphone one foot away, but a reflection of their sound also travels a longer four-foot route by reflecting off the music stand, you'll get a delay between the direct and reflected sounds of around 3 ms—generating a comb-filtering response with its lowest trough somewhere around the 350 Hz region. The severity of the comb-filtering will be weakened somewhat by virtue of the reflected sound being lower in level than the direct sound, but you'll still get sufficient frequency-response undulations to produce a noticeably unnatural tone.

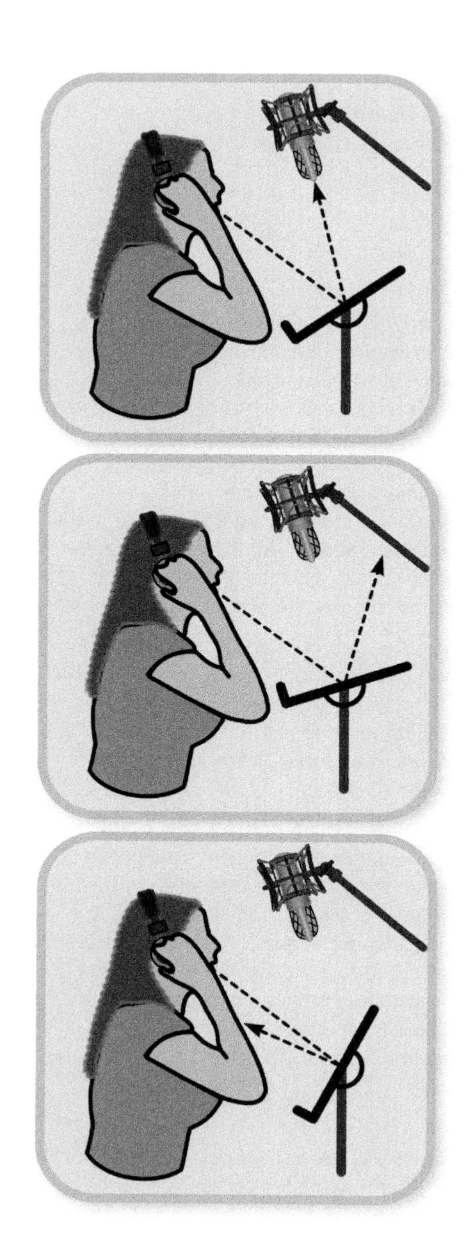

FIGURE 4.12
If a music stand reflects sound from the singer directly back into the mic (top picture), the resultant comb-filtering artifacts can damage the vocal tone. The problem can easily be avoided, however, simply by reangling the music stand (middle and bottom pictures).

If you're using a significantly off-axis microphone orientation on tonal grounds, you should pay particular attention to any discrete reflections that might arrive on-axis to the mic, because not only will they be captured at a higher relative level, but they'll likely be captured more brightly (as mentioned in Section 4.1.2). Extra acoustic absorption may be sensible in this instance.

4.4.2 Resonances

Another concern with reflected sound arises when it bounces back and forth between room boundaries, resulting in room resonances (sometimes called room "modes"). Room resonances at high frequencies manifest themselves as a "flutter echo" or metallic ringing which you'll easily hear if you clap your hands at the singing position, but again those are usually warded off just fine by the padding you've already used to tackle the room reverb. However, low frequencies are much harder to absorb with padding, the result being that the low-end spectral balance in most project-studio recording environments is rather unpredictable—there will typically be frequency hot/cold-spots scattered all over the place, the exact locations of them dependent on the room's specific physical dimensions.

The cheapest way of tackling the effects of lower-frequency resonances on your vocal recording is to experiment with moving the singer and their mic to different locations in the room. Most of the time you won't want to go right up against a wall, because it makes it difficult to control reflections from that surface. It's also unwise to set up the singer or mic within 12 inches of the centerpoint of any of the room dimensions, because these areas will usually suffer most from room-resonance frequency artifacts. In rooms smaller than about eight feet square, however, it may prove impossible to find a sufficiently uncolored tone anywhere, in which case I'd recommend decamping to another room for recording purposes if at all possible. While you can use acoustic treatment to reduce small-studio room resonances, it's not cheap, because lower frequencies need heavier and thicker material to absorb them, so recording vocals in a different space will often be more cost-effective.

VOCAL BOOTHS

For some reason a lot of small-studio operators get a bee in their bonnet about wanting to build a vocal booth, but in most situations I really wouldn't recommend it. The problem is that anything smaller than about six feet square will have such strong acoustic resonances in the vocal range that it'll sound unpleasantly boxy—pretty much irrespective of the amount of acoustic treatment you try to pack in there. If you don't have a separate live room, then you'll get a better sound setting up in the control room and working on headphones, in my opinion. Those who are undeterred, however, should at least make sure no two dimensions of the enclosed space are within a foot of each other, and try to avoid the classic acoustic-treatment mistakes: Don't cover the whole room in acoustic foam (concentrate on the ceiling and the walls behind/alongside the singer so that you don't completely suck the life out of the booth's high-frequency response); don't use the same thickness of acoustic foam everywhere (tiles of different thicknesses will spread the absorption better across the frequency spectrum); and don't rely on general-purpose materials like standard curtains, furniture foam, carpet, or eggboxes, because they're likely to be more or less useless. (And besides, eating that many eggs can't be good for your cholesterol…)

4.5 MONITORING REFINEMENTS

Hopefully it should be clear by now that there's a lot more to vanilla-setup vocal miking than many simple tutorials would have you believe—just sticking a large-diaphragm cardioid condenser mic and a pop shield in front of your singer and hoping for the best is about as smart as choosing a random equalizer preset. By the same token, basic recording guides rarely offer much advice beyond "give the musician a pair of headphones," and while using headphones is certainly the most popular approach for vocal overdubbing, it has plenty of nuances that warrant closer attention.

Closed-back headphones are usually preferred for vocal overdubbing, because that minimizes spill from the singer's foldback mix onto the recording. One of the crucial questions with closed-back headphones, though, is how the singer hears their own voice. There are two main options.

4.5.1 Acoustic Vocal Monitoring

One approach is to ask the singer to slide one headphone off their ear in order to hear their sound acoustically in the recording room. A practical difficulty, though, is that the singer may not hear their high frequencies well enough: They won't reach their ear direct from their mouth, because high frequencies are more directional; and they won't reflect back to their ear because they're so easily absorbed—especially if you've put padding around the singer to reduce room reflections. So if you choose this monitoring approach, it can help if there's some kind of acoustic reflector to bounce their voice directly back toward their ear. By "acoustic reflector" I don't mean anything fancy, because any rigid surface can be pressed into service: a solid music stand, perhaps, or a clipboard gaffered to a mic stand. If you're thinking "that'll make sod-all difference!", try singing something right now and

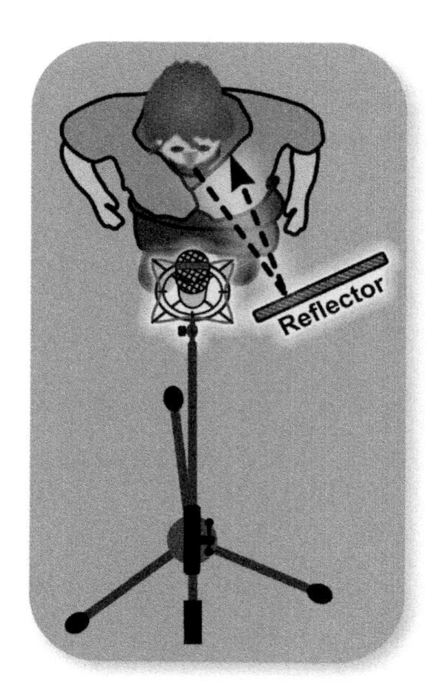

FIGURE 4.13
Where the singer is receiving insufficient acoustic feedback, a small reflector panel can make a big difference—just be careful not to bounce unwanted reflections back into the microphone otherwise you may get comb-filtering problems.

see how the sound changes when you hold this book open a foot in front of your face. (To hell with the librarian!) Of course, you mustn't direct a strong reflection into the mic at the same time, but that's perfectly feasible with a setup such as that shown in Figure 4.13.

The singer may prefer using No Input monitoring (as discussed in Section 3.2.2) in this case, but be prepared for any eventuality, because sometimes they may still want to hear something of themselves in the headphones too. If they decide to leave one earcup completely off, mute that side of the foldback mix to reduce headphone spill—you might need to sum the mix to mono before you do this, though, so that any panning in the foldback mix doesn't cause difficulties. The main downsides of this approach are that the backing mix won't feel as enveloping in mono, so the performer may have trouble "getting inside" the music emotionally; using No Input monitoring can make punch-in recording more awkward; and wearing the headphones one-sided may destabilize them on the singer's head (although most performers will usually be happy to hold them in place by hand).

4.5.2 Vocal Foldback

The second vocal-monitoring option is to feed the vocal signal back into the foldback mix. Our approach so far has been to do this from the recording track's monitor channel, because this allows us to use the Input/Auto monitoring modes, which make punch-in recording so convenient. However, because singers will always hear some of their sound directly, via bone conduction in the skull, "the induced latency will then cause a comb-filtering effect," explains Jay Graydon,[13] "sounding like the vocal is in a fishbowl." This can be quite disconcerting for some performers, significantly affecting their ability to judge performance inflections and pitching. One of the great advantages of overdubbing on analog tape is that (a) you can monitor completely without latency and (b) you can use the Auto monitoring mode for punching in. With digital systems, however, you're almost always forced to choose one or the other:

"One great advantage of overdubbing on analog tape is that (a) you can monitor completely without latency and (b) you can use the Auto monitoring mode for punching in. With digital systems, however, you're almost always forced to choose one or the other."

- Even the best DSP-driven low-latency monitoring will only bring latency down to around 1 ms, which is still more than enough to cause noticeable comb-filtering against the singer's bone conduction, so if you want to use Auto monitoring for punching in, you'll have to sacrifice the foldback sound.

- The only way to avoid the comb-filtering completely is to set up true zero-latency monitoring, using No Input monitoring and then feeding the singer's input signal directly to the foldback mix from the analog side of the recording chain. Of course, there's no simple way of getting that analog foldback signal to mute automatically during playback, which is why you can't use a recorder's Auto monitoring mode under these circumstances. If you've implemented No Input monitoring for the performer by switching off your DAW's software monitoring, then you'll also need to feed the input signal directly to the control-room monitor mix as well, although there's no need to do that in the analog domain if there's a more convenient method in the software.

Despite its complications, the second of these options will always be preferable for me wherever there's adequate analog hardware to implement it, simply because it puts the needs of the performer before those of the engineer.

INPUT/BACKING MIX CONTROLS FOR ZERO-LATENCY MONITORING

Some microphone preamps aimed at the budget DAW recordist include a zero-latency analog monitoring facility. This is basically a mini-mixer which combines the preamplified microphone signal with a backing mix from your DAW, and then feeds that to an onboard headphone preamp. As such, you get just two controls: an input/backing mix knob and a headphone volume knob. Although this configuration seems pretty neat in theory, users with less sensitive headphones (or who are driving two pairs of cans from a single headphone socket) can have trouble getting a loud enough vocal signal in the foldback mix when leaving sufficient recording headroom, no matter how they set the monitoring controls. The simplest solution to this is to try a few other headphones, in case you can find a more sensitive pair, but if that doesn't work, then pass the mic preamp's headphone signal to a second headphone amplifier to boost its output—many budget-friendly headphone amps have a stereo TRS jack input which makes this very straightforward.

4.5.3 Vocal Compression

Right at the start of this chapter I mentioned that vocals have a very wide dynamic range, but this impacts on more than just mic choice. One of the central challenges with producing vocals is that it's often in the interests of the production that the listener can hear their every detail at all times, necessitating a massive reduction in that dynamic range. Trying to implement this in a comparatively natural-sounding manner isn't easy, and one of the key tricks is to squeeze the dynamic range through several different processing stages so that no single stage breaks a sweat. This is one reason why most professionals still record vocals through an analog compressor. Even engineers who usually record most things unprocessed often make an exception for vocals. "I don't add EQ or compression or effects," says Pierre Marchand,[14] for example, "apart from some compression when recording vocals, because they're too dynamic." As processing decisions go, compressing a vocal on its way to the recorder is therefore comparatively low risk.

The main thing to avoid is overdoing it—it's easy to add more compression at mixdown if you undercook it.

There are practical benefits to compressing the vocal input signal from a monitoring perspective too. Firstly, it reduces the levels of louder notes so that singers can monitor their voices louder, making it easier for them to hear the softer intricacies of their performance. That means that they're less likely to oversing, and they'll therefore typically sound more characterful and less strained in the mix. The second benefit is that the compression keeps the vocal levels more stable against the backing mix, which assists in judging the suitability of the vocal tone in context. So even if you're not confident with the idea of compression in the analog domain as part of your input chain, you may still wish to compress the vocal track's monitor channel.

You don't actually need to understand a great deal about compression to use it for vocal recording, as long as you keep your ears open. (The links in this chapter's web resources will give you full details if you really want them.) The main thing to watch is the compressor's gain-reduction meter, which shows how much it's turning down the levels of the singer's louder notes, because as long as that doesn't go beyond 6 dB or so, you're extremely unlikely to do any irrevocable damage to your recording. If there's any Ratio parameter (which determines how aggressively the compressor responds to the loud notes), I'd normally recommend setting it somewhere between 2:1 and 3:1 to be on the safe side, and if there's a switch labeled "soft knee" it's well worth engaging that too, because it'll usually make the processing appear more transparent. If you get separate Attack and Release parameters, then you'll usually be alright setting them to middle of their ranges, but if there's some switch which will set them automatically, then that's likely to be even lower risk.

THE DOUBLE-TAKE TRICK

If you're recording vocals to tape, then ideally you want to keep the levels quite high to minimize noise. At the same time, though, vocal levels can be very unpredictable, and you don't want to scorch the tape on any surprisingly loud notes either. George Massenburg[15] told me a neat trick for dealing with this problem: Route the output of the recording chain to a second track on the recorder, but set its recording level 12–15 dB lower, and then record both tracks at once. That means that if you unintentionally overload the main track, you can be pretty certain that the second track will still be clean.

4.5.4 Balancing and Processing the Backing Track

When recording vocals, most engineers like to have the complete arrangement going. Here's Walter Afanasieff, for instance: "The rule of thumb is, it's got to sound like it's supposed to sound when you're actually cutting the vocals." However, John Hudson[16] highlights an important caveat here: "You've got to be very careful about playing the singer the whole track in the cans. It can be a bit overpowering, and then he or she will start to oversing."

If you're sending a vocal foldback signal to the performer, then its level is critical. "The more you tuck the vocal, the more the singer wants to perform, to hear themselves," says Darryl Swann,[17] for instance, "When it's way out front, they hold back." There's no technically "correct" level, though, warns Rafa Sardina,[18] because his experience is that different singers often have wildly contrasting preferences in this respect. Where either the vocal or the backing track are too quiet, the tuning and/or timing can easily suffer too. Tuning problems may also be exacerbated by latency on the vocal foldback signal, in which case encouraging the singer to use acoustic monitoring (see Section 4.5.1) or making the extra effort to implement true zero-latency monitoring may help. John Hudson[19] offers another tuning tip: "When there's only a basic rhythm track, you normally put a pad in before someone does a vocal, even if you remove it afterwards. It makes the track more comfortable for the singer to pitch to." By extension, an additional rhythmic layer (or even just a metronome click) may help keep a singer's timing focused during moments in the arrangement where insufficient rhythmic cues are coming from other backing parts.

"Tuning problems may be exacerbated by latency on the vocal foldback signal, in which case encouraging the singer to use acoustic monitoring or making the extra effort to implement true zero-latency monitoring may help."

In general, it stands to reason that a great-sounding headphone foldback mix is likely to give rise to a better performance. So given the quality of most small-studio headphones, it can be worth using equalization to compensate for their deficiencies. "I create an EQ curve just for the phones," says Humberto Gatica.[20] "I don't care how good they are, at the end of the day they are these little things that are just blowing volume up your ears." By reducing levels in the 2–6 kHz region, you can make super-loud headphone mixes less fatiguing on the performer's ears, and slapping on a mastering-style limiter preset can also help in this regard by rounding off the transients—a trick suggested by both Stephen Hague[21] and Phil Ramone.[22]

GUIDE-TRACK SPILL

If you're using a guide track with additional "helper" parts which won't appear in the final mix (most often metronome click), it's important that you prevent them breaking through onto the vocal mic as foldback spill. Closed-back headphones will usually take care of this, but it's easy to get caught out none the less. For example, you might have to switch to a one-sided cue mix to avoid leakage if the talent chooses to slip one earcup off their ear, and special super-isolating headphones (effectively industrial ear defenders with headphone drivers in them) may be necessary where a performer wants to hear the guide particularly loud. You may even have to turn down or mute some guide tracks during sparse passages of the arrangement and for end-of-song fade-outs, because those are situations where even the slightest foldback spill can become uncomfortably audible.

4.5.5 Comfort Effects

Minimizing the amount of room sound you capture while recording vocals has many practical advantages, but it does cause one problem where the singer is relying on headphone foldback to hear themselves: The vocal will sound unnaturally dry and exposed. In musical styles that rely heavily on vocal delays/reverbs at mixdown, such a stark presentation may prove quite inhibiting to the performance, so you may be able to make the musician feel more comfortable by applying such effects just for monitoring purposes, without recording them. Some people assert that reverb can also help a performer's tuning, but I have to say I've never found that myself, and remain slightly sceptical about that particular claim.

Adding "comfort effects" is best done using a send–return effect configuration, rather than by inserting things directly into the vocal track's monitor channel. This allows you to keep the control-room monitor mix dry if you wish (perhaps to make it easier to judge the raw recording quality) and also avoids adding any additional latency from digital effects units into the vocal foldback. If you're using Input/Auto monitoring for the performer, here's how to sort out the routing (see Figure 4.14A):

- Use a pre-fader send from the vocal track's monitor channel to feed the effects processor.
- Make sure the effects processor only outputs the effects signal, and not any dry signal—as is standard practice for any send–return effect setup.
- Send the effects-only signal to the foldback mix using a pre-fader send on the effect-return channel.

Where you're implementing a No Input monitoring mode (so that performers can monitor themselves acoustically or use zero-latency analog vocal foldback), here's the simplest way to add comfort effects (see Figure 4.14B):

- Send a signal to the effects processor from the recording path, rather than the monitor path. Ideally, the signal should come from a point in the input chain after any compression, otherwise it'll sound like you're turning the effects levels up whenever the singer lets rip.
- Send the effects-only signal to the foldback mix using a pre-fader send on the effect-return channel.

Whether you implement these sends and returns with an analog mixer, with a DSP mixer (in a multitracker or audio interface), or with a DAW's software mixer is pretty much immaterial. Any latency incurred is unlikely to cause problems, because the most common comfort effects (echo and reverb) are based around delays anyway, and a few milliseconds' more delay shouldn't be noticeable. However, the use of pre-fader sends is a two-edged sword: On the upside, it means that the main control-room monitor mix and the performer's foldback mix remain completely independent in terms of balance, but on the downside, it means that when you change the vocal level in the foldback mix,

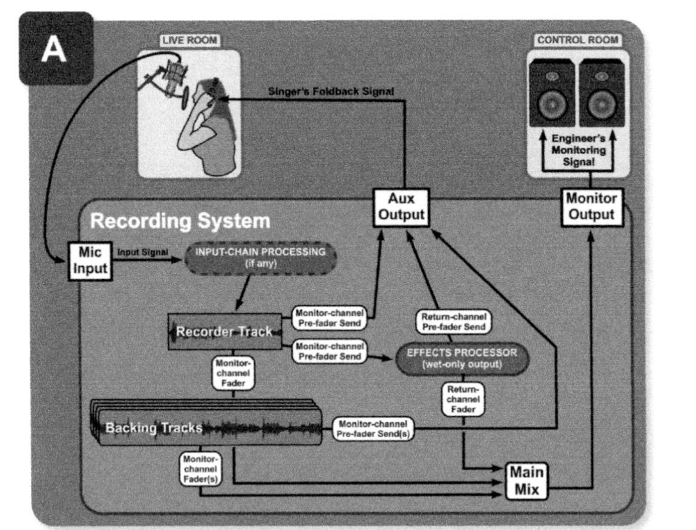

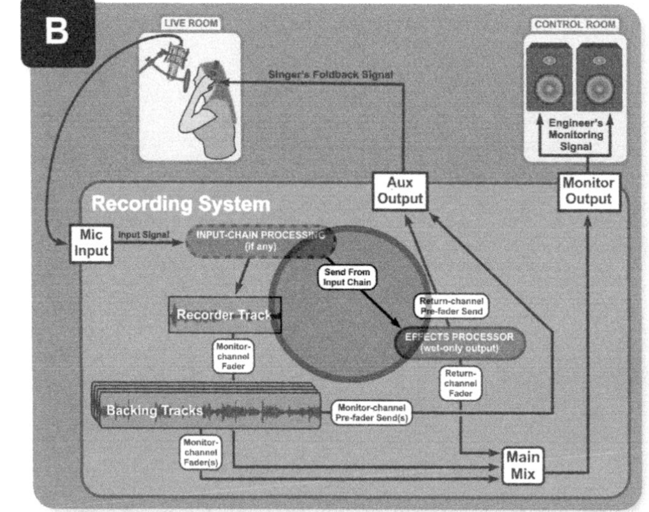

FIGURE 4.14
(A) Implementing vocal "comfort effects" when you're using the recording system's Input or Auto monitoring modes to provide vocal foldback. **(B)** Implementing vocal "comfort effects" when you're using a split-input monitoring system for vocal foldback to reduce latency.

the level of any comfort effects will stay the same. Some engineers are content to work around this by just juggling two different aux sends at once and listening to the foldback mix in the control room to confirm the correct dry/wet effects balance by ear. However, a more elegant solution (especially where you prefer analog zero-latency monitoring) is to use a separate mini-mixer to create the foldback mix. Here's how that works (see Figure 4.15):

- Rather than sending the vocal and backing-mix foldback signals direct to the singer's headphones, you send them to separate channels on the mini-mixer instead, and then feed the vocalist's headphone amplifier from the mini-mixer's main outputs.
- You can then feed the effects processor from a post-fade send on the mini-mixer's vocal channel, and return the processor's effects-only output to another mini-mixer channel. The result: Adjusting the mini-mixer's vocal-channel fader should maintain a consistent balance between the dry vocal and the comfort effects in the foldback mix.

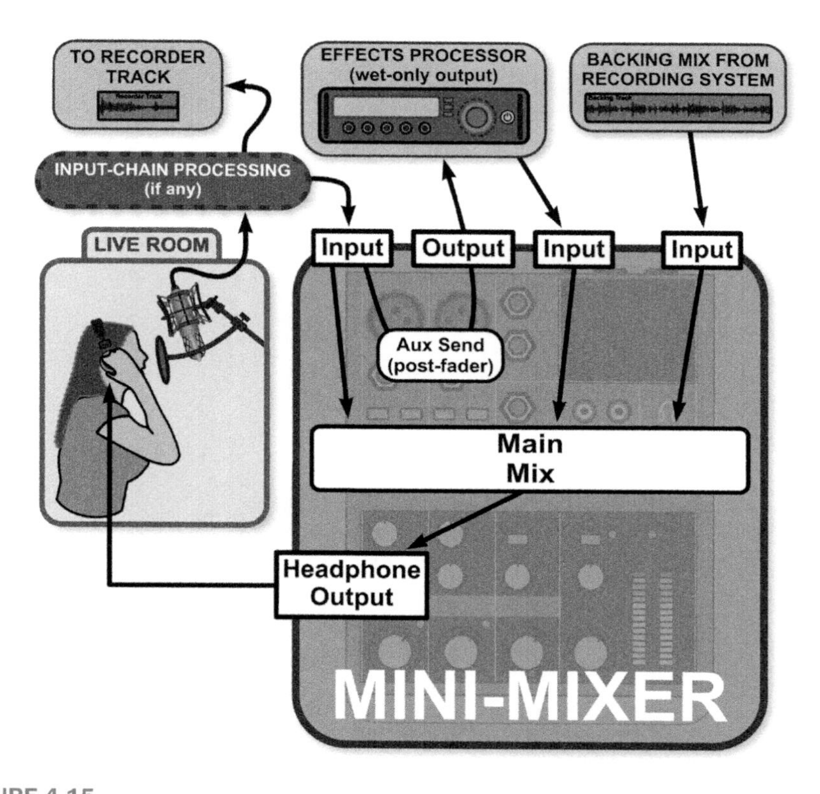

FIGURE 4.15
Using a separate mini-mixer to implement vocal comfort effects. Because the send to the effects is post-fader, the reverb level will remain in proportion if you adjust the mini-mixer's vocal-channel fader.

What's great about this scheme for small-studio users is that you can buy a suitable little analog mixer for about $75 (£50) these days, and it'll usually have a headphone amp built in. If you're lucky, there may also be onboard equalization you can apply to the backing mix as well. Moreover, this approach lets you pull together simple "more me" foldback mixes at extremely short notice if needs be, by skipping the process of building an independent backing mix via pre-fader aux sends and sending the main control-room monitor mix to the mini-mixer in its place. (This won't work if there's any extra latency delay on the vocal heard in the control room, though, because that'd cause comb-filtering against the analog "more me" vocal signal in the foldback mix.)

Now if you're thinking to yourself that all this routing sounds teeth-gnashingly convoluted, then I agree with you! In fact, it's partly because it's such a faff that I normally try to avoid using comfort effects wherever possible. However, Wyn Davis[23] also points out an important psychological effect of keeping foldback effect free: "I think it keeps everyone really honest. If someone wants some reverb or delays it's not a problem, but I try to keep it down to the very least that they'll accept." Phil Thornalley[24] feels comfort effects can actually be counterproductive: "You've got to confront the reality of what you're doing, and a bunch of delays and echoes may be vibey, but may also just add beautiful confusion." Whatever your views on this, though, it's good to prepare for all eventualities, because if it turns out you need comfort effects, then it'll certainly speed up the session if you've already dealt with any routing in advance, found a couple of promising-sounding presets, and checked that the effects signal is reaching the headphones intact.

4.6 ON THE SESSION

Once you've got the gear hooked up and your talent's on the mic, setting levels shouldn't present any additional problems unless you're working with an absolute foghorn, in which case you may find that the singer distorts the microphone's own internal preamplification. Many large-diaphragm condenser mics have a pad switch to deal with this, although it's best left at its unity-gain position unless you actually need it, as with all input-chain gain controls.

Because of the amount of compression typically used on vocals, you should also remain on guard against background noise, wherever it might originate from. "You had to make sure that [Gabrielle] took off her jewelry," noted Johnny Dollar,[33] for instance. "She wears a lot of it, and it makes her jangle quite a bit."

> "Because every voice is different, matching the singer to the right microphone at the start of each project is something that numerous top-name engineers mention making time for."

4.6.1 Refining the Sonics

With any acoustic sound, it's vital that you listen to it acoustically, because that gives you a direct impression of the source itself, and therefore helps you plan and evaluate your microphone

ENGINEER'S QUICKSTART GUIDE: VOCALISTS

As with any instrument, getting the best out of a voice requires good preparation. As an engineer, you can do your bit by keeping the recording room at a reasonable temperature, because the throat tends to lock up when cold. Providing plenty of (room-temperature) water is also a good idea, because singers lose a good deal of water vapor when they sing, and dehydrated vocal cords work less efficiently and tire more quickly. Avoid providing carbonated drinks, not only to avoid unwanted emissions (pardon *me*!), but also because they're usually quite acidic, as are citrus fruit juices. Alcoholic/caffeinated beverages, smoking, and salty snacks tend to be counterproductive (physiologically, at least!) because they all dehydrate your throat, although it has to be said that it's probably better for a singer to be drinking alcohol during the session than the night before, because its dehydrating effects only really kick in after about four hours. Dairy products coat your vocal cords in a way that can cause phlegm to become an unnecessary distraction, and trying to record within a couple of hours of a heavy meal is rarely a good idea either, because a full stomach restricts the breathing—plus the effort of singing may cause reflux, bathing the vocal cords in digestive acid. Eww...

If you have the opportunity to talk to the singer before the session, do your best to convince them to memorize their part, because it makes an enormous difference to the performance. "[Madonna] locks the melody into her head and memorizes the words immediately," says Shep Pettibone.[25] "She doesn't even have to read the words off the paper while she's singing." Young Guru[26] emphasizes the engineering advantages of this when working with Jay Z: "Jay doesn't write his ideas down, he does everything in his head... Once he has the verse memorized, he'll go into the booth to recite it. From a recording point of view, this works much better than someone reading from a piece of paper, and having his mouth tilted to one side." Bruce Swedien[27] has similar memories of working with Michael Jackson: "I don't think I ever saw Michael [Jackson] with the lyrics in front of him. He'd always been up the night before memorizing the lyrics and he sang the songs from memory." But that's not the only tip worth taking from the King Of Pop. "Every day that we recorded vocals," continues Swedien, "he warmed up for an hour beforehand. That made a big difference." While an hour is probably a bit OTT for most small-studio musicians, it's certainly not unreasonable to expect any vocalist who's recording to give 15 minutes of their time to limbering up. If you need a fairly foolproof warm-up routine, try gently humming simple hymn tunes within your most comfortable pitch range: By nature, hymns have a fairly restricted pitch range, so they don't tire you out prematurely with excessively high/low notes.

FIGURE 4.16
The most important maintenance equipment for singers: a glass of water.

Inexperienced singers may not understand how to get the best out of their voices on a session, so here are a few practical pointers:

- Pay attention to the time of day you record, because it can make a big difference. "With some vocalists," comments Stephen Street,[28] "they sound better later in the evening than they do first thing in the morning." You also need to catch the singer when they're in the right state of mind, as Johnny Dollar[29] did when recording Gabrielle's hit song "Rise": "We managed to record at quite an opportune moment: Gabrielle had split up with her boyfriend just before she came in to do the vocal, and that gave the vocal an immediate emotional content. She actually burst into tears in the middle of one take!" Here's Alan Winstanley[30] too, recalling his work with Madness: "Occasionally, we might have to take [Suggs, the lead singer,] to the pub, get a few beers in him to loosen him up, and then go back to do another vocal."

- Taking a sip of water before each take is worthwhile, and swallowing causes less vocal fatigue than coughing if the vocalist needs to clear phlegm. Some professional session singers use a small $35 (£25) steam inhaler before/after sessions and during breaks to keep their vocal cords moist and sinuses clear. While that might seem like overkill for sporadic sessions, it isn't a bad investment if you've got to get a lot of vocals down in a hurry. Just don't use the steamer in the studio itself, because the water vapour can cause condenser mics to malfunction.
- If the performer is raising their shoulders/ribcage and sucking in their stomach when they inhale, they're probably not breathing very effectively. They'll breathe more deeply and with better control if they keep their shoulders down and let their tummy move outwards as they breathe in. If they can't get the hang of this, then a couple of sessions with a good singing teacher would be advisable, to get at least that technique sorted, no matter what style of singing they do. "You won't come out a different person, with a different sound," reassures John Leckie,[31] who is firm advocate of singing lessons, "but... at the end of the day you'll sell more records."
- Standing up while singing will usually promote better breathing by allowing freer diaphragm movement. However, this may not be the overriding concern from a production point of view. "I prefer singers to sit down, to be relaxed, even though there may be certain problems that you get with breathing," comments Neil Davidge,[32] for instance. "You get a presence and a conversational kind of thing... You listen to the record and you feel they're talking to you, singing to you."

techniques. "[With vocals it] can be a little embarrassing," says Mike Stavrou,[34] "but it's well worth the effort. While the singer is in the control room, play the track loud and ask them to sing along for a moment so you can hear their voice in real life... Those few seconds are worth an hour of fiddling with mics and EQ."

Because every voice is different, matching the singer to the right microphone at the start of each project is something that numerous top-name engineers mention making time for, Joe Barresi's[35] approach being pretty typical: "While cutting tracks I'll put a couple of mics out and see what the singer sounds like. On vocal days, I'll then put up a couple of mics again and listen and do a shootout." He also mentions varying the miking per song, as do both Mike Clink[36] and Rich Costey.[37] For example, here's Costey talking about his work with Muse: "In terms of vocal mics, we would switch between different ones according to the song. I'm a big believer in that." Leslie Ann Jones[38] emphasizes the importance of judging such mic selections in context, though: "You can't really hear what a mic is going to sound like on someone's voice if they're not singing along with the [backing]. There are too many characteristics that may not show up. It might sound fine if they're all by themselves, but then you stick them with the track, and it'll sound completely different."

Not that shooting out mics is always feasible, of course, given the importance of getting the right performance. "If a singer is excited and ready to perform," says Jay Messina,[39] "and you take twenty minutes to satisfy your own ego by getting the best possible vocal sound until that person no longer feels like singing, you've defeated your purpose." In fact, with the best singers, you shouldn't count on any setup time at all, says Joe Chiccarelli:[40] "Working with Etta [James], you'd better guess the mic preamp level, you'd better guess the compressor, because chances are from note one it's going to be great." "In the end what really matters is the performance," adds Stephen Street.[41] "I'd rather have a great vocal slightly dodgily recorded than a pristine version of a bad performance. If someone has captured the moment it doesn't matter if the vocal has a few imperfections." One workaround is to arrange a separate shootout session in advance of the actual recording date. "I will try different mics," says Paul Epworth,[42] "but… I always find it works better if it's a totally separate process to actually committing to a vocal performance, because it gets them thinking too much. It's like it switches the brain on, when actually you want the brain off and the heart on."

MICROPHONE PREAMPS

Read a few interviews with high-profile producers, and it won't be long before they'll be extolling the virtues of some high-end preamp or another, or comparing the benefits of different preamps for different mics or sources. As such, it should be pretty clear that this hardware choice makes a difference. However, if you take the time to compare lots of different preamps (3D Audio do some educational demo discs for this), what you'll discover is that the differences between them are much less significant than the differences achievable by changing mics or adjusting mic technique. Because of this, I don't think hugely expensive preamps are the best use of money for many small-studio owners, because money spent on microphones will give you more bang for your buck. As a rough guide, if you add up the total value of your mic collection, I'd suggest spending around a quarter of that money on your preamps. That said, you can get stung at the low end of the market by products which sound unnatural, add excessive noise, or incur unattractive distortion, so you're generally safer spending at least $75 (£50) per preamp, as well as avoiding preamps with valves in them until you're well out of that price range, because they often over-hype the valve character.

If, by hook or by crook, you do manage to interest the singer in doing a mic shootout, then here's my suggested workflow:

- First concentrate on positioning your best-guess mic as well as possible, listening to its sound in the context of the whole production. The advantage of doing this first is that you can immediately skip on into doing takes if the singer begins losing interest, and it also means that you're not subsequently comparing different mics in a hopeless mic position. A good way to audition a lot of different mic positions is to set the mic up at about chin height, and then get the singer to try addressing the mic at different angles, heights, and distances while you listen—moving the singer is a whole lot easier than moving the mic stand! Once you find a promising position, adjust the stand to replicate that position for the singer's natural posture.
- Set up all the other mic contenders as close as possible to your best-guess mic and then do a short take, recording all the mics onto separate tracks. Once you've got that recording, get the singer to join you while you play with the balance of each vocal-mic track against the backing mix and whittle down the options in discussion. As well as any general tonal decisions you may make, listen particularly for whether the lyrics are coming through

clearly at lower fader levels, and whether sibilance becomes overbearing at higher fader levels. Remember that you'll naturally tend to be biased toward louder sounds, so be careful to match subjective volume using the monitor-channel fader for each track.

4.6.2 Takes and Comping

Building up master takes with punch-ins and comping was a major topic in the previous two chapters, and all that material continues to apply when tracking singers, except that the stakes are higher. Lead vocals are usually so important to commercial productions that I'd almost always suggest comping them, and from at least five takes. Just to give a bit of perspective here, it's not uncommon for mainstream chart productions to involve more than 20 vocal takes, and a comping process lasting 8–12 hours, according to Mark Bright.[43] Ben Allen[44] indicates that top-flight artists such as Christina Aguilera may go even further. "Incredible voice, insane work ethic," he elaborates. "[She] sang the song for six hours until it was done—didn't leave the booth once and didn't make a single phone call… That song was probably comped from a hundred different takes! She nailed every single one, but she wants to comp it until she's in love with it… To argue about the validity of it is silly. It's whatever gets you to the destination."

"Anyone other than a professional session singer will start running out of steam within a couple of hours if they're really committing to their performance, so if you need a complete vocal during that time it's important to pace the session astutely."

One thing you have to be wary of, though, is tiring out the singer. Anyone other than a professional session singer is likely to start running out of steam within a couple of hours if they're really committing to their performance, so if you need to get a complete vocal during that time it's important to pace the session astutely, as well as scheduling short breaks at least every hour or so. "There's a kind of crispness to a voice when it's fresh," says Simon Climie.[45] "and sometimes if you've sung too many times it becomes over-compressed and loses that edge—literally, the EQ of the voice changes dramatically."

If you're concerned about the singer's stamina, it's usually more efficient to work on a song in sections, rather than going for too many full-length takes, because you don't want to waste energy rerecording simpler parts of the line more often than necessary. Easier, quieter, and lower-register sections can be good to help the singer get into the swing of performing, and may also provide welcome respite from working on higher-energy phrases. Try to keep the loudest and least restrained sections until close to the end of the session, so that if they leave the singer sounding a bit ragged, it won't affect the vocal tone in the remainder of the song. That said, if there's a quieter section that needs more vulnerability or rawness, then the very end of a session may actually be the ideal time to attempt that, for similar reasons.

Back in Section 3.3.2 I talked about looking for "moments of magic" while recording takes for comping purposes, and nowhere is this more important than for lead vocals. Focus particularly on the starts of verses, where it's crucial to really demand attention, and of course also the run-up to any hook, as well as any memorable pay-off lines. Real nuggets are more likely to occur when the musician is feeling slightly off-guard, which is why the very first take of a song is frequently full of them. Any situation where the singer feels less inhibited is also fertile territory. Dave Jerden[46] regularly records warm-ups, for example, and I remember watching Gerry Kitchingham[47] surreptitiously recording the "rehearsal run-throughs" of less confident vocalists, because he knew that their musical responses would be more instinctive when they thought they weren't yet being recorded—and those takes frequently featured heavily in the final mix as a result.

THE PRE-ECHO TRICK

When recording vocals, it's very common for the first entry of any new section to be substandard, because it takes a few moments for a singer's voice to get going again (and for them to lock onto the right pitch) if they've been silent during the pre-roll. The solution to this is to get them to "pre-echo" the first line a couple of times during the pre-roll, so that their voice is poised and ready for the real entry. You can always edit away the pre-echoes, or even use them as additional take options when comping, if any of them were particularly good.

One of my own favorite tricks for the end of a vocal overdubbing session (once I already have enough takes to comp a respectable master take) is to give the musician a couple of additional tracks they can sing whatever the heck they like on. I'll make it clear that we've already got a master take in the bag, so they've got nothing to lose by trying even the most bizarre ideas that come into their head. Somehow this never fails to unearth one or two absolute gems that really elevate the performance as a whole, and it often brings the session to a close in a really upbeat way too, which is good for general morale. "Assuming the vocalist is not too tired, you'll get some incredible stuff out of that," says Mark Bright,[48] who uses an almost identical tactic himself.

If time is on your side, then another alternative mentioned by Dave Jerden,[49] Clarke Schleicher,[50] Stephen Hague,[51] Steve Levine,[52] and Stuart Price[53] is to comp the vocal from the first session, send a rough mix of it to the singer, and then reconvene for further takes later. "Very often you have to do a comp in order for them to hear how it needs to go," comments Levine.[54] This often results in a significantly more assured second date where the performer feels more confident to take risks in search of greater creative rewards.

When you're selecting vocal takes, don't forget to think about how the singing relates to the lyrics. Mike Clink[55] is just one of the top names who keep the lyric sheet close to hand while tracking for this reason. "I'm trying to see in my mind if the singer is telling me the same story as [they've] written down lyrically," he says. "If I'm not feeling it from a vocalist, I'll tell them so. I'll get them to draw from within, to tell that story." Critical to the transmission of the lyrics, of course, is the singer's diction, so this is something to pay special attention to. In particular, try to put yourself into the position of someone who doesn't know the lyrics beforehand—would they be able to understand them easily? There are many things you can do at mixdown to improve a vocal sound, but it's almost impossible to salvage lyrics that aren't sung clearly. And on a practical note, an emotionally compelling delivery is always more important than perfect timing/tuning, especially now that sophisticated audio editing and pitch-correction are standard features in almost every DAW. (Not that they excuse you from providing the most favorable monitoring conditions for the performer, of course…)

4.6.3 Talkback and Performance Directions

Recording vocals is extremely demanding on the musician both physically and mentally, so it's important to help them all you can. "A guitar player can change his strings or amplifier," explains Tony Platt,[56] "and a drummer can change drums and sticks, but a singer has only got one voice. Psychologically that's not an easy place to be, and it's the most difficult thing that a singer has to deal with. So a lot of vocal production is about putting the singer at ease." In this context, the session-framing techniques I covered back in Section 2.1.4 expand in importance, and it's also usually wise to minimize the number of people in the studio while vocals are being cut. "Lead vocalists always insist that the other band members aren't there," says Steve Levine.[57] "They like to do it on their own, sometimes even with the lights off." A specific example of this comes from Mike Shipley,[58] talking about his work with Alison Krauss: "She doesn't like to have many people in the room when she's recording vocals, so most of the time it'd be just her and me." Phil Ramone[59] takes things even further: "Be careful about soloing a vocal in the control room. It should be done with no guests or musicians in the control room."

LAYERED BACKING VOCALS

One of my biggest gripes about low-budget productions is layered backing vocals, because too many amateur singers seem to think they don't need to perform them with as much gusto as they do lead vocals. In fact, the opposite is the case: Because of the inherent homogenizing effect of the layering process (see Section 3.3.1), you often need to ham up the expression to steer clear of blandness. Overenthusiastically pitch-correcting backing vocals to save editing time only compounds the problem. So don't just toss off the backing parts at the end of the lead-vocal session—take a break and come at them fresh. Jeff Bhasker[60] has another tip, which he used while recording Fun's enormo-hit "We Are Young": "I never let them sing to previous vocal takes. I would always let them sing by themselves, and I'd then add the different vocal takes together. You get a stronger performance that way, because the overdubs are rubbing against each other… It was the rubbing of the vocals against each other that made them sound big."

Another thing to be aware of when layering vocals is that there's a considerable difference in sound between tracking up the same singer and layering several different singers—the former will tend to sound thinner and take up less space in the mix, whereas the latter is more likely to give the thicker texture that small-studio producers more often appear to be seeking. If you've only got one singer available, though, try recording different takes through different mics, a technique that Walter Afanasieff[61] and Billy Bush[62] have both mentioned using on their records.

In a two-room recording configuration, setting up communications is simplified because the main vocal mic doubles as the listen mic. Despite this, however, some producers feel that direct face-to-face conversation is so critical for a vocal session that they'll forgo the technical advantages of a two-room setup and track everything in the control room. "You can react immediately with a simple thumbs up or even by jumping during the take," says Phil Thornalley.[63] "If necessary, you can also sing the next chorus [to] show the singer where they're getting the melody wrong." His own early memories as a young singer lacking self-confidence were also formative in persuading him against the two-room method: "The producers were in the control room. Paranoia would set in. 'Are they ordering pizza, or are they calling around for someone better?' I learned from that experience."

But if you *are* using talkback in a two-room environment, don't let technical inconveniences stop you working your socks off with it. "The worst thing you can ever do," continues Tony Platt,[64] "is stop the tape and have complete silence. The moment you stop the tape you need to be pushing the talkback and making some form of communication. It's a lonely place on the other side of the glass and the first thing the singer is going to want to know when the tape stops, especially in the middle of the song, is how they are doing. It's of absolutely paramount importance to maintain that communication."

"Getting the singer to dance along with the track while singing can completely transform the rhythmic vibe of a take, while upwards/downwards hand motions, raising of the eyebrows, and exaggerated smiling are all well known to influence pitching."

If you're called on to provide suggestions for improving the performance, then Section 3.4.1 should already provide plenty of tools you can use, but I'd like to stress the importance of using posture and movement as a means of influencing the outcome. Getting the singer to dance along with the track while singing can completely transform the rhythmic vibe of a take, for example, while upwards/downwards hand motions, raising of the eyebrows, and exaggerated smiling are all well known to influence pitching. Another key issue to clarify with less experienced rock vocalists is that singing as loud as you can will usually just make you sound thin in the final mix, because individual narrowband resonances and harmonics will dominate over the rest of the spectrum under such circumstances. The secret to achieving a powerful rock vocal recording is learning to sing *as if* you're hammering it out, while actually delivering only a comparatively restrained acoustic level in the room so that the mic captures a well-balanced mixture of frequency content.

4.7 AND ANOTHER THING...

As we've seen, even the vanilla setup presents you with an awful lot of variables to think about, which is perhaps why so many fledgling engineers come a bit unstuck with vocals. However, there are lots of less mainstream techniques that can be more suitable than the plain vanilla approach on occasion, and we'll move on to exploring some of those in the next chapter—once you've had a chance to consolidate what we've learned so far.

CUT TO THE CHASE

- The most popular vocal microphone is a large-diaphragm cardioid condenser, preferably with valves and transformers in the signal path. Although cheap mics can sound great on occasion, pricing remains a good indicator of quality. With cheaper mics, think carefully about the manufacturer's reputation.
- Valve mics will usually have their own power supply, but most other condenser mics will require phantom power. I'd avoid battery-powered and direct-to-USB microphones.
- Set up the mic stand stably, counterweighting it if necessary, and preferably use a suspension shockmount. When arranging the cables, plan for someone tripping over them and try not to defeat the shockmount.
- Certain vocal sounds are quite directional, which helps with positioning the mic and reduces the need for additional interventions against plosives and sibilants. Changing the mic's orientation can provide off-axis frequency contouring, but this may not sound very pleasant with cheaper mics. A variable high-pass filter is extremely useful for controlling proximity effect, giving more leeway in refining the subtler midrange and high-frequency effects of miking distance. Keeping the singer's mouth position consistent makes mixing easier, but be careful not to cramp the singer's style.
- It's usually wise to place sound-absorbing materials around the singer to minimize recorded room reflections and high-frequency resonances, especially if your singer's off-axis to the mic diaphragm. Lower-frequency resonances are more tricky/expensive to control, so try to avoid recording in small spaces where they impact most negatively on the vocal spectrum.
- If a singer wants to monitor themselves acoustically, try rigging an acoustic reflector to help them and be prepared to try different input-monitoring modes. True zero-latency vocal foldback will likely give a better performance, but rules out Auto monitoring on digital systems.
- Moderately compressing vocals during recording can improve the singer's headphone foldback experience and aid mixdown. There are numerous ways of routing comfort effects to the singer if they want them, but the most elegant for small studios is to use a separate analog mini-mixer.
- To perform at their best, vocalists should be well rested and warmed up, and should preferably have memorized their part too. Singers should be well hydrated, so they should avoid diuretic foods/drinks and smoking; sip water regularly throughout the session; and consider using a steam inhaler for extended vocal sessions.
- If you have time, do a microphone shootout to find the best model for each new vocalist, but make a point of listening to the singer acoustically as well, not just through your mics. Find a good position for your best-guess mic first and be sure to audition different mics at comparable levels against the backing mix.
- Voice fatigue is a major concern on vocal sessions, so carefully organize and pace your workflow to reflect each performer's stamina. Maximizing the number of star moments is also a high priority, so don't waste opportunities to record the singer when they're off guard, and try to give the performer the chance to throw caution to the winds for at least a couple of takes.

- Communication is vital for any successful vocal session, so don't let talkback logistics get in the way of that. If the vocalist needs performance direction, suggestions involving posture and/or movement will often prove most effective, and the lyrics can also provide inspiration. Singing very loud typically gives a thin recorded sound, so aim for emotional intensity more than sheer volume, and don't fall into the trap of underperforming layered vocal parts, otherwise they'll sound bland.

Assignment 1: One-Room Dry-Run

This assignment is about preparing both you and your studio setup for the demands of a vocal session. For this you're going to record yourself speaking, because our ears are more sensitive to subtle tonal changes with speech than with singing—and there's no need to worry about a backing track that way either.

- Find about 100 words of text you can read aloud—a bit of *Mr Tickle* by Roger Hargreaves is a good bet, because it's well known amongst linguistics professionals for its wide range of English speech sounds. Deal with the general session preparations discussed in Part 1, and then find the most suitable single-room recording venue you can, with reference to Section 4.4.2.
- Decide on a best-guess vocal mic choice (Section 4.1), as well as a suitable clip/shockmount, stand, and cable to go with it.
- Time yourself as you unfold and set up the mic stand; put the mic into its clip/shockmount and fix that to the stand; and connect a cable between the mic and your recording system's microphone input. If that takes you more than a minute, then I'd suggest putting in some practice to get your speed up! The same applies when packing those bits away: If you can't fold up the stand, coil the cable, and tidy away the mic and its clip/shockmount within a minute, you're in danger of leaving a slimy trail…
- Work out how to rig acoustic absorption around yourself and the microphone for recording purposes, and be very careful with the positioning of any reflective surfaces within four feet of the mic stand.
- Record your chosen text using several different mic positions, ideally including: four inches and twelve inches away, at the level of your lips; above, below, and 20° to either side of your nose, all four positions eight inches away from the mouth; with the microphone rotated so that you're addressing it 30° off-axis. Spend a good half hour comparing these recordings to get a feel for the scale of the differences, and refine the microphone position further, using high-pass filtering or a pop shield to increase your options if necessary.
- If you've not done so already, now's the time to set up a headphone foldback mix so you can hear the input signal while recording. With digital recorders, try to implement analog vocal foldback in some way to get first-hand experience of the subjective difference between zero-latency and latency-delayed monitoring from the performer's perspective. You should also sort out how to compress the vocal input signal, process the backing track for monitoring purposes, and add comfort effects to the vocal in the foldback mix, so that you're on top of your game for Assignment 2.

Assignment 2: Overdubbing a Singer

Following your preparations in Assignment 1, we can move on to recording the performance of a separate vocalist over a backing track.

- Find a friendly singer to practice overdubbing with, and offer them a choice of pre-written songs to sing lead vocals on. (There are some in this chapter's web resources if you don't have anything suitable of your own.) Once they've chosen a song they're comfortable with, provide them with a rough mix (including a guide melody) to listen to before the session and a lyric sheet.
- Choose a suitable venue for recording, whether that's a one-room or two-room setup, and prepare appropriate session-framing, printed materials, and monitoring/communication in advance. Have your best-guess mic already set up in a best-guess position, and make sure any other mics you want to compare are set up on stands, plugged in, and line-checked too. Immediately before the session, check that all monitoring and communications devices are still working properly, and that comfort effects are primed in case they're required.
- When the singer turns up, be ready (as always!) to hit Record the moment they rock up to the mic. If they are amenable to working on sounds first, listen to them singing acoustically and then refine your best-guess mic's positioning before putting up any comparison mics alongside. However, if the performer's stamina or patience begins to flag, be ready to go straight into doing takes.
- Whether you decide to create a master take with punch-ins or (preferably) comping, explain the workflow to the performer. Check whether they can hear their vocal and the backing mix well enough, and ask them whether they'd prefer to have a little vocal echo/reverb in their headphones while performing. Once takes are underway, do everything you can to help the musician achieve the best performance, adjusting the foldback mix (or indeed anything else) as necessary.
- After the session, make notes for future reference, and think back over any problems that arose. After a day or two, listen back to the recordings to re-evaluate the sonics and performances, comping a master take if necessary.
- Before progressing to Chapter 5, try to carry out another couple of sessions like this, preferably with different singers. Try also to do at least one session where you add layered backing vocals alongside a previously recorded lead part.

WEB RESOURCES

On this book's companion website you'll find a selection of resources to support this chapter, including:

- further reading on the technicalities of condenser-mic design, acoustic treatment, zero-latency monitoring, and compression;
- several pre-recorded songs for overdubbing lead vocals, complete with backing mix, guide parts, and lyrics;
- links to manufacturers of mic-mounting and acoustic-treatment products;
- audio demonstrations of vocal dispersion, mic positioning/orientation, acoustic treatment techniques, and different plosive-reduction techniques.

http://www.cambridge-mt.com/rs-ch4.htm

CHAPTER 5

Beyond Vanilla Vocals

The vocal setup I described in Chapter 4 is a great workhorse, especially when you understand how to get the best out of it. However, there are some scenarios where going slightly off the beaten track in terms of studio technique will better serve the music, so in this chapter I'd like to concentrate on why and how you might want to stray beyond the vanilla approach.

5.1 ALTERNATIVE MICROPHONE TYPES

5.1.1 Other Polar Patterns

Useful as directional mics can be, they'll rarely sound as clear and natural as mics which have been designed to pick up sound equally from any angle ("omnidirectional mics" or "omnis"), especially amongst cheaper designs. Al Schmitt:[1] "I keep the vocal mic in omnidirectional. I find the sound richer and more pleasing." Omni mics exhibit no proximity effect and are virtually immune to "popping" on plosives, so you have more freedom to position the mic, plus you get much less level/tone variation if the singer moves around while singing. Phil Ramone[2] and Michael Tarsia both mention the latter in interview: "Sometimes you encounter a singer who just has to move with the music, or even dance," says Tarsia.[3] "That's when I consider using an omni-pattern microphone and carefully place the mic for the most consistent pickup of the vocal performance." Omnis are very insensitive to stand/cable-borne vibrations too, which means they can usually even be handheld without suffering significant handling noise. (It's worth singing slightly past them in this instance, though, to avoid water vapor in the singer's breath from condensing on the diaphragm.)

"Omni mics exhibit no proximity effect and are virtually immune to "popping" on plosives, so you have more freedom to position the mic, plus you get much less level/tone variation if the singer moves around while singing."

Clearly, room reflections are more of a concern with omnis, although the lack of proximity effect does mean you can use mic positioning to counteract this. "Sing up close to the mic," says Keith Olsen,[4] "and all of a sudden the room means nothing." You can further dry up the sound with acoustic padding behind the mic, or by using something like an SE Electronics Reflexion Filter. In my experience, room reflections therefore rarely militate against using an omni mic for vocal overdubbing, so anyone who glibly dismisses omnis as "too roomy" is frequently cheating themselves out of a better sound—particularly anyone working on a tight budget.

The other common polar pattern to mention is "figure-eight" or "bidirectional," which is equally sensitive at its front and rear and has rejection nulls at 90° off-axis. Figure-eights can sound just as good as cardioids on vocals, but they do struggle with singers who move around a lot, for two reasons: Firstly, the frontal sensitivity region is narrower, so the singer only has to move 45° off-axis to drop 3 dB in level and the −6 dB point is only 15° further out; and, secondly, a figure-eight mic's proximity effect is stronger, changing the low-frequency tip-up more for a given front–back movement. Figure-eight mics are also more prone to physical vibration and plosive air blasts.

Some condenser mics incorporate a back-to-back pair of diaphragms, rather than just a single one, and the benefit of these dual-diaphragm designs is that they allow the circuitry to control the polar pattern. Affordable multipattern microphones of this type are usually switchable between cardioid, omni, and figure-eight polar patterns, but more expensive large-diaphragm mics are more likely to offer other intermediate polar patterns, including subcardioid, supercardioid, and hypercardioid (see Figure 5.1), and may even have a smoothly variable pattern control. For vocal recording, what's important is the polar pattern's frontal width (which tells you how careful you need to be about keeping the singer still) and the strength of the proximity effect (which affects how bassy the vocal will sound for a given miking distance). However, proximity effect isn't the only tonal consequence of changing patterns, because you'll also get less predictable smaller-scale frequency variations across the rest of the spectrum.

I mentioned a whole list of classic vocal mics in Section 4.1.4 to demonstrate the professional preference for valves and transformers, but what is also notable about these mics (and indeed the Neumann U87, which seems to be the only non-valve mic that's used as widely) is that they're all multipattern. One of the reasons for this preference is that, assuming you're recording vocals in a fairly well-damped space (as most professionals are), you can access alternate sonic colors just by switching the polar pattern, without recourse to any separate signal processing. "I usually try both cardioid and omni modes with the [mic]," says Alex Clarke,[5] "because there's a subtle difference in the low end between them—you have to decide which one works best for each particular vocalist, and with Tom [Jones] it sounded more open when it was in omni." Be careful, though, to mute the input signal before operating any polar-pattern switches, because these can sometimes send powerful signal "thumps" though the recording chain.

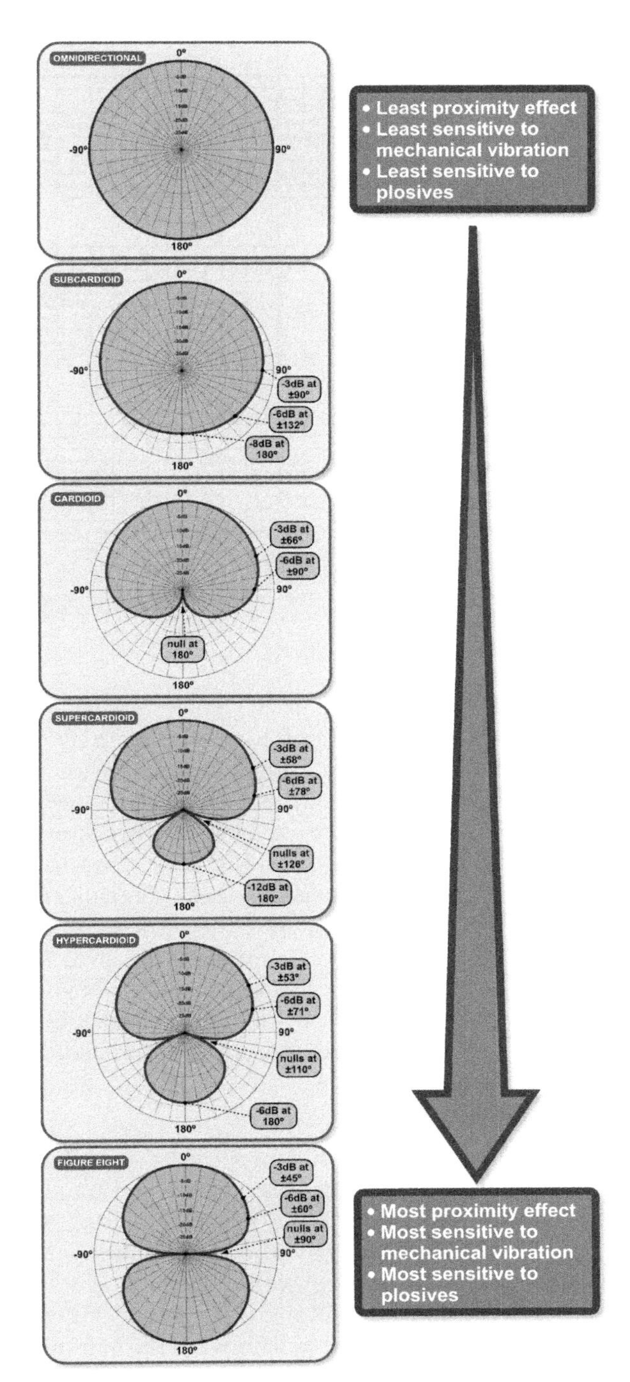

FIGURE 5.1
A range of commonly available polar patterns.

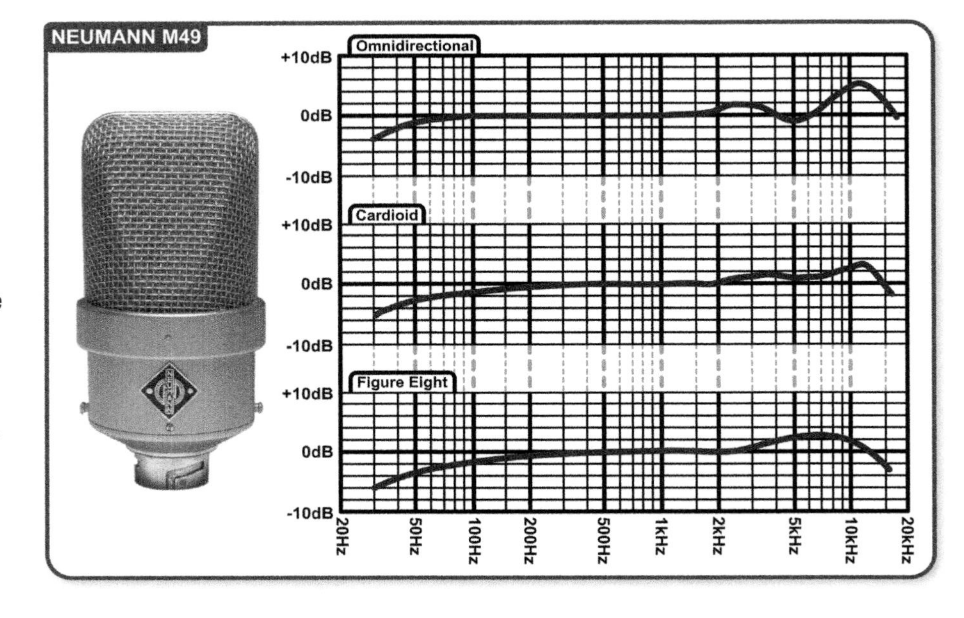

FIGURE 5.2
The frequency response of a multipattern large-diaphragm condenser mic changes as you switch between its different polar patterns. Here, for instance, are the polar plots for three polar patterns of the classic Neumann M49 large-diaphragm condenser mic.

5.1.2 Dynamic Mics

While large-diaphragm condenser microphones are certainly the most common choice for studio vocal recording, there are nonetheless many situations where other designs are preferred. One of the main difficulties is that condensers can be a bit delicate for loud and/or aggressive singers, even when you're using the mic's built-in pad. In these cases, the most common solution is to switch to a dynamic microphone instead, because dynamic mics are usually rugged enough to withstand any abuse and their poorer noise performance ceases to be a concern when the singer's hammering the level. Here's Bill Price[6] talking about recording the Sex Pistols, for instance: "Although we had a beautiful pair of Neumann U47s…when I put one in front of Johnny [Rotten] for his overdubs it sounded awful and died after thirty seconds of being gobbed at due to him using his middle register as loud as he could. So…I put him back on the [Shure] SM58 and thereafter always used that."

But it's not just technical concerns that prompt many engineers to go for dynamic models, because these mics offer a gutsy sound that often suits male rock vocalists better—which is good news for small-studio budgets, because dynamic mics are much cheaper on the whole! The Shure SM58 mentioned by Bill Price is no stranger to studio work: Robbie Adams,[7] Flood,[8] Ann Mincieli,[9] and Stephen Harris[10] have all mentioned that Bono (of U2) prefers it for recording; John Leckie[11] recorded Ian Brown with that mic for The Stone Roses' debut album; Garth Richardson[12] recorded Zack de la Rocha with one for Rage Against The Machine's debut album; Dave Eringa[13] used it for James Dean Bradfield when working on The Manic Street Preachers' biggest album,

This Is My Truth, Tell Me Yours; and Bob Clearmountain[14] tracked some of Bryan Adams' vocals on it for *Reckless*. Its sibling, the Shure SM57 (a very similar design which sounds slightly different because it lacks the SM58's ball-shaped grille) also featured on Anthony Keidis during The Red Hot Chili Peppers' *Californication* sessions (according to engineer Jim Scott[15]), and Eddie Kramer[16] regularly used it for Paul Rodgers of Free and Bad Company.

An important part of the sonic character of the SM57 and SM58 is a strong upper midrange "presence" peak, which helps the sound cut through mixes laden with cymbals and electric guitars, but some engineers prefer Shure's smoother-sounding (and therefore more general-purpose) SM7 instead. "To me, that sounds quite a bit better than your more typical SM58," remarks Rich Costey,[17] and many other engineers have also mentioned it in interview. For example, Joe Chiccarelli[18] used this mic for Jack White of The White Stripes; AC/DC's Brian Johnson went through one during Mike Fraser's[19] sessions for *Black Ice*; Michael Barbiero[20] recorded some of Whitney Houston's vocals with an SM7; and many of Michael Jackson's most celebrated lead parts were tracked with it by Bruce Swedien.[21] Shure mics don't have some kind of monopoly, though, because other affordable dynamic models are also well known for studio vocals. The Beyerdynamic M88, for instance, was for a long time Hugh Padgham's[22] first choice for Phil Collins, and it was also singled out for Bernard Sumner's vocals when Stephen Hague[23] was producing New Order, and for Dion's voice by Eric Schilling.[24] The Electrovoice RE20 is another favorite (Steve Albini[25] is amongst its fans), not least because it employs a unique design that minimizes proximity effect.

Because on-stage vocal mics are so often end-fire dynamic models, another crucial advantage of using one is that the musician can perform with it in the same way they would during live shows. "A lot of people feel comfortable with a handheld," explains Mike Hedges,[26] "even if they are not holding it,

FIGURE 5.3
Some of the most enduringly popular dynamic mics for vocal recording (left to right): Shure SM57 and SM58; Beyerdynamic M88; Shure SM7; and Electrovoice RE20.

but have it on a stand instead. It's more natural, because it isn't a big thing, so they can get as close as they want. This helps the performance, because the vocalist is relaxed." Of course, if the vocalist holds the mic in their hand and is monitoring over loudspeakers rather than headphones (of which more in Section 5.3), you should try to persuade them not to stroll about too much (a few bits of gaffer tape on the floor can provide useful positioning guides here) and to keep facing the speakers so that the microphone's cardioid polar pattern can do its job properly. Some singers have a habit of wrapping their hand around the grille of their handheld mic, presumably in an attempt to shield it from spill, but this is actually counterproductive, because the directionality of dynamic microphones is created using vents at the sides/rear of the head assembly—so if you cover those up you'll get more spill, not less! One final cautionary note comes courtesy of Stephen Hague,[27] and should make sense in the light of what we learnt about vocal frequency dispersion in Section 4.3.1: "Things can get out of control very quickly when sibilant singers use a handheld mic."

THE BEST MIC ISN'T ALWAYS THE BEST-SOUNDING MIC

On occasion the best mic may not be the best-sounding one. Stephen Lipson[28] remembers Paul McCartney letting Ringo Starr use a dodgy-sounding vocal mic simply because he felt that the retro look of it would inspire him to sing better, for instance. Joe Barresi[29] relates a similar studio experience with an Electrovoice 666: "I told the band, 'Hey, this is the mic of the beast—it's the devil mic,' and… I swear, the guitar solo was one take, the piano was one take, and the vocal was one take."

5.1.3 Ribbon Mics

Ribbons were top of the tree quality-wise from the 1920s until condenser designs really began taking the industry by storm during the 1950s. Their understated high frequencies give them a slightly dull tonality to modern ears, but even if you equalize them to reintroduce brightness at mixdown they nonetheless retain a uniquely smooth sonic personality that remains popular to this day. For vocals, what most matters is that the resonant frequency of the mic's diaphragm (a super-thin "ribbon" of metal foil) will be located toward the low extreme of the audible frequency range, which means you don't get that high-frequency "condenser zing"—and sometimes that's exactly what the doctor ordered, for example if your singer has an unpleasantly harsh edge to their voice, or you're simply after the syrupy retro timbre of Doris Day or the Rat Pack. Almost all ribbon mics have a figure-eight polar pattern by the very nature of their design, and many also capture a slightly different tonal balance depending on whether you sing into the front or the rear, so this is something worth experimenting with.

The biggest practical difficulty with ribbon mics is taking good care of them. The ribbon element is quite fragile, so even though it'll happily record extremely loud sounds, it can easily be deformed or torn if hit by a puff of air.

FIGURE 5.4
If you're using budget preamps, active ribbon mics such as the Sontronics Sigma (above left) or Superlux R102 (above right) shown here may give you quieter recordings than passive designs. Alternatively, one of the Cloudlifter boxes (below) from Cloud Microphones can achieve similar noise-floor reduction for passive ribbon models.

With vocals, that means it's sensible to use a pop shield at all times to defend the ribbon against plosives, but you also need to be careful of anything else that might create gusts—ribbons have been known to break just from someone closing the lid of the mic-storage box too abruptly or slamming a door nearby! It's a good idea to store the mic with the ribbon vertically as well, to avoid it sagging over time. Ribbon mics have a large and powerful magnet in them, so be careful to keep the mic away from any small metal objects/filings that might be drawn into the works—this is a good reason to avoid ever putting ribbon mics down on the floor without a dust cover on them. Most ribbon mics don't need power to operate, and feeding phantom power to them may even damage them, so make sure it's switched off unless you happen to be using one of the more recent phantom-powered active ribbon designs.

The final issue with ribbon mics is that they only put out a very weak signal, so you have to push your mic preamp gain close to maximum, at which point cheaper preamps can start getting very noisy. One workaround is to use something like the Cloud Microphones Cloudlifter, a phantom-powered device that connects in-line between the mic and your recording system and provides roughly an extra 20dB of clean gain so that your preamp doesn't have to strain as hard. Alternatively, investigate active ribbon mics, as they typically have much higher output. (Before we go on, let me also clarify my terminology a little. Strictly speaking, ribbon designs are actually a subtype of the dynamic microphones. In this book, however, I've adopted the more widespread colloquial usage I hear amongst small-studio engineers, who generally treat ribbon mics and dynamic mics as totally different beasts. So whenever I have something to say about ribbon mics here, I'll always name-check them directly rather than lumping them in with other dynamic designs.)

SMALL-DIAPHRAGM CONDENSERS?

Small-diaphragm condensers are usually slim end-fire designs (hence their "pencil mic" nickname) with diaphragm diameters of less than half an inch or so. Although such designs often have better high-frequency extension and are less likely to flinch in the face of extreme sound levels, their noise performance usually lets them down a bit by comparison with large-diaphragm models. I don't think this is the main reason they're hardly ever mentioned by professional engineers for vocal recording, though: It's because they don't enhance the vocal sound in the same flattering way, which means that the recorded timbre often ends up sounding rather hard and unappealing. While there are no hard-and-fast rules when it comes to mic selection (small-diaphragm mics were used for many Motown hits, for instance), I think it's probably fair to say that a small-diaphragm condenser for vocals should probably be considered rather a long shot while overdubbing.

5.2 USING ACOUSTIC REFLECTIONS

Although we spent a lot of time in Chapter 4 talking about removing room reflections from a vocal (because that's what people do most of the time), you can also make a virtue of them. Bruce Swedien,[30] for instance, deliberately used an array of cylindrical reflectors (ASC's Studio Traps or "Tube Traps") to introduce a dense and controllable pattern of reflections which enhanced the lead-vocal sound on Michael Jackson's later records, and which also had the effect of keeping the miked sound a little more consistent as Jackson danced around the mic while performing. Another example comes from Steve Bush,[31] talking about recording Kelly Jones for The Stereophonics' debut album: "If we used the small stone room at Real World Studios, all the vocals had a really rich enhanced sound from the room… All the best singing went down in this particular room, because of its sound." This is also part of the reason why so many people like singing in the shower or in the front seat of their car, where there's a hard acoustic reflector right in front of them.

But it's not just tonal enhancement that the reflections offer, because they can also help blend the vocal into the mix. "When you're recording a vocal, you have to think about how you want the vocalist's voice to sit in the track," says Paul Epworth.[32] "If you're doing a rock tune and you've got the singer right up against the microphone, you're going to have to work really hard with EQ to try and make it sound like it sits in the track. Something I learned from John Leckie: You record the vocalist where you think you want to position them in the piece of music, and with a rock singer it's better to record them an arm's length away so that it seems to sit on that scale in the track."

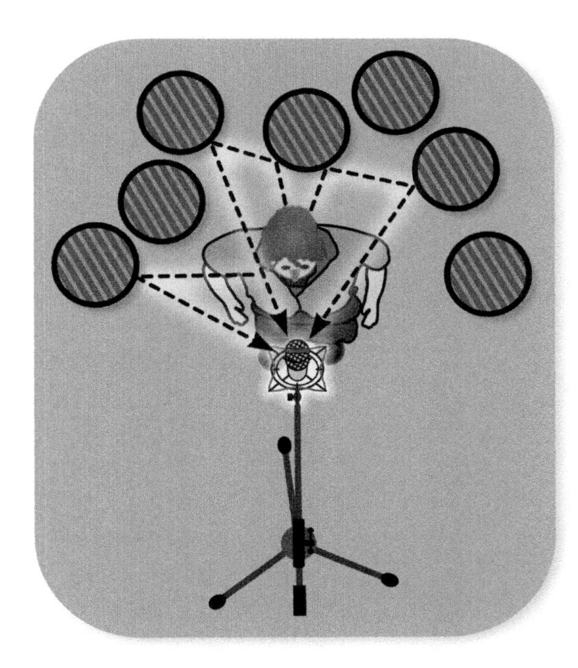

FIGURE 5.5
A group of cylindrical reflectors behind the singer adds a complex pattern of early-onset reflections to the recorded signal. This can provide a subjectively enhanced sound that doesn't change as much when the performer moves. Just a few of the potential reflection paths are illustrated here.

If you'd like to pursue these ideas, however, let me offer a couple of hints. Firstly, a spread of reflections is what you're looking for, not just a couple of isolated ones that'll cause obvious comb-filtering problems, so you're unlikely to get a good result just by dragging a spare plank into an otherwise padded environment. Much better to hunt around for a room that has a more lively sound, and then experiment a little with positioning the singer and mic within that. By the same token, though, you rarely want an obvious reverb "tail," so you may still want to soak up some of the more indirect reflections, especially in larger rooms—this is where Bruce Swedien's setup is so canny, because cylindrical reflectors placed behind the singer within a larger studio room will provide reflections which primarily arrive at the mic within about 10 ms of the dry sound. Room reflections can help thicken up layered vocals too if you get the singer to take a step or two back from the mic for each different take—a well-worn studio trick that Swedien[33] used for Michael Jackson's "Rock With You," for instance.

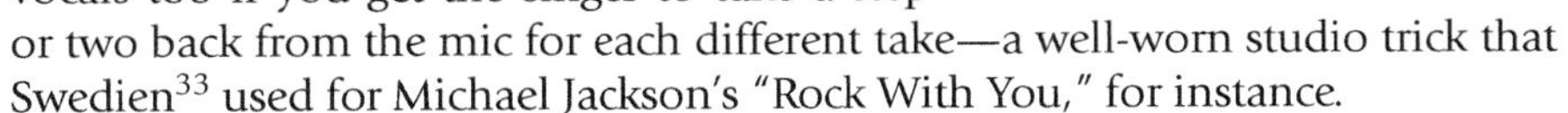

5.3 LOUDSPEAKER MONITORING

Most working engineers will tell you that some singers hate recording with headphones on, but even those who aren't allergic to cans will often give a more emotionally uninhibited performance when monitoring via loudspeakers instead. "Many times I've recorded singers using stage monitors instead of headphones," says Joe Chiccarelli,[34] "and I've even had vocalists sing in the control room while listening to studio monitor speakers. It's better to find an alternative solution than sacrifice what might have been a great performance." Clif Norrell[35] also asserts that speaker monitoring helps singers with their pitching. It's a crying shame, therefore, that this approach is almost universally ignored in small studios, presumably because of anxiety over foldback spill. There's actually a lot you can do to reduce foldback spill to manageable levels, and besides (to quote Eddie Kramer[36]), "if you get a great performance, who cares if there's a little leakage?" The only golden rule is to make sure you don't put anything into the foldback mix that won't appear in the final mix—so you can't have a metronome click in there to help with timing, for instance.

"Some singers hate recording with headphones on, but even those who aren't allergic to cans will often give a more emotionally uninhibited performance when monitoring via loudspeakers instead."

The various methods of implementing loudspeaker foldback are differentiated primarily by how they tackle the foldback spill. Whichever technique you use, it's best to stick to a No Input monitoring mode if you can, so that feedback into the vocal mic doesn't end up comb-filtering the recorded vocal timbre at all. As such, the balance between the backing mix and the vocalist will effectively be set by the volume of the loudspeakers. Where the singer insists on higher-volume monitoring for the general vibe, and this begins to make it hard for them to hear themselves, then try reflecting more of their vocal sound back at them acoustically in the first instance, as discussed in Section 4.5.1.

5.3.1 Speakers in the Mic Null

"Using a fairly directional mic prevents much spill in the first place," says Steve Levine,[37] and this plays into your hands if you're already using such a mic to keep room reflections down. In this respect, though, figure-eight mics will frequently out-perform cardioids, because:

- the cardioid's single null makes it less good at rejecting sound from a stereo pair of speakers;
- figure-eight nulls are, by their very nature, significantly more effective across the whole frequency range;
- the tonality of foldback spill arriving from the back of a cardioid is usually rather nasty, not least because the proximity effect of dual-diaphragm condenser designs is always bidirectional (so it remains strong at the rear even when you switch to a cardioid polar pattern). This is particularly the case with cheaper microphones;
- the figure-eight mic may enable you to place the speakers closer to the singer—alongside the microphone as opposed to behind it (see Figure 5.6).

That said, loudspeaker monitoring seems to be most commonly used in conjunction with dynamic mics, which can't easily be designed to offer a figure-eight polar pattern. Al Stone[38] describes a typical setup here: "We used to do vocals in the control room with a handheld dynamic with the monitors up as loud as you could get without feedback. Then I'd stand right back and [Jamiroquai lead singer Jay Kay would] just perform in front of the desk." Fortunately, however, dynamic models happen to be a little less sensitive to off-axis spill than condensers on the whole, so the rejection should still be pretty good, especially given that on-stage vocal mics deliberately recess their low-end response so that you can sing closer to them without sounding too bassy.

5.3.2 Polarity Inversion Techniques

If your singer can be persuaded to use a mono foldback mix ("Hey, it was good enough for the Beatles…"), one way to reduce spill is to feed that to two speakers equidistant from the microphone and then invert the polarity of the signal feeding one of them. What this means is that as the drivers in one speaker cabinet are moving outwards, the drivers of the other will be moving inwards,

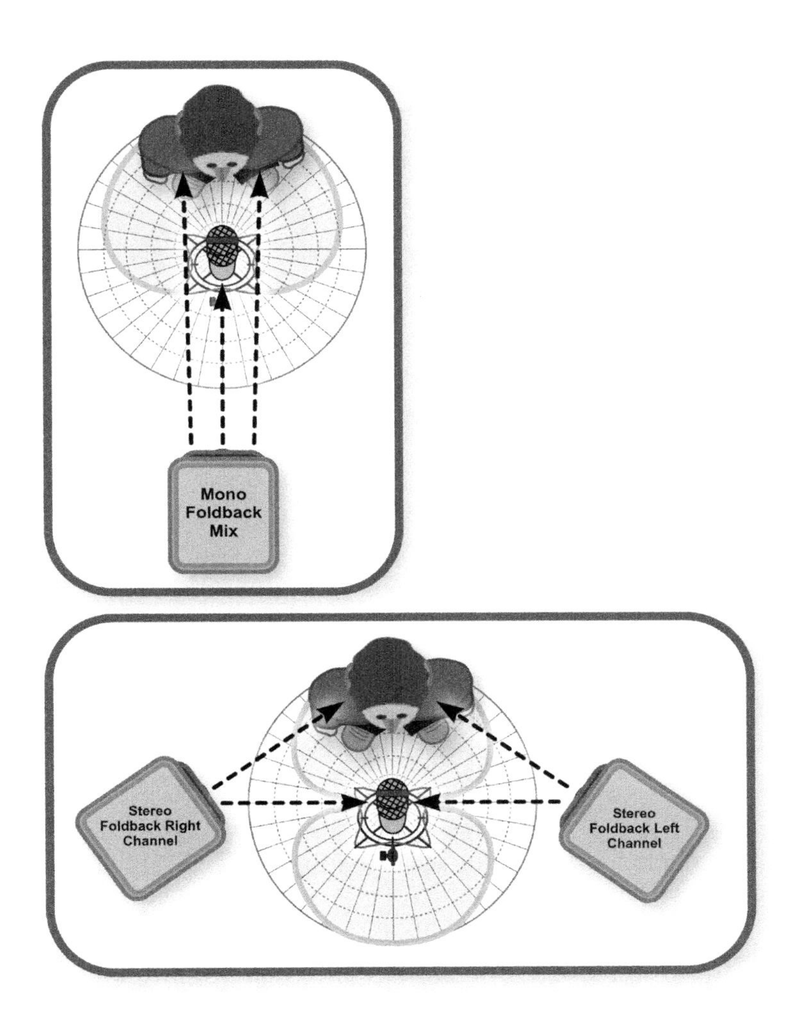

FIGURE 5.6
Placing foldback monitoring loudspeakers to take advantage of the rejection nulls of cardioid (top) and figure-eight (bottom) polar patterns.

creating opposite pressure waves which cancel each other out at the mic position. Here's Jay Messina[39] talking about this approach: "A trick I learned from doing jingles was to put the speakers out of phase while the mic was at the cancellation point of those two speakers, and this allowed Steven [Tyler of Aerosmith] to sing without using headphones and, therefore, with a little more energy and the vibe of being with the band." The single-sided polarity switching used to be easy when most people had passive speakers driven by a separate amplifier, because you could simply swap over the two speaker wires at the back of one of the speakers. Now that so many small-studio monitors have built-in amplification you can't do this, so you'll probably need to invert the polarity of one channel within your recording system. Alternatively, if your speaker connections are balanced, you might use a polarity-inverting XLR adaptor or one of the cables shown in Figure 5.7 instead.

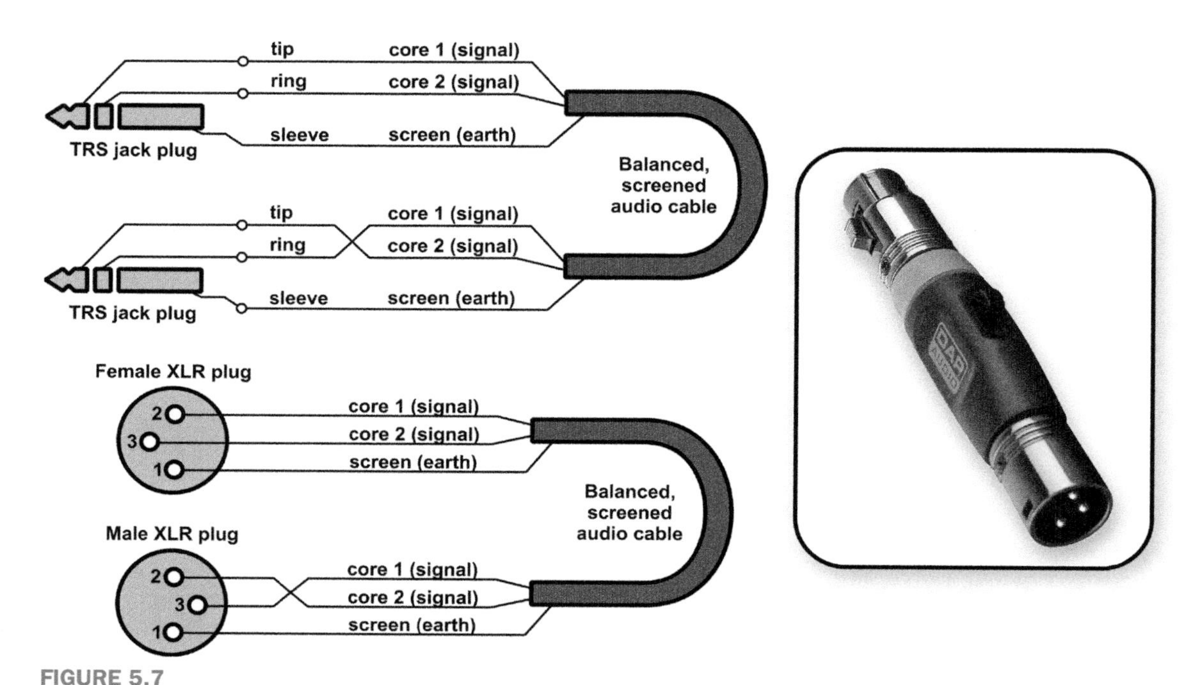

FIGURE 5.7
These two wiring diagrams (left) illustrate how to construct a polarity-inverting balanced audio cable using either TRS or XLR connectors. Make sure to label the cables clearly so they don't get confused with normal ones! An alternative means of inverting the polarity of a signal coming out of an XLR cable is to use an off-the-shelf polarity-inverting XLR adaptor such as the DAP Audio XGA37 (right).

While this approach might seem sensible in theory, to be honest I've never been very fond of it in practice. For a start, only the direct sound from each speaker achieves any significant cancellation, so reflected sound will still be picked up, and the cancellation you do get will mostly be at lower frequencies because it's impossible to match the speaker distances well enough to align shorter high-frequency wavelengths. In addition, the inverted-polarity mono listening experience is rather unsettling, and generally makes you feel like your brain's being sucked out your ear! So I wouldn't recommend this technique as a first port of call for small-studio users.

Besides, there's a much more effective means of using polarity inversion to reduce spill, which also allows a natural-sounding stereo foldback mix. The trick is to make sure you keep the speakers and the mic stationary while recording, and then, once you've recorded all your takes for each section, record one more where the singer simply stands at the mic without making a noise. The waveform of this "cancellation pass" can be polarity-inverted within your recording system and mixed with any of the recorded takes to neutralize their foldback spill. This gives extremely good spill reduction, and can often reduce it to a level you'd associate more with open-backed headphones than loudspeakers. The downside, though, is that any movement of the microphone during takes will reduce the effectiveness of the spill cancellation, so it's clearly not appropriate for hand-held dynamics, and even with stand-mounted mics you should make sure the stand is well tightened to avoid drooping and that you've warned the singer not to touch it. This technique also won't function correctly if you're compressing the recording chain, because the compressor won't react the same way during the cancellation pass.

POLARITY INVERSION OF FOLDBACK SPILL AT MIXDOWN

With any recording that has significant foldback spill, whether from headphones or speakers, the track's signal polarity may have some effect on how this spill interacts with your mixdown balance. So at the very least you should check the overall mix sound with different polarity settings to make sure you've got the best one, or else investigate variable phase-adjustment—something I'll explore properly in Section 7.2.1.

5.3.3 Tackling Loudspeaker Reflections

Whether you use polarity-inversion techniques or not, you'll get less foldback spill if you can minimize sound reflecting from a speaker into the sensitive regions of your vocal mic's polar pattern. This is where placing the speakers closer to the microphone can pay dividends, because the speaker volume then doesn't need to be as high for the same apparent monitoring loudness from the singer's standpoint. Additional acoustic absorption may also be sensible.

5.3.4 Filtering Fixes

High-pass filtering the foldback mix can help avoid problems with a cardioid mic's rear proximity effect, while removing the top octave of the foldback mix with low-pass filtering can significantly improve spill-reduction when using polarity-inversion techniques, without substantially affecting the singer's listening experience—and of course monitor mixes with less high end will also fatigue the ear less at higher playback volumes. Where you're using a directional mic, high-pass filtering the recorded signal also lets you position the singer closer to the microphone (for less spill) without overblown proximity-effect bass boost, especially if you've decided to use a figure-eight polar pattern.

CUT TO THE CHASE

- Omni polar patterns normally give a more natural sound, they're almost immune to proximity effect and handling noises, and they won't punish a singer's movements with big tonal changes. Concerns about capturing excess room reverb are frequently unfounded, because DIY acoustic treatment can be quite effective at vocal frequencies, and omnis can be used closer to the singer with fewer sonic problems anyway.
- Figure-eight polar patterns may cause greater difficulties with mobile singers, because of their narrower frontal sensitivity and strong proximity effect, and they're also more prone to plosive "popping" and picking up physical vibrations than cardioids.

- Varying the polar pattern of a dual-diaphragm mic gives you extra tonal control without additional processing.
- Where condenser mics struggle to deal with high signal levels, condensation (from breath), or spit, dynamic mics may provide a more rugged alternative. However, the main reasons producers use dynamic mics in the studio is for a different sonic character and/or so that the musician can perform as on stage.
- If your vocalist is using a handheld mic, make sure their grip doesn't defeat the mic's polar pattern, and be wary of excess sibilance.
- Ribbon mics need to be handled with care and can initially appear dull-sounding, but they nonetheless have a smooth sonic flavor that can be very useful for vocals on occasion. Small-diaphragm condenser mics, on the other hand, are a rarity on professional vocal sessions.
- Acoustic reflections can enhance your vocal tone if carefully managed: Strong discrete reflections and excessive indirect reflections can both cause problems.
- The performance advantages of loudspeaker monitoring are frequently worth the cost in terms of captured foldback spill. To minimize the spill, make good use of the mic's polar pattern; try polarity-inversion tricks; minimize speaker reflections reaching the mic; and consider filtering the foldback mix or the recording signal.

Assignment 1: Alternative Vocal-Miking Techniques

Repeat Chapter 4's first assignment, but this time:

- try to locate a dual-diaphragm microphone and make a point of experimenting with the tonal possibilities of different polar patterns;
- shoot out a few different mics for your speech, preferably including at least one dynamic, one ribbon, and one small-diaphragm condenser, in addition to your favorite large-diaphragm condenser;
- deliberately try to introduce some "single-bounce" room reflections into the recording, so you can hear the kind of sonic impact they make.

Assignment 2: Loudspeaker Foldback Monitoring

Repeat Chapter 4's second assignment, but this time give the vocalist loudspeaker monitoring in the first instance. Use mic polar patterns, polarity-inversion tricks, acoustic treatment, and filtering to minimize foldback spill. However, make sure you have a fully functional headphone foldback system waiting in the wings in case your loudspeaker monitoring proves unworkable or turns out to be detrimental to the performance.

WEB RESOURCES

On this book's companion website you'll find a selection of resources to support this chapter, including:

- further reading on different microphone designs and the cancellation effects of polarity inversion;
- links to the SE Reflexion Filter and Cloud Microphones' Cloudlifter;
- audio demonstrations of different microphone polar patterns and design types, and the relative effectiveness of different techniques for reducing spill when using loudspeaker foldback monitoring.

http://www.cambridge-mt.com/rs-ch5.htm

CHAPTER 6

Single-Mic Instrument Recording

The time has now come to expand your miking repertoire beyond vocals to encompass solo instruments as well, albeit still using a single mic. So for this chapter's assignments you'll be free to record pretty much any performer you like, with the exception of those who play several different instruments at the same time—things like drum kits and singing guitarists both present additional complications which we'll tackle in later chapters.

6.1 BEFORE YOU REACH FOR A MIC

When miking any instrument, the single best piece of advice is this: Use your ears first. "I always tell guys to go out in the room and listen," says Al Schmitt.[1] "Listen to what the instrument sounds like. Then go inside [the control room] and try to duplicate it. That's what your job is." Bruce Swedien[2] stresses the point, illustrating how this essential studio truth has been handed down through generations of professionals: "I remember [Bill Putnam] saying 'Don't just sit down here in the control room. Go see what it sounds like in the studio and listen to the music.'" Now if this seems like I'm illuminating the blindingly obvious, then great—it means your instincts are already pointing you in the right direction. However, my own experience of hundreds of small-studio sessions suggests that the enlightened are also the minority, so indulge me while I scratch the surface of this subject a little deeper.

6.1.1 Guessing a Mic Position

Beyond checking the instrument's tuning, the first thing you're listening for is roughly where you're going to put the mic. This mic position will inevitably be a guess, and will usually be revised somewhat once you hear the result through your speakers, but the *process* of finding that position will furnish you with a ton of essential information about the instrument and the recording room, and will make it a heck of a lot easier to get the recorded sound you need.

So you just stroll around the room until you find a good-sounding spot, and then put the mic there, right? Unfortunately, this seductive notion actually leads people to inappropriate miking positions more often than not, because of the nature of the human ear's directionality. While the specifics of this are extremely complex, in practice the polar pattern of your hearing system is broadly similar to that of a subcardioid mic, which means that common omni, cardioid, and figure-eight polar patterns will pick up something substantially different to what you're hearing in any good-sounding head location—for example an omni will sound too wet (i.e., it'll pick up too much reflected sound), while a cardioid or figure-eight will typically sound too dry.

"Common omni, cardioid, and figure-eight polar patterns will pick up something substantially different to what you're hearing—for example, an omni will sound too wet, while a cardioid or figure-eight will typically sound too dry."

One solution to this is to compensate mentally: Supercardioid or hypercardioid mics will want to be roughly 50% more distant than your favored listening position to achieve a similar dry/wet balance, whereas cardioid or figure-eight mics need only be about 25% further away. Omnis, on the other hand, can usually afford to be about 25% closer in. The snag there, though, is that miking distance also affects other attributes of the instrument's timbre (not least because of the comb-filtering and room-resonance issues I mentioned in Section 4.4), so the rest of the sound may not be nearly as good in the mentally adjusted position. One way to work around this involves listening in two stages:

1. Find the head position which gives the most appropriate dry/wet balance, without getting too hung up on the quality of the rest of the sound.
2. Estimate what the actual miking distance should be (depending on the polar pattern of the mic you're planning to use), move your head to this distance, and scout around there until you find a position where the instrument sounds its best—while trying to ignore the fact that it'll sound too dry/wet at that location.

Where you're using a cardioid mic, though, there's a trick that can provide a shortcut here: Arrange your hands vertically just behind your ears, perpendicular to your line of sight, thereby blocking some of the sound from behind you and making your hearing a bit more cardioid. You'll look a bit of a twerp, but it does seem to work quite well in terms of allowing you to make a reasonable judgment of both the dry/wet balance and the general timbre simultaneously—and it translates pretty well for figure-eight mics too. However, be careful not to cup your hands or bend your pinnae (the anatomical term for the outer ear flaps) forward when pulling this stunt, because that usually has fairly drastic side-effects for the frequency response you're hearing.

If you have a choice of polar patterns available, this obviously gives you a bit more flexibility in terms of where you place your mic, although the amount of room ambience each pattern captures won't be the only factor to consider, as we've already touched on in Section 5.1.1. Proximity effect will have a big impact for close-miking positions, of course, but when you're miking from more than about a meter away you'll also typically find that an omni mic will give the most extended and natural-sounding low-end response, and that as you shift from omni through cardioid to figure-eight the sound becomes progressively thinner and less realistic at low frequencies. This is one of the big reasons why so many engineers working on classical/acoustic music favor omni microphones for more distant placements. "I like to start with an omni before anything," says Telarc's Michael Bishop,[3] for instance.

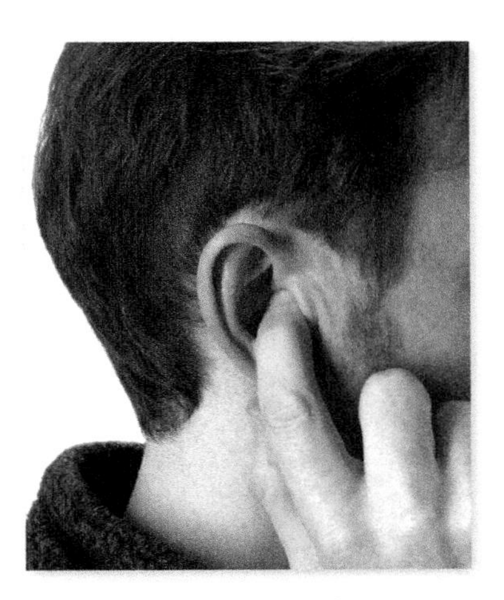

FIGURE 6.1
Although some engineers like to plug one ear while listening for mic positions, the Occlusion Effect may undermine the usefulness of this approach for you in practice.

Some people suggest putting your finger in one ear so that you're hearing in mono (as a single mic does), but I've never had much luck with this myself. The problem is that blocking the ear canal causes something called the Occlusion Effect, which makes any vibration reaching the ear canal by other means much more audible, as well as boosting that ear's perceived low-end response by around 20 dB. Even if you stand motionless to avoid your muscles or breathing apparatus transferring vibrations to the blocked ear via bone conduction, you'll still hear the low-register rushing noise of your own blood flow, and probably the high-pitched "hissing" of nerve activity too. For me, these noises have always proved too much of a distraction when trying to concentrate on the sounds outside my own head! Pressing or cupping a hand over one pinna is unlikely to help either, because some sound will unavoidably leak into the ear canal, and it'll be so tonally colored that it'll skew your perception of what you're hearing with the free ear. So I'd personally stick with the natural two-eared approach if I were you. (What will often help, though, is closing your eyes, because this focuses your attention more on what you're hearing, and makes you less likely to be biased by visual input, as we already discussed in Section 2.3.2.)

If the instrument and recording space happen to be well suited to your needs already, then this kind of listening approach can deliver a pretty good starting mic position in a matter of seconds. It's always a buzz to strike it lucky like this, but every small-studio engineer must face the fact that the job won't be nearly that easy 90% of the time. What usually happens is that you get your head at a distance where you suspect your planned mic polar pattern will pick up a reasonable dry/wet balance, but you simply can't find a sound you're happy with from any angle—either the different elements of the instrument's sound aren't balanced (you get too much breath, bow, or stick noise, say) or the overall subjective timbre of the instrument just isn't as appealing as you'd like.

In this situation, you may be able to improve an unpromising live-room sound with some odd-ball mic technique or by applying signal-processing in the recording chain, but why fight an uphill struggle? "If a recording starts off bad, every other step after that is struggling to make it up," asserts Stuart Sullivan.[4] "I'm not going to say that if it starts bad there's no way anything can get better—that's ridiculous—but with anything you lose at the beginning, you can't get as much back and it gets harder and harder to add to it." "It's not about *getting* a great sound," adds Trina Shoemaker,[5] "it's about capturing a great sound that's already there." It's usually far quicker and easier to address sonic problems before the technology gets involved, and working that way also means you spend more of your time in the live room (where the magic's really happening!) building valuable relationships with the musicians. "I do not like being in the control room during recording sessions," remarks Elliot Mazer,[6] for instance. "I like being with the [musicians]. It gives me a sense of who they are, a sense of what they need in order to get a good performance."

DON'T FORGET THE HEIGHT

I can't tell you how often I've seen amateur (and not so amateur!) engineers walking all round their recording room, intently looking for a good-sounding spot—but without substantially changing the height of their head! This is particularly daft in small studios, where the floor and ceiling are the closest room boundaries for the majority of listening locations, and are therefore most likely to influence the timbre acoustically when you move your head in relation to them. So don't be shy: Crouch down, stand on a stepladder, form a human pyramid... "I'll always get up on a ladder and get up in the air to listen for where the sweet spots are," says Michael Bishop,[7] for instance.

FIGURE 6.2
With a stepladder on hand you can listen to a lot more potential miking positions.

6.1.2 Your Friend, The Musician

If you're not hearing a good sound in the recording room, first ask the player to help you improve it. "I always tell engineers," says Al Schmitt,[8] "your best friend in the studio is the musician. He wants to sound good, you want him to sound good, you're a team." Joe Zook[9] is just one of many who echo Schmitt's view: "Recording instruments is a team sport. Nine times out of ten a great acoustic guitar sound happens when the player makes subtle adjustments...It's all in the hands of the player." This is something I already mentioned with regard to electronic instruments back in Section 2.2.4, but it's even more pertinent where acoustic sources are concerned. Unless you happen to be The Artist Formerly Known As Squiggle, most of the performers you record are likely to be more accomplished instrumentalists than you are, so it's borderline criminal not to take advantage of their experience while you're still at the stage in the production process where the sound can be changed most radically. Besides, a lot of players will feel flattered by your inquiries, and will like nothing more than showing off their arcane instrument knowledge and technical expertise.

Sadly, a lot of musicians these days have become rather blasé about crafting sounds properly at source, because they feel that anything can be fixed in the mix. Yes, it's true that a lot can be done to salvage poor multitrack recordings (as I've demonstrated in *Sound On Sound* magazine's Mix Rescue column), but what I can also tell you categorically is that trying to "fix it in the mix" is extraordinarily inefficient—in most of those Mix Rescue projects a few more hours well spent during tracking would have saved me days of tedious remedial work! To quote Trevor Horn:[10] "The mix is the worst time to do anything." Although accommodating and adapting to the needs of the musician is a recurring theme throughout this book, this is one issue where I urge you to stand firm: Don't let musicians sit on the fence about their sound! "The stupidest thing that any musician can do," asserts Tony Platt,[11] "is to just plug in and play and say 'make that sound good.' I will always say to the guitar player, for instance, 'Is that sound coming out of your amplifier the sound you want to hear? If it isn't, show me what it is and we'll try to get somewhere close to that before we even put a microphone on.' It's a waste of everybody's time to sit there tweaking stuff until somebody says, 'Oh that's good.'"

6.1.3 Tonal Adjustment at Source

Even the most apathetic player will usually come up with a few ideas for changing the sound when prodded. (If not, consider a cattle prod.) However, there are some tactics that are so regularly useful that any engineer should always keep them at the front of their mind (in addition to those discussed back in Section 3.4):

- **Change the Instrument.** This may seem obvious, but the most difficult thing to work around in the studio is the wrong choice of instrument. This is particularly relevant for anything that relies on a resonant wooden body, because very small variations in construction can give rise to surprisingly large sonic differences. Orchestral string instruments, harps, mandolins, and acoustic guitars all fall squarely into this category, for instance, which is why most professional producers lay great emphasis on finding the right tool for the job. "I once asked Daniel Lanois for advice on how to get a good acoustic guitar sound," recalls Pierre Marchand.[12] "His answer was: Get a good-sounding acoustic guitar." Different instruments will often suit different arrangement roles too, as Bob Bullock[13] notes here in relation to acoustic guitars: "The choice of instrument is just as critical as the microphone. A big Gibson J200 gives you a very full, rich sound, which is great for padding the track. Taylor guitars offer a sharper, edgier sound, which is good for licks and solos." But complex resonant characteristics are also inherent to piano soundboards, drum shells, and guitar speaker cabinets, and there again the choice of instrument is frequently paramount—so much so, in fact, that one of the best arguments for a small-studio owner to shell out for commercial studio time is if it gives them access to an extensive collection of instruments.

- **Change the Playing Implement.** Another thing that's often straightforward to change is the implement used to play the instrument. We already mentioned switching guitar picks in Chapter 3, and the design of the mallets/sticks used to hit percussion instruments affects their recorded sound just as much. The same goes for struck strings—even where there's only one set of hammers available, these can sometimes be hardened or softened, and tales abound of people sticking drawing pins into the hammers of an upright piano or sliding a sheet of material between its hammers and the strings.

BACKGROUND NOISE AND SYMPATHETIC RESONANCE

While you're listening in the recording room, keep your ears open for any unwanted noises that you'd rather not record. It's extremely easy to miss persistent repetitive background noises (lighting buzz, ticking clocks, traffic rumble, birdsong) while you're concentrating hard on the sound of the instrument, so make a conscious effort to switch your focus briefly at some point. One specific thing to listen for is anything that's vibrating in sympathy with the instrument you're recording, especially when tracking bass instruments. Unused snare drums are common culprits, their snare wires rattling along sporadically, but luckily these can be silenced in most cases using the "throw-off" lever on the side of the drum, or if necessary a sheet of kitchen towel between the wires and the drumskin. And, speaking of drums, listen out for the rattling of any loose hardware when recording those.

FIGURE 6.3
A scrap of gaffer tape can easily silence most drum hardware rattles—as in this picture, where the mounting lugs (circled) on a kick drum have been immobilized following the removal of the resonant drum head.

- **Change the Tuning.** Drums are frequently retuned on commercial sessions. Not only does tuning provide a wide range of control over the sound's tone and sustain characteristics, but it can also improve the instrument's ability to blend into the mix if pitched resonances are harmonically consonant. Guitars are also frequently used with alternate tunings, and even orchestral string instruments have some leeway in this regard, although I'd stick to fairly conservative downward retunings of any acoustic instrument if the player's in any doubt about whether changes in string tension may damage their instrument. Particular mention should be made of the so-called "Nashville" guitar tuning, where the bottom three or four strings play an octave higher than their conventional pitches. "It has this great, clear, open sound that's not just for Country," explains John Merchant.[14] "If you go back and listen to Keith Richards' guitar on 'Jumping Jack Flash,' that was accomplished using Nashville tuning."
- **Change the Pressure/Angle.** The pressure with which a player applies their fingers, bow, and/or lips to most melodic instruments is a crucial tonal variable, as is the angle of their breath or bowing action. Even the sound of a guitar pick will vary according to the pressure of the player's fingers on it and the angle it hits the strings, as mentioned in Section 3.4.1.
- **Change the Volume.** Almost all instruments will sound tonally different when played at different volumes, and acoustic instruments (just like vocals) often give fuller-sounding recordings when they're not played too loud. "The more quietly you play," says T-Bone Burnett[15] for example, "the less attack there is and the more tone there is. If you hit a guitar too hard, it chokes the note off." (Justin Niebank[16] says almost exactly the same in relation to drums.) Smokier timbres in particular are difficult to get from wind/brass instruments unless you can keep their notes quiet enough so that low-level noise components come through clearly. That said, overdriven electric guitar amps may simply not sound right for rock music unless they're rattling the furniture. Even then, however, smaller amps don't necessarily mean a diminutive sound on record. "At high volume, smaller amplifiers give a bigger sound than big amps most of the time," affirms Chris Lord-Alge,[17] for instance. "If you go back to Duane Allman and Eric Clapton on 'Layla,' they used little amplifiers like Fender Champs, and it sounds pretty damn large."

> "Almost all instruments will sound tonally different when played at different volumes, and acoustic instruments (just like vocals) often give fuller-sounding recordings when they're not played too loud."

- **Change the Damping.** One of the defining sonic attributes of instruments that are struck or plucked is the duration of each note's decay, and there is usually a way to shorten this if you wish by applying some kind of additional damping. With things like harps, guitars, cymbals, and gongs, the player is often able to damp them with their hands as part of their playing technique, but other instruments (most notably drums—see the "Engineer's Quickstart Guide: Drums" box) are less easy for the musician to control in this way, so some additional intervention may be useful.

ENGINEER'S QUICKSTART GUIDE: DRUMS

There's not much you can do to change the basic sound of a cymbal (other than choosing a different one!), but there's masses of stuff that will alter the tone of a drum. The choice of drum head is a crucial tonal component, and it's usually best to put fresh heads onto a drum before important recording dates if you can, and to have a couple of spare heads lying around if your drummer is a particularly heavy hitter. Whatever playing implement is used, the exact point of contact makes a big difference to the timbre, so if consistency of tone is important to a part you're recording (for example, a backbeat), then alerting the drummer to the importance of careful aim can make life much easier at mixdown. If a kick-drum pedal or drum stool squeaks undesirably, you can usually silence it with a quick squirt of light oil.

Tuning drums affects not only their pitch, but also their tone and sustain, so the process is something of a black art, as well as a hot topic of debate amongst performers. Unless you're a drummer yourself, I'd entrust the musician you're recording with tuning decisions. (If you want to do a quick confidence check for yourself, press down gently on the center of the drum head with one finger to damp it, and then tap the drum skin lightly a couple of inches from each lug—normally the tuning should be the same all the way round.) That doesn't mean you shouldn't bring up the subject, though, because although most drums will have a natural shell resonance which determines the range of useful tunings, there'll still be plenty of scope to alter their attack, tone, and sustain within that range.

Another contentious issue is whether, and how, to damp a drum's vibration in the studio. On the one hand, much of the power and harmonic complexity of any drum comes from its complex resonant characteristics, so some players resist using damping on the basis that it robs the sound of its richness and personality. On the other hand, there are lots of situations in music production where you may want to reduce the overall decay tail of a drum, or scotch an overprominent pitched resonance, and applying damping materials is a quick and effective means of achieving that. My advice is to be upfront about discussing the sound with the musician, and let them take the lead when it comes to choosing damping tactics. If they can reshape the sustain characteristics suitably with careful retuning, or by changing drum heads, then great; if not, then you can explore damping possibilities together instead.

FIGURE 6.4
A strip of gaffer tape, folded as shown to leave a small flap poking up into the air, provides a usefully moderate degree of damping—and you can usually get away with using two of those, if required, before you start obviously dulling the instrument's overall tone.

My favorite damping method is to take a six-inch strip of gaffer tape, fold it into a "T" shape creating a one-inch flap in the middle, and then stick that somewhere on the drum skin. This produces a usefully moderate damping effect, and you can usually put one or two of these onto a drum without murdering its high-frequency tone. Bear in mind that the exact positioning of the damping will determine which of the drum's overtones are most heavily targeted, so don't be afraid to pick the gaffer up by its flap and reposition it a few times if you don't get a promising result at first. As a rough guide, damping positions at about 1/4 and 3/4 of the drum skin's diameter will tend to produce the biggest changes, so you can make the effect more subtle if you need by pulling the gaffer more toward the edge. Wherever you put the tape, though, make sure it's not going to be hit, because it'll affect the tone and feel of the drum's attack, as well as making an ungodly mess once the player wears a hole through it. One of the great things about using gaffer tape is that you can damp the underside of single-skinned drums if you need more positioning flexibility, and this is particularly relevant to hand drums of any kind, where the whole drum skin is frequently used for different playing techniques. Where more overtly damped sounds are required, you can add mass to a gaffer patch using a coin or two, or tape down one end of a small bundle of tissue paper to increase air drag.

Gaffer tape isn't the only damping option you'll see in studios, though. Little squishy blue pads called Moongel are popular with a lot of drummers, and they have the advantage of being reusable, although they won't hold to the underside of a drum skin like gaffer can. A cigarette packet or wallet on the snare isn't uncommon either. "[Ringo Starr] always used to have a half-empty soft packet of Lark cigarettes on the top of the snare, which gave it a certain sound," recalls Geoff Emerick,[19] for instance. Tea towels or sheets have even been put over drums on many commercial records, and not just where super-dry disco beats were the aim. Emerick did this with the Beatles too, and here's Andy Smith[20] talking about Paul Simon's album *So Beautiful So What*: "There are several tracks that have a standard drum kit, but Paul usually wanted them to sound a bit different. On many of the tracks...the drummer placed towels over each drum so they'd have more of a muffled quality, leaving more room for the higher-frequency percussion stuff."

Kick drums deserve a special mention as far as damping is concerned. If the instrument's resonant head (the one the drummer isn't hitting) is completely intact, you'll generally get more sustain, whereas using a resonant head with a hole in it, or removing that head completely, alters the sound's decay characteristics. Many drummers will have a towel or small cushion of some type nestling in the bottom of the drum to stop it ringing on too long, but home-studio engineers do sometimes get a bit carried away, packing the thing so full that they end up with a featureless thud which is easily lost in the mix. This makes the drum respond unnaturally for the player into the bargain, which is arguably the greater sin. A folded towel (anchored into position with some kind of weight) is usually fine, and you can moderate the nature of the damping by sliding it so that different amounts of fabric come into contact with each of the two heads. If you feel you need more damping than that, then you'll probably be better served by removing the resonant head in the first instance, rather than just stuffing in your whole laundry basket.

FIGURE 6.5
Here you can see that the snare has both a piece of Moongel and a small flap of gaffer tape applied to it for damping purposes. Any more damping than that, and you start running the risk of destroying the drum's high-frequency tone.

- **Other Tonal Controls.** Many instruments will sound very different when using a mute, and you may even be able to choose between several different designs, each with its own unique timbre. The position of a player's hands can have a big influence too, especially with French horn and harmonica. A guitar amp's built-in equalizer controls are fair game as well ("Let's avoid EQ later, let's use the EQ on your amp," says Bob Weston[18]), although it's poor etiquette to start mucking around with those without asking the guitarist first.
- **Specialty Performance Techniques.** Every instrument has a few non-standard performance techniques that are fairly common currency—things like flute flutter tongue, violin *col legno*, and guitar pinched harmonics. Performers will rarely use them instinctively, though, especially when improvising, so it's sometimes worth asking a musician directly about what specialty tricks might be on the menu.

6.1.4 The Role of the Room

Recording isn't just about the instrument and the mic. The room is equally important, and affects the sound of both, even when using very close mic positions. "The room the amp is in still has a huge effect on the tight-mic sound," says Daniel Lanois[21] with regard to recording electric guitar. "I believe that the amplifier responds to the room, and the two become one." Comb-filtering may detract from the sound if your instrument is particularly close to a hard flat surface, in which case acoustically dampening that (as discussed in Section 4.4) may reduce the unwanted tonal coloration. Cymbals in low-ceilinged rooms are common offenders here, so a couple of square meters of acoustic foam on the ceiling above them can work wonders. You don't need to fix it permanently—just taping it between a couple of spare mic stands will do the job.

FIGURE 6.6 A sheet of acoustic foam slung across the top of an overhead cymbal mic can help absorb reflections from a low ceiling, thereby reducing comb-filtering artifacts in the recorded sound.

It's not always easy to absorb your way out of comb-filtering problems, though, so working engineers will frequently transplant an instrument to different positions in a room (or a different room entirely), where the reflections aren't as strong or where different path-lengths for the reflected sounds transform the frequency-response undulations of comb-filtering from a curse into a blessing. Few sources benefit more from this kind of experimentation than small electric-guitar cabinets, which rarely sound their best when left standing on the floor. This is one of the reasons I keep a couple of collapsible tilting amplifier stands in my own mic-stand bag, but a chair or table can also be usefully pressed into service here.

Comb-filtering isn't the only side-effect of strong reflections, though. Although low frequencies will happily just pass though window panes, doors, and lightweight stud walls, any boundary that's sturdy enough to reflect them will simply reinforce (rather than comb-filter) the low-end response of any instrument placed nearby—the so-called "boundary effect." (In room corners the bass boost becomes more pronounced because of the cumulative effect of two boundaries.) You can use the boundary effect to warm up instruments that are sounding thin (the nearer they are to the boundary, the bassier their sound gets), or move something like an upright piano away from the wall to avoid woolliness.

Room resonances are another crucial concern, which is why you'll see some engineers walking around an unknown recording room clapping (because this produces a giveaway metallic timbre when high-frequency flutter-echoes are present) and making a general racket (to gauge the character of the lower-frequency resonances and reflections more generally). "The place that you like the reflected sound is a good place to start," explains Steve Albini.[22] "A lot of studios are designed to have very little reflected energy and support from the room, and those can be very frustrating environments to record in...Non-professionally designed studios [can] be more flattering acoustically."

Even if you're Barry White incarnate, though, vocal frequencies aren't going to tell you much about resonances for low-end instruments, so an element of suck-it-and-see instrument repositioning will be par for the course with those. "I had a cellist in," recalls Tony Visconti,[23] for example, "and I moved her until I got a good sound. Once I put her in one particular corner, her cello just sang...It's not so much the instrument; the room is very much a part of the sound." Mike Stavrou[24] makes a point of experimenting with the location of low drums. "Have the drummer, or better still your assistant, drag the [drum] around the area while you keep listening as he keeps banging it. The more you listen the more you'll hear. You will find positions where it sounds much thicker, and also dramatically thinner." Bear in mind that it's usually a whole lot easier to move the instrument and player before you've set up mics, headphones, or any acoustic treatment, so do your best not to postpone questions of instrument placement beyond the initial listening stage, otherwise you'll never get round to it.

All the same acoustics issues will also influence the sound at your head position while you're listening in the recording room—and subsequently the sound picked up by a mic there too. However, comb-filtering effects won't be as apparent to your two ears as they'll be to the single microphone, so the

FIGURE 6.7
Using a tilting amplifier stand is a good way to adjust the nature of the comb-filtering caused by sound reflections from the floor. Another advantage is that it brings a standing player more on-axis to their floor-standing cabinet, which usually means they're hearing something closer to the recorded sound while performing.

main things to concentrate on while refining your head position will be the strength of any boundary effect (especially near the floor) and whether the low-frequency resonances at different spots in the room help or hinder your goals. In particular, pay close attention to the subjective balance of different pitches when recording bass-guitar cabinets or upright bass, because getting sufficient low-end evenness with these instruments is one of the biggest challenges of small-room engineering.

FLOOR RESONANCE

The resonance of a studio's floor can affect the low-end sound of any instrument physically coupled to it—especially things like kick drums, speaker cabinets, and upright basses. When first checking out unknown rooms, John Kurlander[25] stamps on the floor to test this. "[If] there's a little bit of give," he says, "the bass response will be a lot more generous than if it feels like a thick concrete slab." One way to combat this interaction is to construct some kind of riser, a tactic Jay Graydon,[26] Jack Joseph Puig,[27] Bruce Swedien,[28] and Shelly Yakus[29] have all mentioned in interview. "It will add clarity to the sound because the amp is no longer coupling to the floor and reinforcing the bass frequencies." explains Puig.[30]

6.2 HOMING IN ON A MIC POSITION

While careful listening in the recording room can find you a ballpark mic position at warp speed, that must always be considered a first guess, especially if you've been unable to investigate a loud instrument's close-miking possibilities by ear without risk of hearing fatigue/damage. The only reliable way to place a mic is to adjust your initial miking position while monitoring the results through your recording system—and you should normally expect to change it several times before reaching a really dependable sound.

6.2.1 Adjusting Your Initial Mic Position by Ear

The ideal studio configuration for comparing mic positions is where the performer is in a separate live room and you've got a willing assistant to move the mic around. That way you can direct proceedings via a sight-line or over talkback, quickly auditioning hundreds of different mic positions through your control-room loudspeakers as the mic's swept around. This is a magnificently swift method which is one of the biggest benefits of overdubbing in a traditional two-room studio environment—but because it's beyond the reach of most small-studio users, I'd like to talk about some alternative approaches too.

If all you're missing is an assistant, then ask the player to fill that role if possible. A lot of instrumentalists can easily move their performance position with relation to the mic, and once you've hunted down a good sound in that way, you can recreate that mic position by adjusting the mic stand relative to the performer's natural performance posture. Where the musician can't easily move the sound source, or you're restricted to a one-room setup, many writers on music-technology matters recommend putting on a pair of closed-back cans and then moving the microphone around for yourself. Eminently sensible as this seems in principle, however, I've always found it disappointingly hit-and-miss. Except with the very quietest instruments, a significant amount of the direct sound will always find its way past the baffling effect of the headphones to skew your perception of the input signal—and if you try turning up the headphones to fight this you'll quickly fall foul of loudness-related subjective biases and ear fatigue. Although the delayed monitor-mix dodge I described in Section 3.2.4 can mitigate some of these problems, that still won't protect you from the natural human inclination to favor louder sounds—keeping the input signal's level constant while you're busy waving the mic around is rarely feasible (unless you have prodigious toe control), so louder mic positions will almost always win out irrespective of whether their tone actually suits the arrangement.

Therefore, where time allows in a one-room situation I always prefer to refine my own mic positions via a series of test recordings, because that means I can make decisions without being misled by direct-sound contributions or subjective loudness differences. If you have two similar microphones, you can make this approach even more effective by routing them to separate recorder tracks and operating them as a kind of tag-team. Here's how:

- Overdub a test recording (against the backing track) with both mics in different positions, and then ask the performer to give you a minute or two to listen back.
- Balance the two tracks to similar subjective levels and compare their sonics.
- Once you've decided which mic signal works better in context, leave that mic where it is and move the other one to see if you can beat it.
- Repeat this process at least a couple of times, and preferably until both mic positions feel equally usable—at which point either just choose one of them for capturing your proper takes, or record both so you can decide between them at leisure after the session.

The big advantage of this tag-team placement technique is that you're always comparing how the mics respond to exactly the same source sound, whereas there will inevitably be performance variations between separate test-recording passes done with a single mic.

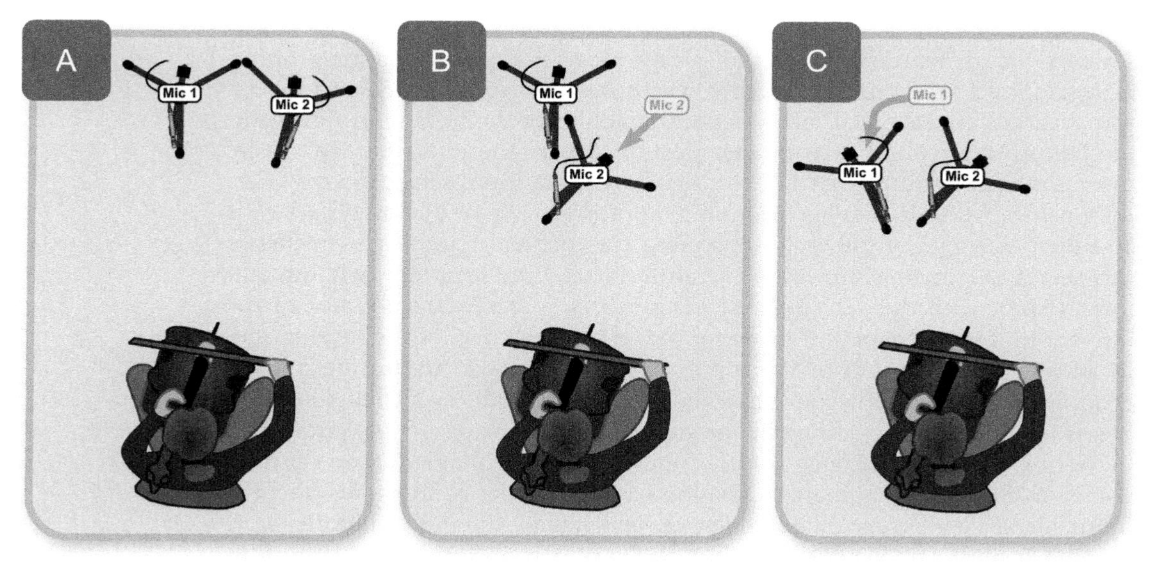

FIGURE 6.8
The tag-team miking method, using two identical mics on cello. Figure 6.8A shows the first two guesswork positions. Let's say that, on auditioning, Mic 2 doesn't sound as promising tonally as Mic 1, but Mic 1 sounds too wet. In response, we might move Mic 2 to address the instrument from the same angle as Mic 1 (for better tone), but from closer up (for a drier sound)—as shown in Figure 6.8B. That might then lead to Mic 2 sounding better than Mic 1 at the next listen, at which point we might choose to move Mic 1 to the same miking distance as Mic 2, but experiment with another lateral positioning, as shown in Figure 6.8C. The idea is to continue this process until each mic signal, while different, is equally appealing subjectively.

6.2.2 Improving Your Guesswork

If you follow the guidelines I've set out so far, finding a decent microphone position for any instrument should be mostly a question of time. However, time is frequently in short supply on real-world sessions, so the big challenge is to speed up this process—primarily by improving the quality of your guesswork, thereby cutting down the amount of time spent on trial-and-error tests.

It doesn't take much Googling to turn up dozens of miking diagrams for different instruments, and it's clear to me that small-studio engineers frequently take their cue from these. You'll find no such list here, though, simply because I don't think that kind of cheat sheet is much use in practice. Indeed, having trawled through thousands of low-budget productions, I'd say that over-reliance on miking "templates" probably causes greater sonic damage than almost any other factor—the very worst recordings I hear tend to be those where someone has clung to a recommended mic position as if it were gospel, despite copious contradictory evidence from their own ears.

"The very worst recordings I hear tend to be those where someone has clung to a recommended mic position as if it were gospel, despite copious contradictory evidence from their own ears."

There's also a misleading assumption underlying generic miking suggestions for any instrument: that there's some kind of "industry standard" sound for commercial recordings. I spend a lot of my time comparing commercial releases (not least for *Sound On Sound* magazine's Mix Review column), and I can tell you there is no such thing. The sound of the same instrument can change considerably between different musical styles, between different artists within a given musical style, and even between different songs by the same artist. Unless you can listen to an audio example of the miking setup described (which you rarely can), you really have no idea whether it might deliver a sound you like, or one which suits the style/arrangement of the production you're working on.

Another thing that really bugs me about per-instrument miking recipes is that they rarely give any useful idea of what to do next (beyond a vague "try moving the mic") if the prescribed positions don't immediately work. Given that the vast majority of initial mic positions won't be optimal (no matter how experienced you are), it's far more important to learn how to use mic *movements* in an intelligent manner to get closer to the sound you want, so you aren't just left floundering around more or less at random. There's also a tendency for published miking suggestions to be based on what top-flight performers use in large professional studios, but many of these rigs don't scale down at all well to small-studio environments where musicians aren't as accomplished, and where gear and acoustics are less forgiving.

THE "RECORDING SECRETS" LIBRARY OF MIC POSITIONS

You're the only person who can decide what sound you like, so no-one can choose a mic position on your behalf. Therefore, as part of this chapter's web resources I've set up a "library of mic positions" which uses audio examples to demonstrate the effects of different placements—use your own ears to judge which might be most promising for your productions! Unlike most other miking demonstrations you may have heard, I've done each set of examples by recording a single performance with multiple microphones of the same model, so that the differences you hear are solely attributable to mic technique.

Enough of the rant, though. Suffice to say that I'd like to follow a different path myself, explaining some of the underlying principles you can use to judge where to mike any instrument (whether or not you've encountered it before) and also how you might sensibly move that mic when it's not sounding up to scratch.

6.2.3 Piggy-Backing

There's no need to reinvent the wheel every time you put up a mic if you can piggy-back on the session experience of others. "If you're a new engineer," advises Al Schmitt,[31] "and you're not sure of anything, and a guy comes in with an instrument that you've never seen before, ask him where to put the mic. He'll say, 'Well, the last time the guy put it here and it was a good sound.' Now you've got a place to start!" This mic-placement tip has much more going for it than anything cribbed off the Internet, because it's related to that specific player and instrument, and may also be well suited to the style you're working on if that musician was chosen expressly for the project. Plus you can ask more questions about the specific context within which that mic position was used if you need.

If you find yourself working in a shared college studio, there's a lot to be gained by speaking to any other students who may be more experienced using it than you are—or, even better, the studio technicians who have frequently spent years observing student follies and working on their own projects during studio downtime. It's for similar reasons that high-profile engineers will frequently consult the resident staff in any unfamiliar studio environment, because they'll have had the opportunity to compare the outcome of dozens of sessions by different engineers and producers in the past. "It helps to talk with the engineers who have worked in the room if you're in it for the first time," says David Thoener,[32] for instance. "They can help with some of the best locations for [different instruments], and this will cut down the time you might use in experimenting."

6.2.4 High-Frequency Beams and Shadows

While talking about vocal miking in Section 4.3.1, I mentioned that high frequencies are more directional and more easily absorbed than low frequencies—information which is just as useful when miking instruments. Reed and brass instruments all create a prominent "beam" of high-frequencies in the direction of their open end(s), for example, and deciding how much of that beam hits your mic gives you powerful control over the tone, especially since the on-axis sound may not be the most desirable. "If you put the mic too close into the bell of the horn," says Shelly Yakus,[33] "the result may seem to be exciting-sounding when you are listening to the horn soloed on the speakers at a loud volume. But when you drop it back into the track, it is going to be this little farty sound."

Any other sound source which uses a horn-style contraption will usually act in much the same way: Bass amps sometimes have horn-coupled high-frequency drivers, for instance, while Leslie speakers use rotating horns to implement their characteristic modulation effect. Speaker cones also exhibit high-frequency directionality, and moving the microphone slightly off-axis is a common means of avoiding unpleasant edginess when close-miking over-driven electric guitar amps. With individual cymbals and gongs, the high end tends to be projected perpendicular to the plane of the instrument in both directions, although a hi-hat refocuses half this energy from its two cymbals through the narrow gap between them, producing an even stronger HF bias in the horizontal plane.

ENGINEER'S QUICKSTART GUIDE: WIND INSTRUMENTS

Let's state the obvious to start with: Players of wind instruments need breath! Even if you aren't a wind player yourself, as an engineer you can still help the musician improve their performance simply by discussing with them where they're going to breathe, and if necessary by using punch-ins and comping to build longer legato lines out of shorter breath-punctuated segments. Another thing to listen for is the player's use of "tonguing," which is when they articulate the starts of notes by briefly interrupting the flow of air through the mouthpiece with their tongue. This is an important aid to phrasing, and (as with string bowing) you can't assume that a performer will instinctively arrive at the most appropriate result straight away, especially if the part in question is still quite new to them.

Mic technique can't do much to downplay excessive mechanical noises from the keys/valves of woodwind instruments, because they emanate from right across the instrument, so if this is a problem you should try to bring it up with the performer, because they can sometimes remedy things by adjusting their playing technique. Finally, remember that wind instruments can be tiring to play, especially when high notes are involved, so try to plan your session workflow to minimize fatigue, much as you would when recording a vocalist (see Section 4.6.2).

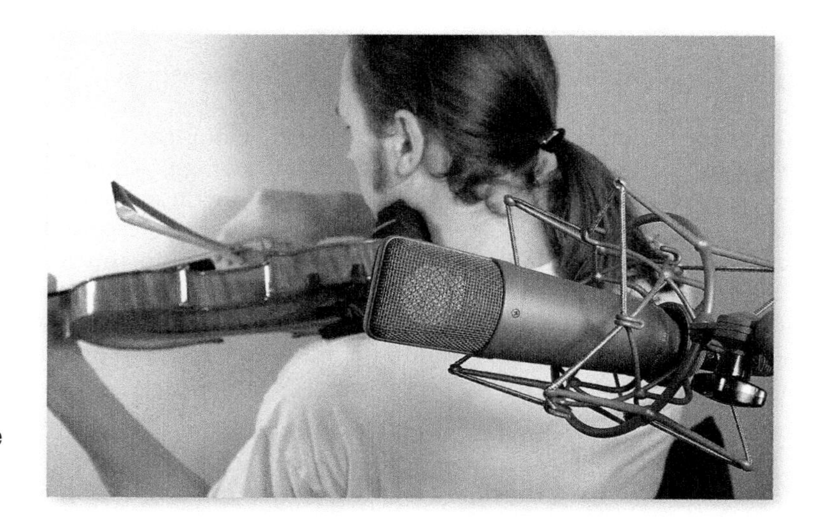

FIGURE 6.9
Miking from behind a violinist can help reduce breathing noises on the recording.

The shadowing effect of the player can have a big impact on the high-frequency content the mic picks up, which is why so many instruments sound brighter in front than behind—the French horn being a notable exception, of course, because of its rear-facing bell. Instruments in the accordion family, with separate left and right reed chambers, may project their high frequencies sideways, however, and harmonica players may also cause sideways HF dispersion depending on how they hold the instrument in their hands. Harmonica isn't the only instrument where you need to be wary of the HF shadow of the player's hands, either. "With the mandolin [I'm] watching where the hand moves," says Bill VornDick,[34] "making sure the mic is placed so it actually gets the instrument and you don't have the masking effect of the hand going in front of the mic." Some engineers like to take advantage of the fact that many of the noisy components of pitched instruments are focused into the upper part of the spectrum. So, for example, the mechanical noises of a tuba are often less prominent behind it, and VornDick[35] also suggests miking fiddles from the rear if you need to tone down unwanted breathing sounds from the player.

With some instruments the shadowing effect of the instrument's casing is another big factor. This is a reason why most engineers miking from outside a piano like to use microphone positions that still "see" the high strings. With grand pianos that usually means keeping the lid open and miking from audience perspective at a height somewhere between the level of the piano frame and the line of the lid, but if the lid has to remain closed for whatever reason, then miking from behind the performer can still afford the mics a valuable glimpse of the high strings. With upright pianos, you can aim a microphone into the open lid to catch the high strings direct, but it's preferable to remove the instrument's whole front panel from above the keyboard instead, affording you greater flexibility to position the mic. (It's usually pretty easy to do this, because upright pianos are designed to have their panels removed for tuning purposes.)

6.2.5 Air Resonances

Most instruments with strings typically have some kind of resonating wooden body that serves to amplify the sound. The resonant characteristics of wood are fantastically complex, and vary a great deal even between quite similar-looking instruments, so it's very difficult to predict which frequencies will be projected in which directions—thus making these instruments quite challenging to record. However, most wooden instrument bodies are also hollow, and the air in the cavity will have a well-defined resonant frequency which will project powerfully from the body's carved vents—for example, from the "F" holes of orchestral string instruments or the central soundhole of acoustic guitars. For the latter, the resonant frequency lies around 90–100 Hz; violin and viola air cavities resonate higher, at around 300 Hz and 220 Hz; and cellos and basses have their air resonances in the 100–110 Hz and 60–70 Hz regions respectively. Choosing how much air resonance to capture is a critical tonal decision, and you should pay particular attention to how it affects the subjective balance of the instrument's lower notes, otherwise it's easy to make a few notes with fundamentals around the resonant frequency dominate over all the others. Single-skinned drums such as bongos, congas, and djembes will also exhibit more "boom" from air resonance if you mike close to the open end.

Naturally, the issue of air resonance influences any mic placements inside the body of an instrument—the most common studio scenario being the rock-music technique of miking the kick drum through a hole cut into its resonant head. In this case, the sound field within the cavity will be peppered with hot and cold spots at different frequencies, a little like a tiny resonant room, and what this means in practice is that small mic-placement changes will result in much bigger tonal shifts than you'd expect elsewhere. I also find that central mic positions are often rather odd-sounding within a drum, which stands to

FIGURE 6.10
Miking an acoustic guitar right in front of its soundhole is rarely a good idea in the studio, because the sound in that position will be dominated by the instrument's booming 90–100 Hz air resonance.

reason given that the spectral effects of the air resonances will be most pronounced there. Internal air resonances also complicate miking inside a closed or semi-open grand piano, and they're another reason why engineers often remove the panels from an upright piano for recording purposes—the sound comes across as less "boxy" that way.

THE BACK-OF-THE-HAND TRICK

Here's a drum-mic positioning trick Chuck Ainlay[36] once showed me, saying he'd learnt it from Ed Cherney. If you hold the back of your hand close to any drum, the hairs there will actually allow you to feel the air vibrations. With a little practice you can use this feeling to improve the reliability of your initial mic-position guesses. Mike Stavrou[37] explains what he tends to look for in this respect: "It's all too easy to find the hot spot up close, but that's not your prime target. Instead, feel for where the vibrations are at their most intense at the greatest distance away from the drum. With practice, you will discover that different locations also vibrate the hairs slower and for longer. This is your target area."

6.2.6 Balancing Instrument Facets

Every instrument produces different facets of its sound from different physical components, so there's a lot you can do to mold your recorded sound simply by modifying the mic's position in relation to these. Much of the tone of reed instruments comes from the open finger holes, for instance, so microphone positions which pick those up fairly equally will typically sound much more natural than placements directly in line with the bell. "I don't ever mic [saxophone] directly coming out of the bell," says Jeff Powell,[38] for instance. "I'll put it off to the side a little bit, to the side where the keys are." With stringed instruments a crucial tonal decision for the engineer is how to balance the "zing" of the strings themselves against the mellower contributions of the wooden body's resonating panels. So, for example, a common mic position for acoustic guitars in country, pop, and rock music is around the twelfth fret where the string component is fairly strong and can be balanced with the body resonances using small horizontal adjustments; but there are nonetheless many engineers (Steve Albini,[39] Al Schmitt,[40] and Jim Scott[41] amongst them) who use positions slightly above/below the soundhole to capture more of the instrument's resonant character, especially in acoustic styles where the guitar can hold its place in the mix without overemphasizing the string brightness. Similarly, the string and fingerboard details of an upright bass can be balanced against the lower-frequency body/air resonances by varying the miking height. "My favorite miking technique was...up near the plucking finger, about six inches above the bridge," says Norbert Putnam,[42] for instance. "This gave you the added attack of a heavily callused finger and the ability to use all sorts of jazzy, buzzy sliding sounds."

Pianos present similar issues, because close-miking the strings favors their bright tone over the complex 100–1000 Hz resonances of the instrument's soundboard—frequently a benefit in busier styles which benefit from a clearer

FIGURE 6.11
You can bring more of a grand piano's strings within range of a single close mic by positioning it where the mid- and low-register strings cross over.

sound, but also rather lacking in natural character and body where the instrument is more exposed. On the other hand, miking from the underside of a grand or the back of an upright delivers a gutsy, retro timbre that may be great for an alternative/indie project, but will usually just add woolliness to any mainstream mix. With an instrument as large as the piano, the physical spread of the strings also becomes a significant factor when you're miking up close, so you have to be wary of overemphasizing the ones the mic's closest to. This is one reason why some engineers choose to close-mike where the higher and lower strings cross over, so that more of the strings are roughly the same distance from the mic.

Playing any acoustic instrument creates some unpitched mechanical noises, and your mic technique can help balance those appropriately against the main body of the timbre. Oftentimes you're just dealing with one main source of mechanical noise that's pretty well defined: the scraping of the bow at its contact point with a cello string; the turbulent breath noise at the flute's embouchure hole; stick noise on a cymbal or hi-hat. In these cases it's mostly just a question of setting the mic's distance and angle in relation to that location—perhaps even moving so that the direct path between the noise source and the mic is obstructed in some way (as mentioned in Section 6.2.4).

Things can get trickier where an instrument generates more than one type of mechanical noise. The most common culprit here is acoustic guitar, where pick noise, string buzz, and fret squeak are all everyday occurrences, and this means that different musical parts may respond best to different mic positions: Excessive pick noise from hard strummed chords might be tamed somewhat by a mic position on the fretboard side of the soundhole, for instance, while moving the mic to the other side of the soundhole may better suit delicate fingerpicking parts where fret noise is a greater concern. Piano can also be awkward in this respect, because of different noises coming from the hammers, the dampers, and the different pedal actions.

MIKING WHAT THE PLAYER HEARS

When looking for microphone positions, don't forget that what the player is hearing might actually be the sound you're looking for, especially if the performer happens to have written the music you're recording. "One of the things I'm really big on is miking from player perspective," says Jon Brion.[43] "Everybody that learns to play an instrument...has learned the balance from where their head is."

With a figure-eight mic, the polar pattern's side-nulls offer an additional option—to reject some elements of the instrument's sound. Again, this can be great for reducing mechanical noises or soundhole boom when close-miking acoustic guitar, but piano hammer noise, the bow scrape of orchestral string instruments, or the rotor whirr of a Leslie cabinet could all be tackled in a similar way.

6.2.7 Spotlighting and Reflections

Many small-studio engineers come unstuck because they try to place directional mics too close, unnaturally "spotlighting" only a small part of a complex resonating body. Part of what misleads people, I think, is that they transfer onstage close-miking practices into the studio, even though these have usually been optimized for minimizing PA feedback howlround rather than maximizing sound quality. "Acoustic guitar sounds terrible with a cardioid mic too close to it," comments Keith Olsen,[44] for example, and many of the home recordings I hear bear this out! Drums can also really suffer when you put the mic too close, because it'll tend to magnify a few of the skin's pitched resonances at the expense of the rest of the drum's timbre—which is probably why low-budget snare-drum recordings so often seem to go "donk!" Mike Fraser[45] doesn't like close-miking cymbals either: "If you mike cymbals too closely the sound just opens and shuts very quickly; you need your mic a bit further away to make sure the sound tails out nicely."

FIGURE 6.12
If you mike your snare drum from this close, you'll be spotlighting just a fraction of its sound-producing mechanics—which usually means seriously misrepresenting the instrument's overall tone.

What's more, it's fiddlier to find a good mic placement when you're spotlighting only a fraction of any instrument, because the sound will change more for small positional adjustments, and will also stay less consistent if the instrumentalist moves around at all while performing. Using an omni mic can alleviate some of these problems where you want a really close-up sound, but in most cases this rule of thumb from Roger Nichols[46] gives a rough idea of the kind of distance where spotlighting becomes less of a problem: "[Place] the microphone the same distance away as the size of the main body of the instrument. The body of an acoustic guitar is about two feet long, so place the microphone about two feet away. A cello is about three feet long, so place the microphone about three feet away."

"It's fiddlier to find a good mic placement when you're spotlighting only a fraction of any instrument, because the sound will change more for small positional adjustments, and will also stay less consistent if the instrumentalist moves around at all while performing."

The big exception to this, though, is when you're miking amplified electric instruments, especially in rock music. "In terms of rock guitar and overdriven guitars," says John Porter,[47] "the closer you can get [the mics], the punchier it'll sound." John Leckie[48] echoes this sentiment: "I always have the mics right up close to the guitar cabinet, literally touching the speaker cloth, and never two feet back." Although other engineers may leave a few inches' gap, this kind of close-miking approach has been used on so many influential records that the listening public are now thoroughly accustomed to it, so many engineers actively choose extreme close-miking even though no-one ever hears an amp from that perspective acoustically. In this context, the dramatic effects of small mic-position tweaks (and the ability to choose between the different drivers in a multi-speaker cabinet) are frequently viewed as a positive side-effect, giving the engineer greater power to design the recorded tone, especially given that the cabinet won't move during the performance and disturb the final delicately honed setup. Bear in mind, though, that there are some big names (such as Alan Parsons[49]) who aren't fans of such close mic placements even on electric instruments, so don't just blindly follow the herd. And then there's Paul McCartney's classic bass sound on *Sergeant Pepper*, where Geoff Emerick[50] miked up his speaker cabinet from 6–8 feet away with an AKG C12 in figure-eight mode. Mitch Easter[51] tried the same approach on REM's debut album: "I'd probably just read that Geoff Emerick miked Paul McCartney from about eight feet away… and, sure enough, it worked!"

The main difficulty with trying to avoid spotlighting is that moving the mic further away or switching it to a less directional polar pattern will inevitably increase the amount of room reverberation it picks up. But that's no reason to just put up with an anemic tone—draw the curtains, sling some quilts around the place, upend the sofa! It's surprisingly easy to shut down a good deal of a room's reverberation without specialist equipment. Another trick is to deliberately reflect more of the direct sound emanating from the instrument back into a close microphone so that it catches a more holistic picture. With instruments such as oboe, clarinet, and soprano sax, for instance, hard flooring will naturally direct some of the strong HF energy from the open end of the instrument up toward frontal mic positions, as will positioning a French horn player with a wall behind them. "I'll put the mics in front," remarks Al Schmitt,[52] "with a

RECORDING ELECTRIC GUITARS ON THE QUIET

Miking up electric guitars can present a problem for home studios in particular, because a lot of guitar sounds rely on a valve amplifier being overdriven to skull-crushing volume. Capturing these kinds of timbres without the neighbors complaining is easy with a digital amp-modeling device (whether hardware or software), and in Section 3.2.4 we already talked about using powersoaks to bring the volume down post-amplifier, but many guitarists feel that both these methods dilute the true sound character of their instrument unacceptably. Two other solutions remain open, though. The first is to use some kind of speaker isolation cabinet, which is essentially a soundproofed box containing a speaker cabinet (or just a single speaker) as well as some means of rigging microphones in front of it. The speaker is fed from your guitar amplifier via a jack socket on the outside of the cabinet, and XLR sockets alongside connect the internal microphones to your recording system.

FIGURE 6.13
A dedicated low-power recording amplifier, such as Cornford's Harlequin for instance, can help solve the problem of trying to mike up overdriven guitar sounds without antagonizing the neighbors.

Although there are commercial guitar isolation cabinets on the market, they aren't cheap, so you can save a lot of money if you build your own instead. (See this chapter's web resources for links to further information about this.) The biggest practical problem with any design is that standing waves within the enclosed space can make it rather boxy-sounding, so if you're going the DIY route, try to make the unit as big as you can to leave plenty of space for sound-deadening acoustic treatment inside, and try to make all three internal dimensions different so that the frequency effects of any remaining resonances are comparatively well spread out in the spectrum. The issue of how to prevent potentially damaging heat build-up in the isolation box also needs thought.

The second alternative for avoiding antisocial noise levels is to investigate one of the dedicated low-power valve amplifiers now available specifically for practice and recording purposes, because these can be "cranked" for their overdrive sound without producing too much real volume in the room. Although these products sometimes look rather unimpressive, they're capable of close-miked sounds which are surprisingly comparable with those of their larger, louder cousins.

reflector behind the horns so the sound will come off of it…or I put them in a spot in the room where they're going to reflect off a wall."

Wooden-bodied instruments respond particularly well to this technique, I find, perhaps because their frequency dispersion characteristics are so complex—by reflecting different perspectives of the instrument toward the mic, you can get a fuller and more consistent sound across the frequency spectrum, and the sound also won't change as much if the performer moves around. The sideways projection of many accordion-style instruments can also be corralled toward a single mic by similar means. A reflective environment will normally flatter high-frequency percussion too by smearing the transients slightly in time—especially with short, sharp sounds like handclaps, claves, and finger clicks.

Using isolated reflections like this doesn't mean adding appreciably to the recorded room reverb, because the overall recording environment can still be kept extremely damped if you wish, but you do need to be aware that combining direct and reflected sounds at the microphone may result in some comb-filtering. This isn't usually a deal-breaker by any means, though—you just have to be prepared to move the player and/or microphone in relation to the reflective surfaces until you find a particular comb-filtering flavor that sounds appealing. Personally, though, I prefer to use bits of hard furniture or hardboard offcuts as impromptu reflectors, partly because multiple reflections from smaller objects tend to cause less audible comb-filtering than a single strong reflection from a wall, and partly because I'd rather move them than the mic/musician if I want to adjust what I'm hearing in the control room.

FIGURE 6.14
You can pick up a more holistic picture of a wooden-bodied instrument's complicated frequency dispersion through a single mic by using rudimentary acoustic reflectors—here just strips of fiberboard propped against spare mic stands.

One final piece of general advice on mic positioning: As when recording vocals, there will be a tendency for any instrumentalist to play toward the microphone, and this may be exactly what you *don't* want if you've deliberately avoided putting it directly in the line of fire of a brass instrument's bell or an acoustic guitar's soundhole. It's equally common for musicians to slowly edge closer to a microphone that you've carefully distanced to avoid spotlighting. Thankfully, it's not usually difficult to deal with this by discussing your intentions with the performer, but even then it often helps if you provide them with some visual guide so that they can check their own position prior to each take. Some gaffer tape on the floor can quickly indicate chair and/or toe positions, but sometimes a more elaborate "dummy mic" setup may prove necessary so that the player has something concrete to point their instrument at. (Or if they're just a dummy.)

SIZE AND PROJECTION

Larger components of any given instrument will tend to project sound further than smaller components, and the same goes for resonant sounds in comparison to non-resonant sounds. I already mentioned this concept in passing in Section 4.3.2 with relation to singers, but it's much more widely applicable with instruments. So, for example, you'll get lots of pick noise and string jangle from an acoustic guitar up close, whereas more distant positions will focus more on the pitched sustain. This principle also underlies many problems that beset low-budget engineers at mixdown: overemphasized bow/breath/key noises on string and wind instruments; distractingly obtrusive stick transients on cymbals, hi-hats, and tuned percussion; and merciless hammer noise on pianos. Moving the mic further away makes these all-too-common afflictions much more manageable—you can always hang up a few extra blankets and/or switch to a more directional polar pattern if you need to keep the recorded reverb levels in check.

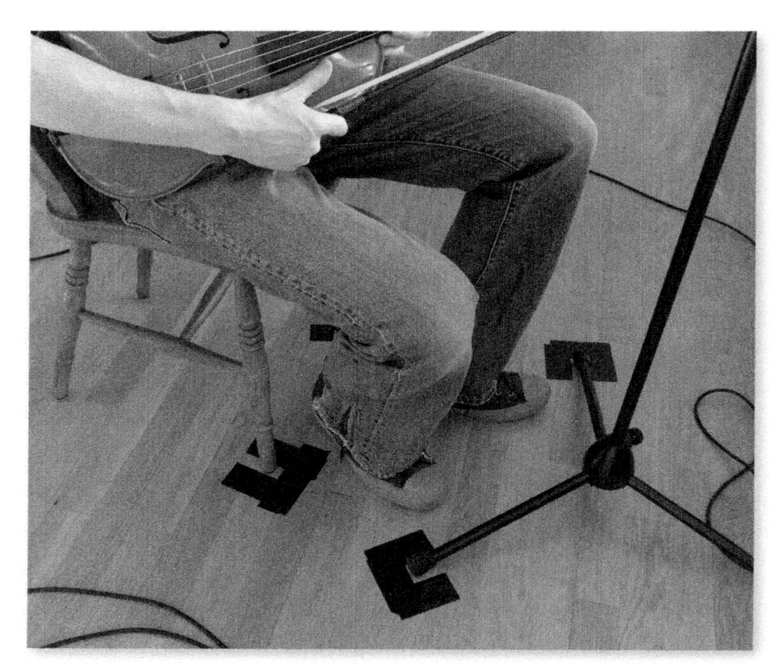

FIGURE 6.15
Using gaffer tape to mark the locations of chairs and mic stands on the floor can help keep the mic position, and hence the recorded tone, more consistent between takes during extended overdubbing sessions.

6.3 CHOOSING MICROPHONES

So far I've studiously ignored the crucial question of how to choose an appropriate mic, because I think you need to appreciate the main principles of mic positioning before you can properly tackle that thorny topic. I already introduced some of the main attributes of dynamic, ribbon, and condenser models in Section 5.1, so let's continue building from there. But allow me to reiterate one vital thing first: The only truly reliable way to decide on the right microphone for any overdub is by listening. Although the following pages will offer a number of generalized suggestions on the subject, the best they can do is improve the odds of making a respectable microphone choice when time is against you, or if you have to put up a mic "blind" and hope for the best while sprinting to keep pace with an artist. When you've got the time, always try to compare at least two microphones on every overdub you do, so that you constantly increase your understanding of each model's comparative strengths and weaknesses.

6.3.1 Large-Diaphragm Condenser Mics

Although large-diaphragm condenser mics are usually the first call for vocals, their high-frequency resonance and typically brightened on-axis response suit some instruments better than others—especially with microphones under about the $750 (£500) mark, which tend to be overhyped in this respect. Mellower-toned acoustic guitars and harp frequently benefit from the extra high-end sheen, but the same tonal modification can easily render orchestral string/brass instruments and metallic percussion rather abrasive, especially at close quarters. The slightly "soft-focus" timbral character and extended low end of large-diaphragm condensers make them popular for capturing bass instruments and kick drums warmly, although close-miking the latter may push some models into audible distortion even where there's an onboard pad switch available. Any instrument's transients will be smoothed a little by a large diaphragm's inertia, especially if they're picked up off-axis, which can give an appealing roundness to plucked strings and tuned percussion, as well as fattening toms and snare drums.

"The real Achilles heel of budget large-diaphragm condensers is the lumpy off-axis response of their cardioid-family polar patterns."

The real Achilles heel of budget large-diaphragm condensers is the lumpy off-axis response of their cardioid-family polar patterns. When recording any acoustic instrument up close, you only get a comparatively small proportion of the sound arriving from directly in front of the mic, which means the mic will significantly misrepresent the natural frequency balance a lot of the time. Once you get more than about 60° off axis the colorations can start turning pretty nasty, which makes recording in more reflective environments more challenging and complicates miking inside a grand piano especially, given that a lot of sound will reach the back of the mic by bouncing off the underside of the open lid. "But surely a cardioid mic rejects sound from the rear?" I hear you cry. Well that's the theory, but actually you'll only get about 15–20 dB rejection right in the null(s) of any cardioid pattern, and plenty of dreadful-sounding reflections will still be coming in quite strongly from either side.

In this respect the omni and figure-eight patterns of a large-diaphragm condenser microphone are usually better behaved (despite sharing some reduction in HF sensitivity at the sides), and the figure-eight pattern's deep side-rejection nulls can help attenuate unwanted background noises or early reflections from nearby room boundaries, especially when recording floor-standing electric-guitar cabs. Finding a good distance for figure-eight mics can be a little tricky to manage, however. Although some producers make a virtue of the strong proximity effect at close distances (notably Steve Albini[53] and Bob Weston,[54] who both use it to bolster electric bass guitar tones), this bass boost and the narrow pickup tend to discourage very close positioning, while more distant placements typically lack low-end weight and may require additional measures to avoid excessive room sound being picked up by the mic's rear sensitivity lobe. As such, an omni pattern is usually my first choice when using cheaper large-diaphragm condensers for instrument close-miking. Even in omni, though, you do still need to be careful, because the large-diaphragm's narrow upper-spectrum directionality still carries a danger of spotlighting, and needs to be aimed carefully to avoid emphasizing undesirable mechanical noises in particular.

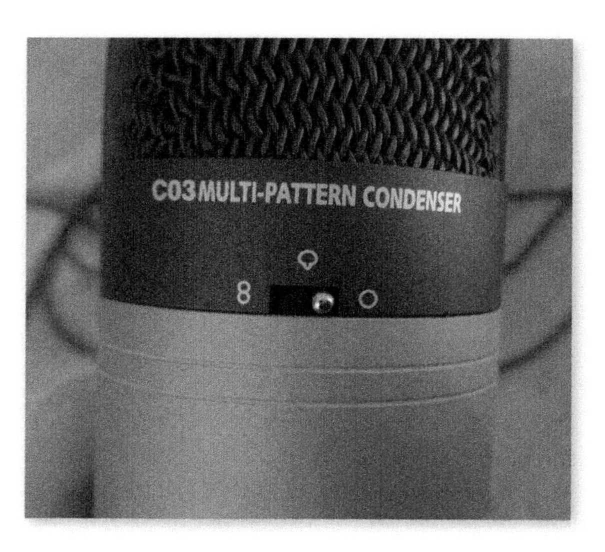

FIGURE 6.16
When using budget large-diaphragm condenser microphones that offer multiple patterns, the omni option will usually produce the most natural-sounding results in close-miking applications, because of its negligible proximity effect and reduced spotlighting.

6.3.2 Small-Diaphragm Condenser Mics

Although small-diaphragm condensers are rarely used for vocal recording, they really come into their own with instruments. Compared with large-diaphragm designs, they can cope with louder sources without distorting and tend to have a smoother and more extended high-end response, partly because the upper resonances of the smaller diaphragm are further up the frequency spectrum where they're easier for the designers to control. They also respond faster to transients on the whole, something that Al Schmitt[58] actively takes advantage of: "For percussion, I'll use an AKG C452 or one of the Schoeps or B&K mics, because these mics are a little faster." In addition, the directional response of small-diaphragm mics is normally much more predictable—you still get some frontal high-frequency boost which wanes progressively as you move off axis, but otherwise the sound remains comparatively natural over a wider range of sound arrival angles. For this reason, such microphones perform well where room reflections are an important part of the target sound, and are frequently the most natural-sounding close-miking choice for acoustic instruments of all sizes, particularly when you're working on a budget and want to use a cardioid polar pattern to minimize recorded room reverb.

MECHANICAL FILTERS

The majority of this book is focused on trying to make recordings that faithfully represent the sounds of the voices and instruments being recorded (albeit with a modicum of flattery on occasion), rather than being about sculpting unnatural timbres for creative ends. I make no apologies for that, though, because my view is that small-studio engineers encounter far greater difficulties when trying to match sounds they're hearing in the recording room than when expressly trying not to! However, on those occasions where you decide to throw realism and fidelity to the winds, one of the most powerful ways to achieve fresh timbres is by using what Tchad Blake[55] refers to as "mechanical filtering," describing it as "anything that alters the sound you're getting into the mic. Something like a pipe or a didgeridoo, metal pipes, metal plates that I would put the metal pipes up to, trash cans, tin cans. Anything you can find that you can put a mic in or on top of." He's by no means the only person doing this kind of thing, either—Roy Thomas Baker[56] mentioned "putting microphones down metal and concrete tubes to get more of a honky sound" when working on the guitar parts for Queen's "Bohemian Rhapsody," while Tom Syrowski[57] talks about "putting an SM57 in a garbage can and putting that in front of an amplifier."

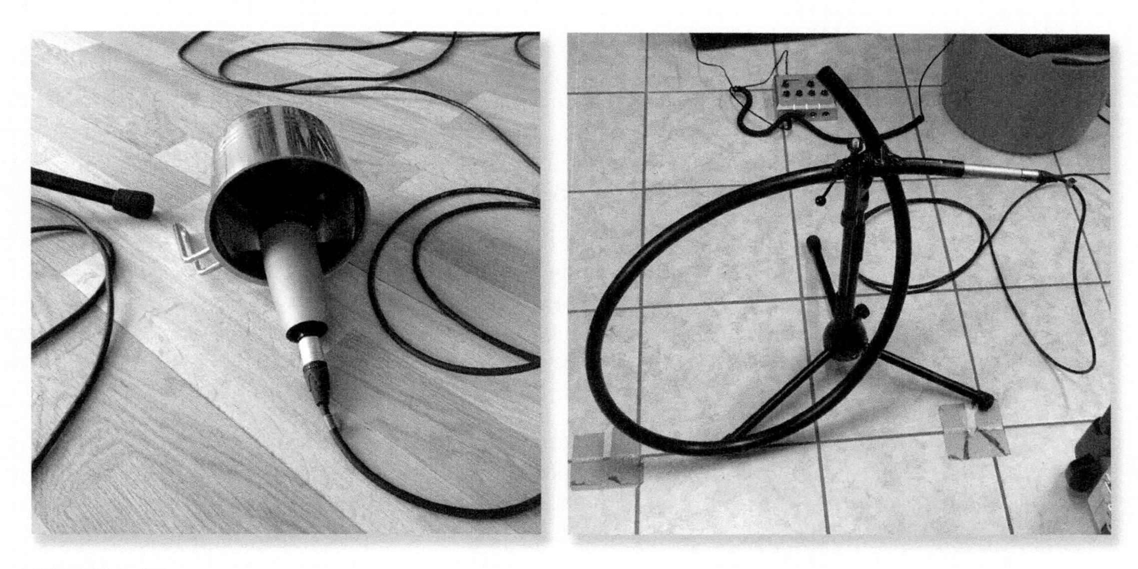

FIGURE 6.17
Some examples of mechanical filters: a large-diaphragm condenser in a metal jug and a small-diaphragm condenser in the end of a length of plastic tubing.

A small-diaphragm's cardioid pattern is achieved in a different way to that of a large-diaphragm, and while the upside is better off-axis sound, the downside is that the polar pattern isn't easily switchable (notable exceptions being Shure's KSM141 and the Schoeps CCM5). Some small-diaphragm models do offer alternate polar patterns, but you usually have to physically unscrew the mic's

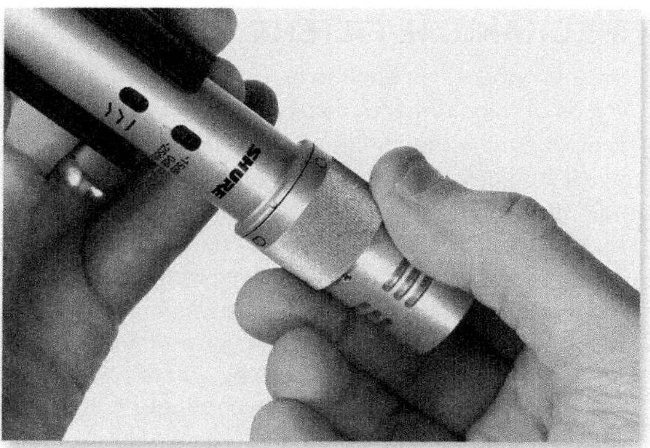

FIGURE 6.18
Most small-diaphragm mics that allow you to change their polar pattern do so by virtue of interchangeable capsules (left). Only very few models, such as Shure's KSM141, are switchable (right).

head unit (the "capsule") and replace it with a different one to implement the change. Another side-effect of this design difference is that small-diaphragm omnis will usually have their lowest high-frequency sensitivity to the rear, whereas large-diaphragm models tend to have theirs at the sides.

The main technical concern with small diaphragms is that their recordings will be a touch noisier than those of large diaphragms in the main. As long as you set up your gain controls sensibly, though, even inexpensive small-diaphragm microphones are still so quiet that you're very unlikely to notice their circuit hiss over the background noise levels in any real-world project studio—particularly when the recording is combined with other parts into a mix.

The harder, clearer tone of small-diaphragm mics is appropriate for a lot of instruments in modern chart-style productions where focused, detailed sounds are the norm, although most engineers still seem to prefer large diaphragms for a less clinical sound on bass instruments, and you do have to be a little bit careful how you angle the mic with strings, brass, percussion, and distorted electric guitars if you're to steer clear of shrillness. Where more laid-back, vintage-style timbres are the order of the day, the clarity and precision of small-diaphragms become less desirable.

ELECTRET OR CONDENSER?

Electret mics are similar in design to condensers, and are stalwarts of project studios because they can often be manufactured more cheaply. They work by storing an electrical charge within the transducer assembly, and early electret microphones got a bad rap for losing this charge over time, resulting in deteriorating performance. The technology has advanced tremendously since then, however, so for all intents and purposes you can treat electrets in exactly the same way you would condensers these days.

6.3.3 Other Condenser Designs

Although almost all large-diaphragm microphones feature circular diaphragms, a few microphone companies use rectangular ones instead, in a similar side-fire configuration. This approach weakens the high-frequency diaphragm resonances, so although you still get the kind of high-end sensitivity you'd expect of a circular diaphragm, the sound is usually smoother in a way that's reminiscent of ribbon designs. The rectangular diaphragm also affects the polar response, giving the horizontal plane a more even tonality akin to a small-diaphragm condenser (assuming that the diaphragm's longer axis is mounted vertically within the mic body, as it usually is). These attributes mean that rectangular diaphragms have a lot to recommend them for instruments such as pianos, strings, percussion, distorted guitars, and brass. However, a side-effect is that the polar pattern isn't symmetrical, and is considerably narrower in the vertical plane, so mounting the mic horizontally is advisable if you want to avoid spotlighting vertically oriented instruments.

FIGURE 6.19
The Pearl Microphones Priority, with its rectangular diaphragm clearly visible through the protective wire grille.

Miniature condenser microphones are routinely used as clip-on mics in broadcast situations, but aren't nearly as common in studio environments because of the unnatural sound of extreme close-miking. However, the ability to fix such microphones directly to an instrument can bail you out on occasion if you're trying to achieve a super-dry recording of a player who moves around a lot during performance. Again, I'd personally recommend going for an omni polar pattern in this scenario so you don't have to worry as much about spotlighting and proximity effect.

A less common condenser design which has interesting applications for small studios is called the "boundary" mic, a microphone element built into a flat disc or panel 4–9 inches across. The idea is that you place the panel on a large solid surface (typically the wall, floor, or ceiling of your recording space) where it uses the boundary effect to boost its output level. This gives it a couple of unique characteristics that can play into your hands if you're recording in small rooms with questionable acoustics:

- The frequency response captured by the microphone is far less affected by the room resonances.
- Reflections from the boundary in question don't cause comb-filtering of the mic signal—in other words, the sound's as uncolored as if the boundary weren't there at all. This is great news if you're recording in a small space, or you want to mike an instrumentalist from above in any room with a low ceiling. It also means that you don't get tonal changes from comb-filtering effects if the musician moves around while playing.

The simplest boundary-mic polar pattern is hemispherical, based around an omni microphone element, which means it has excellent resistance to vibration and no proximity effect. Therefore many engineers put it on the floor because it's easy to reposition there, although it's actually pretty simple to

FIGURE 6.20
Three affordable boundary microphones with hemispherical polar patterns (left to right): the Superlux E304, the AKG CBL99, and the Samson CM11B.

gaffer-tape one wherever you need it. Sticking one to the inside of a grand piano's lid is quite a popular trick, in fact, given that boundary mics usually have very tidy off-axis frequency response, because you pick up a pretty good balance of all the strings from that vantage point. As far as wet/dry balance, a hemisphere polar pattern is probably closest to a subcardioid, which means it "hears" roughly the same wet/dry balance as your ears do. As such, when you're recording in a small room it affords you most of the advantages of a freestanding omni microphone, but without picking up as much room reverb. More recently manufacturers have also begun producing more directional boundary mics with half-cardioid or half-supercardioid patterns for conference and presentation use. However, these often have frequency-response plots which have been mangled to promote speech intelligibility, and their off-axis pickup also isn't as good, so I'd personally suggest sticking to the non-directional variant for studio purposes.

The low-frequency balance of a boundary mic is dependent on the size and solidity of the surface it's placed on, which means it'll sound thinner if you stick it on a tabletop or the side of a gobo—a surface diameter of six feet effectively shelves 6 dB off the low end below about 250 Hz, for instance, and that shelving frequency moves up to around 1.5 kHz if the surface size is further reduced to one foot. This may be a useful tonal tool on occasion, but in my experience boundary mics are often manufactured to sound rather bright as it is, so the traditional mounting approach is invariably my first port of call.

Even though dedicated boundary-mic designs only became widely available in the '70s, some people were already making use of their basic operating principle prior to that by simply laying a small-diaphragm omni mic on the floor (sometimes wrapped in a thin piece of foam) or taping it to the wall. Although this is capable of great results, the mic's diaphragm doesn't get quite as close

PRE-SESSION MIC TESTING

If you find yourself faced with microphones you've not used before, you can save yourself quite a bit of time in the heat of the actual recording session by taking a few minutes to test them out beforehand—if only because you'll be able to check that they actually work! Here's how Mike Stavrou[59] goes about it: "Get one mic lead and one set of headphones. Listen to each mic, one at a time, while you count to ten out loud...This unusual exercise can save you hours of swapping mics later." The great thing about Stavrou's suggestion of using speech as a test signal is that it's the sound us humans are most sensitive to, so it'll show up differences in microphone tonality more clearly than anything else. And, of course, using your voice means you can carry out the tests without having to enlist any outside help.

My main tips for making the most of this process would be to judge the sound from playback, not via live headphone monitoring, so that you're not misled by bone-conduction; and to retest one or two mics you already know for comparison purposes, because tonal judgments made without the benefit of a solid benchmark won't be very reliable. I'd also recommend addressing the microphone from different distances and angles to get an impression of its proximity effect and off-axis response—just remember to say what you're doing so it's clear what you're actually listening to on playback!

Another revealing little test is to jangle a bunch of keys around six inches from the mic. This generates a tremendously complex high-frequency spectrum packed with delicate transients, and as such it severely taxes the clarity and fidelity of any mic's high-end response. High-frequency roll-offs will be very apparent, as will rounded transients, and the more a given mic makes it sound like you're crumpling up a bit of paper, the more it'll struggle with harmonically rich acoustic instruments such as twelve-string guitar, harpsichord, and sitar. (For all you old hippies still out there.)

FIGURE 6.21
Recording a jangling bunch of keys is a revealing test for any microphone.

to the boundary as in true boundary-mic designs, which means that comb-filtering reduces the levels of extreme high frequencies somewhat. You'll get the greatest high-end roll-off for instrument positions perpendicular to the mic, but if you address it at an angle the high-frequency reduction is quite subtle—and not necessarily a bad thing if you're using a cheaper small-diaphragm mic at close range! The high frequency pickup will also tail off toward the rear of the mic, just as you'd expect if it were freestanding, which again may be viewed as a benefit, but otherwise you can get more symmetrical off-axis response by

FIGURE 6.22
Two alternative configurations for setting up a small-diaphragm omni condenser as a boundary mic: If you aim the mic directly at the boundary (left), you'll get a more even, symmetrical off-axis response, whereas laying the mic flat against the boundary will favor sounds in the mic's firing line—especially at high frequencies.

mounting the mic on a stand so that it points directly at the boundary, making sure to leave a minuscule gap so that the air can reach the diaphragm—just wide enough for a credit card to slide through should be fine. (And of course, as in so many other situations, try to use someone else's credit card if you can get away with it!)

6.3.4 Ribbon Mics

We already explored some of the features of ribbon mics in Chapter 5, but there are additional considerations when it comes to using them for instrument recording. One of their most common applications is for rounding off the overly bright sound some instruments deliver at close range. "No instrument benefits as much from the attributes of a ribbon microphone as distorted electric guitar," says Steve Albini,[60] for instance, but things like trumpet, trombone, violin, piano, resonator guitar, cymbals, accordion, and high percussion are all commonly miked with ribbons too. What's more, the transient response of a ribbon is usually pretty good, so you don't lose any attack despite the more restrained tone—in fact, where condensers sometimes seem to overemphasize transients and mechanical noises, ribbons normally respond very naturally in this respect.

"Although condensers can handle most day-to-day overdubbing tasks in a small studio, ribbon microphones are an excellent antidote to the harshness problems that afflict many home-brew recordings."

With less brash sounds, the initial impression from a ribbon can feel a little underwhelming because of the design's high-spectrum roll-off, and this puts a lot of small-studio users off trying one. However, if you reinstate the brightness with equalization (typically increasing the levels above 6 kHz by 4–8 dB) you get a result that's very different from the sound of a condenser mic, and many musicians feel it's truer and more musical-sounding too, especially where pushy chart-oriented sonics aren't necessarily the goal. Although large-diaphragm and small-diaphragm condensers can handle most day-to-day overdubbing tasks in a small studio, ribbon microphones are an excellent antidote to the harshness problems that afflict many of the home-brew recordings I receive for *Sound On Sound* magazine's Mix Rescue feature, so I'd strongly suggest that any studio owner who already has two or three condenser mics should make a ribbon the next addition to their recording arsenal—you can get remarkably capable candidates for well under $500 (£300) these days.

COMBATING MOISTURE AND DUST

All microphones will work better and require less maintenance if you take steps to protect them from air-borne moisture and dust. Keeping one of those packets of silica gel with each of your mics to absorb moisture is wise, and these should be dried out in an airing cupboard or on top of a water radiator periodically to maintain the effectiveness of the desiccant. Mics are better kept in a sealed case when not in use, but if you do leave any set up in your studio between sessions, place a polythene freezer bag over the top of each one to protect it from dust. Just don't seal the bag at the bottom, otherwise changes in the room's ambient temperature may cause condensation within it.

Although finding the right miking distance for any mic with a figure-eight polar pattern requires a little care (as discussed in Section 6.3.1), a ribbon mic can usually be pulled a bit further away from the sound source before its low-end begins to thin out, maybe because of the way its inherent low-frequency diaphragm resonance bolsters the timbre. Even though ribbon microphones can typically capture very loud sounds up close without distortion, you do have to be careful of any part of the instrument that might send damaging air gusts toward the ribbon: the bell of any wind instrument; the speaker cone or bass-reflex port of a loud speaker cabinet; any vent holes in a drum shell, or the hole in a kick drum's resonant head; the moving strumming hand of an acoustic guitarist; the whirling rotary horn/baffle of a Leslie speaker; and the closing action of a hi-hat, which sends out a strong horizontal jet of air that can even cause plosive-style popping on dynamic and condenser mics. A good safety measure is to put your hand into the prospective mic position to feel for any gusts, and it's not a bad idea to use a pop shield with ribbon mics routinely. It can also help if you angle the microphone 30°–45° away from the direct line of fire of brass instruments, bass drums, and speaker cones when you're miking any closer than about two feet away, so that any wind blasts don't hit the diaphragm head-on.

6.3.5 Dynamic Mics

Because dynamic mics are by nature noisier than condensers or ribbons, they're seldom used professionally for quiet instruments or distant miking. Few dynamic mics can manage anything like the same frequency bandwidth or response linearity either (some notable exceptions being the Sennheiser MD441, Electrovoice RE20, and Heil Sound PR40), so capturing the harmonic complexity of delicate acoustic sounds really isn't their forte either. What they really do excel at, however, is adding character and solidity to loud, aggressive, and percussive instruments such as drums, horns, and guitar/keyboard amplifiers.

"What dynamic mics really do excel at is adding character and solidity to loud, aggressive, and percussive instruments such as drums, horns, and guitar/keyboard amplifiers."

It helps that they can take pretty much whatever level you throw at them without overloading, so you can stick a dynamic right inside a kick drum or a trombone's bell without worrying about distortion. In addition, because dynamic-mic diaphragms are comparatively heavy (at least by comparison with those of condensers and ribbons), they naturally round off transients, and in a lot of situations this actually ends up making close-miked drums and percussion sound subjectively meatier—once the attack "spike" becomes less prominent, the body and sustain of the instrument become more of a feature.

Condenser mics generally provide greater fidelity in their presentation of the frequency spectrum, whereas dynamic mics are rather more wayward in their tonality. However, recording engineers throughout history have made a virtue of this by using each specific model's unique colorations to enhance the instrument they're recording, leaving less work to do at mixdown. Indeed, over the years certain microphones have become renowned for certain roles, and while many of these classics are no longer commercially available, three in particular are worth knowing about because they're still current products priced well within range of budget studios:

- **Shure SM57.** Despite the fact that you can pick this up for under £100 ($125), it's without a doubt the commonest choice for two of the most important instruments in modern music production: snare drum and amplified electric guitar. Designed with a low-frequency roll-off to compensate for the strong proximity effect of extreme close-miking, it has two other pertinent tonal characteristics that tend to help in those applications: a little low midrange frequency-response dip which reduces muddiness in the mix; and a broad 2–12 kHz emphasis which lends clarity and "bite."
- **Sennheiser MD421.** More expensive than the SM57, at around $400 (£300), this microphone has a more extended low end without the SM57's low-midrange suck-out. When used close up, the proximity effect causes a bass boost which, together with a hefty high-frequency emphasis from

1 kHz upwards, provides what countless engineers consider a winning combination for kick drum and bass-guitar cabinets—i.e., ample low-end weight, but also enough high–midrange definition so that the kick drum's beater "slap" and the bass guitar's upper harmonics are clearly audible on small mass-market listening systems. The MD421 is extremely popular for tom-toms and electric guitars too, not least because of a five-position rotary switch next to its XLR connector which can reduce the bass level by up to 16 dB below 100 Hz. (Shame the plasticky stand-mounting clip isn't more robust—every commercial studio seems to have a couple of broken ones lying around!)

FIGURE 6.23
Three established classics amongst the ranks of affordable dynamic microphones (left to right): The Sennheiser MD421, the Shure SM57, and the AKG D112.

- **AKG D112.** At around $200 (£150), this distinctive egg-shaped microphone is arguably the most well-known studio kick-drum mic. A resonant cavity within the casing gives it condenser-like bass extension, as well as an extra helping of 100 Hz "beef," while frequency-response bumps up around 3 kHz and 10 kHz really help the drum's beater to carry through the mix—in other words, it'll pretty much give you an instant rock kick-drum sound without EQ. "I haven't used anything else for kick drum for the past ten years," remarks David Luke,[61] for instance. "It sounds closest to what I want. I don't have to EQ it too dramatically; it just gives a good basic sound." Some people also like the D112 for recording bass-guitar cabs, but its heavily scooped midrange timbre isn't quite as general-purpose in this role, and can also compromise the relative balance of different pitches, so I'd personally treat that recommendation with more caution.

There are plenty of other excellent dynamic microphones besides these, of course, but I think a lot of small-studio owners are unfortunately being wooed away from them by the glut of budget-friendly condenser microphones now available. Back when analog tape was in the ascendant, condenser mics had a more important part to play in combating the high-frequency degradations imposed by the recording medium, but nowadays you can afford to have a much more varied mic collection when recording digitally. In my view, there are lots of absolutely terrific dynamic models in the sub-$750 (£500) range that actually represent better value for money than similarly priced condensers in terms of providing extra tonal scope and more musically appealing sonics.

INTERPRETING MICROPHONE FREQUENCY RESPONSE CHARTS

When you don't yet have much experience of what different mics sound like, something you can use to inform your mic-selection guesswork is the mic's published frequency response, which is usually available from the manufacturer's website—although most of the time I prefer using the excellent Microphone Data site at http://www.microphone-data.com, which has a vast collection of frequency-response data carefully formatted for easy comparison. Here are a few things to look out for when scanning through the graphs:

- **Low-Frequency Roll-Off.** The extent of a directional microphone's low-frequency roll-off may influence whether you choose it for close or distant miking. For example, the well-extended low end of Audio-Technica's budget-friendly AT2020 may require significant equalization to keep a natural sound when close-miking, whereas a Shure SM57 will start to sound a bit thin if you set it up more than about a foot away, because the proximity effect won't counteract it's inherent sub-200Hz roll-off. (Bear in mind, though, when comparing technical specifications, that the low-end response curve of any directional mic will depend on the measurement distance, which isn't always explicitly stated by some manufacturers.)
- **Sub-1 kHz Lumpiness.** A lot of pitched instruments have their strongest frequency components in the bottom half of the audible frequency spectrum, and the strength of those frequencies has a big effect on the instrument's apparent overall level in a mix. Microphones with obvious nonlinearities below 1 kHz can therefore cause some of an instrument's pitch ranges to seem out of balance, such that smoothly played melodies become uneven—something I've already mentioned in relation to using AKG's D112 for bass cabs. As a result, I prefer to reserve microphones with sub-1 kHz vagaries for unpitched percussion or any parts that remain within a limited pitch register to avoid making a rod for my own back at mixdown.
- **Vocal Presence Boost.** Many microphones designed with vocalists in mind (such as the dynamic models I mentioned in Section 5.1.2) deliberately offer a presence boost in the 2–6 kHz zone to aid speech intelligibility. Even with vocals this can backfire a little by overemphasizing the sibilance of certain singers, but it can also cause problems with brass instruments and saxophones, which are usually more than strident enough in that region already. As noted above, a lot of producers use the SM57's presence boost to enhance snare drums and guitars, but you do need to be careful not to record too much of your whole production with exaggerated presence frequencies, otherwise you'll face a nightmare trying to get the vocals to cut through all that at mixdown. Indeed, part of the appeal of some large-diaphragm condenser microphones is that they understate the presence region, which nicely complements presence-heavy recordings from other mics.
- **Free-Field Versus Diffuse-Field High-Frequency Balance.** When you point a microphone directly at an instrument to capture predominately its direct sound, this is sometimes called "free-field" miking. By contrast, if you pull a mic away far enough that the recorded signal is dominated by reflected sound arriving from all angles, you're said to be miking in the "diffuse field." To capture the most realistic free-field tone, it's desirable to have a mic with a flat on-axis frequency response at the high end. However, because almost all microphones are duller-sounding off axis, a mic with a flat on-axis frequency response will actually sound too dull in the diffuse field where only a very small proportion of the sound is hitting the diaphragm head-on. Published frequency-response plots typically show only a mic's on-axis characteristics, so if you see a flat response above 5 kHz it can hint that the microphone is better suited to free-field use (i.e., closer positioning), while a strong emphasis in the top two octaves of the spectrum suggests that the microphone has been designed to represent sounds most naturally from a few paces away. (Not that realism is necessarily the primary concern in a lot of studio situations, where unnatural flattery may actually be preferred!)

All that being said, I'd like to stress that a mic's on-axis frequency-response chart provides a rather one-dimensional view of its subjective tone, because its resonant characteristics, the way it reacts to off-axis sound, and its transient response are all equally important in defining its sonic identity. So by all means use published frequency-response charts to help improve your guesswork when time is against you, but do still take every opportunity to experiment with microphones irrespective of whether they "look" suitable, so that you can build up real first-hand experience of their tonal qualities.

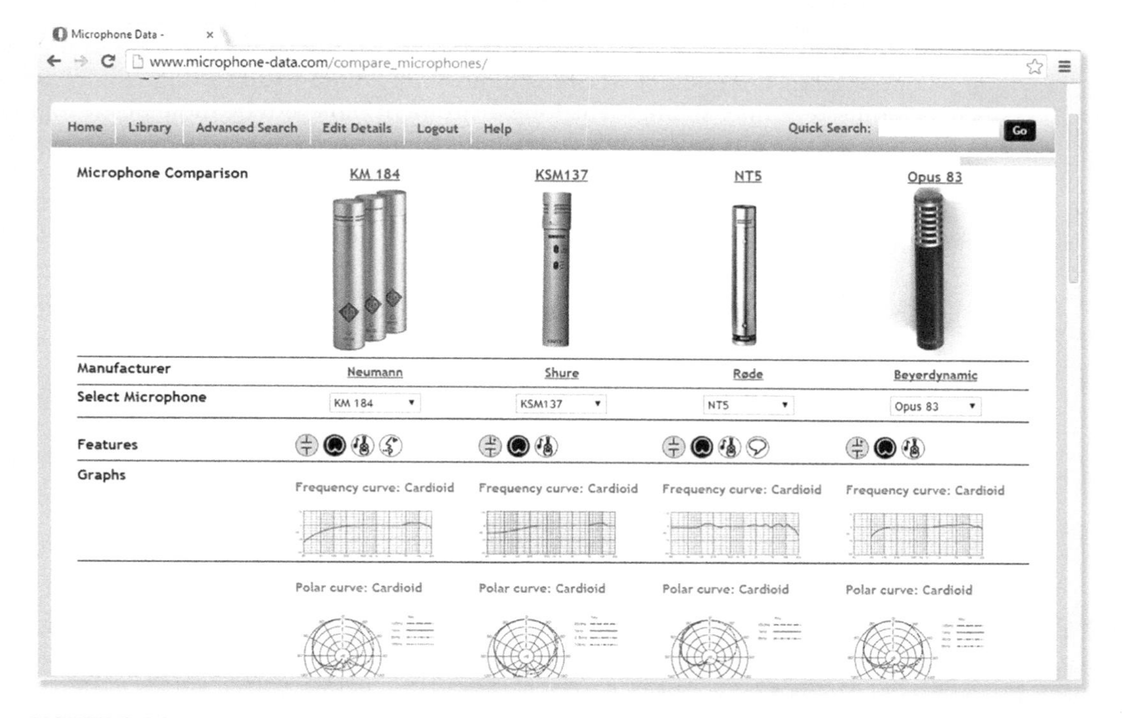

FIGURE 6.24
The Rycote Microphone Data site is a fantastic resource for comparing the specifications of microphones from different manufacturers.

6.4 FURTHER MONITORING TIPS

The assignments in Chapters 3, 4, and 5 should already have given you a good idea of how to set up appropriate monitoring for you, the musician, and anyone else involved in the session, whether you're working in one room or two. However, there are a few little refinements I'd like to add here.

Whenever extreme close-miking of the speaker cabinet is used for amplified instruments, the player may encounter difficulty dialing up appropriate sounds on the amplifier, because the sound they're hearing at their performing position bears little relation to what's being picked up by the mic. This is a big reason why electric guitarists/keyboardists often prefer to perform overdubs while listening to the studio's main monitors, running a speaker cable from their amp to a cabinet miked up in another room. That way any adjustments made to the amplifier settings can be heard in the way they'll be recorded. However, if the performer prefers to play while sitting in the recording room, perhaps to take advantage of feedback from the speakers to the instrument, then it's sensible to try to lift or angle their cabinet so that they hear more of its direct, on-axis sound, because that'll bear a closer resemblance to the mic signal.

You can also run an instrument cable between the control room and the live room, but that's less preferable because you'll have to go into the live room to adjust amplifier settings. The long instrument cable will be more prone to picking up unwanted electromagnetic interference too. "If you use a cable longer than about twelve feet," says Jack Joseph Puig,[62] "the pickups will have trouble driving the line." However, you can work around that if necessary by using a pair of passive DI boxes to temporarily convert the pickup signal to a microphone signal for transmission purposes. Here's how you do it:

- Using short instrument cables, connect the instrument's output socket to the first DI box's input socket, and plug the second DI box's input (yes, input!) socket to the amplifier's input socket.
- Connect the XLR output sockets of the two passive DI boxes together using a suitably long XLR microphone cable. You'll need female connectors at both ends of the cable to achieve this, so either spend $15 (£10) or so on a little female-to-female XLR adaptor or wire up a special cable for the same purpose.

TO EQ OR NOT TO EQ

This book focuses on recording technique rather than equalization (or "EQ"), so I'm not going to go into the details of equalizing here—there's further reading on EQ in the web resources if you need it. It's not that I'm against EQ for some dogmatic reason, it's just that I've observed time and again that small-studio engineers almost always get the best results most quickly if they concentrate on the mic and what it's pointing at, rather than trying to process an uninspiring mic signal. Frank Filipetti[63] is one of many producers who also treat EQ as a last resort: "A change in the microphone or position is worth a lot more than tweaking EQs. [If] you start tweaking EQs too soon, then you're going to miss some obvious things." Robert Carranza[64] adds an important technical concern too, should recorded parts need repairing later in the production process: "I've learned in the past that you want to use very little EQ during recording, because if you later want to drop in a fix, you'll never match it."

That said, there are certainly times when using EQ can be in the best interests of the session workflow. For example, if you've got an otherwise appealing sound from the mic, but it's just a bit too bright/dull, a ten-second fix with EQ may be a more pragmatic solution than spending another ten minutes hunting for an equally nice-sounding mic position with the desired spectral "tilt"—especially when working with ribbon mics, as mentioned in Section 6.3.4. The danger, of course, is that it's human nature to work less hard with mic technique if you feel you can take up the slack with EQ, thereby slowing down the development of your engineering skills. So if you want to improve your chops as quickly as possible, I'd strongly recommend finding some maxim which helps you reinforce your own self-discipline in this respect. Personally, I like to operate a "one band per track" rule: In other words, if I feel the need to use more than one band of EQ on a mic signal, then I take that as a signal to ditch the EQ and change something in the live room instead.

Headphones can cause some practical problems with instrumentalists. With cellists, for instance, the cans may clunk against the instrument's neck during performance, in which case perhaps investigate DJ-style headphone designs which allow one earcup to pivot up out of the way under the headband. Headphones for DJs are usually also a little more stable on the player's head when used one-sided, which is an important consideration given that most instrumentalists can't spare a hand to steady them like singers can.

6.5 THE EFFICIENCY AND ART OF MIC TECHNIQUE

Once you've got to the point of hitting Record, you should be back on familiar territory, and can run the session much as in the previous chapter assignments—just remember to check tuning regularly as you work, keep an eye on your paperwork, and do your best to maximize the quality of the performance using the suggestions in Section 3.4 (which are equally applicable when recording acoustic sources).

I've condensed a lot of information about mic technique into this chapter, but I don't want that to imply that you should therefore be making a huge meal out of choosing and positioning the mic during a real session. On the contrary, the whole point of learning about the underlying principles involved is that it allows you to work more efficiently, homing in on the best sound *within the time available*. On occasion you may have the luxury of faffing about with different mics and placements for days, but in the normal run of things you rarely get more than about fifteen minutes' setup time, because attaining sonic Nirvana is of lower priority than catching the music before the creatives start going off the boil. This is why it's paramount that you've got all your technical setup out of the way in advance, so you can maximize the value of every minute you're given for sound-hunting. If you have to spend ten minutes setting up headphone foldback, say, your only opportunity to tweak the sound may just have gone up in smoke.

"One of the biggest differences between home-brew and professional recordings is that amateur engineers tend to give up *way* too early in the recording room."

It must also be said that one of the biggest differences between home-brew and professional recordings is that amateur engineers tend to give up *way* too early in the recording room. As long as you stay on your toes, fifteen minutes is plenty of time to try a different guitar, switch drum sticks, experiment with half a dozen alternate mic positions, rig up a couple of quilts or reflectors, and maybe swap out the mic a couple of times. Even if each of those stages makes only a minuscule improvement, their cumulative effect will still be significant. So don't just chuck up a mic and then park your backside in the control room twiddling knobs—to be honest, if I see any engineer sitting down before they've actually hit Record, it's usually a pretty reliable indicator that they're not working hard enough on getting the sound right at source! Don't waste a moment of the available setup time to question and refine your initial choices, because even if this doesn't reward you with a better, more mixable sound on the current session, you'll still be gaining invaluable experience that'll improve your results on future dates. No matter how much you think you know about mic technique, there's always room to learn. "I experiment with different microphones, comparing them against a main mic I'm happy with," remarks Steven Epstein,[65] an engineer with more Grammies than fingers. "That's how you evolve, that's how you hone your art."

CUT TO THE CHASE

- Start by listening in the recording room to assess the instrument's sound and find a ballpark mic position (for both dry/wet ratio and timbre), compensating for the directionality of your hearing if necessary. Keep an ear open for background noises and unwanted sympathetic resonance.
- When adjusting the instrument itself, ask the player for help. Common tactics include: switching the instrument or playing implement; changing the tuning, damping, or playing volume; and adjusting the playing technique or amplifier controls. If you can't turn the room acoustics to your advantage by moving the instrument and/or microphone, combat them using acoustic absorption, a riser, or a boundary mic.
- On the whole, the more directional a mic, the less well extended the low-end response, despite the increasingly powerful proximity effect.
- Be wary of prescribed miking "recipes." Instead, try asking the performer or someone else familiar with your studio for miking suggestions, because they're more often useful.
- If you have no assistant to move microphones while you listen, ask the performer to help. Failing that, a series of dual-mic "tag-team" test recordings works well. When deciding where to move a mic, consider the instrument's high-frequency beams/shadows and air resonances, and the location/size of its different sound-producing components. Try to avoid spotlighting one area of an instrument unduly. Reflectors can help recording instruments with unpredictable frequency dispersion, as long as you tweak any comb-filtering to taste.
- Overdubbing electric instruments with the musician and amplifier in the control room and the close-miked speaker isolated elsewhere makes it easier for the player to evaluate their sound. If they're performing in the recording room with a close-miked cab, try to angle/lift the cab so that it fires more toward their ears.
- Large-diaphragm condensers will usually sound less natural for extreme close-miking or more reverberant pickup, because of their on-axis brightness and off-axis coloration, but they're popular for capturing bass instruments with warmth. Most budget models will sound best if you can switch them to an omni polar pattern.
- Small-diaphragm condensers are good all-round performers, although their more clinical high-frequency presentation can sometimes make bright instruments sound abrasive.

- Miniature omni condensers are handy for especially mobile performers, while boundary mics provide omni-like sound quality with reduced reverb pickup, as well as some resistance to room-acoustics problems.
- Don't disregard ribbon mics because they initially sound a little dull—once EQ'd they have a smooth high-frequency sound that's subjectively very appealing and helps tame harsh-sounding sources.
- Dynamic mics will handle the loudest instruments without distortion problems, and are arguably better value than condensers for expanding a small-studio recordist's tonal palette. Three absolute classics (SM57, MD421, and D112) are well within small-studio budgets too.
- Pre-testing unknown microphones with speech can save a lot of session time. Frequency-response plots provide some application clues too—pay particular attention to any low-frequency roll-off, sub-1 kHz lumpiness, presence boost, or general high-frequency lift.
- The hairs on the back of your hand can help you position drum mics, as well as warning of ribbon-busting air blasts.
- Amateur engineers usually give up work in the recording room too quickly—most great sounds require numerous adjustments to the instrument, acoustics, and mic selection/positioning. Although equalization can be useful on occasion, you'll stunt the development of your recording technique if you persistently use EQ as a crutch.

Assignment 1: Pre-Session Mic Tests

This assignment is designed to familiarize you with some of the characteristics of the different microphones at your disposal, and you're going to be supplying the sound sources so that you can proceed with this important exercise at your own pace.

- Collect together a good selection of different mics, preferably including large- and small-diaphragm condensers and a few different dynamics, as well as ribbon or boundary mics if you have any available.
- For each mic in turn, record yourself speaking a few pre-scripted lines of text several times: first on-axis at distances of 1 foot, 3 feet, and 10 feet; and then off-axis at 90°, 130°, and 180°, all at a distance of 1 foot. (Use a pop shield to protect any ribbon mics from plosives.) In addition, record the key-jangling test, and clap a few times from a couple of feet away too. Make sure you're using appropriate gain settings for each sound source, so that you're always getting as clean a recording as you can.
- Put all these recordings onto separate tracks in your DAW and line up the timing of the sections so that you can switch between them easily. Compare the recordings, adjusting levels from the faders if necessary to avoid loudness bias. Once you've had a chance to compare their characteristics, save a copy of the DAW project and ask a friend (perhaps the same helpful soul we met in Chapter 1, Assignment 2) to delete the track names and shuffle the track order while you're out of the room. When they're done, see if you can identify which track is which by ear.

Assignment 2: Overdubbing a Solo Instrument

You now have free rein to overdub pretty much any solo instrument you like, although still over an existing backing track. Where I prescribed many of the session parameters in previous chapters, that's all now wide open, so it's your own responsibility to choose the approach that best supports the performer within your own studio environment. This means you need to make a series of important decisions:

- If there's any choice, which room is likely to give you the best sound? Will you use a one-room or two-room monitoring setup? Who will monitor on loudspeakers, and who on headphones? Will the performer monitor through DAW software, via a low-latency DSP mixer, or in the analog domain? Will you set up any comfort effects for the performer's foldback mix? Do you need any communications system?
- What parts are you going to record? Should you supply the performer with any audio/written materials in advance? What paperwork should you have prepared in advance for your own benefit? Are you going to use punch-ins to build a master take, and if not, then how many takes will you comp from? Do you need any double-tracks or harmonies?
- Who is deciding on the target sound? If it's not you, then how are you going to lock onto the decision-maker's preferences? Which mics will you have plugged up in advance? What absorbers/reflectors should you have set up or ready on standby?

As in previous chapter assignments, make notes following the session and reconsider any problems that arose, returning to the recordings after a couple of days to attend to any comping and to re-evaluate the quality of your results generally. Before proceeding to Chapter 7, try to record at least one member of each of the following groups of instruments. To speed up your progress, schedule more than one instrumentalist (in series) per session:

- upright bass or amplified bass guitar;
- amplified electric guitar/keyboards;
- trumpet, trombone, or saxophone;
- acoustic guitar, mandolin, Dobro, banjo, harp, or harpsichord;
- grand/upright piano or tuned percussion;
- any woodwind instrument;
- any bowed string instrument;
- any drum;
- cymbal, hi-hat, or tambourine.

WEB RESOURCES

On this book's companion website you'll find a selection of resources to support this chapter, including:

- the "Recording Secrets" Library Of Mic Positions, as mentioned in Section 6.2.2;
- audio files demonstrating many of the mic-positioning principles discussed in this chapter and comparing the sounds of different mic designs on different instruments;
- further reading on equalization, drum tuning, and DIY guitar isolation cabinets.

http://www.cambridge-mt.com/rs-ch6.htm

PART 3
One Source, Multiple Mics

In Part 2 we were laying the foundations for robust, general-purpose mic technique, as well as gaining valuable experience of different microphones, miking positions, and acoustics tricks. In Part 3, we're going to head into more specialist territory, by looking at why the highest-profile professional engineers so often record single instruments with multiple microphones simultaneously, rather than just using a single mic as we've done thus far. This is where the subject begins to get quite complex, so it's best approached only once you're fully comfortable with the material in Parts 1 and 2—both in theory and in practice.

CHAPTER 7
Multimiking in Mono

7.1 COINCIDENT MULTIMIKING

7.1.1 Phase-Aligning Multiple Mics

First of all, I'd like to explore "coincident" multimiking techniques, where microphones are set up with their diaphragms as close to each other as possible, the idea being that the sound from the instrument arrives at all the diaphragms simultaneously, and all the recorded signals therefore remain in phase with each other. However, trying to phase-align mics by eye can be tricky, because the exact location of a given mic's diaphragm is frequently obscured by a grille basket (or some such protective apparatus), so most engineers rely on the following time-honoured trick for phase-aligning pairs of microphones by ear:

1. Stick a stationary sound source in front of the mics.
2. Fade up each microphone signal to roughly the same level in your control-room monitor mix.
3. Invert the polarity of one of the two signals.
4. Leaving one mic stationary, adjust the miking distance of the other to achieve the best phase-alignment, which will be where the mic signals cancel each other out most strongly. (They'll never completely cancel each other out, though, because their signals won't be exactly identical.)
5. Remove the polarity inversion applied in step 3.

This should allow you to mix the two mics freely without significant comb-filtering artifacts. To phase-align any further mics, just repeat the process. If you're quick on your feet, you can actually do this from scratch on the session, but I prefer to save time by phase-aligning my chosen line-up of mics in advance—a portable radio provides an excellent sound source for this, and keeps you abreast of the Test Match results into the bargain! Once you've aligned the mics relative to each other, label each one with masking tape (as shown

"Once you've aligned the mics relative to each other, label each one with masking tape so you can easily recover the alignment should it be disturbed accidentally during the session."

FIGURE 7.1
A good way to phase-align coincident multimics prior to your recording session is using a portable radio—remember that the best phase alignment often won't be the best visual alignment! Once you've got the mics in phase with each other, mark their relative positions with masking tape, just in case they get knocked out of line for any reason.

in Figure 7.1) so you can easily recover the alignment should it be disturbed accidentally during the session. With end-fire mics, some people just gaffer-tape the mics together, which has the advantage that you require one less mic stand, but I'm not a huge fan of this because it makes it a hassle to swap out one of those mics later. Phase-aligning multimic rigs in advance can also reveal if any mics or cables are actually wired with inverted polarity, something that's also worth marking in some way to avoid confusion in the heat of a session.

The one disadvantage of using phase-aligned mic arrays is that even the slightest accidental movement of the microphones can significantly damage the recorded sound, so do guard against this. Once your multimiked sound has been given the green light, lock down those mic stands so that they don't sag, and maybe take a picture of the setup on your phone just in case disaster strikes. Having a word with the performer about this issue is a good idea in this context, because they might not think it's a particularly big deal to move the mic slightly while getting out of their chair as long as they put it back roughly in the right place. Make it clear that you need to know if any mic is inadvertently nudged, so that you can recheck the phase-alignment.

7.1.2 Combining Polar Patterns

So what's the point of using coincident mics? Well, firstly it allows you to customize the directionality of the multimic array as a whole. The simplest scenario is where you combine a directional mic with an omni, both mics pointing in the same direction. Balancing the two mic signals against each other then effectively provides any pattern between them—in fact, every one of the cardioid-family patterns if the directional microphone happens to be a figure-eight. (The "Strauss Paket" scheme pioneered by renowned classical engineer Volker Strauss applies the same concept, using gaffered-together omni and cardioid small-diaphragm condensers to create the less commonplace subcardioid directionality.) Advantages of this approach over a single-mic setup are that you can control the pickup response freely from the control room, and that you get access to a full range of intermediate cardioid-style polar patterns between the standardized ones.

Angling coincident microphones relative to each other opens up many other useful options, though. Imagine you have a coincident pair of cardioid mics, both facing an instrument head-on and mixed at equal levels to a single recorder track. The combined pickup of this two-mic array will clearly still be cardioid. However, if you splay the two mics relative to each other, the width of their combined polar pattern will increase.

If all mics had theoretically perfect polar patterns, this "crossed pair" technique wouldn't be big news, because you could just get a wider pattern from a single subcardioid or omni with less hassle. But crossed pairs really shine with real-world mics because they enable you to counteract the high-frequency spotlighting effect of close-miking, given that most mic designs are markedly brighter on-axis. By turning the mics in slightly different directions, the two spotlights diverge, giving good high-frequency coverage across a larger area of the instrument. With cheaper large-diaphragm condensers (the worst off-axis culprits), the effect is particularly striking, but the benefits also extend to well-behaved small-diaphragm mics, albeit in subtler form. In this scenario, it's inadvisable for the "mutual angle" (or "splay angle") between the firing directions of the two mics to go much beyond 90° with large-diaphragm models, otherwise you begin to lose too much high-frequency definition directly in front of the multimic rig, but even with smaller diaphragms there's usually little to be gained, because you'll just get more room reverb without further weakening the spotlighting effect substantially.

The crossed pair's combined polar pattern won't be symmetrical in three dimensions either. The vertical pickup of a horizontally splayed pair will be narrower, for instance, so it will pick up slightly less room reverb than a single mic with the same horizontal pickup width. As with rectangular-diaphragm condensers (see Section 6.3.3), you should consequently take care how the pair's orientation relates to that of the sound source. For example, if you're multimiking an acoustic guitar from somewhere in front, you might splay the microphones horizontally to match the width of the instrument, whereas with a harp you might choose a vertical orientation to take in its height.

Because crossing a pair of mics alters their combined polar pattern, you'll have to account for this when listening for promising mic positions (as discussed in Section 6.1.1) and managing the acoustics in your recording space. In this context, my own impression is that a crossed pair of cardioids or figure-eights roughly approximates the dry/wet balance of a subcardioid, while crossed

MOUNTING HARDWARE FOR MULTIMIKING

Mounting a coincident mic array on a single stand makes it much more convenient to move around without upsetting the phase-alignment. The most common gizmo for doing this is called a "stereo bar," fitting to the end of a normal mic-stand's boom arm crossways so that two microphones can be attached to it at different (and preferably adjustable) positions along its length. You can get good-quality eight-inch stereo bars for under $15 (£10) apiece which will accommodate the majority of end-fire crossed-pair configurations, but such a short bar rarely provides enough leeway to set up crossed pairs with bulkier mics such as large-diaphragm condensers and ribbons. If getting hold of a longer specimen for these situations is a bit pricey for you, check out K&M's 238 Microphone Holder, because this little $15 (£10) metal bracket will let you clamp an additional mic to almost any part of a normal mic stand. In fact, because it'll fix to practically any solid piece of furniture, it's generally a good choice for low-budget studios where there aren't enough proper mic stands to go round, and it's tremendously handy for location recording too—I have a half a dozen in my kit bag, and use them to death! Budget-friendly multimic mounting hardware does have one downside, however: It's often a bit fiddlier to set up than a collection of individual stands, so I highly recommend doing battle with that before the talent turns up.

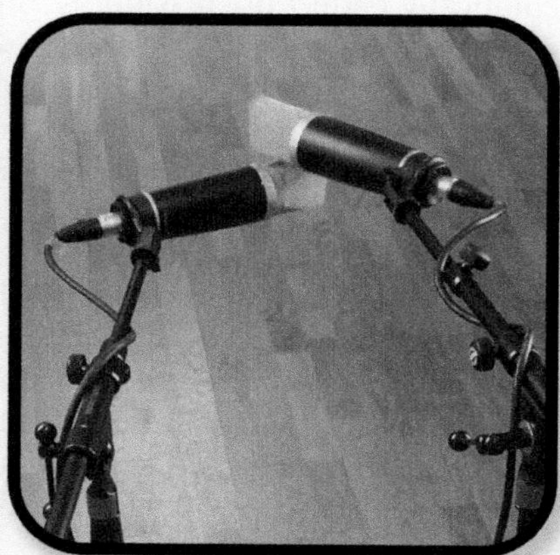

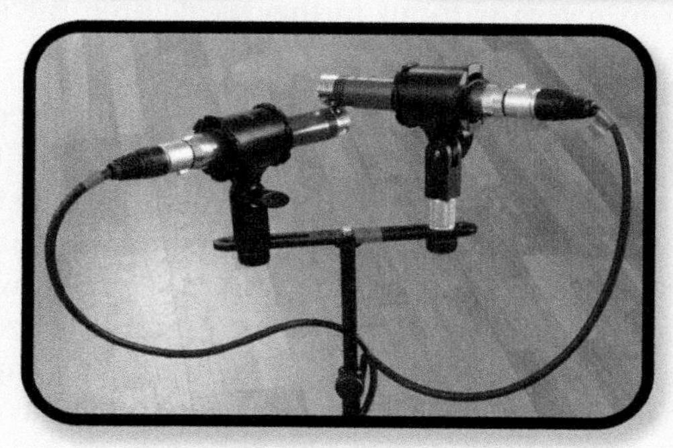

FIGURE 7.2
Here you can see some different mic-mounting approaches for crossed pairs (clockwise from top left): screw-on mic-mounting brackets; separate mic stands; and a short stereo bar.

supercardioids or hypercardioids more closely resemble a single cardioid in this respect.

Crossed pairs aren't just for cardioid-family mics, because crossed large-diaphragm omnis will also benefit from much wider high-frequency pickup at close range than a single mic. However, crossing figure-eight close mics is rarely worthwhile, because the polarity of the rear-facing sensitivity lobes are (by the very nature of the design) inverted with respect to the front lobe—so the rear-facing lobe of the right-facing mic counteracts the frontal pickup of the left-facing mic, and *vice versa*. (Crossed figure-eight mics are definitely useful when you're recording in stereo, though, as we'll see in Chapter 8.)

"Crossed pairs aren't just for cardioid-family mics, because crossed large-diaphragm omnis will also benefit from much wider high-frequency pickup at close range than a single mic."

7.1.3 Balancing Instrument Facets

A crossed pair of directional mics can also balance sounds emanating from different parts of the instrument. So, for example, a crossed pair over the strings of a piano just behind the music stand might have one capsule pointing toward the high strings and one toward the low strings, allowing you to rebalance those against each other easily from the control-room faders. The big benefit here is that it becomes quicker to get a respectable sound directly out of the mics without processing, which is a godsend for fast-moving sessions where you may not have the opportunity to move your mics before takes start going down. Moreover, if you record the two microphones to separate recorder tracks, it becomes possible to change the piano's low/high balance at mixdown too.

Another extremely widespread application of the crossed pair (and a favorite of Nashville engineers such as Chuck Ainlay[1] and Bob Bullock[2]) involves putting the mics in front of an acoustic guitar, roughly in line with the twelfth fret: One diaphragm points toward the fretboard to catch the direct sound of the strings, and one is angled toward the instrument's body for its richer resonances. Andy Johns[3] mentions a similar idea for electric guitars, whereby two SM57s are used on the same guitar speaker at a 45° angle to each other. "Put them together, and it always works…As long as you position them so that they're equidistant from the speaker, they're in phase and one of them will give you all the high end you want and the other one will give you the bottom." With larger speaker cabinets, this technique allows you to blend the qualities of the different driver cones. "They're supposed to sound the same," comments Roy Thomas Baker,[4] "but if you're using a 4×12 cabinet, each of those four speakers may sound different."

The crossed pair can also provide control over a recording's dry/wet balance. For example, Ken Nelson[5] mentions crossing a cardioid and a figure-eight mic for this when recording acoustic guitar: The cardioid faces the instrument to pick up mostly direct sound, while the figure-eight has its null plane aimed at the instrument so that it picks up mostly room sound.

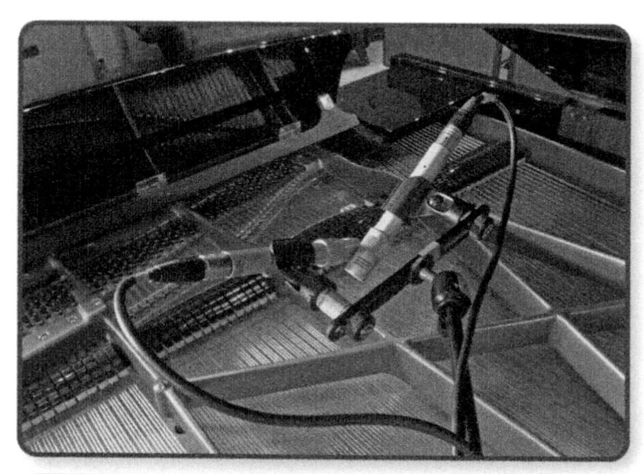

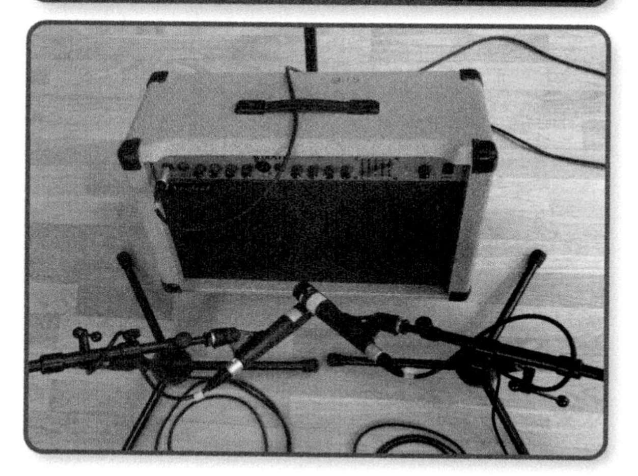

FIGURE 7.3
Crossed-pair mic strategies in action (top to bottom): balancing the high-register and low-register strings of a grand piano; blending an acoustic guitar's string and body tones; and capturing a mixture of two different guitar-amp speakers.

7.1.4 Mixing Microphone Flavors

Coincident multimiking is tremendously popular amongst professionals who like to mix and match the sonic flavors of different mics on a single instrument, with electric guitar/bass speaker cabinets, snare, and kick frequently receiving this treatment in studios. "There's an amazing difference in the sound and coloration you get from adjusting the balance of each of the mics," remarks John Leckie,[6] "and you can get radically different textures depending on your mix of [them]." Chris Lord-Alge[7] is another fan: "Rather than jamming one mic in front of an amp and grabbing the EQ, a couple of mics will give you better tone without the EQ and all the phase problems you get when cranking up the EQ." Again, a big advantage of this tactic is speed, because the sound-blending process is so intuitive: If you put up a ribbon mic and an SM57, you immediately have an idea which fader will give you warmth and which presence, and because the two mics are phase-aligned, those two sound characters will combine in a fairly natural way without comb-filtering artifacts. In addition, I've found that I get a much better recorded sound using low-budget mics if I combine them like this, rather than using them singly, whereas the quality difference between single-mic and multimic techniques seems less dramatic with more expensive hardware.

Making the most of this technique involves selecting contrasting mics, because that gives you a wider sonic range when you balance them against each other. Mic-pairing maxims from various high-profile engineers bear this out: "cheap"/"expensive" (Stephen Street[8]), "bright"/"dark" (Steve Albini[9]), and "good"/"bad" (Jim Scott:[10] "Between the two you can find the ideal sound, and you can get brightness and fullness."). My other tip is to try to get hold of variable high-pass filtering for all the mic channels, because this allows you to moderate the amount of proximity-effect bass boost contributed by each mic, multiplying the number of usable sounds from your mic array without adding much extra operational complexity. In addition, Frank Filipetti[11] makes the important point that adding an extra mic to the mix will also make the combined signal louder, biasing your judgment. "[I] make sure that when I take the mic away, my level is still the same," he says. "That gives me a much truer taste of whether that mix is adding something."

Although often used for studio recording, multimiking is very rarely used on vocals. This might initially seem surprising, given the amount of time and attention lavished on refining vocal sonics by other means, but there are two good reasons for it. Firstly, because upper-spectrum "air" frequencies above 10 kHz are so important for vocal clarity, the phase-alignment of the microphones would have to be matched extremely accurately to avoid dulling them; and, secondly, any movement of the singer off-axis would subtly affect the phase alignment and introduce more-or-less random variations in the high-frequency timbre. Occasionally, though, you might still consider putting up two mics if you're recording someone with the projection of an air-raid siren, in order to insure against your first-choice microphone overloading. In that case, a high-quality dynamic, a small-diaphragm condenser, or even a boundary mic (as John Hudson[12] used for Tina Turner) might serve as a useful safety net.

USING SPEAKERS OR HEADPHONES AS MICS

The transduction principle which most dynamic microphones use to convert air-pressure waves into electrical signals is the same one which (in reverse) allows loudspeakers and headphones to convert electrical signals into air-pressure waves. Because of this, it's perfectly possible to wire up passive speaker/headphone drivers to operate like dynamic mics. As you might imagine, they'll give you a very colored sound, not least because the inertia of the super-large diaphragm provides very little transient detail and typically recesses the high-frequency response, and you're unlikely to get great noise or distortion performance either. However, any of these attributes can become positives within the context of creative multimiking, by providing starkly contrasting tones to complement other more hi-fi mics.

Although a small handful of commercial products have operated on this principle (notably Beyerdynamic's M380, constructed around one of their headphone drivers), the most well-known application of this idea is the hoary old pro-studio trick of wiring up a spare Yamaha NS10 woofer to a mic input and hanging it in front of a kick drum (or indeed any other instrument)—Jim Abbiss,[13] Joe Barresi,[14] Jerry Boys,[15] and Mike Fraser[16] have all mentioned it in interview, for instance. The speaker cone is heavy enough to prevent much in the way of pickup above 500 Hz, and it responds inordinately well to the sub-80 Hz frequency range. However, I suspect the popularity of that particular driver has more to do with the ubiquity of NS10s in commercial studios for mixing purposes (maintenance departments always have a few spare woofers lying around) than it has to do with their specific sound, and other engineers such as Dan Austin,[17] Mike Poole,[18] Eric "Mixerman" Sarafin,[19] and Michael Wagener[20] have all mentioned using other speakers instead. The idea predates the NS10 too, because Geoff Emerick[21] was already using it for Paul McCartney's bass sound on "Paperback Writer" in 1966. Yamaha eventually adapted the idea into a robustly engineered product called the Subkick, but given that it'll set you back around $400 (£300), there's still a good case for soldering up something yourself if you're working on a budget. (See this chapter's web resources for further information on the do-it-yourself approach.)

FIGURE 7.4
A speaker can operate as a dynamic mic if wired up appropriately, although it will typically have very poor high-frequency response.

7.2 NON-COINCIDENT MULTIMIKING

7.2.1 Combining Close Mics

Although crossed-pair coincident miking can expand your polar pickup when close-miking and provide control over the levels of different instrument facets, it still only picks up the instrument from one location. With wooden-bodied instruments in particular, it's difficult to achieve a full picture of the source in this way because the complex body resonances cause such unpredictable frequency dispersion. Close-miking from different angles can therefore pick up a more credible spectral balance. As a mainstay of popular music, the acoustic guitar probably receives this treatment most frequently. Here's Bill VornDick[22] describing his setup, for instance: "I have one [mic] where the neck joins the body, pointed...where the high transients are. Then I have another one that looks down from where his right shoulder is...covering the area in the middle between the wrist and shoulder—that microphone emulates what the guitarist is hearing, and will be deeper in timbre." Notice again how he's deliberately chosen contrasted timbres, increasing the tonal range available to him when balancing the mics in the control room. "You can bring out the high mic or the low mic without having to do anything EQ-wise except high-pass filtering," he adds. Another approach popular with Gary Paczosa[23] and Eric Valentine[24] is to use two mics in front of the instrument, with "one looking down at the strings from the top, and one looking up from below," to quote Valentine himself.

With grand piano, we've already seen that balancing the lower and higher strings is frequently the goal, and Al Schmitt[25] takes this approach by using large-diaphragm omnis behind the music stand, angling the mics to control their high-frequency directionality: "The close mics are usually a couple of feet off the high end and a couple of feet off the low end, kind of at a 45° angle to each other." Another common alternative used by Brian Tankersley[26] involves positioning one mic to cover the low and mid-register strings at the point where they cross over each other, and one mic to catch the higher strings closer to the hammers. Tony Visconti,[27] on the other hand, prefers to place both his multimics pointing toward the soundholes for more wood resonance and less hammer attack, while Ed Cherney[28] and Jay Newland[29] place an additional mic toward the foot of the piano to warm up the otherwise pristine sound of a crossed pair of high-quality small-diaphragm condensers hovering behind the music stand.

Spaced multimics are usually the most convenient choice for upright piano too, whenever the instrument's front panels are removed. By comparison, crossed coincident pairs can be awkward to position, given the player's central seating location. (At least without the aid of a surgical saw.) However, I find that spaced multimic arrays somehow overlay a hint of "honky-tonk" onto pretty much any piano that's not a lovingly tuned nine-foot grand, so I usually recommend coincident approaches for small-studio practitioners multimiking less refined instruments.

The use of multi-speaker cabinets for electric guitars frequently prompts engineers to set up independent close mics for different cones. "It seems a little different than just using one mic," says Steve Churchyard.[30] "It's not twice as good, but it's just mixing the character of two different speakers." And why not take advantage of the different flavors of different mic designs while you're at it? Don Smith[31] mentioned setting up an SM57 and an AKG C452 small-diaphragm condenser for his work with The Rolling Stones, for instance, while Toby Wright[32] and Sylvia Massy[33] have both used a combination of SM57 and MD421.

Kick drums are regularly captured with spaced techniques. One mic is often aimed directly at the beater contact point to achieve better attack definition, with some engineers (e.g., Steve Marcantonio[34] and Tony Visconti[35]) suggesting a position inside the drum and others (e.g., Robbie Adams[36] and Steve Albini[37])

FIGURE 7.5
Spaced-pair multimiking is particularly useful for upright piano where the instrument's front panels are removed, because the mics can be placed conveniently either side of the player.

choosing a position on the batter-head side, particularly where the drum's resonant head is intact. This beater "click" signal might then be balanced with a fuller-sounding mic within the drum or fairly close to the resonant head. Similar approaches are often taken with upright bass too, frequently involving a mic near the instrument's bridge to capture the low power and another higher up to provide mid-range tone—a setup referred to by both Steve Chandler[38] and Bill VornDick.[39]

The main price you pay for the expanded tonal options of non-coincident multimiking is that the mic signals don't mix together as intuitively as you'd expect with coincident arrays, because they're not phase-aligned. In other words, you don't just get a simple addition of each mic's tonal attributes when you blend them—you also get comb-filtering artifacts which will vary in strength depending on how similar-sounding the signals are, and what relative levels you mix them at. You can reduce this comb-filtering somewhat by moving the mics close enough to spotlight different areas of the instrument (thereby making the signals less similar), but this carries with it the danger of a rather unnatural sound. Myself, I think it's easiest to get good results if you first concentrate on carefully positioning a single microphone to catch a fairly solid representation of the whole instrument, because you can then use any additional mic at a comparatively low level for subtle enhancement without drastic comb-filtering penalties.

PHASED ARRAYS

We've already seen how multimiking can be used to adjust the effective polar pickup of a coincident array, but you can also increase the directionality of a non-coincident rig by setting up several microphones in a line, all pointing at your sound source from the same distance—a configuration called a "phased array." Anything arriving at the array head-on will arrive at all the mics in phase, whereas anything arriving off-axis won't be picked up as effectively, because it'll arrive at the mics at different times, resulting in phase-cancellation effects. The more mics you add to the array, the more on-axis sources will be favored. What complicates matters from a practical perspective, however, is that the action of a phased array is frequency-dependent. The basic dilemma is this: On the one hand, widening the array allows the system's directionality to extend further down the frequency spectrum, so larger mic spacings are beneficial from that perspective; but, on the other hand, comb-filtering of off-axis sound becomes more severe for wider arrays, so narrower spacings will provide a clearer tone.

As such, I wouldn't actively seek out phased arrays for directionality reasons while overdubbing, since swapping/moving mics or increasing acoustic absorption in the recording space can achieve similar ends without the comb-filtering damage. However, once you understand how phased arrays work, it provides additional insight into the side-effects of other multimiking approaches. So, for example, the small but unavoidable spacing between the diaphragms in a side-by-side coincident pair will narrow the array's combined polar pattern at high frequencies in the horizontal plane, and it also means you'll probably get clearer off-axis tonality above and below the array than you will from the sides.

Either way, though, you'll never eliminate phase-cancellation between the mics completely, so you have to learn to work with it. For a start, be careful using spaced microphone arrays with instrumentalists who move around a lot, otherwise the comb-filtering response will maraud around the frequency spectrum, giving an inconsistency of tone that's extremely difficult to handle at mixdown. You also need to realize that although moving a mic an inch or two may not have a huge affect on the sound of that mic in solo, it can significantly alter the nature of the comb-filtering effect when multimic signals are mixed. This may act in your favor if you happen to find a pair of microphone positions that you like the sound of, but which comb-filter unpleasantly when combined, because very small mic-position changes can often tweak the phase-cancellation into a more appealing form without upsetting the essential qualities of each separate signal. Any filtering or equalizing you do to individual mic channels will also become less predictable, because those processes don't just affect a signal's frequency response—they also affect its "phase response" (i.e., the phase relationships between its individual sine-wave components), and hence its contribution to any comb-filtering effect.

Because comb-filtering plays such an important role here, a polarity-inversion switch on every mic channel becomes well-nigh essential, because for any given phase-relationship between two microphones, switching the polarity of one of them will swap the positions of the comb-filtering response's spectral peaks and troughs, doubling the number of tonal flavors you can choose from. Where more than two mics are used, the sonic variations multiply according to the number of different polarity permutations available across all the component mic channels. So every time you multimike an instrument with a non-coincident

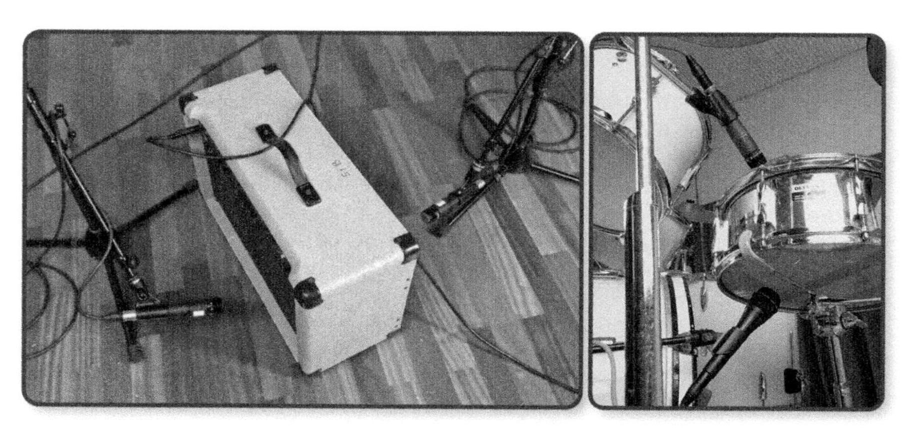

FIGURE 7.6
Here you can see two dual-miking setups where polarity-inversion of one of the mics will usually be advisable: miking an open-backed guitar amp from the front and rear simultaneously; and using top and bottom snare-drum mics in tandem.

array, creating a representative mixed sound in the control room becomes as much about comparing the suitability of different polarity settings as it is about juggling fader levels—something that can take a little getting used to.

Polarity inversion is particularly important if you decide to mike up any instrument at roughly equal distances on opposite sides. This is most common in the case of snare drum, where Jon Astley,[40] Bill Price,[41] and Alan Winstanley[42] are just a few of the engineers who routinely mike it both from above and below, allowing them to decide how much buzz they want from the snare wires. (This is by no means an inviolable standard, though, with John Leckie,[43] Alan Parsons,[44] and Elliott Scheiner[45] all expressing reservations about under-snare mics.) You might also capture open-backed guitar cabinets using front and back mics for a thicker low–midrange tone. "Gerry had an open-back amp that I miked from behind with a Sennheiser 421," recalls Tony Visconti[46] of David Bowie's "Heathen" sessions, "because I found I got more low end by positioning the mic there if that's what was called for. I sometimes also did that with David on his Supro amps, which are open-back too."

Where more nuanced manipulation of phase relationships is required, then you may wish to investigate variable phase adjusters, which provide a whole continuum of alternate tonal colors by shifting a channel's phase response in less straightforward ways. These aren't exactly cheap in the analog domain, unfortunately: The Radial Phazer's good value, for instance, but still won't leave you much change from $350 (£300). However, if you're working in software the good news is that there are much cheaper alternatives available in plug-in form, which are just as powerful (see this chapter's web resources for some suggestions).

Further tonal variations can be accessed by using a digital delay processor to adjust the relative timing of individual signals in a multimic rig. Given that phase cancellation occurs most obviously with timing offsets below about 20 ms, you'll need a delay with decent parameter resolution for this—delay-time increments of 5 ms will only give a handful of different tones compared with steps of 0.1 ms. (Many DAW systems have built-in high-resolution "sample delay" plug-ins which work well for this, but again I've put some plug-in suggestions in this chapter's web resources as well.) Bear in mind, however, that if the effects of your delay processing are audible in the musician's foldback signal, you may seriously tamper with their ability to perform, so I wouldn't suggest using anything more than a few milliseconds of artificial delay for tone-control purposes unless it's easy to set up a separate cue mix without it.

WHICH POLARITY SETTING IS CORRECT?

With coincident multimiking, you'll almost always want to choose the polarity setting which gives the loudest combined sound. However, with non-coincident arrays, there is no "correct" setting, because there will always be some phase-cancellation whichever polarity relationship you choose between the mic signals—so you just have to use your ears! One thing I have noticed, though, is that if I'm having trouble deciding which polarity setting I prefer for a spaced pair of mics, then it's an indication that I should adjust the mic positions. Somehow, good mic positions usually seem to have one polarity relationship that's clearly more attractive-sounding than the other.

7.2.2 Creative Use of Phase Relationships

In some cases, comb-filtering can become a definite virtue, most commonly when miking amplified electric instruments. As I mentioned in Section 6.2.7, close-miking techniques have long been used to sculpt an amplifier's tone into something that scarcely resembles what's heard in the recording room. Coincident multimiking expands on this concept a little, but it's non-coincident mic arrays that really up the stakes, giving you the power to utterly transform the recorded timbre through creatively applied comb-filtering, without recourse to traditional EQ processing. Jack Douglas[47] refers to this as "phase EQ," and suggests that a sound created in this manner will hold its place much more solidly in the mix. "Not only that, it won't wipe out everything else," he adds, "because it will have such a separate and distinct sound character."

The "phase EQ" is easily my favorite guitar overdubbing technique, because it's so easy to experiment with new sounds and textures, and great to do this in collaboration with the guitarist themselves if they're sitting in the control room with you. It's also excellent when you're layering up several distorted guitars, because being able to give each layer its own unique sound tends to create a much thicker and more impressive multitracked texture. However, this powerful method can also be applied to acoustic instruments, something which both Eddie Kramer[48] and Frank Filipetti[49] have talked about in interview. "[You] can create certain [comb-filtering effects] that work to your advantage," remarks Filipetti, for instance, "and sometimes work better than doing it with an equaliser."

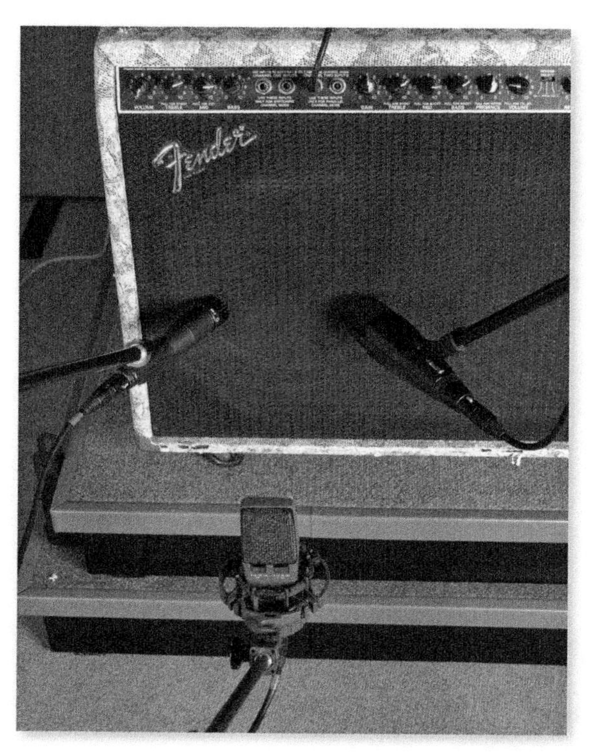

FIGURE 7.7
A "phase EQ" multimic recording rig in action. You can get a vast array of different guitar tones from a setup like this simply by combining the mic signals with different levels and polarities.

With acoustic instruments, though, any mic technique that relies on strong comb-filtering will be heavily compromised if the sound source moves, so more-or-less stationary instruments such as drums, keyboards, and tuned percussion are most likely to reward the effort. Even then, however, the inherent artificiality of creative comb-filtering between the mics may prove a sticking point, simply because most listeners have a strong impression of how acoustic instruments should sound naturally—whereas a much wider range of timbres is fair game with electric instruments because there's not really a "natural" sound people can measure against.

7.2.3 Ambience and Room Mics

Spaced mics also can balance an instrument's close-miked sound with a more distant perspective. In some cases "more distant" just means filling out a spotlighted sound with a full-instrument picture, for example when Al Schmitt[50] mentions putting an ambience microphone up by the curve of a grand piano's lid to enhance his spaced close-mic array. This approach is quite common for "catching the air" around an electric guitar too: Jon Kelly[51] had Neumann U87s close and four feet away on Kate Bush's "Wuthering Heights"; a similarly distanced AKG C414 was used by Robbie Adams[52] to complement a close Shure SM57 for U2's The Edge on "Achtung Baby," while the guitars on Shania Twain's "Up!" were recorded with a comparable setup by Bob Bullock;[53] and Toby Scott[54] used Neumann's KM86 small-diaphragm condenser instead of the C414 for the guitars on Bruce Springsteen's "Born In The USA." On kick drum a similar role is commonly filled by a large-diaphragm condenser, traditionally the classic Neumann supercardioid U47 FET placed a foot or more away from the drum's resonant head, but cheaper products will usually serve pretty well too, as long as they can take the level without distorting.

As with any other non-coincident configuration, the relative polarity/phase/timing of the mic signals will require careful attention, because the more distant mic will still be picking up some direct sound, so you'll need to massage the comb-filtering response to obtain the best tone. Where the aim is to faithfully represent the timbre of an acoustic musician with this kind of multimic rig, some engineers like to use digital delay to time-align the instrument's direct sound in each mic signal, thereby minimizing comb-filtering artifacts. You can do this in the same kind of way we phase-aligned coincident mics in Section 7.1.1, flipping the delayed signal's polarity to search for the

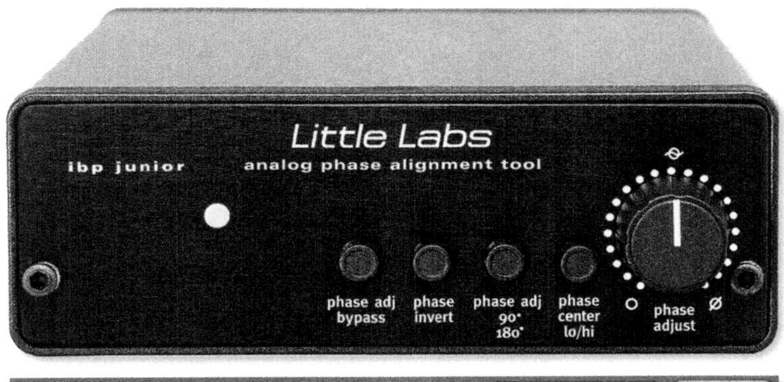

FIGURE 7.8
Analog and digital phase alignment tools: the Little Labs IBP Junior (above) and the Voxengo PHA-979 plug-in (below).

MANAGING EXPECTATIONS

If you're recording your own music, it's entirely up to you to decide what sound you chase when recording any instrument in the studio. The moment you're working on anyone else's material, though, you're dealing with other people's expectations, which usually means doing a bit of research before the session to find out what kinds of sounds they prefer. In this respect it's really useful to get hold of any previous recordings the artist/band may have done, as well as finding out what kinds of music they listen to. Take the time to thrash out what they do and don't like about each of these productions, and which parts of the current project they relate to—it's no use taking your inspiration from a specific single's guitar sound if the band actually like that record for the bass!

Listening to live performances and rehearsals can also be revealing, because they show you how a performer is used to hearing themselves while playing, which can inform the way you set things up on your recording session. If you go to a band's rehearsal room and see the bass player's rig in a corner (incurring serious boundary effect) and the guitarist's combo on the floor firing directly at his knees (putting his ears off-axis to its high-frequency beam), then you can take those sonic preferences into account when molding the amp sounds for those same musicians in a studio environment.

Difficulties can arise, however, where expectations conflict. For example, an acoustic guitarist may complain that your recording isn't capturing the full warmth and depth of their instrument, even though that kind of low midrange might be of no use to the shiny pop arrangement you happen to be overdubbing on. If a little explanation doesn't immediately resolve this, then it's rarely a good idea to argue about it in principle, as I mentioned in Section 3.4—let the musician have their way in the first instance, and once you've got something to listen to, it should become pretty obvious to everyone whether the tone feels right in context.

setting which yields the greatest cancellation. (You can quickly get into the right ballpark by remembering that sound travels roughly one foot per millisecond.) However, the downside of this kind of time-alignment is that the level and timbre of the direct sound (and therefore also the apparent dry/wet balance) become very sensitive to small movements of the instrument, so I wouldn't recommend it unless you're dealing with a stationary source such as a stand-mounted drum or acoustic piano. At least not if you value your sanity at mixdown.

Once you back off (or angle) the more distant mic, such that reflected sound from the instrument begins to dominate over direct sound, comb-filtering begins to impinge less powerfully on the multimic array's combined tone, and the "room mic" becomes more a tool for moderating your recording's dry/wet balance. Despite the ease of adding digital reverberation to dry sounds at mixdown, room mics remain a fixture on many professional sessions. "There's a lot more character to real rooms," explains Husky S. Hoskulds,[55] "and they don't have that flag on them that shouts 'Reverb!' Room sound has dimension to it, because a real room has real dimensions." One of the big reasons why room reverb often sounds better, in my view, is that a real room reflects sound from the whole instrument, whereas an artificial reverb can only work with the limited proportion of the instrument's sound that's picked up by the close mics. (That said, if the room or recording chain are too noisy, then digital reverb may turn out to be a lesser evil.) As long as you try to find a room that suits the sound you're looking for, finding a room-miking position is primarily just a case of using your ears, as discussed in Section 6.1.1, but remember to pay particular attention to the low-end response of what you're hearing, because room resonances can make a big difference to the sound, and the boundary effect may be able to counteract timbral thinning of directional mics used at a distance.

If you decide to try mic placements in excess of about 30 feet, this may affect your mic choices. For quieter instruments, the noisiness of dynamic and ribbon models will eliminate them from the shortlist in most cases. Even if you are able to capture a ribbon cleanly enough for distant miking, however, its muted high-frequency response may be an issue unless you're deliberately seeking out more rounded timbres. When sound travels larger distances, the air itself absorbs some high-frequency energy, and any mic with a duller tonality will exacerbate that effect. (The exact amount of high-frequency damping depends on how humid the air is, but you may potentially lose as much as 3 dB at 10 kHz if you're miking from 30 feet away.) Small-diaphragm condensers are also noisier in the main than large-diaphragm mics, although the downside with large diaphragms is that their colored off-axis tone may cause problems

> "Although omnis are a good general-purpose choice for room mics, in smaller rooms a cardioid or figure-eight mic with its null directed toward the sound source can actually create an illusion of increased room size."

when sound's arriving from all angles. I usually opt for the truer small-diaphragm sound myself, especially when working with cheaper mics, even though that gives me a little more noise.

Although omnis are a good general-purpose choice for room mics, in smaller rooms a cardioid or figure-eight mic with its null directed toward the sound source can actually create an illusion of increased room size. Both Tony Visconti,[56] and Stuart Sullivan[57] mention directing mics toward a wall or window in this situation to emphasize the reflections. "The reflection not only gives me a little more delay," says Sullivan, "it also takes away a little of the edge and smears it so the image isn't quite as clear."

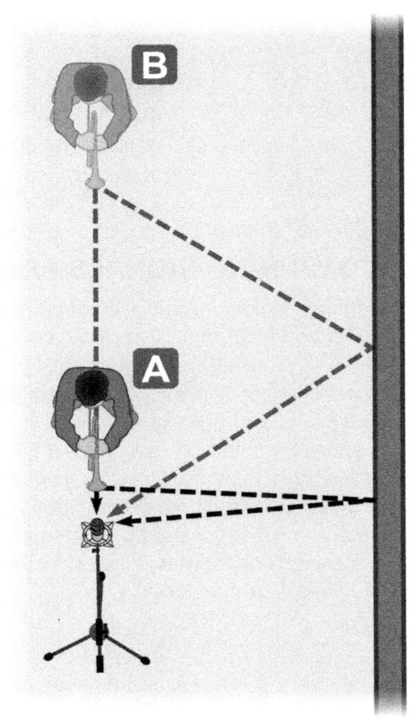

FIGURE 7.9
When a musician is right next to the mic (position A), the distance their direct sound has to travel to reach the mic is only a small proportion of the distance that sonic reflections from the room must travel. As the musician moves away from the mic (position B), these path-lengths become more comparable, so the time offset between the direct and reflected sounds decreases.

Simon Dawson[58] and Ben Hillier[59] take advantage of another time-honored stunt: "If I needed more ambience," recalls Hillier, "I'd just open the door to the corridor and stick a mic down the end. It makes for a more colorful sound." What's particularly good about capturing the ambience separately like this is that you can still keep the close-mic signal very dry and tight—something that wouldn't be as easy if the instrument itself had been put into a more reverberant space. Even in purpose-built recording venues there's much to recommend this approach, and you can hear "non-studio" spaces being corralled into service on many famous records: a concrete loading bay on Brian Adams's "Cuts Like A Knife" and "Reckless";[60] a bathroom at Coast Recorders for Kim Deal's singing on the Breeders' *Last Splash*;[61] and another bathroom (pink-tiled this time!) for Madonna's vocals on "Like A Virgin."[62]

Steve Albini[63] makes another useful small-studio suggestion: "I'll sometimes delay the ambient microphones by a few milliseconds, and that has the effect of getting rid of some of the slight phasing that you hear when you have microphones at a distance and up close...When you move them far enough away, they start sounding like acoustic reflections, which is what they are." However, phase-cancellation isn't the only factor at work here, because increasing the delay between a close mic and a room mic also has the psychological effect of moving the instrument closer to the listener. This is because, in natural listening, the direct sound from a close source will reach your ears much quicker than the reflected sound, whereas the reflected sound of a more distant source will follow the direct sound more closely—as illustrated in Figure 7.9. (You may also hear this perceptual distancing illusion when adjusting timing offsets between the signals of quite closely spaced mics, as discussed in sections 7.2.1 and 7.2.2, but frankly the effect is normally pretty small by comparison with the massive tonal changes incurred by comb-filtering.)

You do have to be a bit wary, though, when delaying room mics, or indeed when placing room mics more than about 30 feet away, because this can perceptually separate the onset of the room mic's signal from that of the close-mic's, creating an audible "slapback" echo effect. While this has a certain retro appeal in some styles, it can quickly become an annoying distraction for the listener, especially in more naturalistic acoustic genres.

COMBINING SIGNALS FROM PICKUPS AND MICS

Although pickup signals are sometimes used on their own for studio purposes (most commonly for electric bass guitar and electromechanical keyboards), they're much more widely recorded in combination with a microphone signal—especially on bass guitar, but also with acoustic guitar and upright bass. Because an instrument's vibrations don't have to travel through the air to be transduced by a pickup system, DI'd signals will always arrive ahead of any signal from a microphone, whether that mic's trained directly on an acoustic instrument or a speaker cabinet. As such, you'll be exposed to the same phase-cancellation issues that crop up between the signals in a multimic array. So if you don't initially get the combined sound you're looking for, try moving the microphone a fraction, flipping its signal polarity, or applying some kind of variable-phase or delay processing before reaching for the EQ. "With bass [there's] usually a DI and a [mic] on the amp," says Andy Johns,[64] for instance, "and then what I do is run it through a delay line so I can pull it back about 3 or 4 milliseconds so it's in phase with the amp."

7.3 MULTI-AMPING AND REAMPING ELECTRIC INSTRUMENTS

7.3.1 Multiple Amplifiers

With amplified instruments, especially electric guitars, there's another situation where multimiking becomes almost inevitable: when the instrument is feeding several amplifiers at once. On the simplest level, a multi-amp approach just increases your tonal options, as Glen Kolotkin[65] remarked while discussing Carlos Santana's "Supernatural" sessions: "He was playing through an assortment of amplifiers at the same time, and by using multiple microphones I was able to get just the right blend." Dave Jerden[66] suggests a more structured plan of attack here, though: "Instead of [having] one sound and one amp [and] using EQ, I'll use an amp basically dedicated to lows, one delegated to midrange, and one for high end. I might use a Vox or Matchless cabinet like an AC30 with a Big Muff—that will give a big bottom end, especially with a Les Paul. Then for midrange I'll use some kind of Marshall, a 50- or 100-Watt lead, or a Bogner. And for high end I'll use a Soldano or another Bogner for a more piercing sound." And it's not just for guitars—Jerden triple-amps bass too.

Eric Valentine[67] and Mike Hedges,[68] on the other hand, use multiple amps to adapt the guitar texture to different sections of the arrangement without having to use layered parts. Hedges, for instance, uses amps with different amounts of drive, so that at mixdown he can fade up the more heavily driven signals for song sections when thicker guitar textures are required, obviating the need for additional overdubs.

Multi-amping can also buy you mixdown flexibility when the guitarist insists on feeding delay/reverb effects directly to the amp. Here's Ken Nelson[69] talking about putting this tactic into action with Coldplay's Jonny Buckland: "Jonny has a Fender Twin Reverb, and he has all these delays going into it, and…I just wanted the option of a bit more dryness. They had another Twin Reverb, a slightly different version, and I said, 'What would be great would be if you could use both amps, one having all your delays going into it, and one that's completely dry.' So every time we'd record both amps." Jez Coad[70] used a similar technique when recording Simple Minds, where three amps were used for Charlie Burchill's guitar parts, the central one dry and the others panned to the sides of the stereo picture layering in effects: "You got the best of both worlds," says Coad. "You got all the power in the middle, and then you got all the lovely, diffused, echoey chorus and any other kind of modulation coming out of the side ones, so we could balance the amount of power to the amount of psychedelia going on!"

7.3.2 Reamping Electric Guitars

A more common means of insuring against Death By Guitar Pedals is to record a clean DI feed from the guitar (usually alongside the miked amplifier signal), so you can reamp it if a cleaner or drier sound is required at a later date. But there's another big reason why reamping is becoming increasingly popular. "It's just such a lifesaver," says Dan Austin.[71] "It means you can keep a really great performance and worry about the sound later." If driven by a desire to avoid boring the talent senseless while you're tending to technicalities (or by the need to record a performer's amp while the neighbors are out!), then I can see the appeal of this method, but it should also be said that a great sound can help inspire a great performance, so I wouldn't advise routinely chickening out of finding guitar sounds during tracking. Also, bear in mind that some players are suspicious of engineers wanting to record DI signals, because they worry you're going to airbrush their unique tonal personality into something unrepresentative at mixdown. Some instrumentalists may also have strong views about the presence of a DI box's circuitry in their signal path unacceptably altering the tone and/or feel of the instrument, and if they do, I wouldn't ever try to force the issue—in the grand scheme of things, a happy player will always sound better!

"A more common means of insuring against Death By Guitar Pedals is to record a clean DI feed from the guitar (usually alongside the miked amplifier signal), so you can reamp it if a cleaner or drier sound is required at a later date."

FIGURE 7.10
An affordable reamping device (such as the Radial Pro RMP shown here) has many applications for small-studio work.

The mechanics of reamping are pretty straightforward, but you won't get the best sound just by plugging a line-level output into a guitar amplifier with

SOME TIPS FOR OVERDUBBING ON LOCATION

There are a lot of instruments that you may be unable to record within your studio space—the neighbors may complain if drums or electric guitars start thrashing away, while getting things like timpani, grand pianos, and church organs up the stairs may be a challenge! And, of course, a lot of budget studios have distinctly dodgy acoustics too. So pretty much anyone doing serious recording on a budget will sometimes have to contemplate the idea of recording on location. Fortunately, this is easier nowadays than it's ever been, because project-studio equipment is so portable—you can realistically carry an entire overdubbing rig on the train, even if you're multimiking. And Mark Howard[76] argues that working in different locations is beneficial both commercially and creatively: "For each project I work on, I do an installation with my own gear.... I set up, I do the record, and I tear the installation down. It's a great way to make records, because there are no budgets anymore these days, and like this I don't have any overheads. You also don't fall into the same routines with always the same drum setup in the same place.... It allows for a lot of creative accidents to happen, and as a result I'm always discovering new sounds."

Planning sessions for evenings and weekends can often get you out-of-hours access to office conference rooms, school gymnasiums, and churches for free if you ask nicely, and background noise levels may well be lower at those times too. To get the best results for overdubs on a budget, I'd suggest investing in the best pair of open-backed studio headphones you can afford, because these will give better monitoring fidelity than closed-back designs—and you can only record as well as you can hear! Although open-backed cans won't isolate you from the instrument's sound in a one-room recording setup, it's usually possible to listen from outside the door on the end of a headphone extension cable if you want to make critical sound judgments while the musician's performing. (If not, have another look at the one-room workarounds in Section 3.2.4.)

Here are few more specific tips if you're considering taking the plunge:

- Try to check out the recording venue beforehand, so you can estimate your gear requirements. Note the positions of power sockets, and pace out the distance between your proposed recorder and instrument locations to work out what cables you'll need. Switch lights on to check for buzzes, and have a careful listen for background noise from things like heaters, plumbing, air-conditioning units, and traffic. If you have to coordinate your session with a building's security staff, confirm opening/closing times and make sure you know where to get/return any necessary keys. Few things can hinder a session more than surly janitors, but if you appear trustworthy and treat them as the important folks they are, you'd be surprised how often they'll go out of their way to help you.
- A simple mains-tester plug is a must for checking that the venue's power supply is correctly wired, as is a Residual Current Device (RCD) or Ground Fault Circuit Interrupter (GFCI) so that no-one gets fried in the event of an electrical mishap.
- After your reconnaissance mission, draw yourself a quick diagram of the setup you're going to use, because this makes it much easier to work out all the cables and stands you're going to need. I find it's helpful to give separate consideration to power wiring, input-chain wiring, monitoring/foldback/talkback wiring, and digital/data wiring—you're more likely to miss something out if you try to tackle them all at once.
- If it's a bit of a trek to the venue, try to build some redundancy into your setup, so that if something flakes out it doesn't ruin the session. In particular, make sure you've got at least one extra cable of every type you're using, a spare pair of headphones, and an extra mic or two if possible.
- Write yourself a full kit list before you go. The first time you do this, it'll feel like a chore, but trust me: It'll be worth its weight in gold. Not only will it stop you leaving anything behind after the session, but it'll also make planning similar jaunts much easier in future.
- A roll of gaffer tape should always be on the essentials list, but watch that you don't stick gaffer on anything it can't cleanly be removed from—at least if you ever fancy recording in that venue again!
- I always take a small pair of locking pliers and a multi-tool with me on location, because I seem to need at least one of them on every job. A torch regularly comes in handy too.
- All battery-powered equipment should be given fresh batteries, and you should also carry spares.

- Old towels/blankets/quilts are tremendously helpful, whether for impromptu acoustic dampening, covering over cable runs, or just holding doors open. Although you can rig up quilts with gaffer, I prefer to fix cheap carabiners to them and then use bungee cords to hang them up, because gaffer always seems to lose its grip just when you're in the middle of a take! (If you really want to kill your singer, then I imagine there are more entertaining methods than smothering them.) A mic stand set up at full height with the boom arm horizontal makes a "T" shape that's good for hanging blankets over.

FIGURE 7.11
When recording on location, a cheap mains tester plug like this one comes in very handy—here it's showing that the socket's live and neutral pins have been connected in reverse.

FIGURE 7.12
It's a lot easier to hang up blankets/quilts for recording purposes if you fix little carabiners at strategic points around the edge.

a balanced-to-unbalanced adapter lead. An instrument amplifier's input is designed to receive an instrument-level signal from instrument-impedance circuitry, so you need something to bring the level down by 30–40 dB and deal with the impedance-matching. There are dedicated reamping boxes that will do this for less than $150 (£100), but you can sometimes get away with just turning down the output level in your recording system and then using a passive DI box in reverse, as we did when extending a guitar instrument cable in Section 6.4.

7.3.3 Other Applications of Reamping

Electric guitars are by no means the only instruments you might consider reamping. "When I was recording with Michael [Jackson] and Quincy [Jones]," remarks Bruce Swedien,[72] "what I did was pick a really good room and record the synths through amps and speakers.… The direct sound output of a synthesizer is very uninteresting, but this can make the sonic image fascinating." Gareth Jones[73] followed a similar approach when creating the dense ambiences of Depeche Mode's "Some Great Reward." Some producers reamp acoustic guitar too, although with more of an emphasis on adding distortion rather than room ambience. "I actually prefer the amplified acoustic sound in some situations," says Daniel Lanois,[74] "because it can give you a little more personality, a little more harmonic distortion." To hear a specific record which uses this, check out Eric Rosse[75]'s production of Sara Bareille's "Love Song," where he felt the reamping "gave the acoustic guitars a little bit more edge."

7.4 THE JOY OF COMMITMENT

While multimiking and reamping offer many operational and sonic benefits, they do also encourage you to defer important sonic decisions until mixdown by recording all the mic/DI signals to separate tracks. I always urge small-studio producers to avoid doing this if they can possibly help it, though. First of all, handling all the separate tracks almost always slows down your workflow: A ballooning track-count makes it more difficult to navigate around the project, more difficult to edit between takes, more difficult to set up monitor/foldback mixes, and more time-consuming to back things up. Plus a computer's CPU and/or hard drive are more likely to run out of bandwidth with huge numbers of tracks, increasing the likelihood of system instability.

If you get into the habit of recording multimiked sounds to a single track (and of recording any reamped audio at the earliest opportunity), it forces you to shoulder the responsibility of capturing timbres that actually suit the production while you're working. Too many people in project studios use "leaving my options open" as an excuse for not deciding what each multimiked instrument should sound like. They put up mics rather aimlessly, without any real sense of the tone they're searching for, and end up with several different shades of "inappropriate" and/or "rubbish" preserved for posterity. And the moment you leave an instrument in your monitor mix with a sound that isn't properly serving the production, it means you're having to assess any subsequent overdubs within a questionable context, so you'll be more likely to misjudge those timbres too.

"Too many people in project studios use 'leaving my options open' as an excuse for not deciding what each multimiked instrument should sound like. They put up mics rather aimlessly, without any real sense of the tone they're searching for, and end up with several different shades of 'inappropriate' and/or 'rubbish' preserved for posterity."

So do everyone a favor: Commit, commit, and then commit some more! "Making records is about making decisions," comments John Leckie.[77] "If you delay those decisions, you pay the price of having to sort them out later. And they mount up, so the sooner you make them, the better." Sure, you'll muck up a decision from time to time, but the worst-case scenario is that you have to rerecord something once in a while—big hairy deal! The worst-case scenario for ditherers, on the other hand, is that every project takes ages and culminates in a mix engineer's sweat-drenched nightmare, which is a much bleaker prospect. If you really can't muster the confidence to jettison individual multimic signals, at least bounce them down to a single track and archive the constituent components somewhere out of sight before trying to overdub any new parts alongside, so that you keep pushing yourself to form an opinion about how the final mix should sound.

That said, if the purpose of your multimiking or reamping approach is to give you dry/wet control over an instrument's recorded room reverb or delay/reverb effects, then playing safe becomes a lot more defensible, because it's not a bad idea to leave those judgments until the mixdown stage. "It's easy to make something more live-sounding or roomy-sounding after the fact," says John Porter,[78] "but it's impossible to make something less roomy-sounding after the fact…once it's in that position in the room, you can't bring it any closer."

CUT TO THE CHASE

- Coincident multimic arrays allow you to blend different mic signals together in an intuitive way, and can speed up finding a good sound. Microphones can be mounted conveniently on a single stand with appropriate hardware and phase-aligned by ear using polarity inversion.
- Crossed pairs of coincident mics effectively provide customized polar patterns, and can also control the levels of different facets of an instrument's sound or alter the recording's dry/wet ratio.
- Arrays combining different mic designs increase your timbral options, especially if the mic tonalities are contrasted and you have access to variable high-pass filtering on each signal.
- Non-coincident miking offers even greater tonal potential than coincident multimiking, but is less intuitive to control because of comb-filtering effects between the microphones.

Per-channel polarity inversion facilities are vital when using non-coincident techniques, and variable phase/delay-adjustment tools are also useful if you need more nuanced control of the sound. Just bear in mind that offsetting mics against each other in time can also affect the instrument's perceived distance.

- Multi-amping electric instruments can increase the range of available tones, as well as offering a way of controlling guitar effects balances at mixdown. If you record an electric instrument's DI signal, you can later reamp it to completely change the amplified sound. Reamping also has applications for synths and acoustic instruments.
- Don't use multimiking and reamping techniques as an excuse for indecisiveness. You're more likely to work efficiently and effectively if you commit to sound decisions as early as you can.

Assignment 1: Coincident Multimiking and Reamping

- Overdub at least one of the following instruments with a crossed pair of mics: acoustic piano, acoustic guitar/mandolin/banjo/harp, tuned percussion.
- Overdub any electric guitar or electromechanical keyboard through an amplifier using a coincident array of tonally contrasted microphones—preferably a condenser, a ribbon, and a dynamic. Take a DI feed from the instrument as well, so that you can experiment with mixing that with the microphones too.
- In both cases, prepare and phase-align the microphone array in advance of the session. Double-check the phase-alignment with the instrumentalist at the start of the session. Record all microphone and DI signals to separate tracks, but make sure you aim for an appropriate mix of them on the session, just as much as you'd concentrate on getting a good sound if you were restricted to using a single mic.
- Following the session, return to the recordings in order to reevaluate and experiment with the microphone and DI signals. In particular, try reamping both mic and DI signals to get a feel for how this adds to your sound palette, and so that you can further experiment with multimiking techniques without pressure from the musician.

Assignment 2: Non-Coincident Multimiking

- Overdub at least one of the following instruments using non-coincident multimiking techniques: acoustic piano, acoustic guitar/mandolin/banjo/harp, upright bass. In addition to a combination of close mics, also record at least one room mic.
- Overdub one of each of the following groups of instruments using a non-coincident array of tonally contrasted microphones, including at least one room mic: amplified electric guitar or electromechanical keyboard; any orchestral string, wind, or brass instrument. Take a DI signal in addition to the microphones wherever possible.
- Keep all mic and DI signals on separate tracks, but aim for an appropriate mix of them on the session, as in Assignment 1.
- Following the session, return to the recordings in order to reevaluate and experiment with the microphone and DI signals. In addition to using polarity reversal facilities, try to get hold of a variable phase/delay-adjustment device to explore the finer tonal shades this makes available.

WEB RESOURCES

On this book's companion website you'll find a selection of resources to support this chapter, including:

- a demonstration of the phase-alignment technique described in this chapter, and audio files showcasing the tonal scope offered by a wide selection of multimiking techniques on different instruments;
- links to affordable multimic mounting hardware and phase-adjustment devices;
- further reading about using speakers and headphones as microphones.

http://www.cambridge-mt.com/rs-ch7.htm

CHAPTER 8

Multimiking in Stereo

To ease the learning curve with mic technique, I've so far only considered recording in mono. The fact remains, however, that anyone with two functioning ears hears the world in stereo, detecting the position of sound sources in a variety of ways. Any noise arriving from one side of the head will reach the nearer ear both earlier and at a higher level, for instance, and our hearing system has evolved to deduce a sound's angle of arrival remarkably accurately based on that basic information. The vast majority of stereo recordings use this fact to generate their sound-positioning illusions, recreating the same kinds of inter-ear time and level differences for the listener via a pair of loudspeakers and headphones. It is possible to create such recordings using just a pair of mics configured in one of several traditional stereophonic arrays, and it's these that I'd like to explore in this chapter.

8.1 COINCIDENT XY ARRAYS

8.1.1 Stereo from Level Differences

One of the simplest methods of stereo miking is to use a coincident mic pair called an "XY" array. In Section 7.1.3, we talked about crossing coincident directional mics to pick out different facets of an instrument, and this concept is easily adapted for stereo purposes. If the mics are crossed so that they favor the left and right sides of an instrument and/or recording room, panning their monitor channels hard left and right will spread the array's combined polar pattern between your loudspeakers to create a stereo panorama. (I'll ignore headphones for the moment, although we'll return to them in Section 8.5.2.)

To be a bit more specific, any sound arriving from directly in front of the array (i.e., on its axis of symmetry) will be picked up equally well by both mics and will emanate from both speakers at the same level, so that the sound will appear to be located at the center of the monitored stereo field. By contrast, any sound arriving from, say, the left side of the array will hit the left-pointing microphone's diaphragm more on-axis, resulting in a louder left-channel

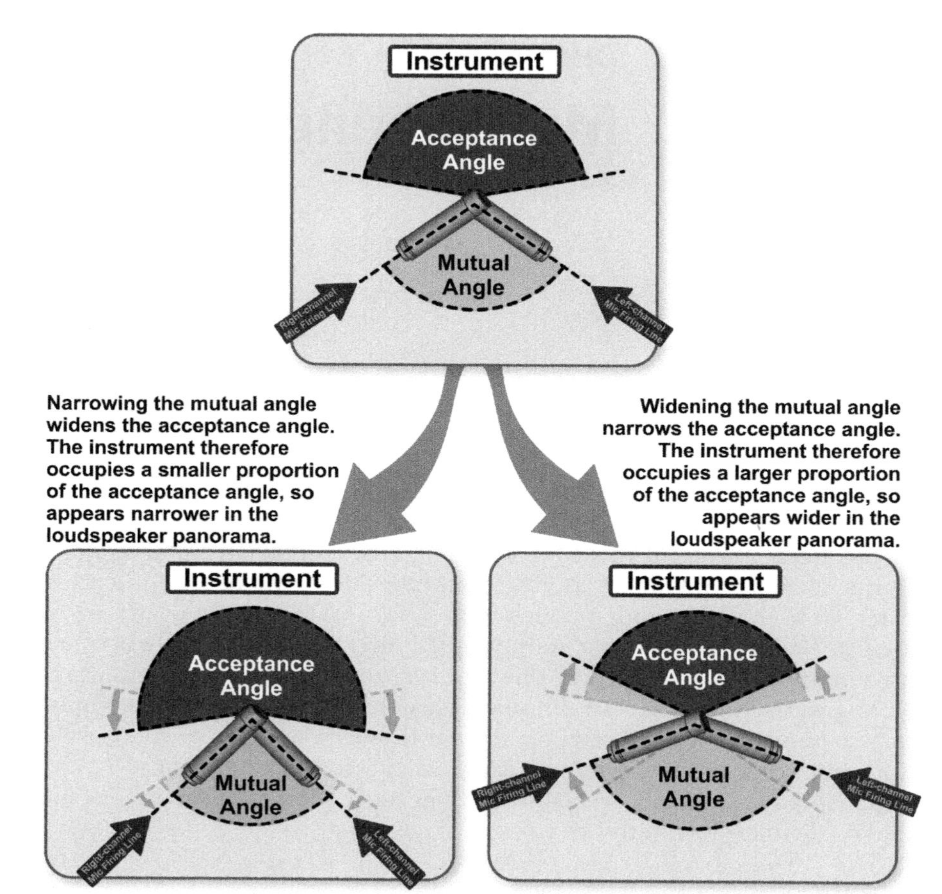

FIGURE 8.1
The correlation between mutual angle, acceptance angle, and playback image width for an XY stereo mic setup.

speaker signal which will position that sound to the left-hand side of the monitored stereo field. Assuming that your monitors are set up in the traditional "equilateral triangle" configuration (described in Appendix 1), every 5 dB of level difference between the mics will move the sound roughly 10° off-center in the stereo playback, and by the time you get to about 15 dB level difference the sound will be pretty much glued to the louder speaker. (Level differences beyond 15 dB can be largely disregarded in this context, because they make very little difference to the stereo positioning.)

Using mics to create a stereo image in this way, via inter-channel level differences, has several important ramifications:

- Increasing the mutual angle between the mics will cause greater level differences for off-center sounds, thereby widening the stereo image. Likewise, decreasing the mutual angle will narrow the stereo image. Another useful way of looking at this is in terms of the "acceptance angle" (or "stereo recording angle"), which is the pickup angle which any specific array will present as a full-width stereo panorama on playback: When the mutual angle is increased, the acceptance angle decreases, as illustrated in Figure 8.1. (Take a moment to make sure you're absolutely clear about what I mean by these terms, because I'm going to be using them a great deal from now on!) Note that some engineers will only apply the "XY" moniker to crossed pairs with 90° mutual angles, but most practitioners (myself included) use it more loosely to indicate any coincident crossed-pair stereo rig.

- Narrowing the panning of the two microphone monitor channels will also decrease inter-speaker level differences, which gives you a simple means of reducing the stereo width from the comfort of your control-room chair. Although this means that each speaker will be playing a blend of both mic signals, this shouldn't normally cause a change in the instrument's subjective tone, given that coincident setups rarely suffer from phase-cancellation issues.
- For a given mutual angle, switching both mics to a more directional polar pattern will increase the inter-channel level differences, widening the resultant stereo image (in other words narrowing the acceptance angle). For symmetrical stereo imaging, two microphones of the same model and polar pattern should be used. Some manufacturers offer factory-matched pairs specifically for stereo recording, which can offer small imaging improvements compared with unmatched mics of the same model and vintage. However, I remain to be convinced that matched pairs actually justify any extra outlay in project studios, unless they're part of a package including additional mic-mounting accessories (such as suspension shockmounts or stereo bars) that you'd otherwise have to buy separately.
- Any crossed-pair configuration that is unable to deliver at least 15 dB level difference between the two mics won't fill the whole stereo field. Obviously that rules out omnis for anything but narrow images, and crossed subcardioids can also only manage 8–10 dB even when placed back to back, so a subcardioid XY rig will always sound a bit constricted horizontally. However, the more directional cardioid-family mics will also fall foul of this if you use too small a mutual angle.

As long as you keep these points in mind, the mic techniques you've already learnt for mono applications should equip you to deal with most of the challenges of placing an XY stereo rig. You've learnt something about preparing instruments and recording environments for tracking purposes. You know how to translate the timbre and dry/wet balance you're hearing in the recording room into promising mic selections and placements. And you can refine your initial decisions efficiently by ear, applying your knowledge of frequency-dispersion principles and drawing on your experience of the different microphone types at your disposal. The only major question that remains, in fact, is how to balance those general considerations of mono mic technique against the need to achieve an appropriate stereo image. Here's my preferred method.

"Make it a priority during your initial listening reconnaissance to decide which mics you're going to use. Having to switch out the mics in a stereo array mid-session for tonal reasons is a hassle, and not just because of the sheer mechanics of it—changing the mics will almost certainly mean having to reassess the miking distance and/or mutual angle of the array."

8.1.2 Choosing the Mics and Guessing the Distance

My first recommendation is to make it a priority during your initial listening reconnaissance to decide which mics you're going to use. Having to switch out the mics in a stereo array mid-session for tonal reasons is a hassle, and not just because of the sheer mechanics of it—changing the mics will almost certainly mean having to reassess the miking distance and/or mutual angle of the array. From a purely tonal perspective, the experience you've built during previous chapter assignments should enable you to make a pretty good mic choice on instinct, but if you fade up your input channels for the first time and discover that luck's not on your side, then don't hang around—swap out those mics as early in the session as possible, well before you think of faffing around with the finer points of stereo imaging.

If tonal considerations alone don't determine which pair of mics you're going to use, then the next most important consideration will be which polar pattern provides the truest off-axis sound, because this will affect the precision of the stereo positioning and the tonal clarity of any recorded room reverb. In this respect, anyone restricted to cheaper equipment will usually find that small-diaphragm condensers give them the best results, and that dual-diaphragm condensers will do best in figure-eight mode.

Once you've taken a punt on which mics you're going to try first, you can estimate an initial miking position based on the wet/dry balance you're hearing in the room and how that relates to the polar patterns you're working with. My own rule of thumb here is to:

- treat crossed cardioids or figure-eights as a single subcardioid, placing the array roughly at my preferred head position;
- treat crossed supercardioids or hypercardioids as a single cardioid, placing them about 25% further away from the source.

ALIGNING XY ARRAYS

With XY stereo arrays, you're typically arranging the microphones to pick up the width of the instrument/room, which means you'll want to position one of the mic diaphragms directly above the other to avoid phased-array side-effects (see Section 7.2.1) in the horizontal plane. If bulky mics leave you with no alternative but to place them side by side, then at least make sure they're facing outwards so that neither mic acoustically shadows the frontal pickup of the other. You can phase-align the mics in an XY pair using the same approach as in Section 7.1.1, so long as you put the sound source directly in front of the array—and once you've phase-aligned once, you'll know how best to arrange the mics against each other for every future session. With simple stereo bars it may not be possible to create a crossed-pair configuration without angling the mics slightly up/down, but I wouldn't lose any sleep over this because it makes very little difference to the sound.

8.1.3 From Acceptance Angle to Mutual Angle

Having a rough idea of the miking distance allows you to make a more informed judgment about what acceptance angle your array is likely to need. When close-miking with XY arrays, it's quite common practice for people to match the acceptance angle to the physical dimensions of the instrument, thereby spreading the instrument across the entire stereo field. This gives the mix engineer complete flexibility to adjust the eventual image width using their pan controls later on. However, where room reflections form an important part of the sound, narrowing the panning of the mic signals at mixdown will narrow not just the instrument, but also its acoustic space, which won't often be desirable. In those cases you need to make a definite decision during tracking about how much of your stereo field is to be occupied by the instrument, and choose your acceptance angle accordingly. When doing this, bear in mind that a wider-sounding instrument image will naturally suggest that the musician is closer to the listener, so it's important to ask yourself whether the chosen width is appropriate for the production as a whole.

A quick scour of the Internet will furnish you with all sorts of mathematical tables, graphs, and software calculators which can generate the theoretical mutual angle required to achieve a given acceptance angle for any flavor of XY array—but what I've noticed is that very few small-studio musicians seem to derive much benefit from them! This is partly because different authorities disagree about the facts, partly because the blizzard of math intimidates the mildly technophobic, and partly because the tools can't easily be internalized for swift and unobtrusive use on the fly—rightly or wrongly, most musicians will feel less confident in your abilities if it looks like you're relying on a bit of software or some printed graph to tell you how to set up the mics! So I'd like to describe a more "broad brush" method of working out mutual angles that even a cicada should be able to remember and apply in practice (assuming it can shut up for long enough to benefit). Idealized mathematical models only ever approximate the operation of real-life mics, so I think there's little point agonizing over theoretical minutiae when you'll always have to tweak calculated acceptance angles by ear anyway.

First I estimate the required acceptance angle in degrees. There are specialized gadgets for this if you've got money to burn, but a simple plastic protractor on the floor is just fine—you only need to be accurate within 10° or so. If you'd like something a little larger and easier to read, there's a PDF file you can download from this chapter's online resources to print out at whatever size you like. Alternatively, if you're caught on the hop, the resources also describe how you can quickly fold a piece of scrap A4 paper into a template for any angle you like. The tripod base of many mic stands may also come in handy here, given that the legs should have 120° between them if equally spaced.

Once you know the acceptance angle in degrees, you can translate that figure into a mutual angle using Table 8.1. Let me explain how the table works. For

ALL-IN-ONE STEREO MICROPHONES

A few manufacturers sell dedicated stereo microphones which feature two independent diaphragms within a single casing, and there are also many portable recorders which now have a stereo microphone setup built in. These are undeniably convenient if you do a lot of coincident miking (especially with bulkier large-diaphragm and ribbon designs) and they also tend to encourage more experimentation with stereo techniques while overdubbing, which I'd consider no bad thing. However, because you do lose out on some flexibility compared with two independent microphones (especially if the angle between the two diaphragms is fixed), I don't think they're particularly good value for money in the small studio unless you regularly record in more specialist situations where setup convenience and portability rank high on your list of priorities.

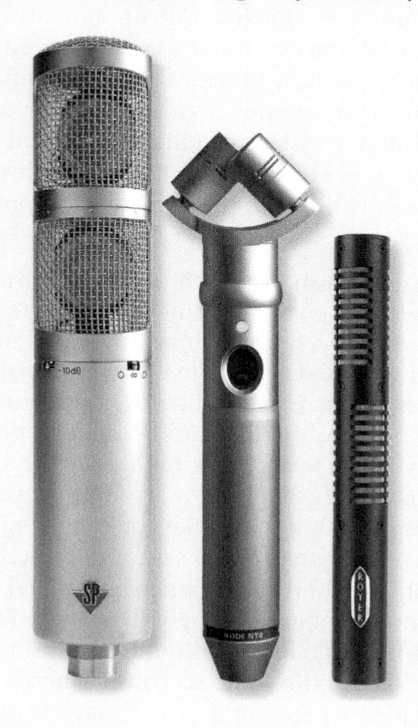

FIGURE 8.2
Some single-body stereo microphones: the Studio Projects LSD2 (left), based around a pair of large-diaphragm condenser capsules; the Rode NT4 (center), which uses two small-diaphragm condenser capsules; and the Royer SF12 (right), a double-ribbon design.

each microphone array, I've suggested three different mutual angle values, each with a corresponding suggested acceptance angle based as much on my own real-world studio tests as on theoretical considerations:

- The central mutual-angle value (in bold face) gives an XY array with a fairly even level balance across the acceptance angle—in other words, sounds arriving from anywhere within the acceptance angle should retain roughly their natural level relationships when heard over stereo loudspeakers.

- The smaller mutual-angle value (to the left of the bold-face figure) is the lowest I'd suggest using with each type of array. At this angle the apparent volume of centrally located sources will usually be increased by a couple of decibels compared with sources at the edges of the acceptance region, and you'll also struggle to get a full-width stereo picture at any lesser angle than this.
- The larger mutual-angle value (to the right of the bold-face figure) is the highest I'd suggest using with each type of array. At this angle the apparent volume of centrally located sources will usually be decreased by a couple of decibels compared with sources at the edges of the acceptance region, and the tonal coloration of central sounds by your microphones' off-axis frequency response will likely become obtrusive at any greater angle than this.

Table 8.1 Suggested Mutual Angles and Acceptance Angles for XY Stereo Arrays

XY Stereo Array	Mutual Angle	Acceptance Angle
Crossed Cardioids	90°...**110°**...140°	200°...**160°**...130°
Crossed Supercardioids	70°...**90°**...120°	170°...**130°**...100°
Crossed Hypercardioids	60°...**80°**...110°	150°...**110°**...80°
Crossed Figure-Eights	50°...**70°**...100°	130°...**90°**...60°

By memorizing just three pairs of numbers for any array (hardly a Herculean labor), you not only have an indication of its usable range of mutual angles and acceptance angles, but also the likely balance side-effects of different setups within that range. For example, if you wanted to create a crossed-cardioids array with a 140° acceptance angle, you might reasonably interpolate between the figures in Table 8.1 to come up with a 130° mutual angle, and would be forewarned that central sound sources might be recessed by a decibel or so compared with sources (or room reflections) arriving at the edges of the acceptance region.

Whether the center-versus-edge balance of a given XY rig is a blessing or a curse will depend on what you're trying to achieve. Whenever the instrument you're recording occupies only a small proportion of the acceptance angle, the relative level of the center region may have a considerable effect on the apparent dry/wet balance, for instance: A center boost will make the recording seem drier, whereas a center cut will make it seem wetter. When close-miking larger instruments, on the other hand, large mutual angles can be useful to compensate for the inherent level dominance of those parts of the instrument closest to the mic array—by recessing the center level, you effectively fade up the more distant instrument components to the sides. This is particularly relevant when close-miking inside a grand piano using crossed cardioids behind the music stand, because a 120°–140° mutual angle can help to avoid overemphasizing the middle-register strings directly beneath the array.

OUTSIDE THE ACCEPTANCE ZONE

Although it's natural to focus the majority of your attention on what's happening in the acceptance zone of an XY setup, what happens outside this region warrants some thought too. With cardioid-family microphones (especially low-cost ones) any sound approaching within about 60° of the array's rear axis will be given a rather unnatural timbre. For single-instrument overdubbing that shouldn't present much of a problem, though, because the instrument will be in front of the mics, and if the polar pattern doesn't adequately attenuate odd-sounding room reflections from the back of the array, they can usually be soaked up with impromptu acoustic treatment. Any parts of the instrument outside the mic rig's acceptance region will just be perceived as coming from the very edge of the stereo panorama.

You need to be warier with crossed figure-eights, however. The biggest problem is that any sound arriving from more than about 20° outside the acceptance zone will start to hit one mic's rear-facing sensitivity lobe (which, if you recall, picks up with inverted polarity), resulting in a polarity mismatch between the mic output signals. Not only does this cause a rather disconcerting "outside the speakers" effect when listening in stereo, but it also gives poor mono compatibility, because polarity-mismatched signals cancel each other out if the left and right channels are summed. I regularly see recording students falling foul of this when close-miking with the classic "Blumlein" stereo array (a crossed pair of ribbon mics at a mutual angle of 90°) without realizing how narrow its acceptance angle is, and it can also cause serious mono compatibility problems when recording inside a grand piano, because of strong sound reflections from the instrument's lid. That said, the fairly narrow acceptance of the Blumlein arrangement makes it well suited to room miking, where the out-of-polarity elements can actually add a pleasant spaciousness to the diffuse reflections without the overall tone collapsing in mono.

Any sound arriving from the rear of a figure-eight XY setup won't suffer from polarity mismatch, because the sound will be picked up by both mics with inverted polarity, but the stereo imaging will be reversed in that situation, so you still have to be a bit careful—strong early reflections arriving at the rear of a figure-eight XY array will effectively overlay themselves on the frontal pickup, and may confuse the imaging. (Hypercardioid mics also have a significant rear-facing sensitivity lobe with inverted polarity, so all the above considerations apply to them too, albeit to a lesser extent.)

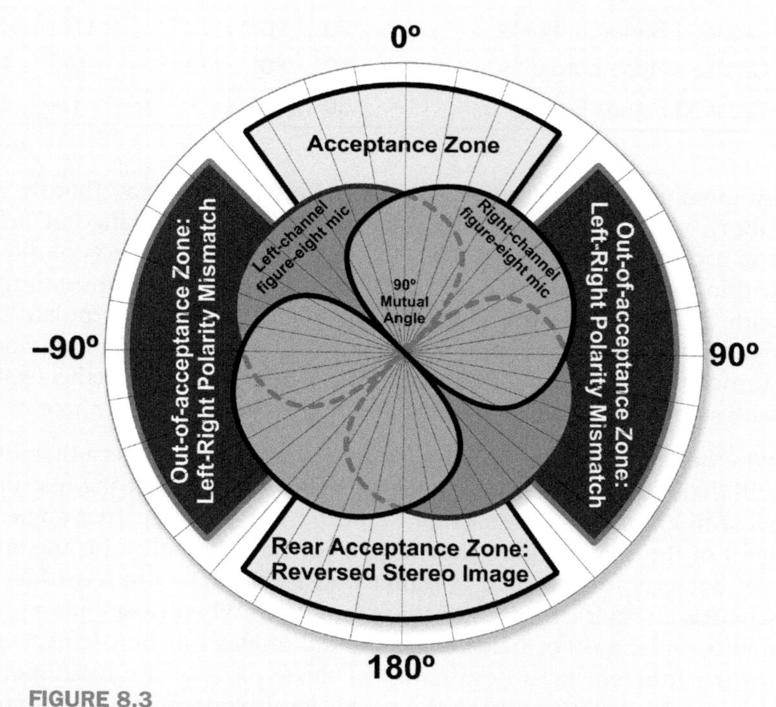

FIGURE 8.3
This diagram shows the front-acceptance, rear-acceptance (with reversed stereo image), and out-of-acceptance (with left–right polarity mismatch) zones for a classic "Blumlein" coincident stereo setup, i.e., a crossed pair of figure-eight mics at a 90° mutual angle.

8.2 COINCIDENT MS ARRAYS

8.2.1 The Middle–Sides Stereo Format

Most of the time we think of stereo recordings as being made up of one signal for the left-hand speaker and one signal for the right-hand speaker, but there's actually another useful way you can encode a stereo recording into two channels: One channel contains the mono elements at the center of the stereo image (the Middle signal) and the other channel contains the elements which appear off-center (the Sides signal). As a sound moves off-center in stereo, its level reduces in the Middle signal and increases in the Sides signal, its polarity in the Sides signal indicating which side of the stereo field it happens to be on.

Given that Left–Right (LR) and Middle–Sides (MS) formats contain exactly the same stereo information (just in different forms), and that it's very easy to convert between them, why bother with MS at all? Well, one big reason is that altering the relative level of a stereo recording's Sides signal will adjust the width of the recorded stereo image—more Sides level will widen it; less Sides level will narrow it. When working in a DAW, Voxengo's freeware cross-platform MSED plug-in is my own go-to LR-to-MS conversion option, but you can also work from first principles on any system that allows you to flip signal polarity, as follows:

- **Encoding LR Stereo to MS Stereo.** Mix the Left and Right channels together and reduce the combined level by 6 dB to create the Middle channel. Polarity-invert the Right channel, mix it with the Left channel, and reduce the combined level by 6 dB to create the Sides channel.
- **Decoding MS Stereo to LR Stereo.** Mix the Middle and Sides channels together to create the Left channel. Polarity-invert the Sides channel and mix it with the Middle channel to create the Right channel.

FIGURE 8.4
If your recording software doesn't have an MS encoder/decoder built in, try Voxengo's freeware MSED plug-in, which is compatible with most DAW platforms.

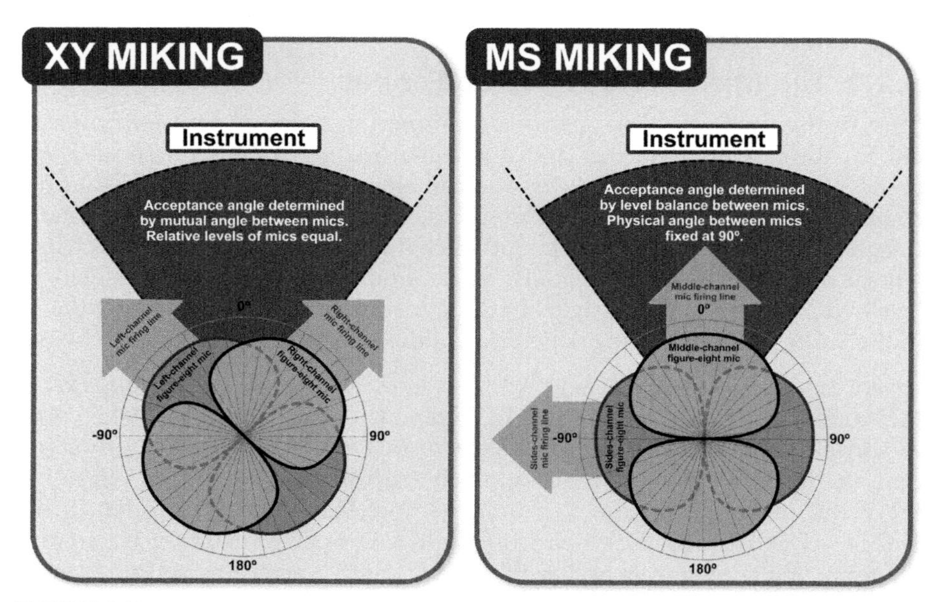

FIGURE 8.5
A comparison of XY and MS coincident stereo mic setups. Note that only figure-eight microphones have been considered here: You might choose a different directional polar pattern for the two mics in the XY setup (as long as they're matched); and you have a free choice of polar pattern for the Middle mic of an MS setup.

In addition, you can record directly into MS-format using a different coincident-miking method. Like traditional LR stereo, this uses a crossed pair, but the Middle-channel microphone is aimed directly along the array's central axis, while the Sides-channel microphone operates at right-angles to it using a figure-eight polar pattern to reject central sound sources (see Figure 8.5). To listen to the recording over your speakers, you have to decode the MS signal to LR format in your monitoring chain. In practice, there are lots of situations where MS rigs can work really well in small-studio situations, so allow me to explain the concept in more detail.

8.2.2 Acceptance Angles for MS Arrays

Although the Sides mic has to have figure-eight pickup characteristics, the Middle microphone can have any directionality you like, and by varying that polar pattern in conjunction with the Sides-signal level you can emulate the stereo pickup characteristics of any of the XY stereo arrays in Section 8.1. Indeed, thinking of MS setups in terms of their corresponding XY rigs is very handy, because it means you can use Table 8.1 to help you choose a sensible acceptance angle. A lot of people getting started with MS microphone techniques harbor the misconception that MS gives you complete freedom to choose whatever stereo width you like, but in reality changing the acceptance angle of an MS rig affects the balance of sounds coming from different positions within the acceptance zone, just as it does when using an XY mic pair.

The correspondence between MS and XY rigs is probably simplest when you're using a figure-eight Middle mic. This MS setup operates in a very similar manner to a figure-eight XY array, for which Table 8.1 suggests acceptance angles in the 60°–130° range. In this case, changing the Sides-mic level of the MS array has a similar effect to adjusting the mutual angle between the XY rig's mics. If you can match the input-gain settings of the two mics, then the 60°, 90°, and 130° acceptance angles can be achieved by offsetting the Sides-channel fader level by roughly +2 dB, −3 dB, and −6 dB respectively. However, gain matching isn't as practical with other Middle-mic polar patterns, so I normally prefer to set my miking distance to give an appropriate acceptance angle first, and then tweak the Sides level by ear until I'm hearing the desired image width coming out of my loudspeakers.

If you use a cardioid-family Middle mic, increasing the Sides level causes a slightly more complicated widening effect, roughly akin to what you'd expect from an XY array if you simultaneously increased the mutual angle *and* made the mic polar patterns more directional. In practice, though, I find that an MS array with a regular cardioid Middle mic seems to react closely enough to a supercardioid XY array that the acceptance angles for supercardioids in Table 8.1 provide a reasonably good guide. In a similar vein, a supercardioid Middle mic gives an MS array which quite closely matches the suggested hypercardioid-pair acceptance angles. For a hypercardioid middle mic, I'd suggest using acceptance angles between 70° and 140°. Using MS techniques with a subcardioid as the Middle mic can lead to difficulties filling the stereo panorama without unduly dipping the levels of sounds arriving from dead center, but the most promising acceptance angles will tend to hover around 130° or so.

An omni Middle mic will cause your MS setup to act like a back-to-back XY array, the directionality of the notional left and right mics dictated by the Sides-signal level. Although you can achieve a full stereo width in this manner by fading up the Sides channel sufficiently, central sounds will then be around 3 dB down in the stereo balance compared with sounds arriving at the edge of the acceptance angle, so using this configuration requires some caution.

8.2.3 Should I Choose XY or MS?

Although it's useful to think in terms of equivalent MS and XY setups to help you choose an appropriate acceptance angle, the two approaches tend to suit quite different applications in practice. For example, the decreasing high-frequency sensitivity of most microphones as you move off-axis to their firing line causes some XY arrays to dull the center of the acceptance zone, whereas the risk with MS rigs is dulling the edges. This means that MS has a lot to recommend it if you're miking your musician from a reasonable distance in order to create a limited-width instrument image within a full-width room sound. However, where you're spreading the instrument itself further across the stereo field using a closer miking position, the wider horizontal HF pickup of a crossed pair may simply catch more of the instrument's upper spectrum.

LOW-FREQUENCY SHUFFLING

Because real-world microphones tend to become less directional at low frequencies, coincident stereo techniques will inherently narrow the low end of the reproduced panorama to an extent. Not only does this degrade the accuracy of the stereo positioning, but also tends to reduce the subjective "warmth" of the final result. For this reason some engineers like to use a low-frequency "shuffler" which compensates for the width loss by boosting the relative level of the recording's Sides signal below 500 Hz or so. This is very straightforward to implement with a combination of MS encoding/decoding and EQ (a couple of specific setups are demonstrated in this chapter's web resources) and boosts of around 3–6 dB would be typical in this application. Do listen, though, for potentially unwanted side-effects of the processing, such as undesirable changes to the recording's dry/wet ratio.

As a brief aside, it's as well to point out that the term "shuffler" is applied to different things by different engineers. What's particularly confusing is that it's also used to refer to an earlier design of phase-shifting network which was developed to enhance the imaging of non-coincident stereo microphone setups. Although this type of shuffler also has its devotees, it's rather more complicated to implement and therefore trickier to justify grappling with in a small-studio context.

That's not to say that MS rigs should be disregarded for close-miking applications, though, because they can really shine under those circumstances too. On a practical level, MS pairs have the advantage that the array's frontal sound (and indeed mono-compatibility) shouldn't suffer as much from high-frequency comb-filtering should the phase-match between the microphones be less than perfect, and tonal changes as the musician moves off-axis to the array are typically less pronounced compared with a similar XY setup. However, the biggest benefit of MS stereo is that it increases the selection of microphones you can use for the Middle pickup, including less directional models. Mike Stavrou[1] explains the thinking: "Some people think that using a cardioid in place of an omni is OK. Not so. In fact, it's a really bad idea.... Using a cardioid means that the left and right extremities of the stereo are created by combining the beautifully flat response on the figure-of-eight with an incredibly dull response of the off-axis cardioid. The end result it a stereo image that loses reality as you stray toward the edges. What's more, when you collapse this picture to mono you lose all the definition from the sides.... You only hear the center." Whereas a crossed pair of omnis or subcardioids will sound narrow, an omni or subcardioid Middle mic can still provide a full-width image within an MS configuration—and the center-versus-edge balance issues of both those MS configurations may also help compensate for the spotlighting effect of the close placement.

> "The ability to adjust stereo width from the control room really shouldn't be a huge consideration when choosing between XY and MS stereo techniques—these days it's pretty easy to apply MS post-processing to any type of mic array within your recording system if you want that kind of control."

Finally, the ability to adjust stereo width from the control room really shouldn't be a huge consideration when choosing between XY and MS stereo techniques. Some people still make a big song and dance about this being an advantage of MS rigs, but these days it's pretty easy to apply MS post-processing to any type of mic array within your recording system if you want that kind of control.

8.3 NON-COINCIDENT AB ARRAYS

8.3.1 Stereo from Time Differences

The XY and MS mic techniques both use level differences between your left and right speakers to create the illusion of stereo positioning. However, there's another fundamental way you can conjure up a stereo image: using inter-speaker time differences. For example, if a sound emanates from the left speaker fractionally earlier than it does from the right one, that sound will appear to come from the left-hand side of the stereo image. The most common way to capture this time-difference information is by setting up a matched pair of mics in parallel with a gap between them, creating the so-called "AB" array shown in Figure 8.6, with each mic being hard-panned to feed a single speaker. Any sound originating from the left-hand side of the AB rig will reach the left-hand mic earlier, and will therefore appear at the left speaker earlier too. Because of the mic matching and parallel placement, level-difference stereo information is eliminated (unless the sound source happens to be very close to the mics, a special case I'll get on to in Section 8.5.2).

A timing offset between the mic signals of around 1 ms is enough to push a sound's position very close to the edge of the stereo loudspeaker panorama, so the time differences involved here are clearly very small. However, as we already discussed in Chapter 7, even sub-milisecond timing offsets between similar signals can trigger comb-filtering when those signals are mixed together. The audibility of this comb-filtering is negligible when the mic signals are being mixed acoustically during stereo playback (which lulls many a newbie into a false sense of security), but it's still vital that you check the mono-compatibility of any non-coincident stereo mic technique during tracking to avoid problems later on. This isn't just about planning for cheap end-user playback systems either, because bad mono-compatibility will also punish you with nasty tonal changes if you attempt to adjust the stereo image of your AB recording during mixdown.

8.3.2 Acceptance Angles, Mic Spacings, and Polar Patterns

As with coincident stereo miking, you'll need to guess a suitable miking distance before you can make any decision on suitable acceptance angles. This should be straightforward, though, given that two mics operating in parallel like this will pick up roughly the same dry/wet balance as one. Adjusting the acceptance angle of an AB array to suit your proposed miking distance will then involve adjusting the distance between the mics—the wider you space them, the narrower the acceptance angle becomes. Again, different theorists disagree substantially about the exact figures involved, so let me make a few

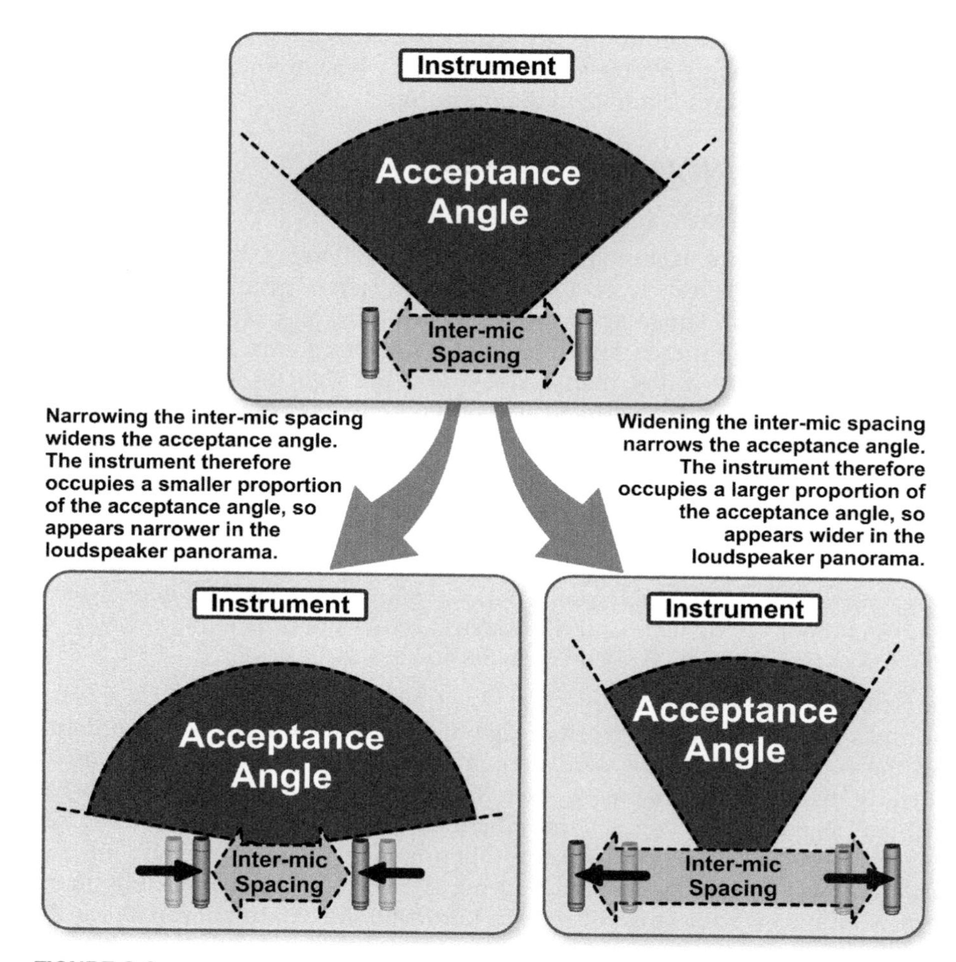

FIGURE 8.6
The correlation between inter-mic spacing, acceptance angle, and playback image width for an AB stereo mic setup.

suggestions in this respect from my own practical experience. Firstly, spacings below about a foot won't really give you a full-width stereo picture. While this may not concern you unduly if room reflections aren't a big factor in the sound, it's a shame to narrow the apparent recording space if you're pursuing a more natural acoustic vibe.

The second thing to say is that most AB recordings I hear coming out of small studios seem to have been recorded with the microphones too far apart, such that most of the instrument's sound comes from the edges of the loudspeaker panorama and the center of the image feels very unstable—a situation that invariably goes hand in hand with poor mono-compatibility. For this reason, the acceptance angles I suggest for different mic spacings are smaller than some authors suggest, and can be seen in Table 8.2. (If you don't have a ruler on-hand, it can be useful to remember that a sheet of A4 paper measures roughly 12 × 8 inches.)

Table 8.2 Suggested Mic Spacings and Acceptance Angles for AB Stereo Arrays

Spacing	Acceptance Angle
12 inches (30 cm)	110°
18 inches (45 cm)	80°
24 inches (60 cm)	60°
30 inches (75 cm)	50°
36 inches (90 cm)	40°

One of the big advantages of AB techniques is that you can choose whatever polar pattern you like for the mics without affecting the quality of the stereo imaging—as long as the patterns are well matched. Omnis are probably employed most commonly, given their typically clearer timbre compared with other polar patterns, but it's a mistake to think that AB isn't possible with other mics. Although figure-eights won't provide the same low-end extension as omnis, so may require low-end EQ compensation, the clarity of their off-axis response makes them otherwise well suited to AB techniques. Renowned classical engineer Tony Faulkner has frequently used this kind of setup commercially, for example. As it happens, figure-eights provide some distinct advantages over omnis in some cases: Firstly, you can place them further away than omnis without drowning in room reverb (an important consideration in light of the narrow acceptance angles in Table 8.2); secondly, the polar pattern's null plane rejects room reflections entering from the sides of the array, where the time-offset between the mics is greatest, improving mono compatibility; and thirdly, in narrow or low-ceilinged spaces the mics will reduce pickup of the typically more colored-sounding reflections across the room's shorter dimensions. The sensitivity of a figure-eight mic will already be roughly 3 dB down once you get 45° off-axis, though, so be aware that this may tamper with the natural balance between sounds coming from the center and the edges of the acceptance region. Cardioid-family mics will also work in an AB configuration, although cheaper ones are best used in drier acoustics where you can count on most of the sound arriving from in front, otherwise nasty colorations imposed on the rear pickup can really pollute the overall recorded tone.

GAIN-MATCHING XY AND AB RIGS

In order to adequately judge the imaging properties of an XY or AB stereo mic array, you should try to match the gain settings of the two mics as closely as you can. Probably the most accurate means of doing this is to set up a sound source at the dead center of the array's acceptance region, and then use the polarity-inversion technique described in Section 7.1.1, but adjusting the relative mic gain settings in step four of the procedure, rather than their positions—the better the cancellation between the mics, the better the gain match. However, such exact gain-matching feels like overkill to me for most small-studio sessions, especially if the musicians have to hang around while you do it. As long as you keep your ears and eyes open while tweaking your mic preamps' gain controls, my experience is that any vestigial stereo lopsidedness can almost always be redressed at mixdown with negligble side-effects.

8.4 COMBINING LEVEL DIFFERENCES AND TIME DIFFERENCES

8.4.1 Designing Near-Coincident Arrays

Although XY, MS, and AB stereo arrays can all generate a stereo picture, there is a distinct subjective difference between the imaging and reverb qualities of coincident and non-coincident techniques: Level differences give a solidity and precision to the stereo imaging that time-difference techniques can't match; but at the same time pure level-difference stereo can feel rather sterile and clinical compared to the involving spaciousness that time differences provide. So it should come as no surprise that recording techniques combining both level-difference and time-difference information are very popular. Some people also find these hybrid methods more intuitive to work with, because they more closely mimic the mixture of inter-ear level and time differences that our own hearing system naturally presents us with.

The most common scheme introduces a spacing between the mics of an XY pair, creating what's known as a "near-coincident" rig. The stereo time differences work cumulatively with the stereo level differences in this case, narrowing the XY rig's acceptance angle. However, if you wish to retain the same acceptance angle, you can just reduce the mutual angle between the mics to compensate—the stereo widening effect of the increased time-difference contribution is balanced by the stereo narrowing effect of the decreased level-difference contribution. The exact way in which mic spacing and mutual angle interact is pretty complex, so once again I prefer to rely on a simplified approximation which can easily be called to mind in the heat of the moment. Table 8.3 expands the information from Table 8.1 to show how the acceptance angles of different XY arrays will be affected by each inch (2.5 cm) of spacing introduced between the mics. So, for example, given that I'd normally use a cardioid XY setup at a mutual angle of 110° to capture an acceptance angle of 160°, I'd expect the acceptance angle to shrink below 100° if I introduced an eight-inch (20 cm) spacing between the mics. While this "back of an envelope" math isn't perfect, it's nonetheless accurate enough to be remarkably useful for guessing initial array setups during the cut and thrust of a real recording job. (Again, a sheet of A4's 12 × 8-inch dimensions can easily be folded to give spacing templates for 2, 3, 4, or 6 inches if you've no ruler with you.)

Table 8.3 Suggested Mutual Angles and Acceptance Angles for XY and Near-Coincident Stereo Arrays

XY Stereo Array	Mutual Angle	Acceptance Angle	Acceptance-Angle Reduction Per Inch (2.5 cm) of Mic Spacing
Crossed Cardioids	90°...**110°**...140°	200°...**160°**...130°	10°...**8°**...6°
Crossed Supercardioids	70°...**90°**...120°	170°...**130°**...100°	8°...**6°**...4°
Crossed Hypercardioids	60°...**80°**...110°	150°...**110°**...80°	7°...**5°**...3°
Crossed Figure-eights	50°...**70°**...100°	130°...**90°**...60°	6°...**4°**...2°

But what's the best combination of mutual angle and spacing? Well, there are a variety of factors to consider. At the most elementary level, the lower the mutual angle you use, the drier (and therefore closer) the recording will sound, because you'll be narrowing the array's combined polar pattern. Where an XY cardioid array will sound roughly as wet as a single subcardioid mic in that location, an AB pair of cardioids will only sound about as wet as a single cardioid (although obviously more spacious on account of the time-difference cues), so juggling the spacing and mutual angle of a cardioid near-coincident array lets you finesse its dry/wet balance within those boundaries. However, trading off mic spacing and mutual angle doesn't make an enormous difference to the apparent balance between sounds at the center and edges of the acceptance angle—for a given microphone polar pattern, widening the stereo image will always tend to recess the center of the acceptance zone, whatever mixture of level and time differences is responsible, so I think you should still be careful using any acceptance angle outside the ranges I already recommended for pure coincident arrays.

I've stressed repeatedly that the mics used in budget-conscious studios often have dodgy off-axis response, and this can cause problems when using larger mutual angles—large-diaphragm models being the worst offenders. If you're using mics at, say, a 140° mutual angle, central sounds will be arriving 70° off-axis to each mic, and each mic's on-axis response will also over-emphasize the brightness of reflections from outside the array's 120° acceptance region. This is where near-coincident arrays can provide a useful workaround, because they let you implement the same acceptance angle with a lower mutual angle. So the aforementioned 120° acceptance angle could instead be created using cardioids with a 90° mutual angle and an eight-inch (20 cm) spacing, putting central sounds only 45° off-axis to each mic and moving the brighter on-axis pickup within the acceptance angle. The result: a more even tone across the acceptance region, and a more natural-sounding room reverb. (For similar reasons, angling the mics of an omni AB array slightly outwards can usefully widen its high-frequency coverage when working with large-diaphragm mics.) However, this has to be weighed up against the issue of mono-compatibility, because widening the mic spacing to narrow the mutual angle will increase the likelihood of comb-filtering problems if you sum the channels.

BAFFLED STEREO TECHNIQUES AND DUMMY HEADS

Beyond the basic XY, MS, AB, and near-coincident stereo techniques, there are plenty of less mainstream methods. One group of these takes inspiration from the way the human ears are separated by the head during natural listening, and therefore places some kind of baffle between a spaced pair of microphones. This introduces high-frequency shadowing effects which add extra level-difference stereo information, sharpening and widening the stereo imaging and adjusting the nature of the reverb pickup.

One of the best-known baffled-stereo rigs is the OSS system, where the baffle is a 12-inch (30cm) absorptive disc—often called the "Jecklin Disc" after its inventor. This is vertically positioned midway between two omni microphones spaced about 8 inches (20cm) apart. Budget recordists quite often experiment with OSS, because building a passable Jecklin Disc yourself isn't that difficult or costly. However, in my experience the technique rarely seems to offer the tonal clarity of a regular AB rig, because of midrange comb-filtering introduced by residual baffle reflections, and the array also seems to dull the top octaves somewhat overall. So if you really want HF level-difference information when working with omni mics, I'd suggest angling them slightly outwards as a first port of call before getting any baffle involved. Alternatively, try mounting your spaced omnis either side of a football instead, right next to the surface of the ball. This should reduce the comb-filtering effects to a small loss of extreme high-end, and although you'll suffer a couple of decibels of upper midrange boost courtesy of the boundary effect (the ball's too small to reinforce frequencies any lower than that), that's easier to compensate for with EQ than heavier comb-filtering. If you have a pair of boundary mics available, a further simple option is to mount them on either side of a large, heavy panel. Again, you may have to add some low-end with EQ, because any manageably sized panel is only likely to provide a useful degree of boundary effect above about 200Hz.

Drawing even greater inspiration from human hearing are systems which use an anatomically modeled "dummy head" (based on Ozzy Ozbourne's, I trust) with mics fitted into prosthetic ears, or else place the mics into the engineer's own ears. The resulting "binaural" recording captures subtle directional frequency-response variations imposed by the human pinnae, and when auditioned on headphones it can recreate uncannily accurate sound-positioning sensations from all around the listener—including from above and behind. However, that realism doesn't translate very well to speaker systems, where the imaging typically becomes rather vague, with an overwide stereo spread. That's not to say that some high-profile engineers don't use binaural techniques in the studio—Tchad Blake[2] and Michael Bishop[3] are enthusiastic proponents, for instance—but given the cost of a proper dummy-head system, or the inconvenience of having to sit out in the recording room with miniature mics in your ears while tracking, I wouldn't really recommend binaural methods for most small-studio folk.

8.4.2 Case Study: NOS or ORTF?

Small-studio recordists lacking the confidence to design their own near-coincident array are often drawn toward a handful of well-known standardized cardioid setups with impressive-sounding abbreviations. In practice, though, this doesn't make their lives much easier, because they still have to work out which to choose and how best to use it. So as a means of consolidating what we've covered so far in this chapter, I'd like to demonstrate how the guidelines I've given above might enable you to make a more informed choice between two of the most well-known: NOS (*Nederlandse Omroep Stichting*), which has a 90° mutual angle and a spacing of 12 inches (30 cm); and ORTF (*Office de Radio-Télévision Française*), which has a 110° mutual angle and a spacing of 7 inches (17 cm).

"Small-studio recordists lacking in confidence are often drawn toward a handful of well-known standardized cardioid setups with impressive-sounding abbreviations. In practice, though, this doesn't make their lives much easier, because they still have to work out which to choose and how best to use it."

Looking at Table 8.3 suggests that 90° and 110° XY arrays will give me 200° and 160° acceptance angles respectively, and that I'd expect the NOS and ORTF mic spacings to reduce those figures to 80° and 104° respectively. Because both these acceptance angles fall outside the "safety zone" for XY arrays, sound sources at the edges of the acceptance region are likely to be favored in the level balance no matter which I use—an effect that will be emphasized by any natural on-axis brightness of the microphone design, because the mutual angle happens to match the acceptance angle quite closely for each of the arrays.

For a more spacious presentation (albeit at the expense of lesser mono-compatibility), I'd choose the NOS rig, given its heavier reliance on time-difference information, whereas ORTF would, conversely, suit applications where slightly sharper imaging were desirable. If my ability to distance the mics from the instrument were restricted, then I'd probably lean toward the ORTF setup on account of its larger acceptance angle, although this might depend on whether the slightly roomier sound of its wider mutual angle proved suitable. However, with budget large-diaphragm mics, the NOS setup might begin to become more attractive irrespective of miking distance if I wanted to maximize the brightness of an instrument (or instrument component) at the center of the acceptance region.

8.5 STEREO MIKING IN THE WILD

So far I've restricted the discussion to traditional stereophonic techniques, where imaging realism is a primary concern. However, high-profile professionals frequently choose to sacrifice realistic imaging when using mics in stereo day to day, so it's as well to consider why and how you might also want to do this yourself.

8.5.1 Tonal Balance Versus Stereo Balance

Where the mics in any close-up stereo array spotlight different sonic facets of the instrument, it's not uncommon for tonal concerns to take priority over stereo imaging. For example, if you had placed a cardioid XY array somewhere around the twelfth fret of an acoustic guitar, you might decide to boost the mic facing the strings for more "zing" (as mentioned in Section 7.1.3), in spite of this stretching the whole stereo picture toward the right-hand speaker. Sure, you could spend a few minutes repositioning the whole mic array to find a zingier overall timbre without skewing the stereo image, but under time pressure most engineers will simply nudge the right-facing microphone's pan control inward to counteract the image shift, just as long as narrowing one side of the overall stereo image in this way doesn't sound too awkward in context.

MIKING FOR SURROUND

Although stereo mic technique might seem a bit involved at times, it's not a patch on the complications of surround-sound miking! For this reason, I'd strongly advise against tangling with surround miking until you've gained a good deal of practical experience with stereo techniques, and have supplemented the practical advice in this book by delving into some of the more technical literature about stereo and surround systems. Because vanishingly few small-studio users ever use surround mic techniques, I also won't deal with them in this book, but if you'd like to pursue the subject further, I've included some links to further reading in this chapter's web resources.

A very similar situation can arise where an instrument is miked up so that its pitch range is spread across the stereo panorama, with low notes on one side and high notes on the other. If you adjust the levels of your stereo mic rig to compensate for a weak-sounding higher register, say, you may find yourself pulling the whole stereo image off-center. As before, repositioning the mics is the ideal solution to this problem, but panning adjustments may be more expedient when the clock's ticking.

8.5.2 Beyond the Bounds of Acceptance

The mutual angles and mic spacings recommended in Tables 8.1, 8.2, and 8.3 are designed to help you capture a full-width stereo picture without seriously misrepresenting the level balance of sound sources within the acceptance region. On occasion, however, an engineer might have good reasons for using unusual acceptance angles in order to rebalance sounds coming in from different angles. For example, narrow acceptance angles can be used with coincident or near-coincident arrays to avoid overemphasizing those parts of the instrument nearest to the array when close-miking—this is something I've already mentioned in relation to piano recording, but it's equally applicable with things like acoustic guitar and tuned percussion, which also present a comparatively wide target to any close-mic setup. (A word of warning, though, when using this concept: If the performer moves a lot while playing, you'll not only get them bouncing around the stereo image, but also ducking and diving in the level balance.)

Stereo room mics may also benefit from the weakened central image of a narrow acceptance angle where they're supplementing signals from separate close mics. This allows the room mics to add maximum spaciousness in stereo,

FIGURE 8.7
Non-coincident close-miking setups such as the one shown here are commonly seen in small studios, because they allow you to catch more of an acoustic guitar's timbral components without picking up too much room reverb. However, if you treat these two mics as an AB stereo pair, the combination of level-difference and time-difference information will yield a very narrow acceptance angle, which means you'll likely have to narrow the mics' panning to avoid a disconcertingly wide stereo image.

without clouding the close-mics' details at the center of the stereo image. You risk unpleasant phase-cancellation, though, if the room mics are ever summed to mono, so you should check for this while recording. If precise directional information is unnecessary (as it often is when room-miking), you can also maximize the reverb pickup within a given room by using the rather odd near-coincident arrangement of a spaced pair of figure-eight mics at a mutual angle of 180°, because this rejects the instrument's direct sound arriving from the front of the array.

"If precise directional information is unnecessary, you can also maximize the reverb pickup within a given room by using the rather odd near-coincident arrangement of a spaced pair of figure-eight mics at a mutual angle of 180°, because this rejects direct sound arriving from the front of the array."

Acceptance angles are also narrowed whenever AB arrays are used in close proximity to an instrument, because the moment the mics begin to pick up significantly divergent timbres they'll effectively be introducing stereo level-difference information between the channels. Even under normal circumstances AB techniques give quite narrow acceptance angles, so it's easy to arrive at an excessively wide stereo image with an unstable center when close-miking, especially if directional mics are used. As such, it's usually necessary to narrow the panning of the microphone channels when using AB rigs up close. "I'm never panning them hard left and right," remarks Gary Paczosa,[4] for instance, when talking about AB-miking fiddles, "I'm only opening them maybe...from eleven o'clock to one o'clock." Fortunately, the increased tonal difference between the two mic signals under

WORKING WITH HEADPHONE MONITORING

Recording with stereo mic techniques is made more difficult for small-studio engineers if they're forced to monitor on headphones while tracking. Although $200 (£150) or so should get you fairly classy-sounding cans from a tonal point of view, no matter how much you spend, they'll still give a very different stereo listening experience than speakers do. For a start, the ±90° positioning of the headphone drivers means that the stereo field covers a 180° angle, three times the normal 60° loudspeaker panorama. In addition, each ear only hears the output of one microphone on headphones, whereas both ears hear both mics on loudspeakers. Together, these two effects not only make it tough to judge the suitability of your mic array's acceptance angle, but also affect your perception of dry/wet balance, because reflected sounds are much more spread out spatially.

The mutual-angle and mic-spacing suggestions earlier in this chapter can help here, of course, but there are a couple of additional workarounds that can further improve the likelihood of workable results. The most useful thing you can do is find a commercial recording that you like the sound of, and then import it into your recording system to use as a yardstick when judging your own mic array's stereo width, dry/wet balance, and center-versus-edge level balance. The other really helpful tool is a vectorscope, a special type of metering display, which is very revealing of stereo width/balance issues that might not otherwise catch your attention. This needn't cost you anything if you're recording with DAW software, because there are a number of free cross-platform vectorscope plug-ins available (see this chapter's web resources for some suggestions).

Several manufacturers now provide affordable software which claims to simulate the speaker-listening experience for headphone listeners, but none of the ones I've heard are particularly helpful in this particular context. For a start, they often alter the overall tone of what you're recording, and some of them add simulated room reverb at the same time, both of which confuse matters. In addition, many of the algorithms are based on averaged data about how the human hearing system works, so they suit some people's ears much better than others. To be honest, I think the traditional headphone stereo experience is no less usable for recording purposes than any of these systems, just as long as you have some commercial reference tracks on hand for comparison.

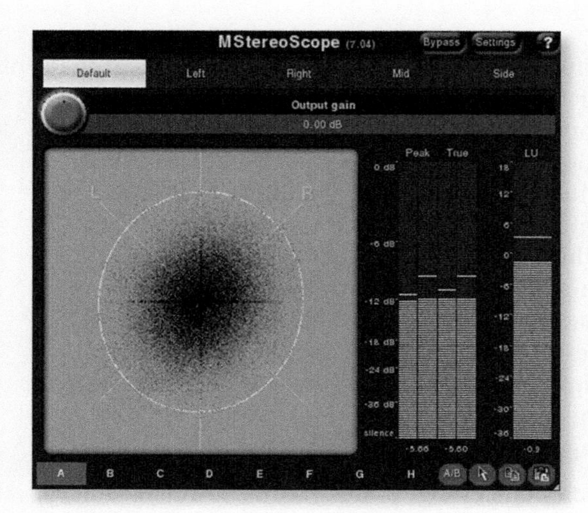

FIGURE 8.8
A vectorscope (such as the freeware Melda MStereoScope shown here) is very handy for checking a recording's stereo width when you're forced to monitor on headphones.

such circumstances tends to reduce the severity of the comb-filtering between them, which makes panning adjustments less damaging to the combined tone than they would be for more distant placements. Some engineers deliberately design their AB setups to take advantage of this. Mike Shipley,[5] for example, mentioned consciously separating the mics by three times the miking distance to minimize comb-filtering when he recorded Alison Krauss's "Paper Airplane," and Bill VornDick[6] has also talked about using that same idea.

Large acceptance angles can also be useful to avoid a disconcertingly wide image when close-miking. If you reduce the array's mutual angle, Sides level, or spacing enough (or indeed narrow the monitor-channel pan controls), then the instrument's stereo image functions more as a width enhancement than as a true stereo picture—an effect that's extremely popular with engineers who want to create full-sounding mixes using only a few instruments. This kind of stereo spread can be heard in Tony Platt's[7] guitar recordings for AC/DC, for instance, as well as on myriad acoustic, jazz, folk, country, and bluegrass productions. "I stereo-mike everything," says Gary Paczosa,[8] for instance, "or at least I use two microphones; it's not always a true stereo configuration." A nice bonus feature of the tag-team positioning trick I suggested back in Section 6.2.1 is that it's usually very good for this kind of subtle image widening if you use your final two mic positions as a stereo pair—the signals will usually end up being quite similar, so the widening remains fairly natural-sounding (as long as you take care with phase-cancellation), and you can also normally afford to mix the mics at fairly equal levels, which avoids lopsided stereo.

8.5.3 Stereo Setups with Unmatched Mics

If you've no choice but to try capturing a natural stereo image without having any mics of the same model, then you'll probably get best results with MS techniques, because even if frequency-response differences between the mics blur the spectrum, at least central images should be solid and the stereo panorama as a whole should remain symmetrical. (This assumes, however, that the figure-eight mic has symmetrical front and back lobes, which isn't always true of ribbon mics.) After MS, a pair of unmatched omnis in AB configuration would perhaps be the best bet, as the subjectively more diffuse nature of pure time-difference stereophony is more likely to disguise any extra blurring of the stereo picture.

However, when you don't need anything beyond a nebulous sense of width from your stereo recording, there's nothing to stop you spreading any of Chapter 7's multimic setups across the panorama using your pan controls—indeed, this practice is extremely popular on a commercial level. Just remember to be wary of mono-compatibility problems when panning any non-coincident rig, especially if heavy phase-cancellation effects between the microphones have been used to sculpt the timbre, as in Section 7.2.2.

8.6 BEYOND STEREO PAIRS

Although we've covered a wide range of stereo recording methods this chapter, that's by no means all there is to know about stereo recording. But more complicated stereo setups usually only make sense within the context of ensemble recording, so we'll take those up in Part 4 once you've had a chance to gain some practical experience of the techniques we've covered thus far.

CUT TO THE CHASE

- Most stereo mic techniques use level differences and/or time differences between the microphone signals to generate the stereo panorama. Pure level-difference techniques (XY, MS) have more precise stereo imaging and (if you're careful) excellent mono-compatibility; pure time-difference techniques (AB) give a more spacious impression and allow you to use a wider range of microphones; and hybrid near-coincident techniques afford greater flexibility to angle the microphones for better dry/wet balance and overall tonality. Baffled stereo or binaural techniques have their uses, but won't provide very good value for money in most small studios.
- Don't choose an MS mic rig just for its ability to control stereo width, because you can do that with any mic array using MS post-processing. More important features of MS are its typically brighter and more solid central images; its ability to use an omni Middle mic; and its symmetrical stereo image even when using unmatched mics.
- You can decrease the acceptance angle of an XY array by increasing the mutual angle or using more directional polar patterns; of an MS array by increasing the Sides mic's level; and of an AB array by increasing the spacing between the mics. With near-coincident arrays, the acceptance-narrowing effects of mutual angle, mic polar pattern, and inter-mic spacing accumulate.
- All stereo arrays have a range of acceptance angles for which they'll give their most even level balance across the acceptance region, but there are situations where tampering with center-versus-edge levels can be used creatively. Some arrays won't give you a full-width image at all, although this may play into your hands if you're overdubbing dry sounds.
- Don't assume that AB recordings have to be made with omni mics; other patterns can also be very useful. Coming in very close to an instrument with AB will add level-difference information, narrowing the acceptance angle, in which case panning adjustments may be required to avoid an overwide stereo image.
- Swapping out the mics in a stereo array can be a pain, so do your best to decide which mics to use before messing around too long with stereo-specific setup issues. Dedicated stereo mics and factory-matched mic pairs are rarely a good investment when you're working on a budget.
- Stereo miking techniques are best judged on loudspeakers, but if you have to monitor on headphones then use reference tracks in conjunction with a vectorscope to improve your guesswork.

Assignment 1: Close-Miking Instruments in Stereo

Record at least one of each of the following groups of instruments in stereo: acoustic piano, acoustic guitar/mandolin/banjo/harp, tuned percussion; any monophonic orchestral instrument; any amplified electric guitar or electromechanical keyboard.

- For each session, make a point of comparing two different types of stereo array, preferably one each of the following: XY versus MS; XY versus AB; and AB versus near-coincident.
- Feel free to choose whichever mics you'd like for each stereo configuration in pursuit of the most appealing result, evaluating each rig for general tone; stereo width; dry/wet balance; and mono-compatibility.
- This assignment will be most revealing if you record both mic pairs simultaneously, but on separate tracks. Following the recording session, compare the two stereo arrays in detail, and take the opportunity to experiment with MS post-processing on each one to get a feel for how they respond.

Assignment 2: Stereo-Miking Instruments with Room Reverb

Record any solo instrument you like, but from a distance which allows you to take advantage of the recording space's reverb—if the room reflections in your normal recording space sound a bit dodgy, consider recording on location for this assignment.

- For each session, make a point of comparing two of the following stereo arrays: XY, MS, AB, and near-coincident.
- Feel free to choose whichever mics you'd like for each stereo configuration in pursuit of the most appealing result, evaluating each rig for general tone and stereo width (both of the instrument and the reverb); dry/wet balance; and mono-compatibility.
- If possible record both mic pairs simultaneously, but to separate tracks. Following the recording session, compare the two arrays in detail and experiment with the possibilities of MS post-processing.

WEB RESOURCES

On this book's companion website you'll find a selection of resources to support this chapter, including:

- audio demonstrations of a variety of stereo setup issues;
- a downloadable stereo protractor printout in PDF format, as well as instructions on how to fold your own angle templates using plain A4 paper;
- some suggested implementations of low-frequency shuffling for coincident pairs;
- further reading on XY, MS, AB, near-coincident, and baffled-stereo techniques, and on creating click tracks;
- links to on-line stereo setup calculators and freeware MS processing plug-ins.

http://www.cambridge-mt.com/rs-ch8.htm

PART 4
Multiple Sources, Multiple Mics

By now it should be clear that single instruments can be overdubbed in a multitude of different ways, and that the working practices of different professionals can be understood as their response to the specific music, performers, environment, and equipment they have to work with. You'll also doubtless have developed your own preferences while working through the chapter assignments—but you should also have an insight into the comparative strengths of many alternatives, so you'll always have somewhere to turn in situations where your favorite tricks draw a blank. And nowhere does this adaptability have greater value than when you're capturing several instruments/voices at a time, whether that's a large ensemble such as an orchestra, big band, or choir; a smaller chamber/folk/rock group; or simply any musician who performs multiple instruments at once, such as a rock band's drummer or a singing accordionist.

As I see it, there are two basic approaches toward capturing any group of instruments/voices in stereo:

- **Dominant Array.** Use a primary mic array (whether that's a single mic or a complex multimic setup) to achieve a faithful overall representation of all the sound sources. If that picture then requires adjustment for corrective or creative reasons, this dominant array may be supplemented by additional "spot" or "accent" microphones, although the main array typically still dominates the balance. This approach is good for natural-sounding ensemble recording, hence its popularity in the classical world, but it's also extremely useful for more mainstream commercial styles (especially when recording drum kits), and is very well suited for small-studio engineers who need to get respectable results in a hurry and don't have many mics to play with.

- **Peer Arrays.** Use a separate microphone array for each instrument (or group of instruments) and create the desired ensemble balance by adjusting the mix level of each array. None of the arrays can be seen as dominant here, as each is usually responsible for capturing the bulk of the sound of its associated instrument(s). This approach can be used for any type of music, but is more commonly encountered in popular styles where the engineer wants greater control over the balance and timbre of individual ensemble components. The downside is that peer-array recording generally requires more gear and more setup time.

Of these two methods, the first presents the gentler learning curve, so we'll begin there. However, let me first just clarify my use of the term "mic array" here. Although this phrase usually implies a multi-mic setup of some kind, in the remainder of this book I'll be assuming than an array can also be just a single mic, just so that I can talk about the general functions of mic arrays without having to write "mic or mics" the whole time!

CHAPTER 9

Ensemble Recording with a Dominant Array

9.1 USING A SINGLE-MIC DOMINANT ARRAY

To home in on the bedrock techniques of ensemble recording, let's start by considering the most "no frills" dominant array: a single mic capturing everything in mono.

9.1.1 Ensemble and Venue Reconnaissance

Organizing musicians for an ensemble recording can be a massive logistical headache, so sessions frequently have a strict schedule and you'll be under time pressure. Therefore good guesswork remains paramount if you're to avoid squandering precious minutes on setup chores. At the very least, find out the ensemble lineup ahead of time and scrutinize any available information about the recording venue if it's unfamiliar to you. However, there's no substitute for first-hand experience, so get yourself to a rehearsal/concert to hear the players in action, and sneak into the recording venue in advance to check it out. Pay special attention to the following questions:

- **How do the performers set themselves up for rehearsal and/or performance?** Every player will be accustomed to hearing a certain balance of the ensemble, and may rely on close visual contact with the other performers too, so this should be reflected in your session planning. If in doubt, I suggest following this advice from Leslie Ann Jones:[1] "Find out how people normally set up and try not to deviate from that, because that's what they're most comfortable with." However, any differences you notice between the rehearsal and performance layouts may indicate which instruments can be repositioned in the studio without unduly discomfiting the players, so keep your eyes peeled for those.
- **Does any instrument (or group of instruments) either overpower the others or get lost in the balance?** Professional musicians will intuitively find an appropriate balance amongst themselves without much help from the recording engineer, but if you don't get to work with that caliber of performer very often, then be alert for any instruments/voices which may

need assistance in maintaining a sensible level. For example, many amateur choirs have one or two weaker sections (or perhaps a single hurricane-force soprano!), while student orchestras are frequently underpowered in the strings/harp department or else overwhelmed by the brass. Outside the classical sphere, acoustic guitars and upright basses can find it difficult to cut through (even piano may struggle in larger ensembles such as big bands), while any low-register drum can easily eat a small acoustic ensemble for breakfast. Where the "ensemble" is a drum kit, less accomplished players may hammer the hi-hat at the expense of the other drums. "With a really good drummer," says Jack Endino,[2] "you just hang a mic over and it's amazing, no matter what he's playing. You don't get drummers like that too often. You get drummers who are just wailing on the hi-hat and they've got a little wimpy snare drum." Whatever imbalance you hear, try talking to the performers about it, because they may be able to solve such problems before the mics get involved. "If you talk to the drummer," says John Hampton,[3] "you'll find out that he may be able to live with the ride in a place where it is more easily heard. A lot of drummers are like that." Gordon Raphael[4] illustrates this with a splendid anecdote about working with The Strokes's drummer Fabrizio Moretti: "To get a proper 'drum machine' gated snare… we had to put the hi-hat on the other side of the drum kit, and that was Fab's idea: 'Maybe if I put the hi-hat four feet away from the snare mic, I can still play it with another hand and keep time, but there won't be any bleed.' That was a great move, because we had a coherent, live sound, not an overdubbed hi-hat."

- **Do any instruments sound horrid?** Thin-sounding violins, banjos, or acoustic/resonator guitars? Abrasive electric guitars or horns? Honky piano? Anemic bass? Any kind of melodica? Again, query things like these with the players at the earliest opportunity. But if your suggestion of ritually sacrificing the melodica falls on deaf ears, then just make a note of the issue so you can plan to tackle it by other means.
- **Does the recording venue restrict the layout of instruments, musicians, or microphones in any way?** The space requirements of ensemble recording often force small-studio engineers to record on location, but this can incur other physical restrictions. A narrow room may require a large ensemble to be arranged with an unnaturally large distance between front and back, for instance, while a low ceiling may militate against using mics overhead. Are there any risers, pews, steps, doors, or balconies which might hamper the placement of large mic stands or heavy instruments (e.g., pianos)?
- **What are the acoustics of the recording venue like?** If the room doesn't sound terrific, then you'll want to investigate ways of minimizing the recorded reverb. However, on those happy occasions where the recording environment sounds like it might enhance the sound, you'll be more concerned with achieving a good dry/wet ratio for all the instruments in the ensemble. Also, consider how you might use any areas of the room that have different reflective properties, and explore the acoustic effects of

moving large objects in the room (carpets, curtains, rows of seats/pews) or opening/closing doors, windows, and partitions.

- **Is anything in the venue usable as an acoustic absorber/reflector?** Soft furnishings such as rugs, bedding, sofas, and clothing racks can serve as rough-and-ready absorbers, while things like whiteboards, large sideboards/wardrobes, and upended tables can make great reflectors. It's always a wise idea to have a selection of absorbers and reflectors on hand for any ensemble session, but there's no sense in lugging bulky things like that to a venue unnecessarily.
- **Have the ensemble or venue been involved with any previous recordings?** If you've gone as far as making contact with the performers or venue owners, then you might as well find out if they've witnessed any relevant recording setups, whether successful or not. Given the number of variables involved in ensemble recording, don't be shy about fishing for suggestions—if nothing else, the responses you get will give you a good idea how confident the musicians are likely to be in the studio, and how suitable the venue is likely to be for recording purposes.

Laden with the fruits of this research, you're ready to start making some educated guesses about suitable initial setups for your recording.

CRITICAL DISTANCE

If you move far enough away from an ensemble, there comes a point where the level of direct sound from the ensemble matches the level of reflected sound from the environment—the so-called "critical distance." In large recording spaces you can use the critical distance as a way of estimating initial miking distances, so some engineers like to measure this aspect of unfamiliar venues during their pre-session investigations. The normal way to do this is by measuring the level of a broadband noise source (placed where the front of the ensemble is likely to be) at different points in the room with a portable SPL meter, but you can also make a reasonable stab at it using a detuned radio, an omni mic, and any reasonably wide-range digital level meter. Within the critical distance the measured sound level will decrease 3–5 dB for every doubling of the miking distance, but beyond the critical distance you get very little change at all—i.e., the measurement stays pretty constant. (For more detailed measurement instructions, check out this chapter's web resources.) Common consensus is that sensible initial miking distances tend to cluster around a third of the critical distance for omni mics, and a half of the critical distance for cardioids, although clearly you need to develop your own personal tastes here.

It's fair to say, however, that most practicing engineers are content to judge initial miking distances solely by ear, so don't feel obligated to work in terms of critical distance if you're already fairly happy with your own intuition. Nevertheless, if you think the measurement-based approach might come in handy on some occasions, then it's worth doing dry-run critical-distance measurements for a few recording sessions before you need to use such data in earnest, so you develop a personalized sense of how your own preferred miking distances relate to the data.

9.1.2 Which Mic?

When shortlisting promising contenders amongst your mic collection, all the guidelines in Part 3 continue to apply, but with the proviso that groups of contrasting instruments are unlikely to be well served by mics which strongly color the recorded tone, since the chances that the coloration will flatter several instruments at once are pretty slim. Dynamic mics are rarely useful for mixed ensembles for this reason, and I would in general favor mics with the flattest possible frequency response, both on- and off-axis (which usually means leaning toward pricier models). Don't discount ribbons on those grounds, though, because although they're typically duller sounding than condensers, and lose some low end when used at a distance, they may otherwise be very well behaved in the frequency domain and should respond well to broad corrective EQ boosts at the frequency extremes.

9.1.3 How Wet Overall?

As the next step, I suggest asking yourself how much room reverb you want to capture overall, because the issue of dry/wet balance probably imposes the greatest restrictions on your mic placement. This is where the venue reconnaissance can really help. Firstly, it tells you whether the room reverb's tonality is likely to be useful at all—if not, then you'll have to work hard to eliminate it. And, secondly, it indicates roughly what range of miking distances will give you a decent dry/wet ratio should you choose to make use of the natural reverberation.

Other than that, your main tools for controlling reverb pickup will be just the same as when working with single instruments: shifting the mic position, switching its polar pattern, and adjusting the room acoustics. However, the first two of these strategies offer less leeway when you're working with ensembles. For example, you may struggle to get a single mic close enough to an ensemble for the right overall dry/wet ratio without overemphasizing the levels of the closest individual instruments in the balance. In a similar vein, trying to dry up an ensemble's sound by switching to a more directional polar pattern may mean that the mic's off-axis response seriously misrepresents the level or tone of some players. In this respect, a figure-eight mic with groups of performers arranged on either side may outperform a cardioid facing a single group of performers, by virtue of the cleaner off-axis timbre. (Using a figure-eight mic in this way also provides the considerable ancillary advantages of improving eye contact between the performers and allowing you to place foldback speakers in the mic's deep side nulls if you don't have enough headphones for all the musicians.)

Because ensembles reduce the power of pure mic technique in this way, the importance of dealing more directly with reverb issues increases, so strive to get maximum value out of acoustics modifications in the first instance. Reducing excessive (or just dodgy-sounding) reverb is the most common requirement on low-budget sessions, although the very act of putting lots of musicians in one place can actually help with this. "I always felt that the sound in any studio was always better when the room was filled with people," remarks Larry Levine,[5] "regardless of whether you could get isolation or not." The DIY absorption measures we've touched on in previous chapters can get you a long way in smaller rooms too, but when you're in a space capacious

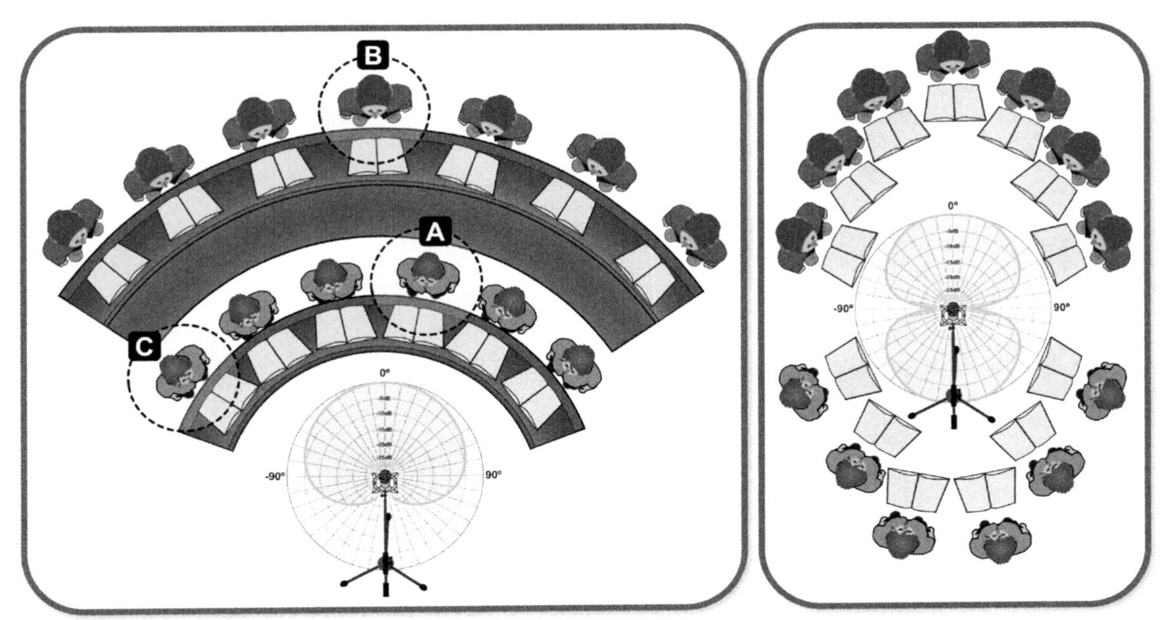

FIGURE 9.1
The recording setup in the left-hand diagram will struggle to capture all the singers in the ensemble with a uniform dry/wet balance: Singer B will sound more reverberant than Singer A because he's further away from the mic; and Singer C will also sound more reverberant than Singer A because the mic's cardioid polar pattern attenuates her direct sound. By contrast, the right-hand diagram would be more likely to achieve the same dry/wet balance for all the singers by moving them out of the pews and switching the microphone to a figure-eight polar pattern: The singers are in a single row all around the mic, and the laterally placed voices are moved closer to the microphone to compensate for the polar pattern's side-rejection.

enough to hold a large ensemble, a given stock of improvised acoustics materials won't be able to treat as large a proportion of the room's reflective surfaces, so clustering your available baffling close around the microphone instead will usually prove more efficient. (The danger there, though, is that high frequencies will be absorbed much better than low frequencies, rendering the remaining reverb pickup rather woolly sounding.) Your on-site reconnaissance plays a vital role here, of course, because moving furniture and carpets/curtains may make a huge difference. Be aware, however, that heavily damped acoustics may bring problems of their own, by making it hard for the performers to hear each other well enough while performing.

9.1.4 What Level for Each Instrument?

Once you've got several instruments coming through one mic, you can't easily adjust their levels after recording. This makes it vital to capture a good ensemble balance during the session, so plan for this as soon as you've considered issues of overall reverb pickup. Again, mic technique only offers limited

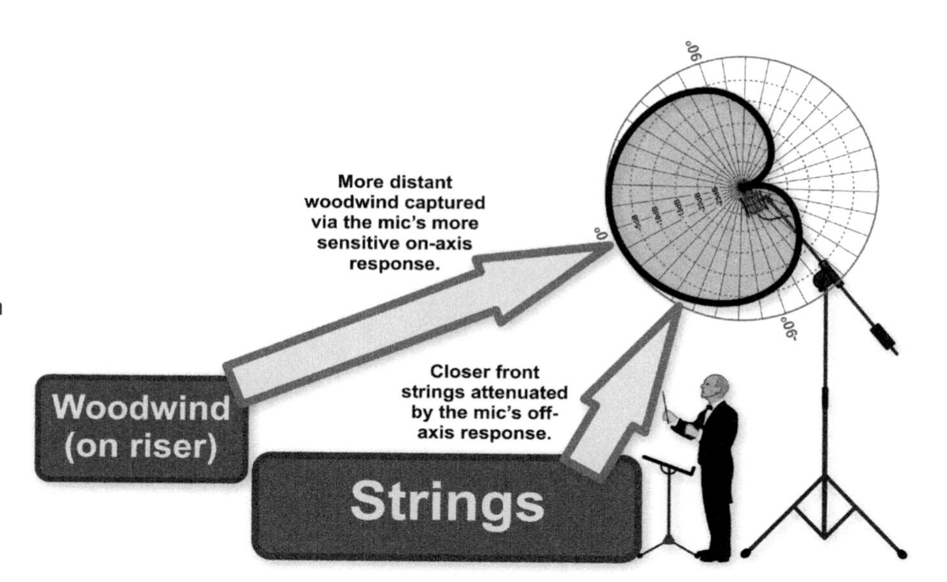

FIGURE 9.2
Miking an orchestra from a close frontal position can overemphasize the level of the strings in the balance. This diagram shows how you might compensate for this problem to a degree by using a cardioid polar pattern angled to favor the woodwind section.

control here. For instance, changing any instrument's apparent level by adjusting the mic-to-instrument distance will also alter that instrument's apparent dry/wet ratio, and possibly incur proximity-related timbral side-effects too. On the other hand, angling a directional mic to rebalance the dry sounds of instruments will also affect their dry/wet ratios, and may cause tonal damage to some of them as well.

As a result, I think a mic's type/placement is incapable of substantially correcting poorly balanced performances, although it does still offer solutions to a couple of common technical difficulties when you're trying to capture the balance "as played":

- If a mic's polar pattern introduces balance problems, then changes to miking distances can compensate. So where, say, a supercardioid mic makes an ensemble's off-axis instruments too quiet and wet, bringing those instruments closer to the mic can counteract this by boosting their direct level.
- If balance problems are caused by unsatisfactory miking distance, angling the polar pattern of a directional mic can compensate. For instance, if concerns about overall reverb pickup force you to mike from a position too close to some instruments (as is frequently the case when working with large groups), angling a directional mic's polar pattern can compensate for this by making closer sources both quieter and wetter. This is a very common tactic with orchestral recording, because it allows a mic to be placed in front of the ensemble, but aimed toward the wind section to avoid overstating the level of the nearest string players or rendering them unconscionably dry sounding, as shown in Figure 9.2. (Mic positions above the ensemble can help with this too, but a low ceiling or the lack of a sufficiently tall mic stand may prevent this approach in practice.)

Because a single mic offers so little independent control over individual instrument levels, any ensemble imbalances you hear in rehearsal (or anticipate from a glance at the instrument lineup) are invariably better addressed by other means. Since most practical remedies involve the cooperation of the players, however, the main question at the planning stage is whether the reflection/absorption characteristics of the room (or any additional DIY treatment) can be used to improve the balance. A reflective surface behind rear-firing instruments (most notably French Horns) is often desirable, for example, but any instrument that's having trouble projecting can be assisted in a similar way by hard-surfaced rear reflectors, and also by removing carpet from in front of and beneath them—even with apparently "forward-firing" instruments, additional early reflections will still provide significant level reinforcement. By extension, putting overloud instruments on carpeting with absorptive panels behind them may provide some balance correction, and in more extreme cases (such as when a rock drum kit is playing alongside acoustic instruments) you might even build a roofed booth around the back and sides of the player to more severely curtail the total energy they send out into the room. Any acoustic manipulation like this won't just alter an instrument's balance, though, because it'll also affect the timbre—the high-frequency content of the instrument will be most affected in each case, with reflectors tending to brighten the tone and absorbers tending to dull it.

9.1.5 What Tone for Each Instrument?

If the ensemble already sounds gorgeous, your primary goal is to avoid tainting that sound during the recording process, which is why classical engineers so often prefer the naturalness and clarity of high-class small-diaphragm omni microphones for ensemble capture. In the less rarified atmosphere of lower-budget sessions, though, both the ensembles and the microphones tend to be less than ideal, so it pays to think quite carefully about workarounds.

Where certain instruments sound unappealing during rehearsal, and the players themselves aren't able to remedy things sufficiently at source, try using your knowledge of instrument dispersion characteristics to advantage, particularly high-frequency beaming effects. So if the viola in a string quartet sounds a bit muffled, turn the player slightly, thereby directing more of the instrument's forward-thrown high-frequency spectrum toward the microphone; or if the tenors and basses in a large choir are being obscured by the sopranos and altos standing in front of them, investigate ways of lifting them higher, perhaps using temporary risers or a dais/step already in the venue. By contrast, with an overbright trumpet you might place a little acoustic absorption directly in front of the performer to soak up some of those searing high frequencies emanating from the instrument's bell. Philip Hobbs[6] did something similar for the harpsichord when recording John Butt's award-winning 2012 interpretation of Bach's *Matthew Passion*: "John directs from the harpsichord, so it has to be in the middle, right under the [dominant array]. So it's a challenge to stop the whole project sounding like a harpsichord concerto. We take the lid off the instrument, but put a bit of board over the middle of the soundboard just to get rid of some of the immediate 'ping' of the sound, so it sounds present enough, but nicely diffuse."

Where the ensemble sounds fine in the flesh, you have to be a bit careful that the extra on-axis brightness inherent in most mic designs doesn't misrepresent instruments directly in front of the mic. In fact, in situations where players are seated all around an omni mic, there is a good argument for orienting its firing line vertically, such that all instruments address it similarly off-axis. (The main drawback of this approach is that it tends to brighten the recorded reverb, so some additional acoustic absorption may be necessary, particularly in smaller rooms.) The same applies with cardioid-family mics, where firing the microphone a little over the top of the players may match captured instrument tonalities better across the width of the ensemble. Sometimes, however, a mic's on-axis HF boost can be turned to your advantage, especially when using omni mics, which won't change the overall level of off-axis instruments very much: If an instrument needs brightening, place it on-axis; whereas an instrument which is too biting could be placed off-axis where the mic's pickup is dullest (i.e., typically at the rear for small diaphragms, or 90° off-axis for large diaphragms).

"In situations where players are seated all around an omni mic, there is a good argument for orienting its firing line vertically, such that all instruments address it similarly off-axis."

9.1.6 What Depth for Each Instrument?

The next major issue is how far away each instrument should appear relative to the others—in other words, what it's subjective depth should be in the front-back perspective. This is something that is primarily determined by each instrument's dry/wet ratio, although tonal brightness does provide a psychological cue to the listener as well. The big problem is that changing an instrument's reverb level independently of its overall level is something that microphone technique is all but powerless to do when the whole ensemble is coming through one microphone—you may be able to adjust a player's dry/wet balance by changing their distance and/or angle relative to the mic, but either way you'll also change the instrument's apparent volume compared to other ensemble members. Given that good balance and an appealing tonal quality are usually deemed more important than a completely natural depth perspective, most pragmatic engineers would tend to prioritize balance/tonality concerns over depth issues wherever only one mic is being used.

9.1.7 Sonic Adjustments on the Session

Pondering the above issues should bring one or more likely-looking mic tactics into your sights, around which you can plan the technicalities of the session itself. As always, try to get all the equipment set up and tested long before any performers arrive, so you can concentrate on making musicians comfortable and ensuring that instruments are all well prepared for the recording. With amateur ensembles, don't be afraid to insist that small sections of the group (or even single individuals) tune separately, because few things make a recording sound cheaper than sour intonation. Where electronic tuning devices are being used, it's not a bad idea to use the same device for all the instruments, so you don't fall foul of differences in tuning reference between units.

The moment you clap ears on the mic signal, refine your initial setup guesswork by racing through the questions in Sections 9.1.2–9.1.6 again:

- Which mic?
- How wet overall?
- What level for each instrument?
- What tone for each instrument?
- What depth for each instrument?

Because these questions proceed roughly in order of importance, you're likely to work most efficiently if you concentrate on mic choice/directivity, miking distance, and global acoustic treatment issues before transferring too much attention toward per-instrument concerns. Having more than one mic primed can really speed up the selection process, as can the presence of a willing assistant to move microphones under talkback control. When assessing the overall reverb capture, you'll find decisions a whole lot easier if you've set up your monitoring system to switch easily between your mic signal and the playback of some commercial reference tracks, so I'd recommend configuring some way of doing that in advance. If you're in any doubt with reverb, though, I'd err on the side of "too dry," because that's easier to remedy at mixdown than "too wet." Don't forget to critique the tonality of your captured reverb too, because the timbre of reflected sound can be one of the major reasons to reconsider your choice of microphone and/or polar pattern. Once you get into per-instrument concerns, it's very easy to fixate on one or two instruments while overlooking others, so where any ensemble has more than a half-dozen sound sources, it pays to keep a paper list of the lineup close to hand. Working through this list methodically (first for level, then for tone, then for depth), helps you feel more confident that nothing important has slipped under the radar. ("Heavens—I can actually hear the banjo! How careless of me…")

The most effective fix for balance problems is simply asking certain performers to play quieter/louder (or asking their musical director to do it for you), but you might also request that someone change their instrument or instrument settings; that more/fewer doublings are used in the arrangement; that a soundbox lid be opened/closed; or that amplifier levels be adjusted. Although all these methods of implementing balance changes may have tonal side-effects, they shouldn't impact too much on the ensemble's depth perspective, whereas if you start shuffling an instrument's position/angle, you'll also affect its subjective distance from the listener in the recording.

Of course, in some cases the link between level and depth can work in your favor, because repositioning a solo instrument to emphasize it in the balance

"Repositioning a solo instrument to emphasize it in the balance was a staple technique prior to the advent of multitracking, with instrumentalists encouraged to step closer to the mic for solo passages. Many small-studio recordists seem to have a blind spot toward this kind of 'performer choreography,' but it remains tremendously handy when working under space/budget restrictions."

will usually also pull that instrument forward of the main ensemble by reducing its dry/wet ratio—a staple technique prior to the advent of multitracking, and one which was regularly used dynamically, with instrumentalists encouraged to step closer to the mic for solo passages. The rise of the large-scale multi-room studio during the 1970s caused this kind of "performer choreography" to fall out of fashion, which means that many small-studio recordists seem to have a blind spot toward it, but it remains tremendously handy when you're working under space/budget restrictions. If you're going to use it, though, try to mark each musician's different position (perhaps with gaffer on the floor) so that they're precisely repeatable between takes—and keep your ears open for squeaking shoes!

For single-mic ensemble recordings, trying to enact per-instrument tonal tweaks using EQ is pretty much a non-starter—again, this is something best addressed by the performer. However, if you've already established a decent level balance, you have to be a bit careful that a player's timbral adjustments don't upset that. For example, many instruments are louder along their frontal axis, so rotating them won't just achieve a different sound color. In this respect, the way the ensemble and microphone interact with the room's acoustics becomes quite a useful variable. In other words, moving the whole ensemble (together with the microphone) around the recording space can significantly change how the room's acoustic resonances filter each instrument's tone, but shouldn't have an enormous effect on the level or depth of each source. For finer timbral control, a single instrument might be moved into another sector of the mic's polar pattern, perhaps in conjunction with a change in miking distance to maintain the level balance. Changes to the miking height will also give tonal adjustment, because of the way inevitable room resonances affect the captured frequency-response at different miking positions (see Section 4.4.2), as well as allowing creative use of boundary-effect bass boost as the mic approaches the floor or ceiling. In this context, fairly moderate repositioning can have a surprisingly big tonal impact without usually incurring too many level/depth-related side-effects—unless you're miking really close to the ensemble, adjusting the height of the dominant array by a foot (30 cm) or so shouldn't make a huge difference to the distance between each instrument and the mics.

Once you've found an appropriate dry/wet ratio for your recording as a whole, and also a satisfactory level and tone for each separate instrument, your ability to tackle any remaining depth concerns without sacrificing some other hard-won aspect of the sonics may be very limited—unless you start adding additional mics, as we'll see in Section 9.3. As such, the art of single-mic ensemble recording is all about deciding what (if anything) you can afford to trade in for better depth presentation within the context of your specific production.

9.2 USING SIMPLE STEREO DOMINANT ARRAYS

Now let's consider the additional complexities introduced when we substitute a stereo setup in place of our single mic, thereby increasing the number of questions that need answering while planning and running the session. Here's the expanded list—again, tackling these questions in order provides a pretty sensible workflow, although there is inevitably some interaction between them:

- Which mic?
- Which stereo method?
- How wet overall?
- How wide overall?
- What level for each instrument?
- What tone for each instrument?
- What depth for each instrument?
- What image for each instrument?

In stereo our options for microphone positioning immediately become more restricted, because of the need to place different sources appropriately within the stereo image, and you may also be forced to bring the microphones into a more central room position, where tonal fluctuations on account of room resonance are typically more severe. Where microphone directionality is responsible for creating stereo level-difference information, you can't then use it as freely to rebalance or tonally reshape individual instruments either.

9.2.1 Balance, Tone, and Depth Across the Acceptance Region

The issue of center-versus-edge balance (which we first mentioned back in Section 8.1.3) will usually become a more critical issue in ensemble settings, so ideally you should stick closely to the setup suggestions in Tables 8.2 and 8.3, and then try to arrange your musicians at equal distances from the mic rig. Bear in mind that mutual angles which misrepresent the balance of signals reaching the array will also misrepresent the ensemble's apparent depth: Overwide mutual angles will not only attenuate central sources, but will also make them appear wetter (and therefore more distant) than they are; while mutual angles which are too narrow will cause similar problems for instruments at the edges of the acceptance zone. You may be able to compensate for these issues by changing the ensemble layout in some cases, but there's no sense in making a rod for your own back if you can help it.

Sometimes the chosen ensemble layout will itself cause level/balance challenges. For example, instruments at the center of the acceptance region may be considerably closer to the mic rig than those at the edge, either because of physical layout limitations within the recording venue or because of the sheer size of the ensemble. The most extreme situation you're likely to encounter in this regard will be where your performers are set up in a straight line across the acceptance zone, which will give you roughly 3 dB more level from central musicians than from those 60° off-axis to the array, in my experience. If you're using an AB rig, there's precious little you can do about this (short of supplementing the rig with additional mics, which we'll get to in Section 9.3), but with directional mics the wider mutual angles in Table 8.3 may help you out by recessing central sources, provided that you can still fit everyone into the narrowed acceptance angle. You'll be placing greater demands on the off-axis linearity of your mics by doing this, though, so it may prove a struggle to maintain consistent tonality across the ensemble and to avoid overbrightening the room reverb under those circumstances, especially with large-diaphragm condensers. As such, I'd treat this as a fairly high-risk strategy when working on a budget, and test the tonality of the proposed stereo rig in advance to see if it's likely to be viable. (The easiest way to do this is by recording yourself walking across the ensemble setup while speaking, given how sensitive we humans are to the timbral nuances of speech.)

Where your ensemble setup and mic selection allow you some choice between different arrays while retaining a decent level/depth picture, you should try to take advantage of this to design a rig which faithfully reflects the timbres of instruments at different locations. This is where the smoother off-axis response of omni or figure-eight mics can be of assistance, but the narrower stereo acceptance angles these mics yield in typical XY, MS, and AB setups may cramp your style in more reverberant spaces, unless you can find some way of rigging acoustic absorption behind the array. Whichever mics you're using, though, don't forget to account for their on-axis brightening effect. Obviously this is easiest with AB omni pairs, where the mic angle has least effect on acceptance angle, but near-coincident pairs can also be usefully reconfigured by trading off mutual angle against mic spacing, as discussed in Section 8.4.1. Moreover, any of the stereo arrays we've encountered so far can be tilted upwards/downwards to adjust the recorded tone by putting the entire ensemble slightly off-axis, and this can be particularly useful for avoiding overbright central sources when using an MS rig. Although tilting any directional stereo array like this has the potential to narrow the effective frontal acceptance angle in extreme cases, this shouldn't be a cause for serious worries if you keep within a ±45° angle.

9.2.2 Combining an Array's Front and Rear Pickup

"Less experienced engineers often overlook the additional setup possibilities offered by stereo mic techniques that pick up equally from both front and rear."

One thing that less experienced engineers often overlook is the additional setup possibilities offered by stereo mic techniques that pick up equally from both front and rear. For example, in music which combines choir and orchestra, it's common in live performances for the singers to be set up behind the orchestra on risers, but this causes problems if you want to mike the ensemble

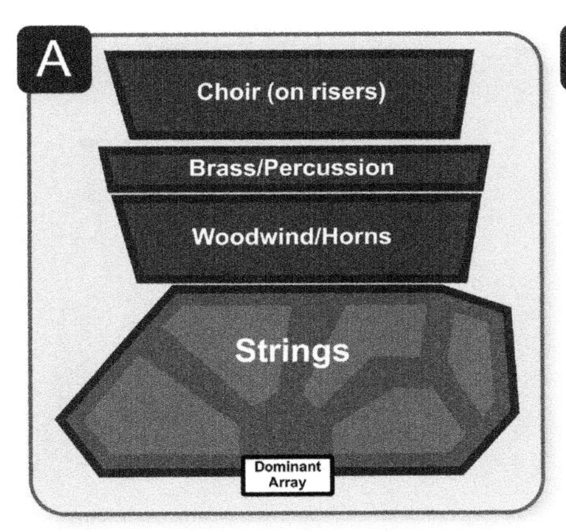

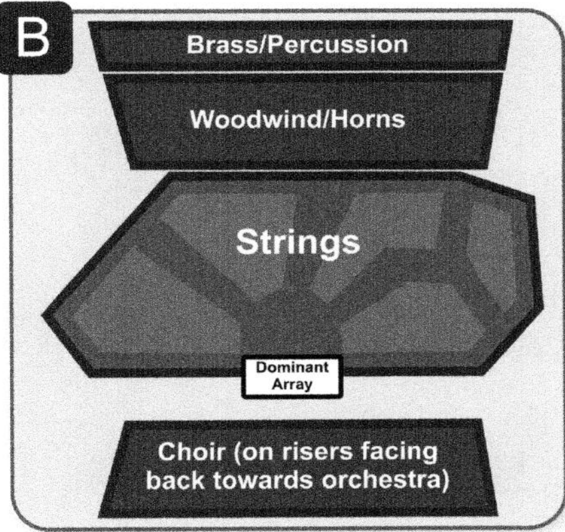

FIGURE 9.3
The ensemble setup shown in Diagram A is common in live performance, but in a recording situation the choir may end up sounding rather soft and distant if you're trying to pick up the whole ensemble with a simple dominant array behind the conductor. One solution to this problem is to stand the singers on the other side of the dominant array, facing the orchestra, as shown in Diagram B—assuming, of course, that your choice of dominant array has suitable rear pickup characteristics.

from the front while recording: The choir will usually appear both too low in level and too far away. However, if you depart from the standard concert layout by moving the singers to the other side of the mic array (i.e., facing toward the orchestra from behind the conductor, as shown in Figure 9.3), you can place them much closer in to achieve better definition, assuming of course that you're using a mic technique with adequate rear pickup—spaced omnis or figure-eights, say.

The same dodge will also work with crossed figure-eights (whether in XY or MS configuration), but you'll need to account for the array's reversed-image rear pickup when placing the musicians. Do take care to keep instruments well clear of the out-of-polarity regions on either side of the rig, though, otherwise imaging clarity will be poor in stereo and your ensemble balance will take a big hit in mono.

Another option here is to suspend any stereo array above the ensemble, pointing at the floor—just so long as you take the necessary measures to stabilize tall mic stands so that they don't topple onto the players! The players can then be positioned on either side of the array, effectively stacking the players vertically, as far as the mic array is concerned. There are a few caveats here, though. Firstly, a lot of instruments are designed to project their most balanced tone forward rather than upward, but lowering the array to capture the frontal dispersion will cause stronger coloration of the ensemble's tonality by directional mics' off-axis response. Secondly, the small physical spacing between coincident mics will cause some comb-filtering effects for instruments above and below the array's normal on-axis stereo plane—a subtle effect, admittedly, for slim-line

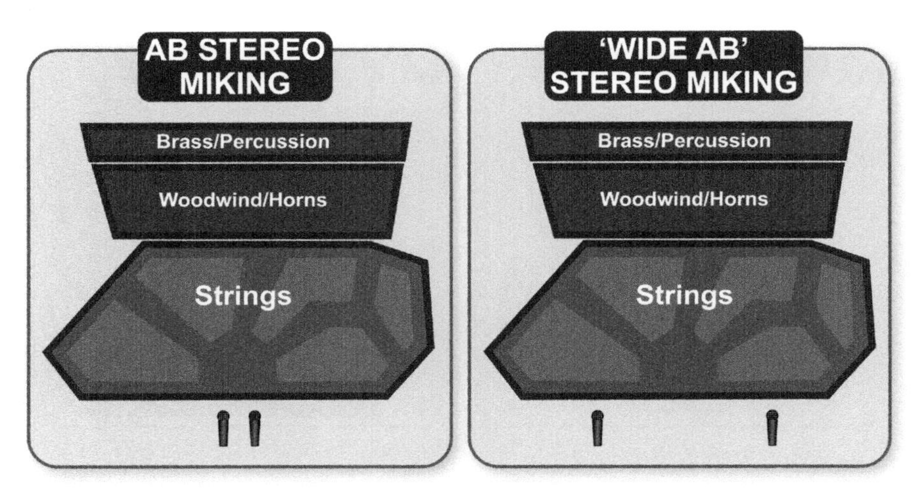

FIGURE 9.4
The traditional AB stereo setup in the left-hand diagram typically gives a more realistic stereo spread, but favors central instruments in the balance; whereas the "wide AB" setup in the right-hand diagram gives a more uniform balance across the width of the orchestra, albeit at the expense of an unnaturally overstretched stereo picture.

end-fire mics, but bulkier large-diaphragm or ribbon pairs may lose a significant amount of high end. And, finally, you need enough height in the room to avoid tonal problems from ceiling reflections when miking overhead like this.

9.2.3 Wide AB and Close AB Arrays

Although in most cases I'd suggest following the spacing and acceptance-angle guidelines in Table 8.2 when using AB stereo arrays, balance or tone concerns arguably override those of imaging realism in some cases. For example, you'll get a better balance between the instruments of a physically wide ensemble by separating the two mics so that each one sits midway between the ensemble's center line and its edge—the "wide AB" setup shown in Figure 9.4. Although the downside of this is an unnaturally stretched stereo image which places most of the instruments at the panorama's edges (a situation some describe as a "hole in the middle"), you might prefer that to a beautiful stereo image with a dodgy balance—and you may even be able to compensate for the overwide imaging by narrowing the panning, as long as you make sure comb-filtering between the mics isn't too objectionable and you don't mind a loss of subjective width in the room reverb.

Trying to achieve a dry ensemble sound with traditional AB miking can result in similar image-stretch compromises, because of level differences introduced between the mic signals when working at very close range (as discussed in Section 8.5.2). Fortunately, using panning adjustments to weaken the hole-in-the-middle sensation is usually less damaging here, because the narrowed reverb pickup will be less noticeable, especially if you're planning to add artificial reverb at mixdown anyway. However, it is very easy when using a close-up AB rig to end up with lots of instruments well off-axis to the array, where the inter-mic time differences are largest, so do check carefully for mono-compatibility.

While the unnaturally sides-heavy stereo image of both these types of non-traditional AB setup may not suit more purist genres, it's as well to point out that it can actually serve modern mainstream productions better in some contexts. For example, wrapping overdubs such as percussion, strings, and atmospheric ambience noises around more important central sounds (e.g., vocals, solos, kick, snare, bass) by effectively overwidening them can help prevent all those different elements from obscuring each other in a busy mix.

9.3 MULTIMIC DOMINANT ARRAYS IN MONO AND STEREO

It's frequently possible to catch a respectable ensemble sound with just a simple mono or stereo setup, especially with smaller groups of instruments, which is great news for cash-strapped recordists or those sessions where you're racing against the clock. However, it should also be clear by now that this kind of recording can be fraught with compromise, and often involves trading off various desirable sonic attributes against each other to some extent. In more demanding recording situations, therefore, many engineers resort to building a dominant array using more complicated multimiking techniques, so let's have a look at the thinking behind some of these.

9.3.1 Coincident Multimiking

Using a coincident multimic array for mono ensemble recording offers many of the same advantages that it does for single instruments, as described in Section 7.1. For example, the broader pickup of a crossed pair's combined polar pattern (especially at high frequencies) catches the edges of wider ensembles better in situations where you only have access to directional mics. You can also carefully aim the polar-pattern sensitivity lobes and rejection nulls of coincident multimics at different sections of the ensemble to give considerable balance control from your faders. Figure-eight mics really come up trumps in this respect, one popular application being to arrange a crossed pair of them vertically in front of a singing guitarist so that one mic points at the voice and the other at the instrument—if you carefully aim each mic's plane of rejection you can frequently achieve quite astonishing separation between the two mic signals, in spite of the small distance between the two sound sources.

Another interesting scenario is where you capture one lead vocal and two backing vocals on the same crossed figure-eight pair—the lead singer 45° off-axis to both mics, and each of the backing vocals on-axis to one mic's rear sensitivity lobe. In mono, you can rebalance the two mics to shift emphasis between the two backing vocals, but without affecting the lead-vocal level; you can use just one mic to eliminate either of the backing vocal lines; or you can invert the polarity of one of the mics and mix them together at equal level to cancel out the lead line, leaving just the backing vocals. Alternatively, if you want to use the rig in stereo, adjusting the array's Sides level with MS processing allows you to adjust the stereo width of the backing vocals relative to the lead.

Mixing an omni mic with a similarly aligned directional mic to create intermediate directivities (as in the Strauss Paket, for example) has interesting applications too. Not only does it give you control over your recording's overall dry/wet balance when using two-mic mono or four-mic AB stereo arrays, but it also allows you to get a wider range of useful acceptance angles from XY or MS stereo rigs. For example, if you combine carefully phase-aligned forward-facing omni and figure-eight mics to create your Middle-channel signal, you can mimic any flavor of two-mic MS rig simply by rebalancing the resultant array's three mic signals. (In fact, if you want to get really freaky you could even use complementary EQ settings on the two Middle-mic signals to effectively create different stereo rigs for different frequency ranges—a concept at the heart of the Schoeps Polarflex system, for instance.) There's also nothing stopping you from using coincident multimiking (in either mono or stereo) just to blend the timbral characteristics of different microphone types, as discussed in Section 7.1.4.

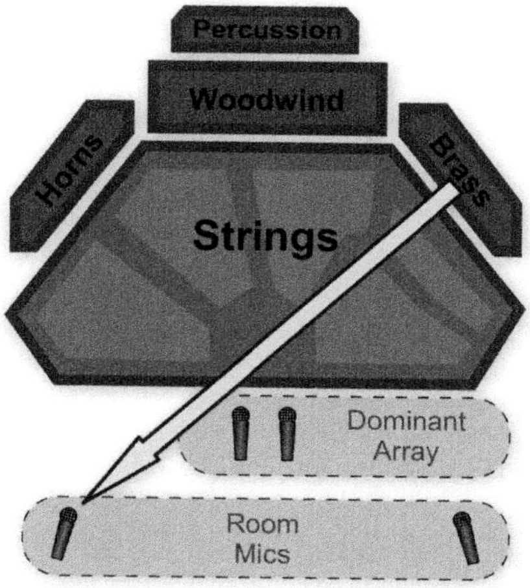

FIGURE 9.5
Wide-spaced room mics can introduce conflicting stereo cues if there are any strongly directional instruments firing across your ensemble—as you can see in this diagram, where an orchestra's brass section is arrayed on the angled section of a rear riser.

9.3.2 Dual Arrays for Dry/Wet Control

The first reason to expand the dominant array with extra non-coincident microphones is to give yourself more control over the dry/wet ratio of the recording at mixdown, perhaps to work around physical mic-placement restrictions within the recording venue. The basic idea is to use one mic rig to catch a comparatively dry sound from the ensemble, and then to supplement that by mixing in room reverb from an additional rig dedicated to the purpose. We already considered the fundamental ramifications of room miking back in Section 7.2.3, and we also assessed the suitability of various two-mic stereo arrays for this kind of application in Chapter 8. The main thing to add here is that using directional mics for the reverb pickup with their nulls pointed toward the ensemble offers two advantages for acoustic styles that trade on stereo realism. Firstly, it allows you to achieve a good degree of wet/dry contrast between your two mic pairs without putting the mics too far away (thereby avoiding a distracting "slapback" effect); and, secondly, with wide-spaced AB room arrays it also reduces the likelihood that highly directional instruments (e.g., trumpets and trombones) firing across the ensemble from one side might cause conflicting directional cues when the two mic pairs are combined, as illustrated in Figure 9.5.

The question of image width also needs a little extra thought. The crux of the matter is that if your dry pair presents the ensemble as stretching right across the panorama, this creates the impression that the listener is positioned quite close to the musicians—after all, in real life an ensemble sounds wider the closer you get to it. Therefore, if you take a full-width dry image and add reverb to it using room mics, the question of where the listener feels they're located in relation to the orchestra can get a bit confused—the dry pair's width implies that they're closer up, but the overall reverb level implies they're further away. Choosing a slightly wider acceptance angle for the close array would be the preferred solution to this, but narrowing that mic pair's panning may also do the job as long as your chosen stereo setup doesn't punish you with excessive comb-filtering. (And speaking of comb-filtering, do check the phase/polarity relationship between the different dry and wet arrays to make sure you're getting the most pleasant combined tonality.)

9.3.3 Balance and Depth Corrections

One of the big problems with trying to record an ensemble with a single mic array (whether mono or stereo), is that placing microphones close enough to achieve an appropriate overall dry/wet ratio may unduly bias the capture toward instruments at the front and center of the group—leaving instruments at both the sides and the rear sounding too quiet and too distant by comparison. As such, it's very common for professional recordists to plant additional arrays close to those underrepresented sections, thereby bringing them up in the balance and forward in the depth perspective.

A popular strategy for defining the sides of an ensemble in this way is to place additional "outrigger" mics flanking a central array, effectively picking up the full width of the group by stitching together the pickup from three narrower sections. If you're purely looking for level/depth compensation, then you'll want to place the additional mics to achieve the same kind of overall dry/wet ratio you're hearing on the central array. A line of three very similar mics across the front of a choir or orchestra is a common application of this idea (a setup that was apparently the bedrock of Jack Renner's classic Telarc sound, according to Michael Bishop[7]), and if the outriggers are hard-panned for stereo, then the central mic provides the useful service of counteracting the hole-in-the-middle effect that would otherwise arise if a widely spaced pair had been used alone. However, it's important to realize that if the spacing between any of the three microphones exceeds three feet (90 cm) or so, the stereo imaging will still be significantly stretched toward the locations the mics are panned to. So although this three-mic setup avoids the hole-in-the-middle phenomenon when listening back on loudspeakers, you may still end up with "holes" at ±15° off-center (as in Figure 9.6B). Different engineers disagree about how much of this kind of image-stretch is acceptable, but if it offends your own ears, then try increasing the number of mics to allow more conservative inter-mic spacings.

Of course, the moment you have more than two spaced microphones creating your stereo picture, some of them will have to be summed, so the timbral impact of inter-mic phase and polarity relationships won't just be felt in

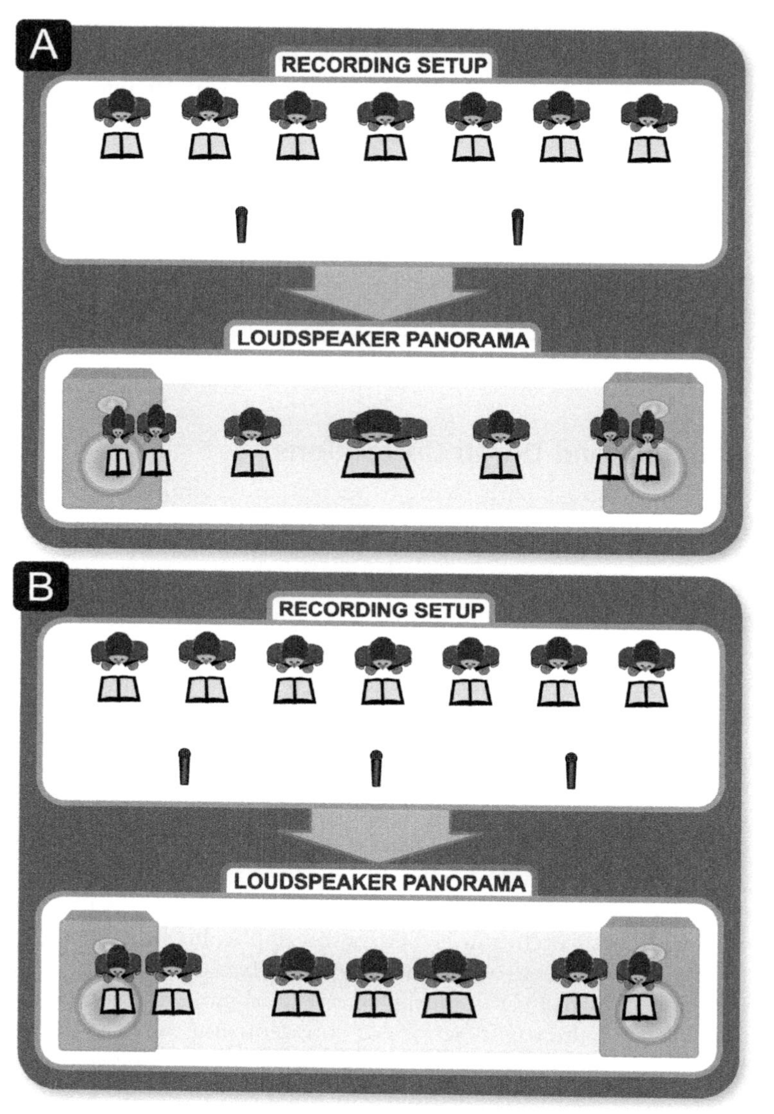

FIGURE 9.6
If the widely spaced microphones in Diagram A are panned to the stereo extremes, your stereo panorama will suffer from "hole in the middle"—i.e., most of the singers will seem bunched up at the edges of the panorama on playback, and the central singer's image will be stretched and destabilized. Using three mics instead, as shown in Diagram B, can help avoid the worst hole-in-the-middle problems, although weaker image-stretching artefacts may still remain audible.

mono any longer. (As such, the techniques we used to manage comb-filtering in Section 7.2.1 will be equally useful here, although I'd be a little wary of trying to delay the outriggers to match the central array when working in stereo, because this can generate rather confusing stereo impressions.) You may find that your multimic rig starts to exhibit phased-array characteristics too (see Section 7.2.1), boosting on-axis instruments and comb-filtering off-axis room reverb, so don't be surprised if you end up with a slightly drier sound than you'd expect when using this multimic "curtain" method. Irrespective of how many mics you use, in practice you may find that the outermost mics need a little more gain than the inner ones. This is because the ensemble section in front of any of the inner mics is also being picked up as spill by mics on both sides, whereas mics at the edges only have supportive bleed coming from mics on one side. Angling the outriggers slightly outwards can make sense too, because spreading their brighter on-axis response further across can improve the consistency of high-frequency coverage across the ensemble.

Outriggers are frequently used in conjunction with a stereo central array, but if the central mics are directional designs then you'll probably get a better match for that pair's overall dry/wet ratio by using subcardioid or cardioid outriggers than omnis, as mentioned in Section 8.1.2. The stereo imaging of the combined setup needs a little thought too, because it's very easy to end up with an overwide stereo image from your central array when miking a wide ensemble up close, and this will only be exacerbated if you mix in hard-panned outriggers. This is something you actually hear quite frequently on commercial releases, so it's clear that some engineers and producers aren't too worried about it, but if it bugs you as much as it does me, then my advice is to insure you're using a wide enough acceptance angle for the central array, or if that's not possible for some reason then consider narrowing the panning of the center pair's mics a little. In fact, if you're planning on mixing in additional room mics, you might also want to pan the outriggers slightly inwards too, so that the ensemble as a whole doesn't create a stereo width at odds with its apparent distance from the listener on playback (as mentioned in Section 9.3.2).

The most common means of correcting balance/depth problems for instruments at the rear of the ensemble is to set up a second array in the midst of the ensemble, targeting those ranks of instruments that need additional definition. As far as stereo recordings are concerned, a wide variety of XY, MS, AB, and near-coincident rigs have been used commercially in this role, and the multimic curtain approach we've just touched on can also function perfectly well. I'd steer clear of crossed figure-eights myself, because when they're surrounded by instruments they'll usually pick up lots of things way off-axis, leading to troublesome polarity mismatches between the stereo channels (see Section 8.1.3). There is also a strong argument for each stereo rig using a similar blend of inter-mic level-difference and time-difference information to create its stereo illusion, to avoid the incongruity of having one group of instruments with razor-sharp XY/MS-based positioning set against another with softer-focus AB-based positioning. Plus, it stands to reason that you should try to match the spread of the instruments between the pairs too if possible. The off-axis response behavior of mics used for the supplementary array will be critical, whatever rig you choose for it, so those on a budget would be well advised to stick to small diaphragms if cardioids are unavoidable, and otherwise to favor

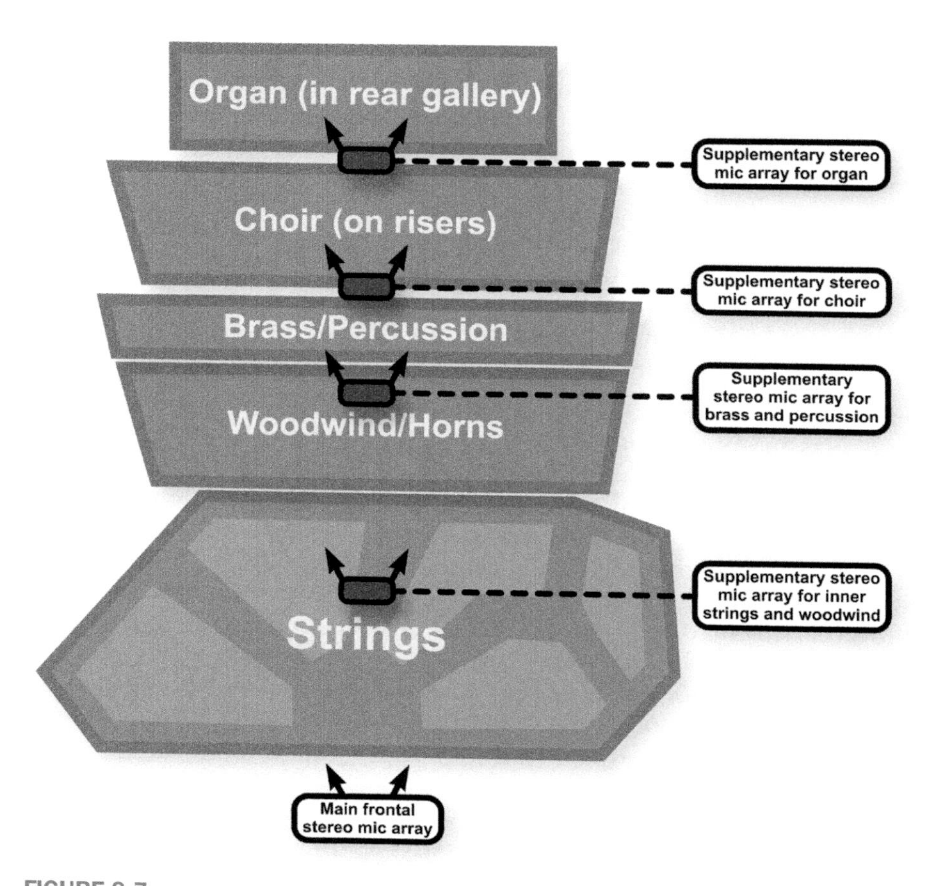

FIGURE 9.7
A dominant array system for capturing a large-scale concert work with orchestra, choir, and organ might incorporate a frontal mic array for the primary ensemble pickup working in conjunction with a series of supplementary mic arrays designed to compensate for the frontal array's depth misrepresentation.

omnis and figure-eights—remember that the latter can be used in parallel for non-coincident stereo without the polarity problems of a crossed pair.

Because the main purpose of the supplementary array in this context is to correct for depth misrepresentation on the front array (rather than errors of musical balance), you shouldn't need to put these additional microphones too close. Just try to get roughly the same dry/wet balance for the rear-section instruments on those mics as you've already obtained for the front-section instruments coming through the front array, which means you should be able to mike from far enough away to capture a reasonably natural timbre. Once you've found a suitable mic position for the rear rig:

- Check that its stereo image matches that of the front pair. The best way to compare the rigs is to visualize the stereo position of different instruments while toggling between the two pairs using your monitor-channel mute buttons. If possible tweak the mics to address any problems here, but if this isn't possible, then adjust your channel pan controls or make use of MS-based stereo width adjustment.

- Find a good phase/polarity relationship between the front and rear arrays. Personally, I prefer to start off by delaying the rear array to roughly compensate for the distance between the front and rear mics, based on my little "foot per millisecond" rule of thumb (or the slightly more accurate "meter per 3 ms" version)—this usually helps give a more natural timbre and avoids rear instruments being pulled right up to the front of the group. Then I'll grab the rear array's fader and try mixing it in slowly, listening to the entire ensemble (not just the rear instruments!) for unwanted tonal changes. Ideally, you'd hope to arrive at phase/polarity settings for the rear array that would allow you to mix it in without changing the tone of the front array at all, but frankly that's a pipedream, because phase-cancellation will always occur to some extent. Hence, a more realistic goal is a tonal change that's as benign as possible.
- Balance the rear array sensibly—which usually translates as "lower than you think!" A mistake a lot of small-studio recordists make is trying to position every instrument in an ensemble at the front of the mix's depth perspective. The purpose of mixing in a rear array is to avoid the back ranks of your ensemble sounding unnaturally far away, not to create the illusion that the musicians are all piled on top of each other in some sort of group hug!

"The purpose of mixing in a rear array is to avoid the back ranks of your ensemble sounding unnaturally far away, not to create the illusion that the musicians are all piled on top of each other in some sort of group hug!"

In the normal run of things a single rear pair should be ample for most of the ensembles that small-studio engineers are likely to meet, even when working on location, but that's not to say that more might not be called for in exceptional cases. A massive concert work like *The Dream Of Gerontius* might warrant three or four additional pairs in certain venues, for instance: one for the inner strings and woodwind; another for the brass and percussion; and separate arrays for choir and organ.

9.3.4 Hybrid Stereo Arrays

Adding extra microphones to a two-mic stereo setup can also let you blend selected characteristics of the basic XY, MS, near-coincident, and AB arrays into a hybrid listening experience. This is one reason why supplementing a coincident central array with wide-spaced outriggers can work nicely, because it adds spaciousness that would otherwise be lacking. Choosing omni models for these outriggers is also popular, because that provides the improved low-end extension of a traditional AB array, but it does mean the outriggers will pick up more reverb than the directional central rig, so you have to decide how you react to that. If you're concerned about correcting balance/depth misrepresentation across the width of the ensemble (as in Section 9.3.3), then it's tempting to move the outriggers closer to the ensemble to dry up some of the reverb, but this risks imbalancing and distancing the rear sections of the group instead, and also has the potential to muddle the stereo imaging. A more common attitude, therefore, is to embrace the omni outrigger's extra reverb as a central part of its appeal (and even emphasize it using slightly more distant placement), such that the purpose of the outriggers becomes less about correcting balance/depth problems, and more about gaining some independent control over the overall degree of low-end warmth, spaciousness, and reverberance.

Probably the best-known hybrid stereo technique is the Decca Tree, a setup developed over the course of more than a decade by a group of pioneering engineers at the Decca record company. The exact setup they used varied a great deal during the label's early years, but had settled into a fairly standardized rig by the beginning of the 1970s: three Neumann M50 microphones suspended above the front row of the ensemble in a triangular configuration, the leading mic 2–6 feet in front and panned centrally, and the remaining two mics 3–9 feet apart and panned to opposite sides of the panorama (see Figure 9.8). Most professional users of this system scale the physical dimensions to match the size of the ensemble, while maintaining roughly a 2:3 ratio between them—values of around 4 feet (1.2 m) deep by 6 feet (1.8 m) wide are fairly typical for a regular classical orchestra, for example. Where a Decca Tree is being used on its own, the side mics would usually be panned hard left/right to create a full stereo picture, but where outriggers are also in use, many people like to narrow the Tree's panning a little. The mics are almost always angled slightly downward toward the ensemble and set up at the same height initially, usually via a special "T"-shaped bar on a single tall mic stand—preferably one built like a tank! You'll sometimes see one or two of the mics being lowered a little on any given session to adjust the balance between the three sectors of the ensemble being covered. As with the simpler three-mic "curtain" arrangement in Section 9.3.3, a lot of users fade the flanking mics up a little more than the center mic, but this is very much a question of personal taste—some engineers use a very low level purely as a subtle "center fill," while others (such as Eberhard Sengpiel[8]) advocate keeping all three mics at identical levels.

Part of the unique character of the classic Decca Tree is related to the unusual design of the specific microphones used. The Neumann M50's head basket contains a small-diaphragm omnidirectional transducer mounted into the surface of a golf-ball-sized sphere, an arrangement that generates an overall high-frequency lift above about 3 kHz, as well as making the mic significantly more directional at higher frequencies. What this means is that when three M50s are set up in a Decca Tree formation, with the side mics angled about ±45° outwards (as is typical), the high-frequency directionality produces level-difference stereo information, sharpening the stereo imaging; the high-frequency coverage remains fairly even over a wide acceptance angle; and the omni design's extended low end and clean off-axis response in the lower half of the

FIGURE 9.8
A traditional-style Decca Tree setup in action. The microphones are TLM150s, Neumann's modern transformerless update of the classic, but now discontinued, M50.

spectrum outperforms fully directional mics. In addition, the wide spacing of the side mics provides good ambient spaciousness; the central microphone guarantees a certain baseline mono compatibility; and the slight acoustic delay between the center mic and the trailing pair gives the whole array a subtle "building out from the center" character which a lot of listeners like.

The bad news about the classic Decca Tree for most small-studio folk is that Neumann discontinued the M50 long ago, so it's now a collector's item, and most modern mics designed specifically for the same application tend to be eye-wateringly pricey—especially when you want three of them! That said, it is still possible to adapt the same concept to get good results with more general-purpose microphones. One workaround suggested by Ron Streicher[9] is to substitute large-diaphragm omnis for the M50s, which makes a lot of sense given the unavoidable HF directionality of such designs. Another option is to choose directional mics instead, and then supplement their comparatively dry and bass-light sound with other mics, such as omni outriggers perhaps. Alan Meyerson[10] used this idea while recording one of the *Pirates Of The Caribbean* film scores, in fact, suspending a trio of figure-eight ribbon mics right over the conductor. Using ribbons like this has a lot to recommend it with amateur orchestras too, given that overhead miking positions can easily make lesser-quality string sections sound scratchy. One word of caution here,

"It's possible to adapt the Decca Tree concept to get good results with more general-purpose microphones. Using ribbon mics has a lot to recommend it with amateur orchestras, given that overhead miking positions can easily make lesser-quality string sections sound scratchy."

though: Directional polar patterns will more readily cause unnatural image-stretch in this kind of configuration, so you may need to narrow the mic spacings to compensate for this—listen especially for what's happening (or not!) in those critical mid-left and mid-right panorama locations.

Another of Ron Streicher's[11] adaptations of the Decca Tree concept is to replace the leading mic of an omni-based setup with an MS stereo rig, effectively blending the pure level-difference stereo information from the coincident rig with the spaciousness, low-end extension, and high-frequency coverage of a three-point non-coincident arrangement. If you'd like to try this for yourself, then I'd suggest using a cardioid Middle mic, because that should give you a sensible acceptance angle without skewing the center-versus-edge level balance too much. The difficulty with using a cardioid center mic, though, is that the MS pair will almost certainly pick up significantly less reverb and low end than the omnis. If you find that this pulls the center of the ensemble too close to you over loudspeakers, then one response might be to switch the omnis for cardioids, thereby matching the sonics more closely across the rig, and then to move the whole array a little higher and apply some EQ to address the loss of reverb and low end.

The delicious whiff of mystique surrounding the Decca Tree "trade secret" means you'll find no shortage of Internet debate about its original usage and potential adaptations, but rarely do such discussions suggest any downsides of the concept. In practice, creating a respectable balance and depth impression from an overhead miking scheme frequently requires the array to be raised 9 feet (3 m) or more above the floor, and that's not easy to do securely on the cheap—plus the recording venue must be large enough to accommodate such mic placements without comb-filtering from ceiling reflections noticeably coloring the overall timbre. As with simpler omni AB rigs, the classic Decca Tree arrangement will suffer from phase-cancellation when summed to mono, but I find it's also more prone to sounding over-reverberant under those conditions. The Decca Tree is also highly dependent on room acoustics, and John Kurlander[12] and Tony Faulkner[13] advise caution with overly dry and overly wet rooms respectively. Faulkner also has concerns with placing performers directly underneath the array, where the leading microphone's center-fill function is weakened: "There can be problems when doing concertos because soloists close to the main array can generate unstable or overwide stereo images."

And while we're speaking of Tony Faulkner, he has suggested another interesting hybrid-stereo alternative: a pair of omnis at a mutual angle of approximately 90°, spaced roughly 26 inches (66 cm) apart, and between them a near-coincident pair, usually subcardioids at a similar mutual angle, spaced

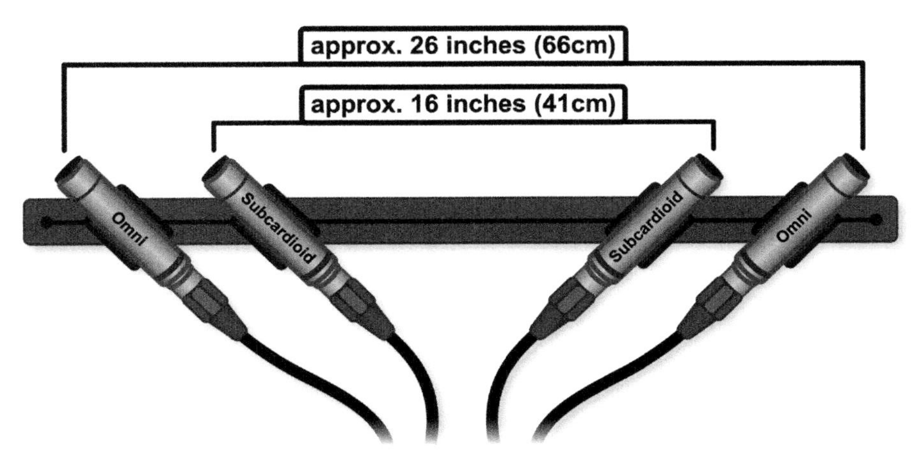

FIGURE 9.9
An example of a four-mic phased-array stereo setup.

roughly 16 inches (41 cm) apart. On the face of it, this might appear to be just another attempt at blending the sonic characteristics of two different stereo rigs, but there's more to it than that. The microphone diaphragms are also carefully set up in a line so that sound from the center of the acceptance zone arrives at all of them simultaneously, which means that they operate as a phased array (see Section 7.2.1). As such, the whole shebang becomes more directional, and can therefore be placed further from the ensemble (for a given overall dry/wet ratio), with the result that the group's balance and depth can often be captured more naturally. The main trade-off is mono compatibility, particularly at the high end for instruments on either side of the ensemble, so take time finessing the array placement and mic spacing, and I'd recommend mixing one of the pairs at least a few decibels lower than the other, to soften the tonal impact of any comb-filtering. You may also wish to experiment with smaller mic spacings if you're presented with an especially wide ensemble, because the aforementioned setup only gives an acceptance angle of around 60° in my experience—although narrowing the spacing will also weaken the phased-array effects by shifting them further up the frequency spectrum.

9.4 ADDING SPOT ARRAYS

The primary aim of a dominant array is to capture an ensemble performance holistically, warts and all. Clearly, some flattery is possible via sensitive mic selection/positioning, but none of the techniques we've talked about above will bail you out if a barbershop group's bass singer is struggling to project his lowest notes; if you're trying to give a big-band's rhythm section more definition on record than they have in the live room; if an orchestra's clarinets sound dull; or if a jazz pianist isn't differentiating between their rhythm and solo parts sufficiently. For all those jobs you need spot mics/arrays which pick up a specific instrument/section with more isolation, allowing you to control its level, tone, and depth independently of the rest of the ensemble—not only in the static balance, but also dynamically throughout the musical timeline if necessary.

ENGINEER'S QUICKSTART GUIDE: DRUM KITS

Back in Chapter 6, I already discussed some of the variables involved in preparing individual drums for recording, but there are a few extra things to mention about pop/rock drum-kit setups. When you're recording them with the dominant-array method, you'll get the best results if all the component instruments are already naturally balanced, because that means you don't have to rely too heavily on your (inevitably less natural-sounding) close-placed spot mics.

The first thing to be wary of in this respect is damping individual drums too much, because it's easy to imbalance the kit that way. This is especially risky with snare drums when you're dealing with less experienced drummers, as they always seem to hit the hi-hat and cymbals too hard as it is. If you need to adjust hi-hat or cymbal levels, the best way is to swap them out for alternatives—thinner ones will usually be quieter, as well as brighter-sounding. However, the high-frequency dispersion patterns described in Section 6.2.4 give you some additional control here, depending on how you place your mics in relation to each instrument's high-frequency beams. With toms, get the drummer to play round each tom in turn to listen for their relative levels and sustains. The trickiest balance problems to solve at mixdown are where the level and timbre of the drummer's hits aren't consistent enough, so if you notice this I'd recommend chatting it over with the performer—sometimes small adjustments to the kit's layout can make certain instruments easier to control.

Whenever any drum in the kit is hit, all the other drums resonate in sympathy to a degree, and the snare wires will rattle along a bit too. Some engineers seem inordinately harassed by this, while others (myself included) are more of the opinion that such peripheral noises are an important part of the whole sonic package. "Trying to get rid of those rings and resonances is sort of standard practice, but I've never followed that advice," remarks Steve Albini,[14] for instance. If you do decide you want to reduce some of the resonance, the first thing you can do is remove any drums that aren't actually being played for the part you're recording. Beyond that, sympathetic resonances and snare-rattle can be reduced a great deal with careful tuning, and you'll almost always get a better overall sound if you try this approach before getting all heavy-handed with the damping.

9.4.1 Managing Spill

The biggest complication when using spot arrays is managing spill. Let's suppose you're recording a choir with an underpowered alto section, and point a spot mic at those singers to correct the level imbalance. If that mic picks up loads of spill from the tenors standing in the row behind, you'll undesirably overemphasize those voices in the process of boosting the altos—and if the mic's off-axis response is highly colored, or you're unlucky with the phase/polarity relationship between the spot and the dominant array, you may also seriously harm the overall tenor timbre. Avoiding this kind of problem involves controlling the amount and quality of spill on each spot array, something that places heavy restrictions on your mic technique.

The most basic spill-management tactic is to maximize pickup from those instruments your spot array is ostensibly pointing at (which I'll call the

"Focus'), while minimizing pickup of everything else in the recording room (which I'll call the "Backdrop"). To this end, you're usually forced to place spot mics much closer to their Focus instruments than you'd ideally like, and to use directional polar patterns to reject Backdrop sounds—both of which requirements make it tougher to capture appealing timbres. For example, the Focus may suffer from proximity-related problems such as bass boost, spot-lighting, and overemphasized mechanical noises, while the Backdrop tone may be adversely affected by the vagaries of cardioid-family off-axis pickup. Furthermore, multimic workarounds, which are great for overdubbing single instruments, can come seriously unstuck in ensemble situations—the comb-filtering inherent in non-coincident methods can wreak havoc with the Backdrop tone, while the broadened directionality of some coincident techniques may struggle to reduce Backdrop levels sufficiently.

So the best advice I can give here is to tackle spill problems as much as possible by acoustic means, because this'll give you more leeway to select and position your mics for a better recorded tone. Try adjusting the ensemble layout, perhaps, increasing the physical separation between quieter spot-miked instruments and troublesome sources of spill, as well as directing the firing lines of louder instruments away from nearby mics. Here's Shawn Murphy[15] talking about this concept within a film-scoring context: "If you seat the orchestra properly and work on the balance in the room, by and large you will have lots of isolation—no matter what dynamic people are playing at...You don't have to worry about excessive leakage because [each] instrument is acoustically sitting in a place that is complementary to the part it is playing ... [It's a question of] how tightly you pack them in, how far back you push the percussion or brass from the main mics, and recognizing the radiation pattern of the instruments."

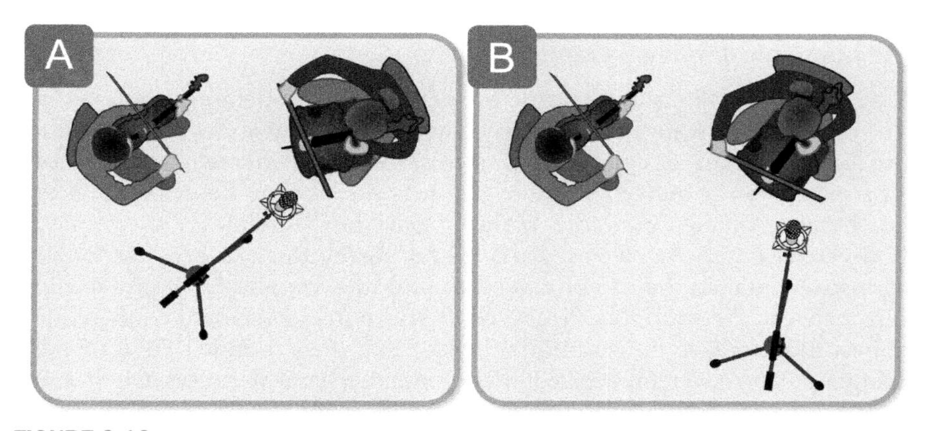

FIGURE 9.10
In Diagram A you can see a cello being spot-miked. If the cello tone sounds fine through this mic, but the violin in the Backdrop sounds horrid, you might improve the situation by rotating the whole cellist+microphone setup as shown in Diagram B—the violin moves more on-axis to the mic, where the tone should hopefully be a bit clearer, while the mic's position relative to the cello remains all but unchanged.

Using a two-sided dominant array (see Section 9.2.2) can really help out here, because it allows you to increase the distances between imbalanced instruments within the recording space, but still have them apparently sitting next to each other in the stereo panorama because of the array's superimposed front and rear pickup. Risers can increase the physical separation between neighboring instruments too, without making a huge change to the sound in the dominant array. Where your spot array is catching a great Focus tone, but the Backdrop sounds awful, a trick Mike Stavrou[16] suggests is subtly adjusting the location/angle of the instrument *together with its spot mic* to improve just the spill sound without unduly affecting the Focus (see Figure 9.10).

"You're usually forced to place spot mics much closer to their Focus instruments than you'd ideally like, and to use directional polar patterns to reject Backdrop sounds—both of which requirements make it tougher to capture appealing timbres."

Acoustic treatment can be your friend too. You might carefully arrange reflectors to bounce more sound into the mic from just the Focus instrument(s), for example, or use absorbers to obstruct paths of direct/reflected sound from the most prominent Backdrop sources. Shielding "unused" sectors of the spot array's pickup pattern isn't a bad idea either, the most common application of this idea being the "tunnel" of blankets or drum cases/shells that rock engineers regularly construct in front of a drum kit to protect an outside kick-drum mic from cymbal spill (as Butch Vig[17] did when recording Dave Grohl's drums for Nirvana's *Nevermind*). On a smaller scale, little handmade acoustic baffles can fulfill a similar function with solo instruments where spill-reduction requirements aren't quite as extreme. The most frequent application here is probably keeping hi-hat spill out of a snare-drum close mic: Bruce Swedien[18] has a homemade wood-and-mu-metal panel, for example, while Darryl Swann[19] uses a pop shield with a ski cap over it.

With the best will in the world, though, spill considerations usually prevent you setting up spot arrays with complete freedom, so let me share a few tips for finding an acceptable compromise between spill rejection and tone when you're on a budget. Figure-eight mics are unsung heroes in this context, because of their powerful plane of rejection (for loud sources of spill nearby), but otherwise clean off-axis sound. Again, this can be great for singing instrumentalists, but do be careful to find the best phase/polarity relationships between the spot mics if they aren't set up in a coincident configuration. Ribbon models have the additional benefit that their rounded HF tone helps compensate for overemphasized highs and mechanical noises when close-miking, especially with brass and strings. Small-diaphragm designs are usually preferable over large diaphragms as well, because of their smoother off-axis frequency response—so much so, in fact, that they become a much less controversial choice for lead vocals within an ensemble context than they would be for vocal overdubs (as discussed in Section 5.1.3). Dynamic spot mics designed for on-stage applications can be excellent at rejecting spill, and can even work with more acoustic sources, because the natural sound in the dominant array

FIGURE 9.11
One way to achieve excellent separation between the vocal and instrument spot mics on a singing acoustic guitarist is to use figure-eight polar patterns, angling each mic's deep rejection null to reduce the unwanted Backdrop signal. Just take care with the phase/polarity relationship between the spot mics—not just the relationship between each spot and the ensemble's dominant array.

helps disguise their tonal colorations. Boundary mics are very versatile in ensemble contexts too, given that they can be mounted on walls, floors, and gobos right next to an instrument without reflections from the nearby surface incurring undesirable comb-filtering.

In pursuit of maximum spill reduction, some people suggest placing mics inside instruments, a strategy borrowed from live sound-reinforcement. I'm not too keen on this in recording situations, though, simply because the mic signal then requires heaps of EQ to render the Focus listenable. True, kick drums sometimes rely on this approach in rock styles, but inside-miking usually makes pitched instruments sound so awful that the cure ends up worse than the disease. If everyone can hear the Focus instrument(s) in the room, then you shouldn't need ridiculously low spill levels on that spot array—normal close-miking, instrument placement, and acoustics dodges should provide ample control. Where you're really fighting spill on something like upright bass, harp, or acoustic guitar/piano, consider it a message from some higher power that you probably need to rethink your working methods! For example, any instrument that's barely audible on the dominant array should perhaps be moved closer in, or else replaced later with an overdub. Alternatively, it may be time to question the suitability of that instrumental line-up, or to consider moving some instruments into separate isolation rooms (of which more in Chapter 11).

By the same token, taking a DI feed from an instrument can also provide a spill-free Focus, but it won't give you a huge amount of control over that instrument's balance within the ensemble mix unless the DI signal can stand on its own sonically. This is certainly feasible with electric guitars and basses where reamping and/or digital amp emulation are available, but the piezo-electric DIs on acoustic instruments rarely offer anything like the tone-quality you'd hope for, so I'd consider them a last resort for anything but fairly subtle balance adjustments.

A SPOT OF SMOKE AND MIRRORS

Besides their audio contribution, don't overlook the psychological effects of spot mics. For some musicians, their presence can be intimidating, whereas other performers may suffer bruised egos if they're denied that special attention—indeed, putting up an impressive-looking spot mic in front of a temperamental soloist may significantly improve the quality of their performance even if you don't switch the thing on! And, of course, where any musician's movements are causing distracting sonic inconsistencies on the recording, a dummy spot mic may help to anchor them more firmly into position.

9.4.2 Stereo Imaging Considerations

If your dominant array delivers a convincing stereo image, then you don't want any spot arrays messing with that. The best way to guard against this is to repeatedly toggle the Mute buttons of each spot array while listening for image-shifts in the monitor mix. With single-mic spots, stereo alignment can be carried out with simple pan adjustments, but with stereo spot arrays, panning is only part of the process, because they capture some stereo information directly. My recommendation here is to set the spot array's acceptance angle according to the width of the Focus, not the width of the entire ensemble. Although this means you'll likely have to reduce the array's image width using your pan controls to match that of the dominant array, it usually makes it easier to find a good match in practice. Just be wary of overemphasizing out-of-acceptance Backdrop pickup in this case if you're using wide mutual angles with a coincident or near-coincident stereo spot rig.

However, matching the position of each spot system's Focus with the imaging in the dominant array isn't the end of the story either, because Backdrop pickup on each array can cause imaging side-effects too. For example, if you place a spot mic on the cello of a string-quartet recording, heavy viola spill on that mic would have the effect of pulling the viola toward the cello's stereo position. Faced with such a situation, you have to decide whether there are any further steps you can take to reduce the viola spill in the Backdrop—perhaps by baffling one side of the spot mic, moving the spot mic closer to the cello, or switching to a more directional polar pattern. In some cases, though, there may be little you can do to combat the image-shift without excessive damage to the ensemble timbre, in which case the stereo anomaly simply becomes the price you pay for extra control over the cello.

However, you can sometimes reduce Backdrop-driven image-shift by using stereo spot arrays in place of single mics. Imagine if you spot-miked our cello in stereo instead, such that it appeared in the center of the spot array's full-width stereo image and the viola spill appeared to be emanating from well off to one side. If you gave both these mics the same pan setting (in other words, treating them like a mono multimic setup) to match them with the apparent positioning of the cello in the dominant array, you'd get the same kind of viola image-shift as with a single mic; but if you offset the panning of the two mics equally either side of that position-matched setting, the viola would shift back toward its natural position. This kind of trick is particularly handy where soloists are placed at the front of an ensemble (for example with a lounge jazz singer or a classical concerto soloist), because it gives you much more freedom to ride the spot array's level dynamically during the mixdown without modulating the recording's overall stereo width. However, you do need to exercise some care with this technique if the soloist moves a lot, otherwise their proximity to the stereo spot array may exaggerate those movements to the point of absurdity in the loudspeaker panorama.

"When you're miking a broad source, such as a choir, a multimic curtain of spaced directional microphones will usually give you a better combination of Focus balance and Backdrop rejection than any traditional two-mic rig."

As far as choosing specific stereo methods for spot arrays is concerned, the recommendations I made when discussing depth-perspective compensation within a dominant array (see Section 9.3.3) still hold true, but I'd also advise against using any crossed figure-eight rig when you're miking close in amongst the players, because of its reversed-image rear pickup. When you're miking a broad source, such as a choir, a multimic curtain of spaced directional microphones will usually give you a better combination of Focus balance and Backdrop rejection than any traditional two-mic rig.

9.4.3 Spotting for Tone

Most of the time spot mics are used simply to adjust an ensemble's natural balance, which puts the onus on the engineer to avoid unwanted timbral changes occurring when spot-array faders are moved at mixdown. Typically, this means:

- matching the tone of the spot array's Focus instrument/section with the sound heard in the dominant array, preferably with minimal EQ so that you avoid mangling the tone of Backdrop instruments;
- finding a polarity/phase relationship between the dominant and spot arrays which renders inevitable comb-filtering effects between them as pleasant as possible—both for the spot rig's Focus instruments and for any instruments prominent in its Backdrop signal.

Spot rigs can also perform a more tonal role, though, if you deliberately design/process the spot array to correct or enhance the timbre of specific ensemble instruments (although this necessarily reduces the usefulness of such a spot system for pure level adjustment). For example, many engineers like to retain the transient snap of percussion instruments at the rear of an ensemble (a swing band's drum kit, say, or an *avant garde* orchestral percussion section) and low frequencies often need tailoring in small-studio recordings to compensate for unhelpful room resonances and lesser-quality instruments/amplifiers. Shawn Murphy[20] is circumspect about the benefits of heavy-handed

LOW-FREQUENCY CORRECTIONS FOR SPOT ARRAYS

Although I suggest trying to avoid equalization when recording ensembles with the dominant-array approach, the very nature of spot-miking can introduce low-end aberrations which even purists will use EQ to correct. For example, close-miking with directional mics will often incur considerable proximity-effect bass boost, especially with figure-eight polar patterns, and directional mics are also more sensitive to unwanted physical vibrations transmitted through their mic stands. A variable high-pass filter can deal with both these issues pretty swiftly.

The other problem, though, is that any spot mic's Backdrop signal will tend to have an overabundance of sub-500 Hz frequencies, because anything higher up the spectrum is dispersed fairly directionally by most instruments (as we saw in Chapter 6), and is more easily absorbed by physical obstructions in transit. The off-axis dulling effect of most mics (especially cardioid-family ones) only compounds the issue. As a result, it's very common for engineers to shave a little low end off their spot mics. A high-pass filter isn't that useful in this role, because its attenuation increases as you move down the spectrum, so a type of EQ called a "low-frequency shelving filter" (or "low shelf" to its chums) is a better choice, because it cuts everything below its nominal frequency by a similar amount (determined by a gain control). I normally set a shelving filter to around 100–300 Hz for this application (because almost all EQ designs have a residual effect on frequencies a little above their nominal frequency) and wouldn't suggest using any more than about 3–4 dB of cut for this reason unless you're looking to thin out the spot array's Focus timbre into the bargain.

processing, though: "If you crank a lot of EQ on the individual instruments," he says, "it alters the characteristic of the leakage, so you're fighting the fact that in order to make the direct sound right on a certain mic, you've got badly EQ'd leakage to deal with." Because of this, drastic EQ also makes spill reduction more critical (with all the sonic compromises that may entail), so I'd suggest chasing your target sound as hard as you can with careful mic technique before dialing in any processing. "Usually the cure for something not sounding right," adds Murphy, "is to take the EQ off."

Tweaking polarity/phase relationships between the spots and the dominant array is another valid tonal tool, albeit one that's heavily reliant on the instruments remaining stationary—fine perhaps when you're recording a rock drum kit and three electric guitar/bass cabs, but much less advisable on acoustic sessions. Whatever means you use to sculpt per-instrument tone in a dominant array setup, however, the naturalistic dominant-array capture will inevitably restrict your creative scope. Where you crave greater artistic license, the peer-array method explained in Chapter 10 becomes more appropriate.

9.4.4 Dealing with Depth Perspective

Because spot arrays are usually drier-sounding than the dominant array, changing the level of any spot mic may also change the dry/wet ratio (and thereby the apparent distance from the listener) of its Focus instrument/section. In some cases this may be exactly what you want—it's certainly part of the appeal of drum-kit spot mics for mainstream mix engineers who want to cram the kick and snare right up the listener's nose! However, assuming that your dominant array already gives the entire ensemble a fairly representative front–back depth, you probably don't want your spot array's Focus instrument(s) to shimmy closer to the listener whenever you fade up those mics—especially if you're planning to correct ensemble imbalances dynamically as the music progresses.

"Because spot arrays are usually drier-sounding than the dominant array, changing the level of any spot mic may also change the dry/wet ratio of its Focus instrument. In some cases this may be exactly what you want—it's certainly part of the appeal of drum-kit spot mics where you want to cram the kick and snare right up the listener's nose!"

One solution to this is to add artificial reverb to each spot array, keeping the dry/wet ratio more consistent as faders are moved. Ideally I like to do this in the monitor mix while recording, but you can also perfectly well leave it until mixdown, so I won't delve into the thorny subject of reverb processing in any great depth here. If you want to dip a toe in the water, though, here are the most salient points:

- Feed the reverb from post-fader sends on your spot-array monitor channels, and make sure the effects processor is set so that it doesn't pass any unprocessed signals (i.e., make sure it's set to "100% wet").
- Take care choosing an initial preset, avoiding unnatural-sounding emulations of vintage devices such as mechanical plate/spring reverb units. Try to choose stereo reverb presets which seem to match the acoustic characteristics of the recording room (as captured via the dominant array)—this can take quite a bit of patch-surfing, so stick at it!
- Find out how to adjust the length of the reverb's decay, because this is the most common parameter you're likely to want to change.
- In practice it's usually easier to work with two reverbs than one: Use an "early reflections" or "ambience" patch without any discernable effect "tail" as the primary means of pushing the spot array's Focus instrument(s) backwards in the depth perspective; and use a longer "room" or "hall" patch to increase the sense of size and sustain for Focus instruments if necessary.

But artificial reverb may not adequately distance the Focus instrument(s) if your spot-mic signals precede the dominant-array signals in time (as discussed in Section 7.2.3). With smaller ensembles that are set up within about 10 feet (3 m) of the front of your dominant array, this effect will often prove weak enough to ignore, but with larger ensembles it usually pays to delay the spot arrays to roughly compensate. I say "roughly" here for several reasons: Firstly, it's impossible to compensate accurately for both Focus and Backdrop sources at the same time; secondly, with ensemble recordings it's rarely straightforward to judge exact timing offsets using the polarity-inversion trick in 7.2.3 or by comparing waveform displays, because dominant-array and spot-array signals will be quite different in character; and, thirdly, you won't be able to exactly

THE 3:1 RULE

An oft-quoted principle of ensemble miking is the "3:1 Rule," which states that a spot array should be placed three times closer to its Focus instrument(s) than to any Backdrop instrument. The assertion is that this keeps spill from Backdrop instruments roughly 10 dB below the level of the Focus, so that it doesn't cause too much mischief at the mix. What a lot of people don't realize about the 3:1 Rule, though, is that it's based around the following assumptions:

- That the instruments in question project all their frequencies equally in all directions. No instrument does this, as we saw in Part 2.
- That the Focus and Backdrop instruments are equally loud. This is rarely the case—indeed, the whole point of using spot arrays in many cases is to support Focus instruments that are quieter than those around them.
- That the mics pick up all frequencies equally in all directions. Even the best omni microphones don't do this, and definitely not cheap directional types.
- That the mic signals are balanced at identical levels. This is only likely when the two instruments already balance correctly—but if you're using spot mics at all, then it's often because there are balance problems.
- That you're recording in mid-air, without a floor, ceiling, and walls reflecting sound back into the mics. Admittedly, this one might apply to you if your studio is in an anechoic chamber. Or a hot-air balloon. Anyone?

But it's not just these issues that call the 3:1 Rule into question for me, its also that I don't think there can be a "correct" spill level for any spot mic, because the amount of spill is just one variable amongst many that influence your mic-positioning decisions, and each engineer will have their own opinion regarding how much comb-filtering between mics is acceptable. Leaving the theory aside, though, I just don't find the 3:1 Rule much practical use in real recording situations. Granted, it maybe offers beginners the flimsiest of guidelines for initial mic setup, but I wouldn't suggest giving the 3:1 Rule much credence once you're on the session, because it's only your ears that can tell you whether you've got too much spill or not.

Incidentally, some engineers define the 3:1 rule differently: that the distance between any two spot arrays should be at least three times the distance between either spot array and its respective Focus instrument(s). To my mind this variant is even less useful, because the amount of spill picked up then depends much more heavily on the specific layout of the performers—for example, if you were independently spot-miking two musicians, following this version of the 3:1 rule would give you much less spill for mic positions between the musicians than it would in setups where either musician was between the spot mics.

time-align off-center spot arrays to any dominant array that incorporates wide-spaced mics such as outriggers. Hence, my preferred method is to drag a tape-measure from each spot rig over to the dominant array's center-front point and use that to approximate the delay amount.

If the array's Focus happens to be something that doesn't move around (a drum kit, guitar amp, or piano, say), then I'll stick with the same "meter per 3 ms" estimation I used in Section 9.3.3, because that's where the tightest phase-match is likely to be found. However, where a spot array's Focus is something that moves around during the performance, then there's a good argument for using slightly longer delay times instead. To understand why this is, imagine you've miked up a big-band with a simple XY dominant pair and have then set up a spot mic for the saxophone, delaying the spot mic so that its sax signal is phase-matched very closely with the sax signal in the dominant pair. This will give you a clear saxophone sound with minimal comb-filtering. However, if the saxophonist destroys the phase-alignment by moving, you suddenly get a big tonal change as the mics start properly comb-filtering against each other. By contrast, if you delay the saxophone's spot mic *past* the point where its sax signal phase-matches with the dominant XY pair, you'll get a certain amount of comb-filtering the whole time. In other words, when the saxophonist moves, we hear a selection of different comb-filtering variations, which won't be as strong a subjective effect as the transition between unfiltered and filtered tones (as we had when the spot mic was set for a tight phase-match). How much additional delay you use in this scenario is open to debate, and will depend on tonal considerations of course, but I'd say 10–25% is a pretty sensible range.

"One side-effect of close-miking that can't be corrected by artificial reverb or delay-compensation is the unnatural emphasis that's often given to a Focus instrument's high frequencies, especially where musically unrelated mechanical noises are concerned."

One side-effect of close-miking that can't be corrected by artificial reverb or delay-compensation is the unnatural emphasis that's often given to a Focus instrument/section's high frequencies, especially where musically unrelated mechanical noises are concerned—a particularly troublesome problem with bass-register instruments such as bassoons and upright basses. Equalization can be a godsend here, so don't be afraid to apply several decibels of high-frequency shelving cut to any spot array for this reason. "High frequencies make something sound closer," explains Reinhold Mack,[21] "and less high frequency tends to [move things] further back."

Before we leave the issue of front–back perspective, it's worth pointing out that depth realism is the aspect of commercial recordings that is most commonly jettisoned when compromises have to be made. For example, instruments will often hurtle merrily backwards and forwards in recorded film scores as a byproduct of extreme rebalancing of thematic material or dramatic sound effects, or because of changing perspectives in the visual imagery. Pre-digital classical recordings are replete with unusual depth perspectives too, partly because unusual ensemble layouts (harps by the conductor, say) were commonly used to improve balance, and partly because tools for compensating for spot-array depth distortion (e.g., delay/reverb processing and multitrack recording/mixdown) didn't become widespread until the 1970s. Even on modern recordings, the logistics and timescale demands of location recording, especially for broadcast, will frequently militate against the use of detailed depth-adjustment techniques on spot mics. So if you need to get results in a hurry with limited gear, depth presentation is probably the aspect of your recording which will best tolerate any unavoidable corner-cutting.

WHAT'S IN A NAME?

One thing that confuses a lot of people when learning about ensemble recording is that a number of the terms used to describe miking techniques apply more to a mic's positioning than to its actual purpose in the mix. Foremost amongst the culprits is the term "outriggers." For example, in an orchestral recording:

- the microphones designed to supplement a central stereo setup by picking up the edges of the ensemble better (as discussed in Section 9.3.3) are usually called outriggers. These form an integral part of the dominant array, and are often selected to match the central rig's dry/wet ratio and overall tonality. They are unlikely to be processed or delayed independently.
- omni outriggers might be used to bolster the low end and spaciousness of a central stereo setup that's reliant on directional mics, again as an integral part of the dominant array (see Section 9.3.4). These mics will already sound wetter than the central array because they're less directional, but some engineers will increase this wetness, giving the outriggers more the character of room mics. Again, this kind of outrigger would rarely be processed or delayed independently, but is more likely to have its level tweaked (subtly!) during playback to adapt the apparent width and reverberance of the space to different orchestral textures.
- sometimes the outriggers might be operating as string-section spot mics, rather than forming part of the dominant array. In this case directional polar patterns and closer mic placement would make sense, and you'd be much more likely to use unmatched mics or apply processing such as EQ, delay, and reverb to tackle timbre and depth concerns. If the purpose of these spot mics was for balance control, then it wouldn't be unusual for their faders to be ridden over a range of 6dB or so during the music.

This isn't the only ambiguous term, though, by any means. Some of what might be called "spot mics" within an ensemble may actually be integral to the dominant array, as a means of redressing the overall depth-distortion of a main stereo rig placed at the front of the ensemble (a technique we talked about in the latter half of Section 9.3.3). In a similar vein, the "overheads" in a drum-miking setup might be the dominant array for miking this one-player ensemble, but they are just as often treated as cymbal spot mics.

The upshot of this terminological tangle, as far as students of microphone technique are concerned, is that you have to be very careful how you interpret advice from other engineers. So if someone recommends a certain brand of overhead mic, say, on the basis that they like it for cymbal close-miking, you might not want to give that suggestion too much weight if you prefer using your overheads as a dominant array to capture the sound of the whole kit.

9.4.5 How Should my Spot Mics Sound?

Because there are so many factors involved in choosing and using spot mics, there's a lot of confusion amongst the small-studio community about how they should actually sound. The first thing to set straight is that any assumption that a certain spot mic (say a drum kit's snare-drum mic) should always have a certain sound is unhelpful, to say the least. What's important is how each spot array reacts *when combined with the dominant array*—and indeed with any other spots. In this respect, the biggest trap you can fall into is passing judgment on spot arrays when they're soloed. Soloing a spot array can be useful for homing in on a broadly appropriate sound, because you can hear the effects of setup and processing tweaks in isolation, but it's vital that you treat what you hear in solo mode as provisional, postponing final decisions until you can scrutinize how the spot array interacts with the rest of the ensemble setup.

As when setting up a dominant array, there are a lot of different things to listen for when judging the effectiveness of any given spot array, so it helps to structure the task by asking yourself questions very similar to those listed at the start of Section 9.2:

- **How does it affect the level?** Where you want to rebalance the Focus instrument/section, check you can do this sufficiently without also imbalancing Backdrop instruments. If you can't, you may need to reduce the spot array's spill levels or adjust your polarity/phase settings.
- **How does it affect the tone?** If your spot system is for rebalancing purposes, then toggle the array's Mute buttons to check for unacceptable tonal damage of either Focus or Backdrop instruments. (Remember that the degree of comb-filtering will depend on the spot array's mix level, so this judgment should be made within the context of an appropriate range of fader settings.) Where your spot array is designed to adjust the Focus timbre, you'll want to insure this isn't at the expense of unintended consequences for Backdrop instruments. Dealing with this question satisfactorily may involve a combination of miking tweaks, polarity/phase adjustment, and EQ.
- **How does it affect the depth?** Should toggling the array's Mute button bring the Focus closer, and if so by how much? Does the Focus's front–back position vary too much if the spot array's fader is ridden in real time? When you're not happy with what you hear in these cases, you might turn to compensating delays, artificial reverb, and/or EQ for a solution.
- **How does it affect the imaging?** Toggle the array's Mute buttons to check for stereo image-shifts, adjusting your stereo setup and pan controls accordingly. Don't just listen to the Focus, though, because Backdrop instruments may also move noticeably, depending on spill levels.

9.4.6 Small is Beautiful

The beauty of the dominant-array recording approach for budget-conscious engineers is that it allows you to capture even quite large-scale ensembles with comparatively minimal gear, but simple setups may retain their attraction even where money is no object. Here's Michael Bishop,[22] for instance: "I always look for the imaging across the orchestra that lets me feel where each musician is on stage. Use of spot mics pretty much destroys that." Whether or not you subscribe to this view, or indeed any audiophile notion that simpler recording setups inherently sound better, the speed with which you can achieve results by streamlining your approach is another important benefit of smaller-scale systems, leaving you more time to deal with the musicians on your sessions.

Keeping the technological paraphernalia to a minimum is frequently a question of psychology too, as Tony Faulkner[23] explains: "Keep it simple and you'll make friends with a lot of orchestral players who do not enjoy a [spot mic], stand, and cable in each of their faces while they are trying to play… You'll also probably make friends with the conductor if you use fewer mics, for at least two good reasons. First, if he/she sees a lot of microphones he/she will suspect (correctly) that you'll be messing after the event with his/her careful balance. Second, if the stage is full of stands and cables it is a pain in the neck for the players and especially the conductor to get to/from their place if they are walking through the string section."

"Even though amateur ensembles and quirky recording spaces frequently justify multimic approaches, most small-studio engineers get better results by erring on the side of simpler setups which allow them to concentrate more on the mic placement, the acoustics, the musicians, the performances, and the music."

So overall, I'd urge you to resist the temptation to put up more mics just because you have the hardware. Even though the amateur ensembles and quirky recording spaces typically used for low-budget projects frequently justify multimic approaches, my firm opinion is that most small-studio engineers get better results by erring on the side of simpler setups, because this allows them to concentrate more on the mic placement, the acoustics, the musicians, the performances, and the music.

9.5 SOME CASE STUDIES

Ensemble recording can be a bit daunting for newcomers, because there are so many things to weigh up when planning and executing a session. So what I'd like to do now is bring the whole discussion back to earth by walking you through the thought processes I might follow when tackling a couple of different dominant-array recordings using typical small-studio resources.

Let's set the parameters first. We'll say I have the following motley selection of mics to work with:

- One Shure SM58 and two Shure SM57s. These cardioid dynamic mics typically cost less than $120/£100 each.
- A pair of SE Electronics SE1A small-diaphragm cardioid condensers (under $400/£300).
- The Cascade Victor, one of the new breed of budget-friendly ribbon mics, priced around $200/£160.
- AKG's D112 cardioid dynamic mic (under $200/£150) and a C414B XLS large-diaphragm multi-pattern condenser mic (around $1000/£1600).
- A $150/£100 large-diaphragm cardioid mic of questionable quality and with no published specifications. We'll not name names, but I'm sure you know the kind of mic I'm talking about…

In addition I have six mic stands, a small stereo bar, a handful of screw-on mic-mounting clamps, a couple of thick old quilts, and a few random scraps of acoustic foam. Sound familiar? Right, let's get to work!

9.5.1 A Drumkit

I've been asked to capture a fairly natural-sounding drum sound for a laid-back indie-rock production, and I decide to use the dominant-array approach, although I'm aware from listening to releases by some of the band's influences

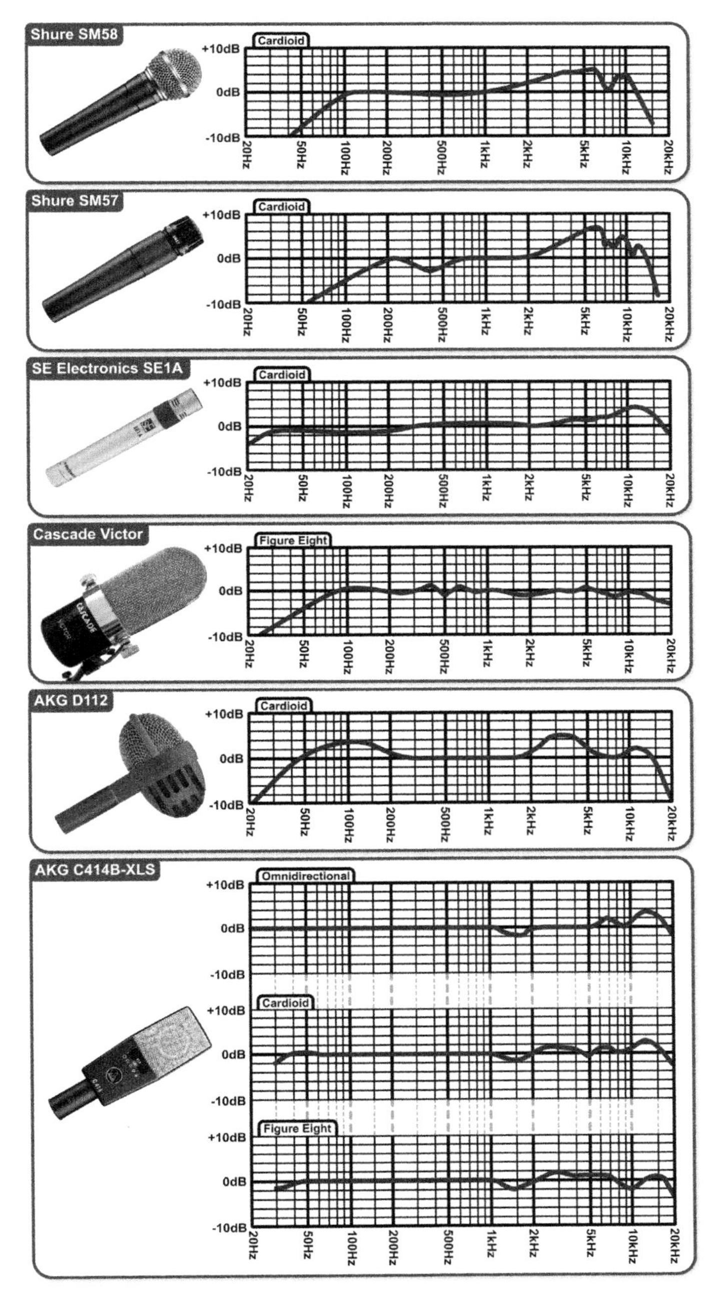

FIGURE 9.12
Some technical information about the specific microphones featured in Sections 9.5 and 10.3.

that I'll probably need to use spot mics on all the drums to produce the required presence in the mix. Turning up to a band rehearsal, I see that the kit setup is fairly conventional, with the odd extra cymbal, but that the hi-hat's set quite low, so I ask the drummer whether he'd be happy raising it at all if we want to reduce hi-hat spill on the snare close-mic. We also chat about whether a slightly deeper snare tone might work better, and agree that tuning it down a little seems to make it fuller-sounding. Because of the rehearsal room's low ceiling and carpeted walls, we plan to record one weekend at a local warehouse that has a more spacious and reflective environment. The drummer's not keen on recording to a click, but I ask the band to agree tempos in advance so that we can at least start each take consistently, increasing the likelihood of us being able to edit between takes.

I want time-difference spaciousness from my dominant array, so I plan to put my only stereo pair (the SE1As) in a non-coincident setup, treating the snare as the center of the stereo picture (i.e., putting it equidistant from both mics). I'll set up a single stand just behind the drummer with its boom roughly horizontal, and then I can use screw-on clamps to fix the two mics in position. To balance the snare's high frequencies and overall level with the cymbals (which will be closer to the mics) I decide to point the cardioid polar patterns directly at the drum as an AB pair, rather than going for any kind of near-coincident setup. To supplement this stereo rig, I'll try putting the ribbon mic on the other side of the kit facing the kick drum, to try to compensate for inherent kick-drum weakness/distance in the overhead pair (see Section 9.3.3); to catch more of the shell-tone from the other drums (in other words, multi-miking those instruments from different angles to catch more of their frequency dispersion, as discussed in Section 7.2.1); to fill out the harder timbre of the small-diaphragm condensers with the smoother ribbon flavor; and to provide a more solid center to the spaced-pair stereo image. For spot-miking individual drums, I decide to go for dynamic models across the board for their chunkier sound, knowing that I should be able to rely on the overheads for general realism: The D112 goes on the kick; an SM57 on snare; another SM57 covering both rack toms; and an SM58 on the floor tom. The C414B I'll use as a room mic (clipped to a chair, given that all six stands are already accounted for), because it's quiet and has polar-pattern control, and I'll keep the cheapo condenser in reserve in case I need another mic for anything else.

On the session, the first job once the drummer arrives is to walk around with his kick and snare to find a promising-sounding location, whereupon we set up the kit. The lower rack tom seems to ring longer than the others, so a little retuning is required there, and a strip of gaffer helps zap one rather odd metallic resonance from the snare without sucking the life out of it. Balancing the SE1As and ribbon against each other proves the most critical part of the mic setup. First I try to get a suitable balance and depth across the kit. Here the AB pair seem to work pretty well, but I need to reposition the ribbon mic several times to get the kick, toms, and cymbals reasonably evenly balanced. At that point I refine the stereo width a little (with reference to some commercial releases, because we're working on headphones) while checking the AB pair's mono-compatibility, and then see whether finer mic positions and further phase/polarity tweaks can improve the tonal combination of the three mics. During this whole process, it becomes clear that the hi-hat is coming across a little too strongly on the dominant array, so I do a test recording and ask the drummer to have a listen. He doesn't have any other hi-hats with him, but says he may be able to adjust his playing to deal with this, and I decide to hang my quilts on the nearby wall on that side of the kit as an additional measure. I allow the dominant array to be a little on the dry side, so that the room mic will give a bit of control over the kit's apparent dry/wet level at the mix. The C414B is set to its figure-eight pattern so that I can use the null to reject the direct kit sound, and then I look for a spot in the room where I get a good low–midrange body to the drums, while avoiding excessive hi-hat pickup.

All this takes some time, but pays off with a pretty creditable dominant-array capture that needs comparatively little level from the spot mics. In order to get passable snare and tom sounds, I keep their close mics at least 4–6 inches (10–15 cm) away, with the rack-tom mic placed between the two toms to catch them equally. By moving the hi-hat a little higher, I'm able to gaffer some acoustic foam on one side of the snare close mic to take some edge off the hi-hat spill, just in case hi-hat balance concerns re-emerge later. Although the kick drum's resonant head has a hole cut in it, I nonetheless mike from just outside the drum to get a less "caricatured" sound, refining the amount of beater click captured through the hole by angling the mic. I don't bother delay-compensating any of the spot mics, given that a comparatively upfront drum sound is in keeping with the genre, but I do make sure to experiment with polarity inversion in each case for the most appealing tone. While doing this I first check each spot microphone in solo to listen for any Backdrop instrument that's obviously awful-sounding—an activity that demands some adjustment of the floor-tom spot mic because of its cymbals Backdrop.

If I'd found myself pressed for time, I'd probably just have guessed the room-mic and spot-mic positions, knowing that the final production sonics would depend mostly on the dominant array, and that the side-effects of any heavy remedial mixdown processing would be confined to the low-level spot signals. In fact, *in extremis* I'd probably even have got away with postponing stereo-width and polarity/phase adjustments of the dominant-array mics until mixdown if I'd had to hit the Record button very early on in the session, as long as I'd been able to record all those mic signals on separate recorder tracks. Not that I'd suggest taking either of those shortcuts out of choice, of course, because the session time saved wouldn't be worth the disproportionate increase in mixing workload unless it resulted in an utterly blistering drum take.

9.5.2 A Chamber Orchestra

As a second case study, let's imagine I wanted to record a college orchestra playing something like a Mozart symphony—nothing very large-scale, mostly just strings and woodwind, plus a couple of French horns, a couple

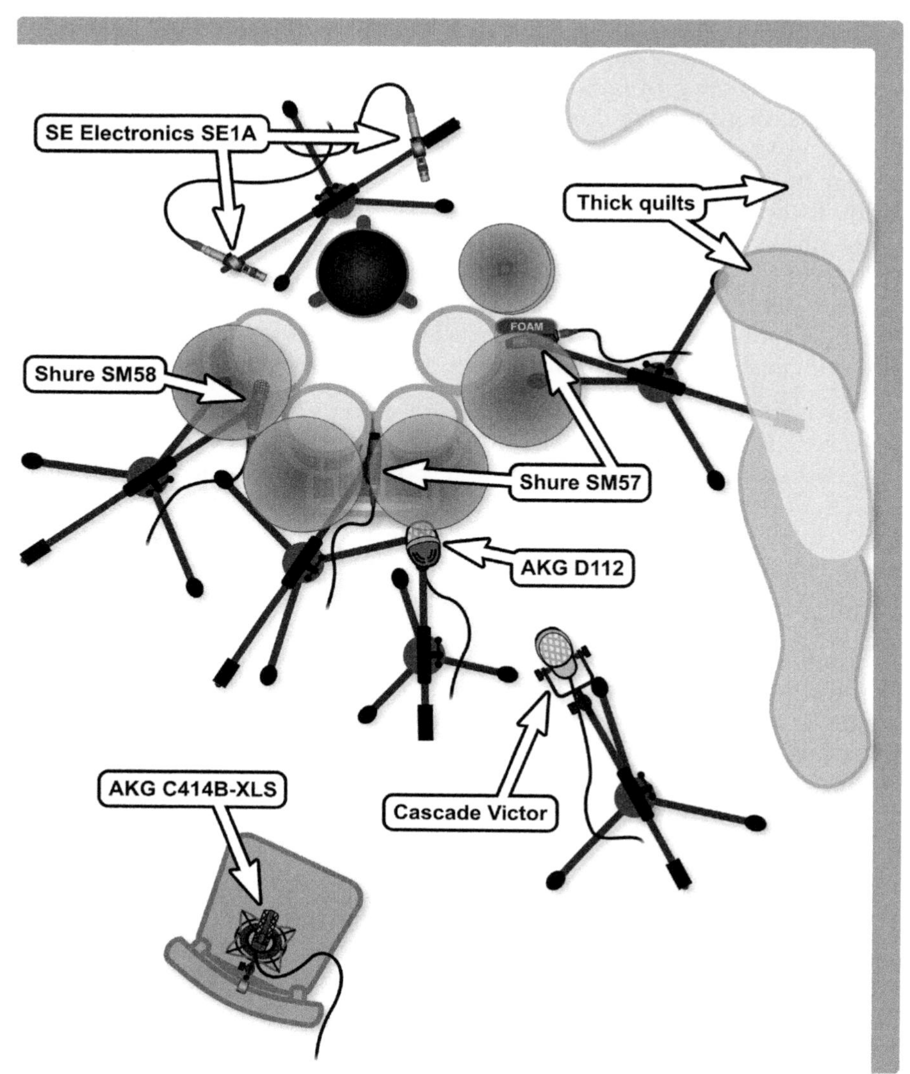

FIGURE 9.13
A diagram illustrating this chapter's first case-study setup, as described in Section 9.5.1.

of trumpets, and some timpani. I've heard them play the piece in an end-of-term concert, and will be recording them in the same room: the college's main assembly hall. Without the audience in the room, its gymnasium-grade reverb gets rather overwhelming, so I suggest moving the orchestra off the dais/stage (where they performed the concert) and setting them up in the hall, so I can then close the heavy stage curtains to dry things up a bit. I also plan to sit the orchestra with their backs to the curtain so that directional mics pointing toward the back of the orchestra reject reverb from the live end of the hall.

Although the SE1As are my only mic pair, I'm concerned the small-diaphragms will give the strings in particular a rather scratchy sound, so I do a pre-session test comparing them against an MS rig comprising the C414B-XLS in subcardioid mode and the figure-eight ribbon mic—the latter array isn't as hot on imaging (even after some high-end lift for the ribbon mic), but feels much smoother in tone, so I go for that. However, the coincident setup lacks spaciousness, so I add the SE1As into the dominant array as wide-spaced outriggers to combat that, with the intention of setting them a few feet further back from the MS pair to give a super-wide and more ambient signal that I can blend in as a subtle supplement.

By jury-rigging two boom arms onto the same mic-stand base, I'm able to raise a stereo bar with the MS rig above 7 feet (2.1 m) into the air while retaining reasonable stability (with the help of a few sandbags!) and have also asked the caretakers about borrowing some of the lightweight stacking tables that are sitting at the back of the room in case I need more height. Each of the outriggers has its own stand, which leaves three stands left over for spot-miking—although none of the remaining mic choices fills me with tremendous enthusiasm in that respect. In the end, I decide to set up the cheapy large-diaphragm condenser on the flute and oboe, as they had a little trouble projecting at times during the concert, but otherwise I resolve just to do as well as I can by rejigging the ensemble itself.

On the session, I find a reasonable dry/wet balance on the MS rig at around 5 feet (1.5 m) behind the conductor, and decide against raising the stand on a table so that the strings stay strong in the balance. However, adjusting the level of the MS rig's ribbon mic for a suitable width seems to distance the woodwind at the center of the picture rather too much, so I switch the C414B to its cardioid polar pattern instead, which makes for better center-versus-edge balance for the given acceptance angle. While moving the mics slightly backward to restore my previous overall reverb level, I also experiment with a couple of alternate miking heights to see what the tonal ramifications are. Listening to the outriggers, I decide to raise them up on the tables and angle them roughly parallel to the floor in order to warm the strings a little (by virtue of the mics' off-axis response). Mixing the two pairs together isn't initially a roaring success (overblown width and some nasty comb-filtering between the mics), but I persevere, narrowing the spacing a couple of times until I manage to get something that combines pretty well with the MS rig at moderate mix levels and doesn't sound horrid in mono.

While refining the dominant array in the hall, a couple of ensemble problems become increasingly apparent. Firstly, the timpani sound rather strident and overbearing, so I have a quick word with the conductor, play her a snippet of test recording to demonstrate what I mean, and she then decides to investigate some alternate mallets. While this is going on, I attend to the second issue, upending a couple of tables behind the French horn player (one of the two players has called in sick!) to reflect more of its high-end into the main mics—the drapes behind it were making it sound rather muffled in the dominant array, and the single instrument now needs some help in the balance anyway.

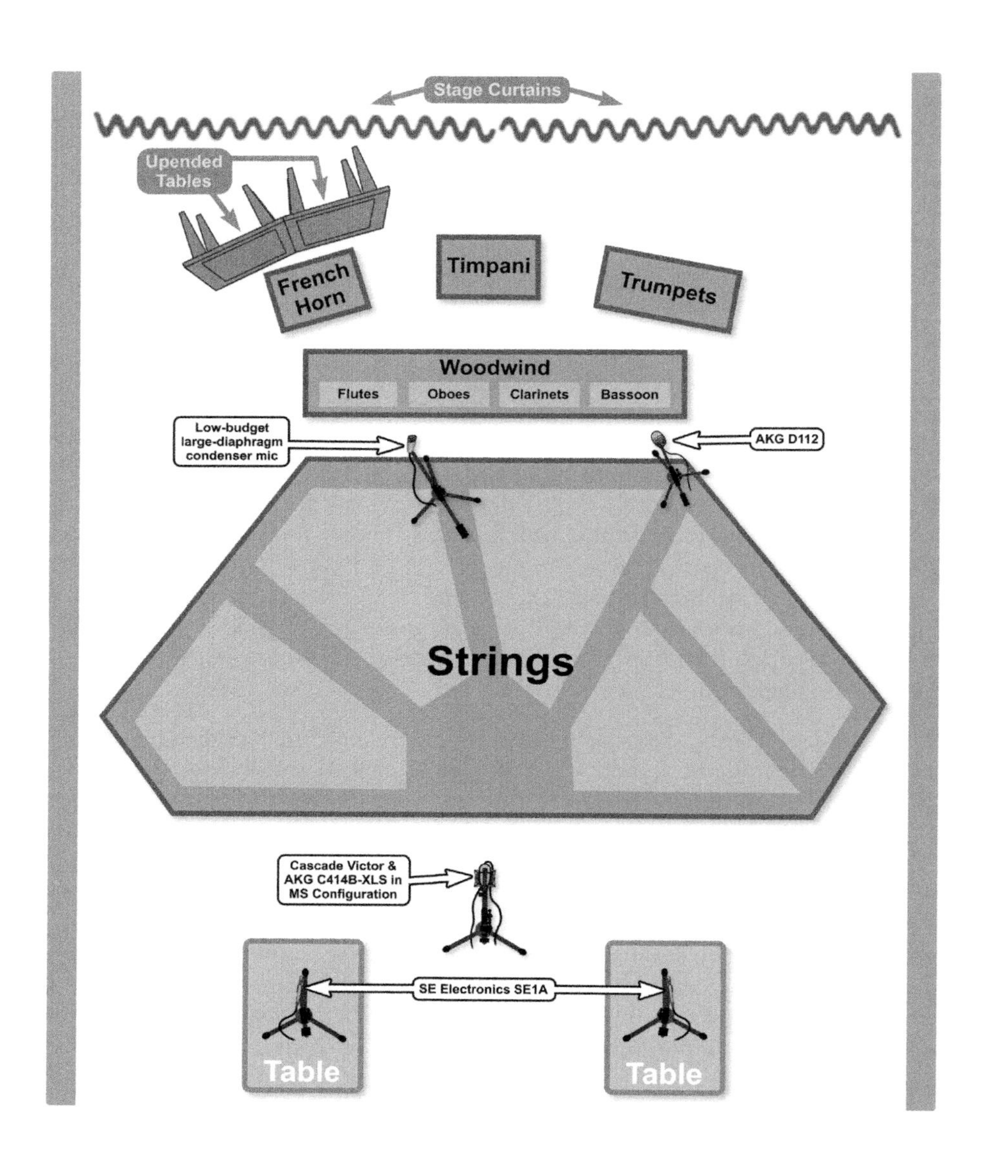

FIGURE 9.14
A diagram illustrating this chapter's second case-study setup, as described in Section 9.5.2.

While I'm over there, I move the spot mic a little closer to the oboes, but angled more toward the flutes, to try to get a better level/tone balance between the instruments. Another quick test recording confirms that these dodges have made a useful improvement, and we also decide to change the angle of the trumpets slightly, so as to mellow their tone on the MS rig.

At which point we need to start recording, so I have to do a couple of remaining tweaks to that spot mic in the gaps between the first few takes, and I also throw up the D112 spot on the bassoon as well—not a tremendously refined choice, but good enough for a bit more low end once I've cut away some of the highs with EQ.

9.5.3 The New Orthodoxy

So what can you take away from these case studies? Well, the first thing to highlight is that you're unlikely to find either of the above recording configurations described as a "standard studio setup," by any stretch of the imagination. That really shouldn't be surprising, though, given that neither of them has much in common with any standard professional recording situation—the way Abbey Road deploy a fleet of vintage Neumanns to record the London Symphony Orchestra has precious little bearing on how I best use a handful of budget mics to serve an amateur ensemble playing mediocre instruments in a school hall! Getting decent results in project studios is all about unorthodox measures, because most orthodox techniques were developed with expensive professional setups in mind.

"The way Abbey Road deploy a fleet of vintage Neumanns to record the London Symphony Orchestra has precious little bearing on how I best use a handful of budget mics to serve an amateur ensemble playing mediocre instruments in a school hall! Getting decent results in project studios is all about unorthodox measures, because most orthodox techniques were developed with expensive professional setups in mind."

I'd also like to make it absolutely clear that these case studies are entirely hypothetical, and not designed to be some kind of template for your own small-studio recording. What I'm trying to illustrate is how planning and executing an ensemble recording session is about suiting your working methods to the facilities you have and the immediate needs of the session. For example, I'd expect to completely rethink Section 9.5.1's drum-miking setup if I traded in the Victor ribbon mic, D112, and SM58 for an extra C414B-XLS. Or if I had to set up in a living room. Or if one of the small-diaphragm mics died on me during setup. Or if the combination of the small-diaphragm condensers and the ribbon just happened to sound terrible, however many good experiences I'd had with it in the past. Or if the drummer hit his cymbals too hard. Or if I only had four mic inputs on my audio interface. Or if…There's no end to the variations, and ultimately it's the way you adapt your recording technique to each of them that primarily determines the quality of the recording. If you only ever stick to someone else's tried-and-trusted setup, you'll never actually work out how to get the best results from your own limited collection of small-studio gear.

Finally, when resources are scarce, do at least use all the tools at your disposal, not just the studio gear. It doesn't cost any money to change the ensemble layout, to ask musicians to adjust their playing, or to use spare bits of furniture as impromptu acoustics devices, so there's no excuse for overlooking these things when you're short on cash—especially as they're often the quickest route to big improvements anyway.

9.6 MONITORING AND SIGHTLINES

I've already underlined how important it is that individual performers can hear what they need while overdubbing, and that's no different in an ensemble situation: The better any group of musicians can hear each other (and any backing cue mix you're using), the better they'll play together. If you have the hardware for it, there's nothing stopping you giving every musician their own headphones with an independent headphone mix, in which case the previous chapters have already covered everything you need to know—just scale everything up! Honestly, though, that kind of monitoring setup is the last thing I'd recommend for the majority of small-studio setups, for several reasons:

- It's tremendously gear-hungry: Four musicians will need four separate auxiliary sends, four separate outputs from your recording system, four independent channels of headphone amplification, four suitable pairs of headphones—and probably various mains/headphone extension leads as well. I'd rather spend more money on mics...
- It's a big logistical headache, leaving you less session time for sorting out the microphones. If you're flying solo on the engineering side (which most small-studio engineers are), you need additional complications like you need a hole in the head!
- Headphones can easily hinder communication between members of an ensemble both during and between takes. "When you have the musicians in the same room together without headphones," remarks Bones Howe,[24] "they tend to balance themselves better than any engineer can."

So my advice is to try to keep monitoring arrangements as no-frills as possible when working with ensembles. One of the best ways of doing this, in my opinion, is to use a recording setup that allows each performer to hear themselves and the rest of the ensemble naturally within the recording room (i.e., acoustically), because this means you can usually get away with using only a single backing-track cue mix for No Input monitoring. If the musicians are using headphones, they can set their own backing-track level relative to what they're hearing in the room, so all you need to do is keep a careful ear out for unwanted headphone spill (especially from any click track). Where there aren't enough headphones for a larger ensemble, you can often use speakers instead without too many problems (as discussed in Section 5.3). Failing that, distribute the available headsets to the conductor, rhythm players, and/or section principals, and ask them to lead the rest of the group—as Reinhold Mack[25] did when he ran out of headphones recording orchestral overdubs for ELO: "I gave headphones to the lead guys; all the good ones who I knew could actually play."

On a practical level, though, getting such a stripped-back monitoring scheme to work may require additional effort in the live room. Repositioning and angling musicians, instruments, or amplifiers can make a huge impact here, and setups where musicians are facing each other often work very well in this respect by taking advantage of the forward directionality of most instruments—and indeed of the human ear. Reflectors can make themselves useful by bouncing additional sound from quieter instruments back to the

FIGURE 9.15
Here a lorry's wing-mirror mounted on a mic stand maintains the sight-line between a singer and an organist in a situation where a back-to-back performance layout made sense for recording purposes.

performer or other musicians. Jack Miller[26] recalls using this technique when working with Duane Eddy: "We had these baffles that hung on the wall that we could turn. One side was live, one side was dead. We could bounce stuff around. If the rhythm guitar was [saying], 'I can't hear the drums,' like a mirror we bounced the drums toward him." This trick can also encourage over-forceful players to rein in their levels by making them more aware of their own volume—especially in the case of loud forward-firing instruments like trumpets and trombones. Where a singer is struggling to hear themselves, cupping a hand behind one ear can really help, as can blocking an ear with a finger to amplify bone-conducted sound via the Occlusion Effect (see Section 6.1.1).

Equally important to most ensemble playing is the issue of sight-lines, so this should always be at the back of your mind. One thing a lot of people forget is that you can use mirrors to maintain sight-lines in tricky situations—for example in an organ loft, or where two conductors are standing back to back so that groups of performers can face each other. Cheap lorry wing-mirrors are available for under $30 (£20), and can easily be gaffered to spare mic stands for accurate positioning. A webcam or closed-circuit TV system can also expand your options, so long as the display doesn't have too much of a time-lag.

Naturally there will be times when ideal "acoustic monitoring" remains elusive despite your best efforts, but that still doesn't mean you have to have everyone monitoring everything on headphones, because it's usually much easier to provide targeted foldback just for those who need it. One common scenario is where one or two members of the group aren't hearing a certain part well enough. There's no need to give them a full separate headphone mix in this case, because you can just feed a little of the relevant spot mic to their cans, assuming they can still hear everyone else acoustically. Similarly, if you're recording a full rock band, the vocalist will usually be inaudible in the room against the drums and amplified instruments, so vocal-only headphone foldback may be all that's required to supplement what's going on acoustically. Again, where headphones are in short supply you could substitute speakers in both these situations, and floor wedges designed for onstage applications are quite nifty for this sort of thing. "About fifty percent of my projects use monitor wedges as opposed to phones," comments Clif Norrell,[27] for instance, "primarily due to the fact that many of my projects ... need to have a gig atmosphere approximated, in order to capture the most natural performance."

Clearly, any of the above suggestions may throw your carefully prepared recording setup into disarray, especially where foldback spill on the recording mics risks unwanted tonal coloration or feedback howlround. But don't expect too much sympathy from me! Great performances will always be in the best interests of the production, and the very reason for learning a wide range of different miking techniques is because they allow you to adapt to the unique monitoring needs of each new group of performers.

Finally, if headphone foldback is unavoidable, make sure you dedicate enough time to getting each of the different cue mixes right. "Amazingly bad things can happen to even the best players when the headphone mix is all screwed up," warns Wyn Davis.[28] "I don't think you can pay enough attention to that part of it," agrees Brian Ahern[29] wholeheartedly, and he also shares a great time-saving tip: "What I usually do is set up the phones prior to the session with a pre-existing basic track tape that has similar instrumentation. That puts me that much further ahead in the game when the musicians walk in." Don't assume that the musicians will alert you to problems here either. "Sometimes I go out [into the live room]," says Jeff Powell,[30] "and the phones are completely distorted. So I will sneak out there and tweak it up a little bit and all of a sudden they can hear. It amazes me how little importance some musicians place on headphone fidelity and mixes."

CUT TO THE CHASE

- The dominant-array approach is good for natural-sounding ensemble recording, and makes excellent use of limited time and equipment. The dominant array captures a picture of the whole ensemble at once, and its setup and balance typically remain static once recording has begun. Although a dominant array can be just one mic or stereo pair, it will often include other mics to enhance the overall recorded tonality; to allow the array's directivity to be changed more flexibly from the control room; to give control over recorded reverb levels at mixdown; to correct misrepresentations of the ensemble's balance/depth by close-placed mic rigs; and to blend appealing features of different stereo miking methods together.
- Spot mics can be mixed with a dominant array to enhance or correct selected instruments/sections of the ensemble. These mics must avoid catching excessive/ugly spill; must support the dominant array's stereo imaging; and may need processing to avoid upsetting the ensemble's natural depth perspective.
- Do some research about the ensemble and venue before the recording session, and preferably check both out in person. Make a note of any preferred ensemble setups; any

balance/tone problems between players; and any concerns with the acoustics or physical layout of the room. If possible, try to find out about setups that may have worked in the past, and keep your eyes open for anything on site that might operate as impromptu acoustic treatment.

- You can structure the process of planning and setting up a dominant array by using the following series of questions, roughly in order. For the whole ensemble: Which mics; which stereo method; how wet overall; and how wide overall? For each instrument: What level; what tone; what depth; and what image? Because of the limitations of pure mic technique, answering these questions is usually best done by manipulating the acoustic environment and by working with the musicians to adjust the instruments, ensemble layout, or performance.
- When refining the sound of each spot array, ask yourself in turn how it affects level, tone, depth, and imaging within the ensemble. Only make final judgments about spot arrays when you're hearing them mixed with the dominant array.
- The real-world directional characteristics of microphones are critical when miking ensembles. As such, think very carefully about how any mic's on-axis and off-axis response will color incoming sounds, not just about the positioning of rejection nulls. Budget cardioid-family microphones should be treated with particular caution in this respect.
- Although most small-studio engineers won't have the resources to try a classic Decca Tree arrangement, its underlying operational principles can be adapted quite successfully for those working on a budget. Phased-array rigs are a good choice for cost-effective large-ensemble dates, because their natural balance and depth capture tend to require less correction from supporting mics.
- Be careful how you interpret the terms "outriggers," "spot mics," and "overheads," because in each case they can refer to mics which serve very different purposes for different engineers.
- One of your top priorities when recording ensembles is to make sure that the musicians can hear and see each other properly, even if that makes your recording job more difficult. Rehearsal-format or face-to-face ensemble layouts are often preferable in this respect, and acoustic absorbers and reflectors (whether purpose-designed or improvised) can help with this. Mirrors provide a cheap way of working around sight-line problems.
- Try to keep both recording and foldback schemes as simple as possible. An ensemble session is usually a race against time, so minimizing the technical workload means you can concentrate more on what gets results quickest: dealing with the musicians, the instruments, the ensemble layout, and the acoustics.

Assignment 1: Ensemble Recording with Just a Dominant Array

Record two or three different ensembles using a mono or stereo dominant-array approach, but without recourse to spot mics. You're free to choose any size of ensemble, so there's no need to wait around for big groups—even a singing instrumentalist will provide useful experience. You may use whatever recording equipment you wish, but make the effort to research and plan your approach well in advance and do your best to get maximum mileage from adjustments to the performance, instruments, ensemble layout, and acoustics—by hook or by crook!

Assignment 2: Dominant-Array Ensemble Recording with Spot Mics

Record a few different ensembles in stereo using a fully fledged dominant-array approach, complete with spot mics as required. You should record a drum kit in this way at least once, and also do your best to find a large ensemble (big-band, orchestra, or vocal chorus for instance) you can try your hand with too. When you've planned each job, make a point of asking yourself whether simplifying the setup might actually serve the final production better, bearing in mind your anticipated technical workload during the session.

WEB RESOURCES

On this book's companion website you'll find a selection of resources to support this chapter, including:

- audio demonstrations of a variety of dominant-array setup issues;
- further reading on multimic ensemble recording techniques, critical distance, and the fundamentals of reverb use;
- links to information on the Schoeps Polarflex system and all the mics used in Section 9.5's case studies.

http://www.cambridge-mt.com/rs-ch9.htm

CHAPTER 10

Ensemble Recording with Peer Arrays

Although the dominant-array recording method can be extremely quick and efficient at capturing all sorts of different ensembles, it works best when preserving a natural acoustic event for posterity. What it doesn't allow is a tremendous amount of control over the balance or timbre of any instrument from the control room, whether during tracking or at mixdown: You can't process the dominant array without affecting most of the ensemble at once; you can't fade down an instrument that's too loud in the dominant array; and you can't use a spot mic to drastically change any instrument's tone without upsetting the ensemble balance. Fundamentally, a dominant-array recording is what it is, and trying to fight against its inherent nature will usually end in tears.

Which is why some engineers prefer to use a peer-array approach instead, removing the dominant array from the equation and relying on separate microphones for all of the ensemble's musicians/sections. Ditching the overall pickup means you can sculpt the tone of each instrument and rebalance everything much more flexibly, and it also allows more creative positioning of players and acoustic treatment to improve performances and manage spill. The downside, though, is that it's arguably the most complicated type of recording to pull off successfully. Part of the reason for this is that any single array within your peer-array setup may range in complexity from a single mic (discussed in Part 2) to a complex multimic dominant-array rig with accompanying spot mics (discussed in Chapter 9), as in the following hypothetical band setup:

- **Peer Array 1:** A lead vocalist singing into a handheld dynamic mic.
- **Peer Array 2:** An acoustic guitar multimiked with a crossed coincident pair of condensers.
- **Peer Array 3:** An electric bass guitar recorded simultaneously through an amp (via a single mic on the cabinet) and a DI box.
- **Peer Array 4:** An electric guitar tone created by blending together three different close mics and adjusting their phase relationships.
- **Peer Array 5:** A drum kit captured with a stereo-pair dominant array, spot mics on every drum and cymbal, and additional multimics for snare and kick.

THE FOCUS AND BACKDROP OF A PEER ARRAY

I'm going to continue using my terminology from Chapter 9 here: i.e., the Focus of a given peer array is what the array's mics are ostensibly directed at, whereas the Backdrop is everything else these mics are picking up besides the Focus instrument(s). However, a little clarification is required, given that we can now nest dominant arrays within a peer-array setup. The overall Focus of a dominant array is the entire mini-ensemble it's capturing, which means the full drum kit in the above example setup, even though that dominant-array rig may comprise spot mics which each have their own single-instrument Focus.

Driving a setup like this is no mean feat. Not only must you handle every one of the individual peer arrays confidently in its own right, but you also have to manipulate all the spill that arises between them. A lot of small-studio engineers try to make life simpler for themselves here by just declaring all-out war on spill in general, heaping up masses of acoustic treatment and using super-tight mic placements to maximize isolation between instruments. However, this kind of recording is absolute hell to mix, because it just provides you with a bunch of spot mics—or, to put it another way, a dominant-array-plus-spots rig without the best-sounding bit (i.e., the dominant array itself)! The real trick with peer arrays is not trying to eliminate spill completely, but turning the spill to your advantage. In our imaginary full-band setup above, for example, you wouldn't necessarily break your back trying to totally shield the acoustic-guitar and vocal mics from spill. Instead you'd work out how to set up those mics so that their Backdrop signals would actually enhance the sound of the rest of the band within the mix.

"A lot of small-studio engineers try to make life simpler by heaping up masses of acoustic treatment and using super-tight mic placements to maximize isolation between instruments. However, this kind of recording is absolute hell to mix, because it just provides you with a bunch of spot mics—or, to put it another way, a dominant-array-plus-spots rig without the best-sounding bit!"

Once you adopt this kind of mindset, the interdependence of your peer arrays can become an enormous asset, because each one is effectively using all the others as supplementary multimics, thereby capturing a much more holistic view of each instrument. So the electric-guitar cabinet in our example wouldn't appear in the mix solely via its own fairly close-placed three-mic peer array, but could be significantly enhanced by more than a dozen other mics nearby, albeit ostensibly pointing at other instruments (see Figure 10.1)—they might pick up useful alternative views of its frequency dispersion, say, and/or act as room mics to adjust its apparent depth in the mix. Here's S. Husky Hoskulds[1] talking about this concept: "For example, if I'm recording drums, piano, and acoustic guitar, I'll often run out while people are getting sounds and move the piano mic further away to match the [drum] ambience in the guitar mic,

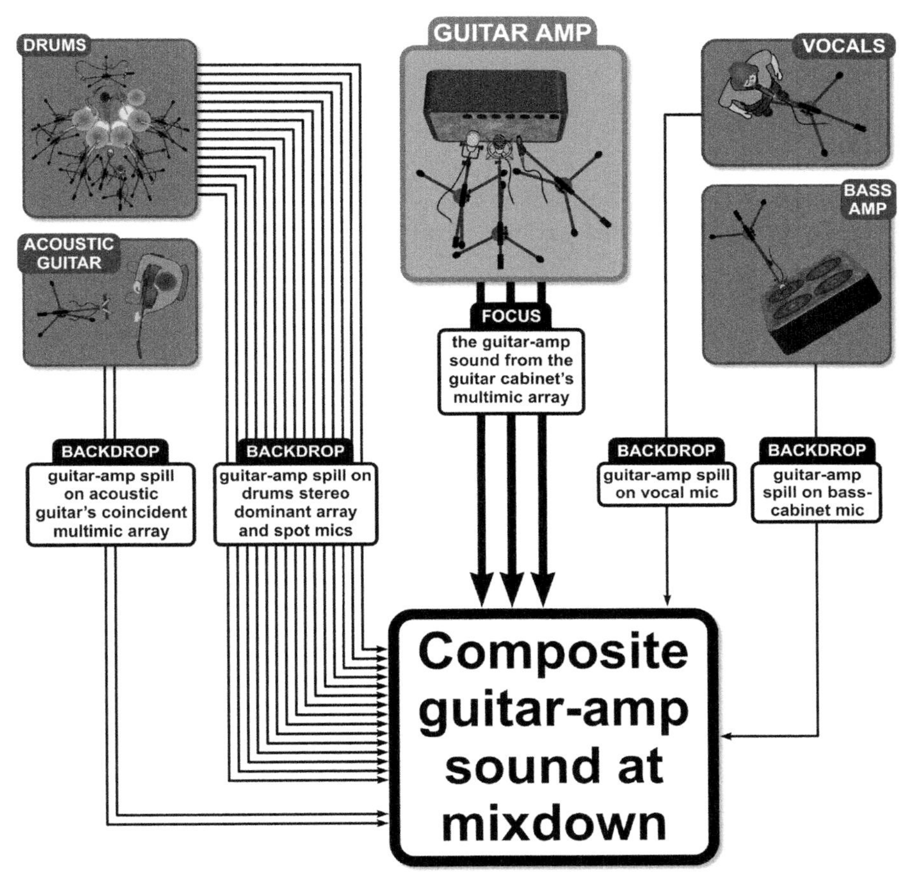

FIGURE 10.1
In this hypothetical peer-array ensemble recording of a five-piece band, the final mixed sound of the electric-guitar will comprise the Focus sound from its own peer array combined with the Backdrop electric-guitar sound coming through all the other open mics—in this case four separate peer arrays for drums, acoustic guitar, vocals, and bass amp.

and then pan them left and right. That way the piano and guitar mics actually include half of the drum sound. A lot of the organ stuff you hear on [Solomon Burke's] *Don't Give Up On Me* is well set back in the mix, and I got that sound by using the spill from the guitar mic in the next room."

While this kind of juggling act can be terrifically intense work from an engineering perspective, it also offers exceptional rewards if you can judge it just right—a recording where every peer array sounds much richer in the mix than it does in solo, where individual instruments achieve an almost holographic three-dimensional quality, and where mixdown begins to become a mere formality. "Leakage is your friend," affirms Frank Filipetti,[2] "and what makes the sound real and live and wonderful."

10.1 PLANNING THE SESSION

Planning any ensemble recording involves reconnaissance work, whatever capture method you're using, so everything we discussed in Section 9.1.1 still applies. There are additional considerations to take on board when using peer arrays, though, mainly because there may be a considerable difference between what's happening acoustically in the recording room and what you actually want coming through the control-room monitors. Therefore:

- You have to make an executive decision about what kind of recorded sound you're after. If you've been following the assignments throughout this book, you should already have some practice at developing a sonic "vision" in your head and then chasing it with your mic technique, and doing this for a complete ensemble is a natural extension of this. If you're struggling to define the parameters of what the artist wants to hear, then grill them about what music they like listening to and how that relates to what they want to record, as I suggested in Section 7.2.3.
- You have to compare your vision for the recorded outcome with what the ensemble is actually likely to sound like acoustically. This is where confirming an instrument line-up and attending a rehearsal are both so useful, because they tip you off if any instruments will be significantly out of balance with the others, and whether any of them will need serious engineering work from a timbral perspective. Without this kind of information you could totally misjudge your initial setup, thereby wasting a whole bunch of session time.
- You have to work out how to transform the ensemble's acoustic performance into the envisioned control-room mix. As usual, the more you can match the reality to your vision before any microphones get involved, the fewer engineering difficulties you're likely to encounter, but whatever mismatch remains must then inform your session planning.

RECORDING DRUMS: DOMINANT ARRAY OR PEER ARRAY?

The most common situation where small-studio users will have to choose between dominant-array and peer-array approaches is when recording drum kits: Do you try to pick up the whole kit with a dominant array and then add definition with spot mics, or do you give all the different kit components their own peer arrays? Unfortunately, confusion over the basic method being used undermines many of the budget multitracks I hear. For example, I can't tell you how often I see people trying to record a full drum kit with only four mics (for budget reasons) by putting a mic in the kick drum, a super-close mic on the snare, and wide-spaced overheads a foot or so above the left and right groups of cymbals—a rig which doesn't make much sense however you view it:

- Looking at it as a dominant-array setup, where the overheads would normally serve as a stereo pair to capture the whole kit, the placement of those mics is going to give far too much cymbal level in the balance, dull and distant snare/toms, and a whacking great hole in the middle of the stereo image.
- Looking at it as a peer-array setup, all the elements of the kit should be the Focus of at least one array, so where are the hi-hat and tom mics? It's also important that the Focus instruments of individual peer-arrays sound pretty good, but here the kick and snare mics are both placed in very unnatural-sounding positions—you might get away with that for kick in some styles, but not for snare.

FIGURE 10.2
This small-studio drum setup can't decide whether it's a dominant-array or a peer-array approach: If you take it as a dominant array with a snare spot mic, then the overheads are probably too high to get a representative overall kit sound; whereas if you take it to be a peer-array rig, then where are the close mics on the toms?

In general, a dominant-array approach tends to work best in small studios, because it's easier to get a respectable kit sound that way. "When you move the mics in closer and separate the kit into nine tracks," says Phill Brown,[3] "that's usually when all the problems start. Then it doesn't sound anything like a real drum kit." Dominant-array recording doesn't typically require as many mics either, and only two of them have to be particularly hi-fi. To get best results out of a peer-array setup, on the other hand, you really need to have individual mics for each instrument (although miking groups of cymbals and toms can work), so you'll usually be forced to use considerably more gear. And if you're placing peer arrays close enough to allow serious independent tone/balance control, it begins to become advisable to multimike some of the drums as well (particularly the snare, given the snare wires underneath), in order to combat timbral spotlighting—which only expands the complexity of things further.

What you've read so far in this book will guide you toward getting a decent sound from each peer array in isolation, and Chapter 9 has already suggested which mic choices and miking techniques tend to work best within an ensemble-recording environment. So all that remains to explore from a sonics standpoint is how to position the peer arrays (in conjunction with their associated Focus instruments) and any acoustics measures to make the best of the spill situation in your recording space.

10.1.1 Acoustic Balance and Depth Perspective

The first thing to ask yourself is whether the ensemble's acoustic balance is going to be problematic. You see, although the peer-array system gives you the power to adjust level balances after recording, acoustic imbalances can also cause problems with the ensemble's apparent depth perspective. For example, if the electric guitar is too loud in the room, you might reduce the level of its peer array in the mix to compensate, but that won't affect the level of electric-guitar room reverb coming through all the other peer arrays (see Figure 10.3). The result is that the electric guitar will sound wetter, and therefore more distant, than it would have if you'd not had to rebalance it with your faders. Likewise, a quieter instrument that hardly features at all in the Backdrop signal of other peer arrays will come across more dry and upfront if faded up. "The hardest thing to manage in terms of bleed," affirms Steve Albini,[4] "is if you have really quiet instruments and really loud ones playing in the ensemble—like if you have a violinist playing with a rock band."

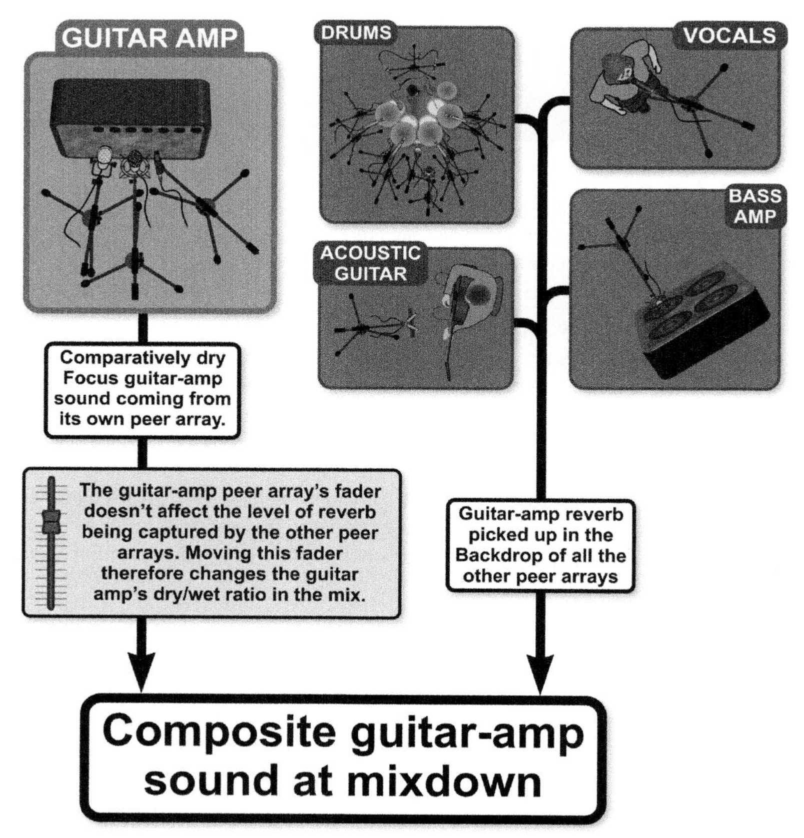

FIGURE 10.3
Although rebalancing your peer-array mic signals can give you significant control over the ensemble balance, such fader adjustments may also mess with your mix's depth perspective.

This effect can work in your favor sometimes, if you happen to want the quieter instruments at the front of your mix and the louder instruments in the background, but otherwise it's probably the biggest reason you'd want to try reducing spill between peer arrays. We talked about various ways of doing this using mic polar patterns and acoustic devices in Section 9.4.1, but I'd like to add that you should try to be as targeted as you can with absorption, rather than just baffling everything off. If the vocals and acoustic guitar are too quiet in our setup, then target your resources toward getting sufficient spill-level reduction on their peer arrays, rather than worrying too much about cordoning off the electric-guitar and bass cabs as well. Alternatively, if our five-piece band were playing entirely unplugged instead, the drums might be the only thing that was out of balance, in which case it'd probably be more efficient to cut down spill by putting gobos around the drums, rather than trying to shield the four other peer arrays individually.

Old-school engineers such as Bones Howe,[5] Wayne Moss,[6] Al Schmitt,[7] and Armin Steiner[8] offer very similar advice regarding spill: Be careful of combating it by increasing the distances between the ensemble instruments. "The closer [the musicians] are, the tighter they are, the better things sound, and the better they play," says Schmitt,[9] for instance. "[Some engineers] would try…moving them farther and farther away from each other. Well, the drummer would then play louder and louder! So I learned that moving guys closer together made them play at a more comfortable level. It worked better for me and for them."

"Be careful of combating spill by increasing the distances between the ensemble instruments. When instruments are set up close together, the Backdrop of all the peer arrays sounds drier, whereas when you place the instruments further apart the Backdrop signals begin to operate more like room mics, making the instruments washier."

When instruments are set up close together, the Backdrop of all the peer arrays sounds drier too, which means it fills out the timbre of each instrument without distancing it as much; whereas, when you place the instruments further apart, the peer-array Backdrop signals begin to operate more like room mics, making the instruments washier without enhancing the dry timbre as much. Alan Parsons[10] illustrates this rather counterintuitive concept in relation to the challenging task of capturing live drums and piano at once. "That's always a nightmare," he says, "but the way around it is to move the piano as close as possible to the drums, which is contrary to what you'd think.…It's the time delay that's the problem, not the actual separation."

SMALL ROOMS FOR BIG DRUMS

Nick Launay[11] has a tip for people wanting powerful drum sounds. "The misunderstanding about recording rooms is that to get a big sound you need a big room. I find it absolutely the opposite. Some of the biggest drum sounds on earth have been recorded in very, very small rooms." John Leckie[12] echoes this sentiment, adding that "it doesn't have to be cramped, just a normal-sized room with an 8- to 10-foot ceiling."

10.1.2 Anticipating Processing Requirements

Even where an ensemble is well balanced overall, spill can still cause you problems if you're planning to heavily process individual peer arrays for any reason. If we wanted to use high-frequency EQ boost to brighten the acoustic guitar in our five-piece band setup, for instance, that might overemphasize spill from the hi-hat or add an unpleasant edginess to the electric-guitar timbre. Or if we wanted to compress the lead vocal in that line-up, we might find that the compressor's automatic gain changes caused distracting variations in the balance and/or depth of other instruments by modulating the vocal mic's Backdrop level. And the same would apply if we wanted to do any drastic vocal fader riding at mixdown.

In either of these cases there might be a good argument for additional baffling of certain instruments, effectively sacrificing some of the timbral support and ensemble cohesion that spill can provide in favor of increased mixdown control. However, it's as well to mention that side-effects of heavily processing one peer array might also be compensated for by adjustments to the others—turning down the hi-hat spot mic in response to the aforementioned acoustic-guitar EQ, say, or equalizing the electric guitar's peer array. Unnatural compression "gain-pumping" of Backdrop signals is also par for the course in many commercial rock styles. This is why you need to have a clear idea of the kind of product you're hoping to end up with, because processing side-effects that are acceptable in one style won't necessarily be tolerated in another.

10.1.3 Stereo Considerations

Part of what makes good peer-array recordings seem so three-dimensional in stereo is that spill between the mics prevents any individual array ever appearing as a narrow point source at mixdown, even when it's recorded in mono. So, for example, if I panned all the electric-guitar multimics in our band-recording setup dead center, you'd still get some stereo width on account of electric-guitar spill in the drum kit's stereo overhead mics. Most of the time this is a benefit, because the listening public isn't that fond of super-narrow stereo images on the whole, but you do have to be a bit careful sometimes to avoid one peer array contradicting the natural picture presented by another peer array's stereo mic technique.

Let's say, for instance, that I positioned the acoustic guitar in our band setup to the left-hand side of the drummer, such that its mics picked up lots of snare and hi-hat spill (see Figure 10.4). If I then panned those guitar mics to the opposite side of the panorama, the spill might risk dragging the snare and hi-hat onto the wrong side of the stereo image. Alternatively, imagine that the acoustic guitarist were in front of the drum kit facing the drummer (see Figure 10.5). If I panned the acoustic guitar's crossed-pair peer array to create a stereo picture of that instrument, the stereo positioning of the drums in the array's Backdrop might then contradict the stereo picture presented by the drum kit's own peer array.

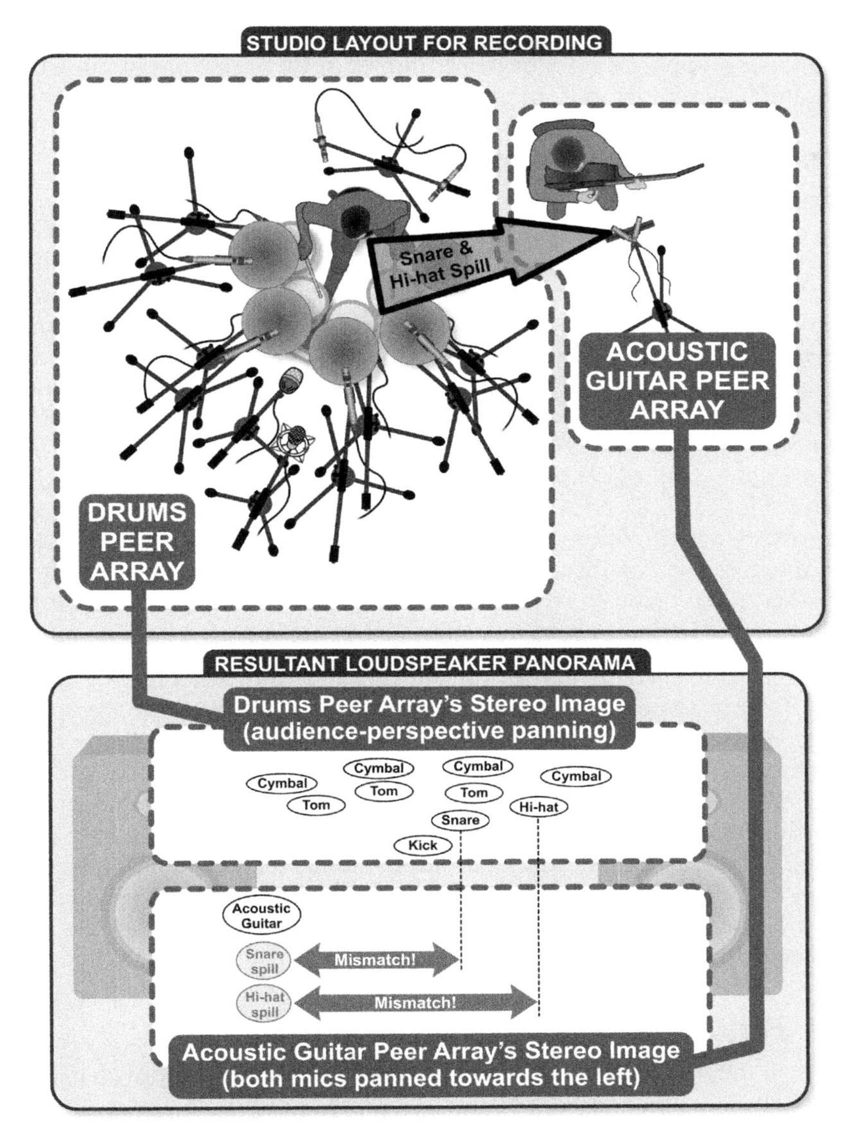

FIGURE 10.4
If your acoustic guitarist is playing alongside the drummer (as shown here), then spill from the hi-hat and snare may cause stereo contradictions in the mix if the acoustic guitar's peer-array mics are panned unwisely.

Although you can solve some stereo imaging problems by simply tweaking pan controls and swapping the left and right channels of stereo mic rigs, I prefer to try to deal with things like this at source if possible, planning an ensemble layout from the outset to roughly match the desired stereo panorama of the final mixdown. So if I wanted to pan the two guitars in our five-piece band left and right respectively, while keeping the bass and vocal in the center, I might place the guitars on opposite sides of the drummer in the recording room, with the singer and bass cabinet in front of the kit.

10.1.4 Session Layout for the Performers

Whatever your concerns about the effects of spill on the sonics, however, your needs as an engineer should (as always) remain secondary to those of the musicians. "Don't ever forget," reminds Mike Stavrou,[13] "[that the musician's] job is more difficult than ours. We just have to capture his performance. If our recording setup ruins his ability to perform, we're not helping." So whenever you take any steps to adjust spill, you always have to question them from a performance perspective. Yes, you might prefer to put gobos up all around a jazz quartet's upright bass to protect that peer array from drum spill, but if that means the player can't see/hear the other players well enough to stay in time, then you've clearly put the cart before the horse. In addition, I think it's also the engineer's responsibility to think creatively about whether deviating from an ensemble's instinctive rehearsal/concert layout might help the players see and hear each other more easily. For instance, although facing groups of musicians toward each other doesn't tend to work that well within a concert situation, it can dramatically improve visual communication within a group for recording purposes.

As in Section 9.6, I'd suggest using the simplest foldback system possible—and preferably none at all if you're laying down an ensemble performance from scratch without a guide/backing track. That said, headphone monitoring is more often required in peer-array setups, because putting acoustic baffles around some players to counteract spill is more common, and that can make it very difficult for those musicians to hear what's going on outside their little cocoon. If you do end up running multiple independent foldback mixes, however, my main advice would be to consider setting these up via a physical mixer or control surface, rather than with mouse-driven software controls, so that you can work more quickly.

10.2 BUILDING THE ENSEMBLE SOUND

Because all the microphones in a peer-array system interact, the only reliable way to assess the suitability of what you're recording is to mix the arrays on the fly while setting up for tracking. As you might imagine, handling all those interdependent Focus and Backdrop signals simultaneously can be a complete brain-fry, not least because it's so easy to damage one facet of the ensemble sound inadvertently while working to improve some other aspect. So let me share my own plan of attack for this combined setup/mixing process, which helps make the task a little less daunting and the decision-making more logical—as well as organizing the workflow to keep backtracking and second-guessing to a minimum.

You need to start off with all your monitor channels muted, the goal being to build up the full-ensemble mix one peer array at a time. There's a bit more to it than that, though, so let's walk through the whole scheme step by step.

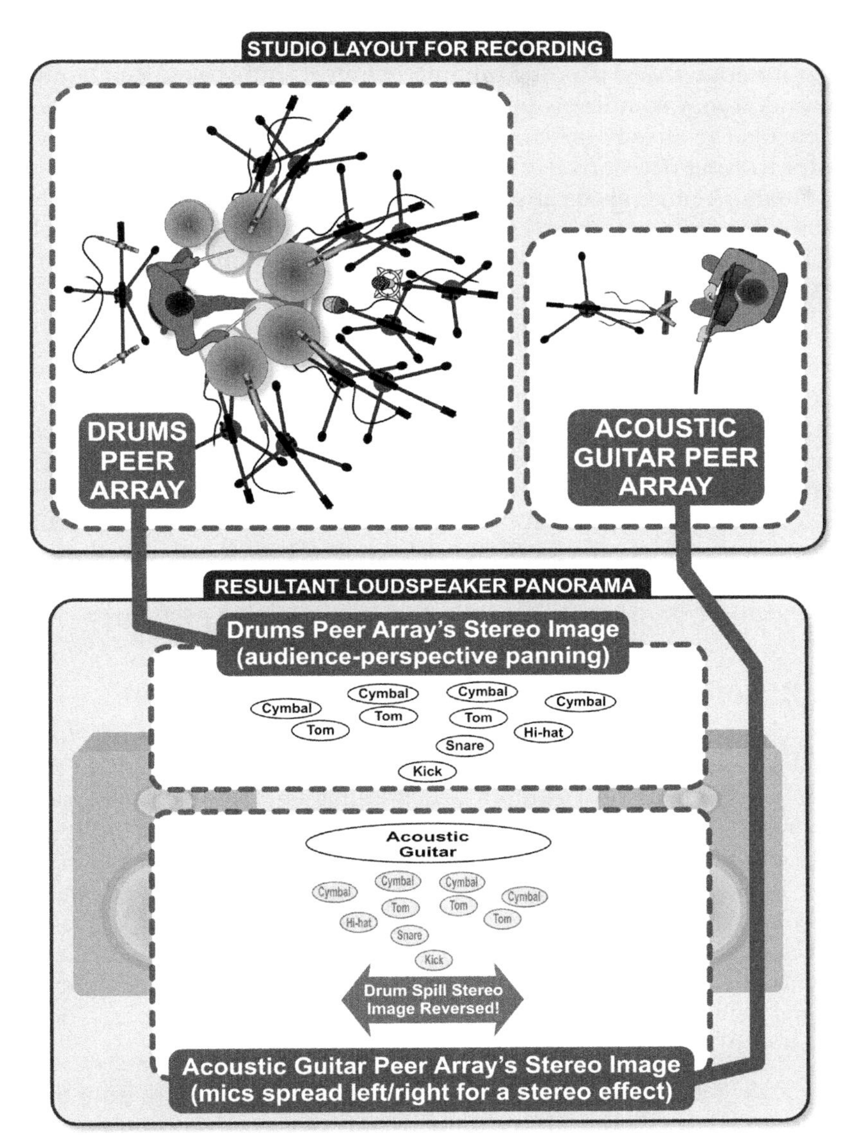

FIGURE 10.5
If your acoustic guitarist is playing facing the drummer (as shown here), then the stereo positioning of the drumkit Backdrop in the acoustic-guitar peer array may contradict the stereo positioning of the drumkit peer array's Focus.

10.2.1 Choose a Peer Array

Firstly, choose which peer array you're going to unmute first. Ideally, you should start work with those peer arrays which pick up the most spill, because your sonic decisions regarding the Focus instrument(s) on all subsequently added peer arrays will be better made within the context of that spill. Or, to look at it another way, if I only unmute my most spill-soaked peer array right at the end, it may completely upset the carefully wrought sonics of most of the instruments I've already mixed, forcing me to backtrack on a load of my prior setup/processing decisions. Fortunately, peer arrays with lots of spill are often well suited to being mixed early on, because they're usually the ones pointed at quieter Focus instruments, and the sound of quieter instruments won't change as much as the mix progresses because there'll be less spill from them on other peer arrays.

In some situations, however, the difficulty of moving a peer array or its Focus instrument(s) may cramp your style. This is most likely when you're working with pre-installed microphones or an immovable instrument such as a chapel organ, but session logistics may also militate against moving a grand piano or drum kit in many cases, and there may be musical factors to weigh up too—for example, the need to keep rhythm-section players in close contact with each other for the sake of a tight groove. Where layout restrictions reduce your ability to adjust the sound of a particular peer array, getting it into the mix early on makes sense, because it provides you the maximum opportunity to compensate for its sonic shortcomings by adjusting the remaining peer arrays.

10.2.2 Refine the Peer Array: Focus and Backdrop

Having chosen your first peer array, concentrate on getting its Focus to sound as good as you can. You should be on familiar territory here, whether the peer array is a single mic or a complicated dominant-array-plus-spots system. To recap briefly: Your job will be to decide on the Focus's overall dry/wet ratio and stereo width (remembering to check mono-compatibility) as well as the relative level, tone, depth, and imaging of each individual Focus instrument. At this point in the mixing process you've got a pretty free rein in terms of how you decide to sculpt the sound—reposition the instruments, adjust the acoustic environment, change mics and miking positions, tweak polarity/phase relationships between channels, add processing in the control room… So make the most of it!

Once you've got a good Focus sound, shift your attention away from that and concentrate on any Backdrop instruments you can hear, running through the same list of sonic decisions regarding those too (overall dry/wet balance and width; per-instrument level, tone, depth, and imaging), but within the context of their status as spill signals. So let's say the acoustic-guitar microphones in our hypothetical five-piece band had the most spill on them, so I'd started my rough mix with those. Having achieved a nice acoustic-guitar tone, I'd then listen to the guitar's XY peer array and consider how it was picking up the other instruments. Looking at the big picture first, overly wet spill may make the final mix too washy, whereas overly dry spill may not give much ensemble cohesion; overly wide spill may cause mono-compatibility problems, whereas overly narrow spill may cause the amount of room reverb to be rather uneven across the stereo image. On a per-instrument level, and taking the drum kit as an example:

- Is the level of drum-kit spill appropriate? Does the level balance between the different drums and cymbals support my vision for the final mix? Tackling these questions now makes it less likely that drum spill from the acoustic-guitar peer array will cause balance/depth difficulties when I'm trying to add the drum kit's own peer array to the mix later on.
- Is the tone of the drum-kit spill appropriate? If the snare drum's unpleasantly honky, the kick drum bloated, or the cymbals harsh, this'll make it harder to get great drum sounds overall.
- Is the imaging of the drum-kit spill appropriate? We considered some possible problems you might meet with this in Section 10.1.3.
- Is the depth of the drum-kit spill appropriate? For instance, if the spill sounds like it's coming down a tunnel from Colditz, you'll struggle to get an upfront sound for the drums—assuming that's what you happen to be aiming for.

Seeing as we're in a questioning mood, you'd also be perfectly justified in asking the following: How the hell am I supposed to know what "appropriate" means in each of these cases? Unfortunately, there is simply no way of knowing for sure about this until the whole mix is up and running, so you just have to make the best guesses you can and resign yourself to spending extra time revisiting Backdrop issues on some of the peer arrays later on if it turns out you've misjudged things. This is one area where experience counts for a lot, because the better developed your instinct for how each peer array's Backdrop should sound, the less you end up having to retrace your steps in your quest for a complete ensemble mix.

"Your options for altering the Backdrop will be constrained by the desire to maintain the sound of the array's Focus, so you should prioritize methods which don't disturb the relationship between the peer array's mics and their Focus instrument(s)."

It's also worth pointing out here that your options for altering the Backdrop will now be constrained by the desire to maintain the sound of the array's Focus, so you should prioritize methods which don't disturb the relationship between the peer array's mics and their Focus instrument(s). In other words, your first ports of call should be adjusting and/or moving the Backdrop instruments, or changing acoustic treatment. Here's Joe Chiccarelli,[14] for instance, talking about doing this on a jazz session where they couldn't prevent drum spill on the piano mics: "We moved the piano around the room and found a place where the drum leakage was mostly short, early reflections. This actually thickened up the drum sound a bit and was a pleasing effect that didn't compromise the piano track. On cuts where the pianist didn't play, I still left his mics open because it actually helped the drum sound." If you feel compelled to move your peer array on account of the Backdrop, then try to do this without changing the spatial relationships between that array's mics and its Focus instrument(s). If none of those options work, and you realize that you need to change your miking strategy for that peer array, then be sure to backtrack properly and check all aspects of the Focus with the new mic setup. It may be that you choose to trade off a little Focus quality because of Backdrop concerns, but bear in mind that it's the Focus signals that will often have greatest impact on the final mix, so I'd usually give those precedence if I were you.

Before moving on to the next step, mute the first peer array, unmute the next one in line (referring again to the recommendations in Section 10.2.1), and repeat the steps above to achieve a sensible Focus and Backdrop for that as well. However, take care that the techniques you use to modify the sound of the second peer array don't interfere with the good work you already did on the first. Again, if this proves impossible, you must reassess the first peer array's Focus/Backdrop thoroughly before continuing.

SPEAKER FOLDBACK SPILL

If you're using speakers for foldback purposes, don't forget to consider their impact on each peer array's Backdrop signal. In most cases speaker placement and/or baffling will combat foldback spill, but where it's unavoidable, consider equalizing the foldback speaker's feed signal to reduce any strong tonal colorations captured on the recording. If you're in really desperate straits and the foldback speaker is being driven from your mixer's monitor channels, you could try recording a cancellation pass, as discussed in Section 5.3.2—in other words, feed the foldback speaker with playback, while recording its spill through all the mics on a new set of tracks. Polarity inverting those tracks should reduce the spill.

10.2.3 Refine the Meld: Focus and Backdrop

Hopefully it shouldn't be too tricky to get reasonable Focus and Backdrop sonics for your first two peer arrays, at which point you'll be ready to try combining them. Although most people would instinctively call this process "mixing," I prefer to use the word "Melding" instead, because getting a good combined sound out of peer arrays often means making adjustments to the recording setup too. Your goal is to evaluate and refine the Meld (i.e., the combined setup and mix of both peer arrays) in exactly the same way you evaluated and refined each individual peer array. In other words, first concentrate on the Meld's collected Focus instruments, and then on its combined Backdrop signal, scrutinizing overall dry/wet balance, overall stereo width, and per-instrument level, tone, depth, and imaging.

If you've been conscientious in Section 10.2.2, the Meld will rarely hold any surprises in terms of overall wetness/width, and per-instrument balance will usually fall into place once you've decided on the overall level of each peer array in the mix. What can be really tricky, though, is trying to get a good tone for the entirety of the Meld, because of the way comb-filtering effects inevitably impact upon the sound when the different peer-array signals mix together. Some small-studio engineers fall into the trap of immediately EQ'ing lots of individual mics to fight the comb-filtering, but this is a recipe for frustration. For a start, the EQ won't work as effectively as you might hope, because it'll change the phase response (and hence the character of the comb-filtering) as well as the frequency response—in other words, you'll find yourself hunting a moving target. This typically encourages overprocessing, leading to level/tone side-effects which demand yet more corrective processing, whereupon you start chasing your tail. So don't take the bait! If you're happy about the sound of each peer array in isolation, but don't like the way they combine in the Meld, equalization really isn't the best answer.

The trick is to accept that comb-filtering between the peer arrays will be unavoidable, and then to direct your energies into grappling with that directly, manipulating its tonal flavor until it suits your needs. As you'll know from previous assignments, this is not a very predictable process, so the key to success is lots of experimentation. Given the time pressure on most ensemble sessions, try to get the best out of quick control-room fixes (such as polarity-inverting or phase-adjusting one entire peer array relative to the other) in the first instance, reserving recording-room phase-adjustments (such as adjusting the relative positions of the two peer arrays) for situations where the simpler tactics don't bear sufficient fruit.

USING DELAYS FOR PHASE-MANIPULATION OF PEER ARRAYS

I explored the subject of using delays for phase-aligning multimic rigs back in Sections 7.2.3 and 9.4.4, but this concept also has a bearing on peer-array recording setups. As far as your individual peer arrays are concerned, I don't think there's any new information to add to what you already know—in other words, feel free to phase-align any of the multimics within a given peer array if that gets you the Focus sound you're looking for, but be wary of phase-related damage to the Backdrop. However, I don't think there's very much to be gained by trying to time-align whole peer arrays against each other.

To understand why this is, imagine the following very simple two-peer-array scenario: a trumpeter and a saxophonist sitting next to each other, each with a single mic pointed at them. In this situation, the trumpet mic will pick up delayed saxophone spill, and the saxophone mic will pick up delayed trumpet spill. If you electronically delay the saxophone mic to time-align its Focus with the trumpet mic's Backdrop, the saxophone mic's Backdrop will slip further out of time-alignment with the trumpet mic's Focus. In other words, phase-aligning the spill of both instruments is impossible.

Leaving the idea of phase-alignment to one side, though, what about using delays simply as a phase-manipulation tool to massage the tonal quality of the comb-filtering between peer arrays? In principle I have nothing against this, as long as you keep the delays below about 5 ms—longer delay times may start to have unwelcome knock-on effects on the ensemble's depth perspective, as mentioned in Section 9.4.4.

While comb-filtering will probably be the main tonal issue to address at the Meld, the tonal aspects of your Backdrop guesswork will also be put to the test. The main tip-off of a problem in this department is if some tonal problems persist despite your efforts with polarity/phase work. Returning to our imaginary five-piece band setup, let's say that I'm trying to Meld the vocal mic with the acoustic-guitar mics, but find that the guitar tone is plagued by a persistent boxiness which isn't apparent on its own peer array. This is the moment to quickly Solo the vocal mic again and check whether that boxiness might actually be coming from its Backdrop. If it is, then I'd be inclined to finesse the sound of the guitar spill on that mic in some way, before recommencing the Meld. Some kind of acoustic treatment would probably make most sense, since that would disturb the phase-relationships in the Meld the least, but it's vital to realize that pretty much anything I do to change the sound of an individual peer array has the potential to affect every aspect of the Meld. In this case, for example, it would be very easy to miss some unwanted tonal side-effect of my acoustics fix on the vocal sound while I'm concentrating my mind on the guitar. The only fail-safe attitude is steely self-discipline: Whenever you change anything about a peer array, don't skimp on fully rechecking the Meld!

If you've properly considered the stereo imaging of each of your peer arrays in Solo mode, there should be little else to do at the Meld than just toggling the second peer array's Mute button(s) to check for image shifts. Problems with the depth perspective, on the other hand, can take more work to sort out. Again, this is where your Backdrop guesses for each individual peer array will be put on trial: Do all your Focus instruments seem to be in the correct front–back positions in the mix? If they aren't, then the best tools at your disposal will be those which have the least effect on other aspects of the Meld, for example changing the acoustic balance between the instruments (if possible), getting busy with any acoustics devices you have available, or adding artificial reverb. As before, if you have to reposition the instruments or change anything about the peer arrays themselves, don't overlook the possibility of unforeseen consequences in the Meld.

You may already have had difficulties getting the right Backdrop sound for the two peer arrays individually, because of overriding concerns about the Focus instruments in each case, but it gets even more challenging when you start Melding peer arrays together. So although you should certainly do the best you can with it, don't beat yourself up too much if you have to make some Backdrop compromises, because it's pretty much an inevitability given the complexity of the endeavor. The main thing is to try to minimize those compromises and restrict them to less important aspects of the production—which for many engineers means favoring issues of overall dry/wet ratio and per-instrument level/tone over questions of per-instrument imaging/depth.

RECORDING SAFETY SAMPLES

Whenever you record an ensemble containing percussion instruments, it's wise to spend a few minutes at the end of the session recording a few good clean hits of each one, through all the mics in the ensemble rig. This effectively gives you a kind of "get out of jail free" card if you discover at mixdown that the percussion mics have caught undesirable spill. Spill-free cymbal samples can be a godsend for uptempo pop/rock productions, for example, if headphone click-track spill becomes embarrassingly audible through the drum overheads as a song's final note decays. (Er, or so I've been told...) There are now plenty of ways to repeatedly retrigger a sample in sync with a live-played track (see this chapter's web resources for some suggestions), and this is another thing that can come in very handy—for example, if your snare close-mic signal has picked up too much hi-hat spill, replacing the signal with a retriggered close-mic sample can bail you out big-time.

10.2.4 The Ghost of the Next Array

When you're convinced that the Meld has reached the point of diminishing returns, you can choose a third peer array to work on. Don't unmute it just yet, though! Instead, just concentrate on the combined spill from its Focus instrument/section across all the mics in the Meld so far—what I call the "Ghost" of the third peer array. If you don't like what you hear, and weren't able to remedy this during the Meld without unacceptable damage to the first two peer arrays' Focus instruments, then first try adjusting/moving the new Focus instrument(s) and/or altering the acoustic environment (as discussed in Section 10.2.2), as this is least likely to undo the work you've already done on the previous peer arrays. If this doesn't reap the rewards you're hoping for, however, then you have to think about potentially compensating for the Ghost's remaining deficiencies by adapting the third peer array's Focus sonics.

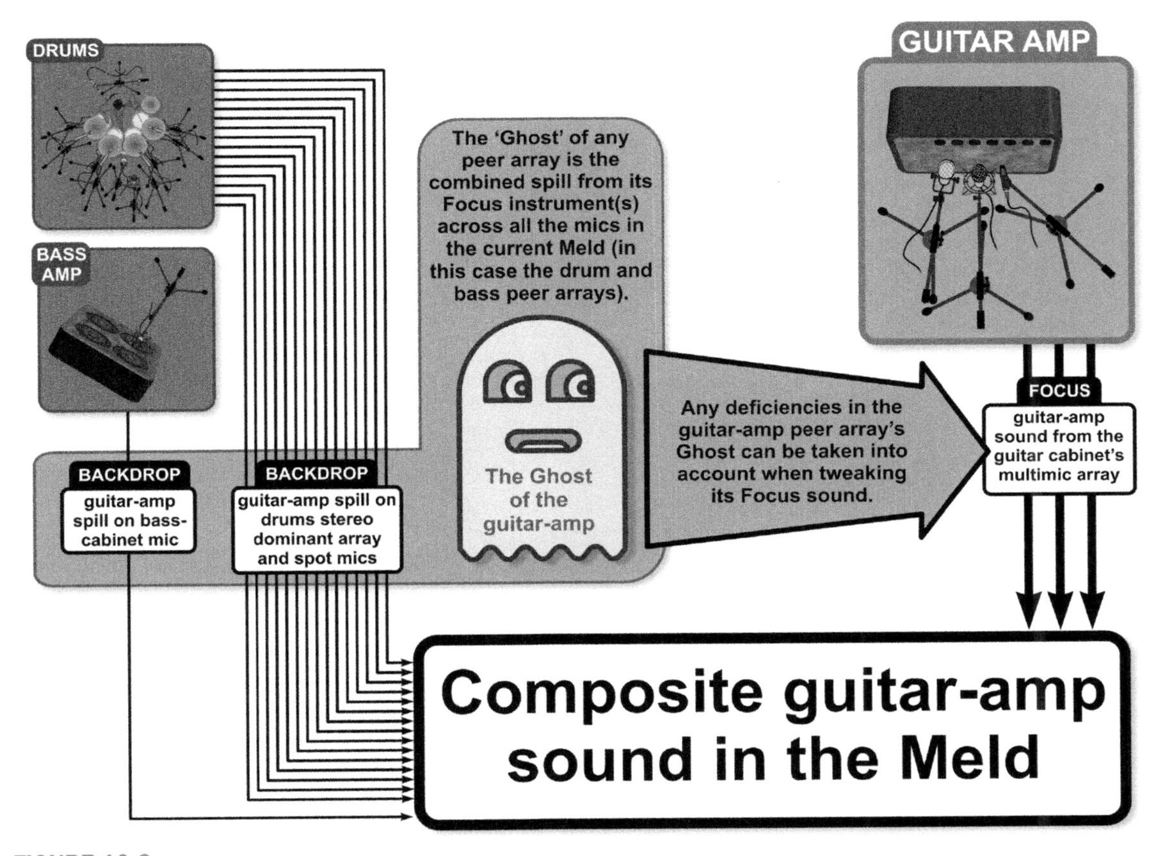

FIGURE 10.6
An example illustrating the concept of the Ghost. Here I've already melded two peer arrays (drumkit and bass amp) and have chosen a third (guitar amp) to work on. First I'd address the sound of that peer array's Ghost (i.e., the guitar amp's combined spill on all the drumkit and bass-amp mics), and then take any unavoidable deficiencies into account when tweaking its own Focus sound.

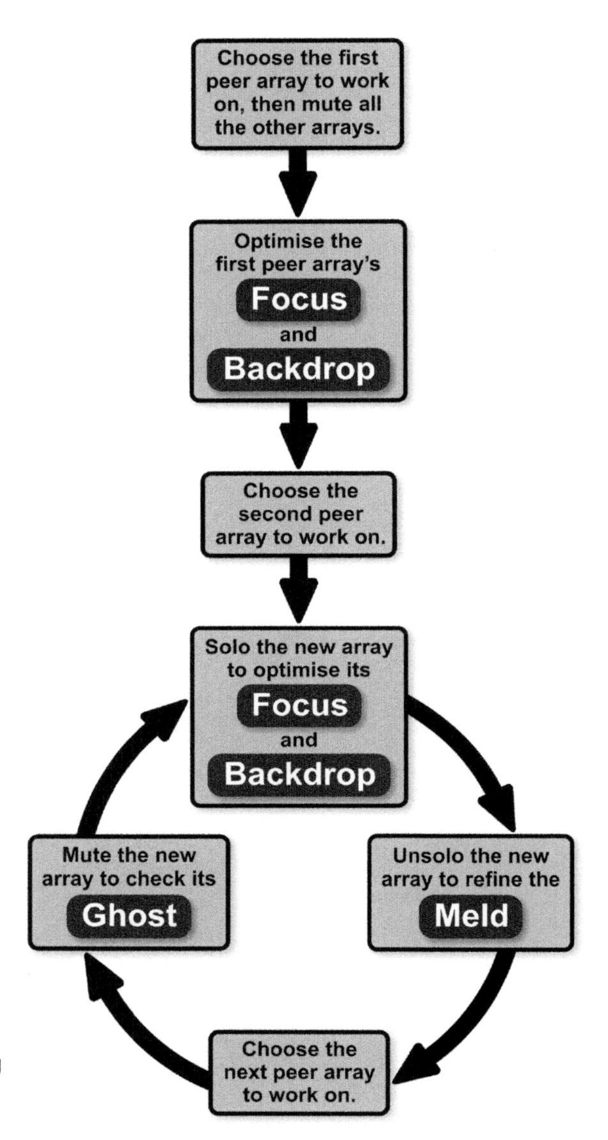

FIGURE 10.7
This block diagram shows the basic steps involved in constructing a full ensemble recording from multiple peer arrays.

So, for example, if you can't avoid some rather hissy cymbal spill when Melding the vocal and acoustic-guitar microphones in our five-piece band setup, you might subsequently consider cutting some of those same cymbal frequencies from the drum-kit peer array, perhaps by selecting different cymbals, altering your mic choice/technique, or dialing in some EQ on the overhead-mic mixer channels. Or let's say you wanted to bring in the electric guitar as the third array instead, but there was more electric-guitar spill in the Meld than you'd have liked. You might respond to that by moving the electric guitar's peer array closer to its amplifier cabinet, giving it a drier sound. Doing this kind of thing helps reduce the likelihood of big surprises at the *next* Melding stage, and therefore speeds up your workflow. The main drawback with such compensatory maneuvers, however, is that if you change the ensemble balance dynamically at mixdown, you get greater tone/depth side-effects, so I'd recommend using this dodge as sparingly as you can get away with.

10.2.5 Turn the Handle...

You're now ready to work on the third peer array's Focus and Backdrop (as in Section 10.2.2), and then Meld it with the previous two peer arrays (as in Section 10.2.3). Subsequent peer arrays can be dealt with in a similar manner, as summarized below and illustrated in Figure 10.7:

- Choose the peer array, but leave it muted to assess its Ghost, making a mental note of any shortcomings.
- Solo the new peer array to refine its sound, checking both Focus and Backdrop instruments.
- Meld the new peer array with all previously Melded peer arrays, checking both Focus and Backdrop instruments.

The more peer arrays you use, the more complicated the Melding process becomes, so it pays to be as methodical as you can in terms of giving each and every Focus instrument enough attention. As suggested in Section 9.1.7, keeping a checklist of all the ensemble instruments next to the mixer makes it less likely you'll forget to run your ears over one.

10.3 SOME MORE CASE STUDIES

As in Chapter 9, I'd like to use a couple of thought-experiment case studies to help illustrate the nature of the workflow when using a peer-array system. As before, I'll restrict myself to the equipment listed in Section 9.5, and I'll choose a fairly simple situation to start with: A student singer-songwriter has asked me to record an intimate ballad she's written, and which she'd like to sing while accompanying herself on the piano.

10.3.1 A Singing Pianist

The most suitable instrument available turns out to be a mid-size grand lurking in her college's performing arts department, so we both toddle over to have a pre-session gander. The room it's in is a multi-use rehearsal space covering an area of roughly 20 feet × 30 feet (6 m × 9 m), but with a typical domestic ceiling height of around 8 feet (2.4 m)—in other words, large enough to escape the worst room-mode colorations, but not high enough to avoid a fairly constricted reverb sound. The room's pretty bare, with slatted blinds hanging over the windows along one side, so options for adjusting the acoustics are pretty thin on the ground, and there's also no separate space to use as a control room—looks like I'll be working on headphones, then! The piano is in fairly good nick despite heavy use, and reasonably well in tune, but doesn't have casters, which inclines me against trying to move it around. At least it's not right up against a wall. It's rather bright sounding, such that it overwhelms the singer's breathy delivery when I ask her to do a quick run-through of the song for my benefit. I also notice that she's using the piano's music stand, which means I'll need to mic her slightly from one side to keep her sight-line clear. In light of that, I make a mental note of the fact that she naturally appears to sing a little to the left.

Anticipating that I'll want to use dynamics processing and/or level automation on the vocal line, I decide to mic the singer fairly close, with one of my quilts intercepting the room sound from the sides and rear of her sitting position. I figure that my large-diaphragm condensers will probably be best suited to the breathy vocal delivery, and decide to do a little shootout between the AKG C414B XLS and the other bargain-basement mic at the start of the session. For the piano sound, I'm in something of a quandary. On the one hand, I'd like a stereo piano sound with some time-difference spaciousness, but on the other hand the only matched pair of mics I have are quite hard sounding, and I'm concerned that they'll exaggerate the instrument's brightness, especially at the moderately close miking distances I'd like to use to keep the unflattering room reverb manageably low. My tactic will therefore be to set up the SE1As as a wide AB pair just outside the piano casing (rather than right up by the hammers), and then to use the Cascade Victor ribbon mic perhaps a little closer up to provide some tonal warmth and avoid too much of a hole in the middle of the stereo picture. (The second quilt I'll arrange around the backs of those mics to give me a bit more reverb reduction.) I briefly entertain the idea of closing the piano lid to round off the high-end timbre, but quickly abandon that train of thought, anticipating the difficulty in getting a natural-sounding piano tone that way.

USING ROOM MICS IN A PEER-ARRAY SYSTEM

In Chapters 7 and 9, we investigated using room mics to control a recording's overall dry/wet ratio from the faders. It can also be useful in peer-array ensemble recording, if you deliberately capture a slightly under-reverberant sound from the peer arrays and then supplement that with some kind of room-mic array which delivers a fairly balanced ensemble reverb. However, if the ensemble you're recording doesn't balance well acoustically, room mics will only tend to exaggerate the kinds of depth-presentation problems I described in Section 10.1.1—in other words, the depth disparity between quiet and loud instruments will just increase, because the quiet instruments create less reverb. Artificial reverb is therefore usually the best solution under these circumstances, because you can apply it more liberally to the instruments that sound driest. (One alternative, though, which is quite popular when recording rock drum kits, is to set up a PA system in the room to supplement the room sound: For example, Bill Price[15] used this technique when recording The Rolling Stones's *Tattoo You*, Steve Churchyard[16] used it for The Pretenders' *Learning To Crawl*, Dave Jerden[17] used it for Alice In Chains' *Dirt*, and Dave Bottrill[18] used it for Tool's *AEnima*.)

I wouldn't really advise using room mics as part of individual peer arrays either, unless you're willing to make a fairly firm commitment to the reverb level the moment you Meld that specific array—the room mics will pick up so much spill from the whole ensemble that you may seriously disrupt the ensemble sound by trying to change their level later on.

On the day of the session, I'm able to get into the room an hour or so early to set up and test the gear. By fixing both SE1As to a single stand using screw-on mounts, I only need four stands in total for the mics, so the other two I use to rig one of the quilts into a V-shaped booth behind the piano stool. The other quilt I manage to suspend from the ceiling-tile lattice with bungee cords. With 20 minutes still in hand, I decide to make use of the time and do a few test recordings (showcasing my spectacularly meager keyboard skills) to see what the guesswork-based piano peer array sounds like in practice. It's a good job I do, as well, because the ribbon mic doesn't deliver a particularly thrilling sound, even after three or four position changes. Deciding to rely mostly on the small-diaphragm condensers instead, I pull them closer together into a more traditional AB spacing and then experiment with their height, angle, and spacing in search of a relatively smooth sound with reasonable mono-compatibility. While I'm in the middle of doing that, the singer shows up, so I immediately change tack and suggest she warms up a little so we can check out how she sounds on the two vocal mics.

I've set up both mics a little to the singer's left side and slightly below her nose level, my reasoning being: that I'll probably get a brighter vocal tone there; that I shouldn't get comb-filtering from reflections off the music stand; and that she's less likely to move off the mic if she needs to look where her hands are going. A pop shield is fixed to one of the stands around 8 inches from the mics, and I talk to the performer about trying to keep close to the pop shield, but without singing downwards into the mics. While she gets used to this, I quickly check the vocal-mic signals. Both of them give slightly overblown proximity-effect bass boost, but are already catching enough spill that I decide against moving them any further away. Reaching for my high-pass filters instead, I roughly match the low end on both mic signals by ear and then hit Record so that we can both have a listen. Another surprise awaits: The cheapy mic wins hands-down! There's just something about its high-frequency spectral contour that seems to bring out the airiness and clarity in the singer's voice, whereas the C414B leaves her sounding rather spongy. I suggest giving the C414B's hypercardioid pattern a go too, but that only makes her too sibilant into the bargain.

What concerns me about the cheapy mic, though, is that its Backdrop piano sound is pretty nasty—gutless in the midrange, and with a harsh upper-midrange edge that clearly won't do the production sonics any favors. I try dealing with the upper-spectrum harshness first by gaffering a piece of acoustic foam to the C414B's stand and positioning it behind the vocal mic, which does seem to help somewhat. Then I try rotating the microphone a little further to the side of the singer, and after a couple of repositionings manage to tame the Backdrop tone somewhat without unduly affecting the vocal sound or making the singer's performance position too awkward. I end up with a little more piano spill overall on the vocal mic, but its timbre feels a whole lot less objectionable.

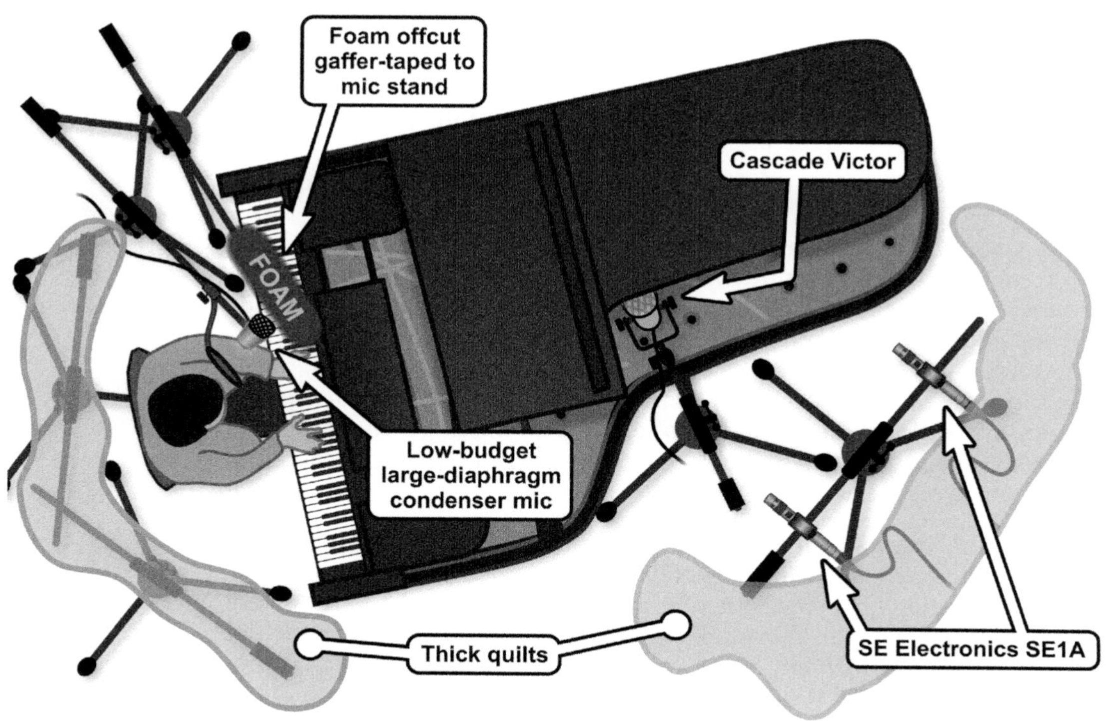

FIGURE 10.8
A diagram illustrating this chapter's first case-study setup, as described in Section 10.3.1.

Now I explain to the artist that I'd like to take ten minutes or so to get the piano sonics together, and ask her just to play the piano part while I do this, so we don't tire her voice unnecessarily. Although I'm working on headphones in the same room, my recording system is set up a fair distance away, so I'm able to make some judgments about the sound while she's playing. Returning to the task of refining the SE1A positions, I end up moving them further away from the instrument than I'd originally anticipated in order to minimize tonal edginess. This means they start picking up a touch more room reverb than I'd like, so I rehang the quilt closer around the mics to make them a bit drier again. In addition, the mics seem rather light on low end at this distance, so I try adding in a little of the closer ribbon mic (checking for the warmest-sounding polarity setting). This is of some use, but I can't fade too much of it in without clouding the midrange tonality, so I end up routing the monitor channels of all three mics to a group channel in my recording system and then boosting that by 3 dB at around 100 Hz using a low shelf EQ, as well as further mellowing the brightness above 3 kHz with a couple of decibels of high-frequency shelving cut.

Now I ask the performer to sing and play at the same time, so I can listen to the vocal Backdrop on my piano peer array. As suspected, the voice doesn't feature particularly strongly in the Backdrop signal, but what spill there is appears to be fairly neutral sounding, albeit as muted in tone as you generally expect with off-axis pickup. The Ghost's image is a little off-center in the stereo panorama, but I don't really want to jeopardize the piano tone by moving its mics around any more, so resolve to pan the vocal mic a touch if any compensation is required later on.

Combining the two peer arrays for the first time is a disaster, but flipping the polarity of the vocal mic immediately restores a more sensible tone, so I record a short section and suggest the singer take a short break while I listen critically to the Meld. A touch of muddiness has crept into the piano sound, and that's only marginally improved by reducing the ribbon mic's level, so instead I insert a variable phase-adjuster into the vocal-mic channel, which makes a bigger difference—and which affects the vocal tone very little, given the lack of vocal spill on the piano peer array. The stereo offset of the vocal Backdrop on the piano peer array doesn't turn out to be particularly problematic in practice, so I keep the vocal mic itself panned centrally. Moving the quilt closer in behind the piano mics appears to have avoided distancing the piano too far, but I do blend the vocal a little better with it by adding a dab of short artificial reverb to that mic signal in my monitor mix. At this point I take the opportunity to quickly reference a couple of commercial voice-and-piano recordings I've already imported into my recording system (checking with a vectorscope to confirm that the stereo width's appropriate), and realize that I can ease up my initial SE1A high-frequency shelving cut a fraction and bring the ribbon mic even further down in the mix balance.

At this point I figure it's time to start doing takes. The sound's not perfect, but good enough now that I'm pretty sure we won't get substantial improvements even if we spend the rest of the session faffing around with the setup—in other words, we've reached the point of diminishing returns. Knowing the spill situation on the vocal mic, though, I do my best to alert the singer of any swallowed syllables in her vocal performance, because I know that my ability to fade those up without affecting the piano sound may be a little restricted.

TALKBACK SPEAKERS AND SLATING

Even if you're recording an ensemble without recourse to a foldback mix, it can still be useful to have little loudspeakers out in the recording space for talkback purposes. These don't need to be anything fancy—in fact, little multimedia speakers have a number of advantages in this role: They're cheap; they're usually designed to be plugged directly into a headphone socket, so you can use your existing headphone foldback hardware to drive them; and they're usually designed to require no earth connection, which prevents earth-loop problems. Do be careful, though, that such speakers don't add appreciable background noise in the room when they're not being used for talkback.

An ancillary advantage of having talkback speakers in the recording room is that you can use them to start each take with an announcement about what's about to be recorded—a technique known as "slating." Although in theory your note-taking during a session should make slating redundant, I think it's a very good idea to do it anyway. Written notes can easily be lost or seem confusing in retrospect, whereupon slated take numbers are invaluable. I also sometimes like to slate all the recorder tracks prior to the session, by recording myself speaking into each mic, identifying what it is and where it's supposedly pointed. Not only does this help out if audio files get mislabeled later on in the production process, but the process of checking over each mic also frequently exposes ridiculous setup gaffes you'd rather weren't witnessed by a room full of musicians!

10.3.2 A Jazz Trio

Now let's say the same singer comes back to me and wants to record some jazz standards, adding a drummer (playing with brushes) and a bass guitarist to the line-up. How would I rejig the same venue/gear situation to deal with this? Well, first of all, I'd try to take advantage of the work I'd already done on the previous session, and set up the vocal mic and the piano's small-diaphragm condensers roughly how they were before—slinging the vocal "booth" from the ceiling lattice as well, though, to free up those two mic stands. Given that the piano mics are a stereo rig, I'll place the bass and drums in the room so that they appear roughly central within the Backdrop panorama, and I'll face the drum kit toward the piano to maintain good eye contact between the singer and drummer. The big question is where to place the bass cab, and the following thoughts go through my mind:

- The bass player is likely to perform best when close to the drummer, but he needs to hear himself too, so I'll avoid cab positions behind the piano which would make that difficult.
- The singer may need to be able to hear the bass part for her pitching, so I don't want to put the cab on the other side of the drum kit.
- If I put the cab underneath the piano, the casing of the piano will help reduce bass spill on the piano's peer array.

As a result, I decide to put the bass player a little to the left of the drummer (and to the right of the singer), with his cabinet under the edge of the piano in front of the kick drum, pointed toward the bassist.

Now to decide on mics. Given the less-than-stellar room sound and limited acoustic absorption available, I'm pretty sure I'll want to bring the mics in fairly close and stick with directional models for the most part, but the main question is whether or not I should recycle the idea of a ribbon+C414B stereo MS pair I discussed in Section 9.5.2. The problem I have with this approach, though, is that it doesn't leave me with any very suitable mic for the bass guitar—the D112's highly contoured LF response is unlikely to work well with more wide-ranging jazz lines, while the SM57 and SM58 are both likely to be light on low end. So I abandon the idea of miking the drums in stereo, and instead rely on the drum spill in the piano mics to give the drum kit some stereo width. I guess that some high-frequency definition on the drummer's brush work wouldn't go amiss, so I figure I'll allocate the C414B XLS as a single overhead mic, and then use the ribbon down in front of the bass cab with its null plane rejecting the piano and kick drum—taking care to protect the microphone from potential air blasts from the kick drum and bass cabinet. I imagine that a kick-drum spot mic will probably be required to fill out the overhead sound, so decide to use the D112 there. The kick-drum has no hole in it, and I feel that removing the front head would kill the more natural and resonant tone required for this style, so I resolve to mike the instrument from the outside. With the piano and bass cabinet in such close proximity, I'm a bit cagey about using this mic around the front of the drum, so decide to go for a position around the beater side instead (where those other instruments will be shadowed more), and then angle the mic to regulate the degree of beater noise. The final mic stand, the Shure dynamic mics, and the second quilt I'll keep on standby just in case I need them…

SOME BAND REHEARSAL TACTICS

Rehearsing and working out arrangements with bands is an art in itself, but there are some time-honoured preproduction tricks. Paul Worley,[19] for instance, likes to work out his arrangements in a rehearsal room, so that all the musicians know exactly what they're doing before they get into the studio. "It means that you don't have to waste your time in the studio having discussions about structure, what instruments to play, what amp to plug in, because you've already figured all that out…Each musician gets a chance to figure out what *not* to play as well as what to play, and to work out when it's their turn to step up."

Gareth Jones[20] values this initial preproduction stage for other reasons: "We all get to know each other in a relatively low-stress environment, so when we get into the recording studio we all know where our strengths and weaknesses lie and there's already a level of trust that's been developed with the band. In the studio, that band is more willing to take what I'm saying seriously if we've done a lot of preproduction, because they've worked out that some of my ideas are worthwhile—and that's really important."

Another great little trick comes courtesy of Jack Douglas:[21] "I always tell them they're holding back in rehearsal, even if they aren't. Even if they're at rehearsal playing their absolutely best, I will say, 'Well, when we get into the studio, you can really let go.'" And if you're actually rehearsing in the studio, Tony Brown[22] recalls a clever approach used by Don Was: He'd only rehearse the band section by section, concentrating particularly on any tricky transitions. Once all these smaller sections had been rehearsed, the band would then invariably nail their recorded performance within the first couple of takes. Producers such as Neal Avron[23] and Gareth Jones[24] regularly make little demo recordings during the preproduction to help focus arrangement discussions, and this also serves to remind everyone of decisions they've agreed on once proper recording commences. James Ford[25] cautions against making such recordings too elaborate, though: "It's that demo-itis thing. I [don't] want the rehearsal-room recordings to be too good, in a way, because you can kind of get attached to them."

I arrange to get into the venue a couple of hours before the musicians do, and ask the singer/pianist to turn up twenty minutes before the rest of the band. I also warn everyone that we're going to need at least an hour's setup time once they've all arrived, in order to get a solid recorded sound. When the singer arrives, we set up the Focus, Backdrop, and Meld for the vocal and piano peer arrays as well as we can, within the constraints of a slightly closer piano-miking position—I don't want the quilt in front of the piano this time, because I'm relying on the piano mics to fill out the drum kit's stereo picture. While the other players are warming up and tuning, I have a listen to what each sounds like as a Ghost in the existing Meld. The bass amp isn't that powerful, so its Ghost is fairly understated, but the drums are (as expected) producing a fair bit of spill. Although the kick and snare don't fare too badly in terms of sonics, the piano mics appear to be making the cymbals rather harsh and unnatural sounding. Moving the kit a little off-center seems to help, as do some small adjustments to the angling and spacing of the mics, but there's only a certain amount that can be done without undermining the Focus sound of the piano's peer array.

Once I've done as best I can with that, I recheck the piano array's Meld with the vocal mic, and make a mental note that the drums Ghost is still sounding a touch harsh. In response I try positioning the C414B for a smoother sound, but its high-frequency presence makes this a little tricky, and drum spill picked up by the D112 spot mic also makes it hard work to find a clear tone for the whole kit—unless I put the spot mic so close that it picks up a very strange sound from the kick itself. Trying to Meld this drums peer array with the vocal and piano mics doesn't go particularly well either, with cymbal harshness remaining a problem, and colored spill in the drum-array Backdrop clouding the piano sound. I ask for two minutes' hiatus to have a listen to a test recording, but become increasingly convinced that we're on the wrong track using the C414B as the drums overhead. Replacing it with the ribbon mic gives a much less abrasive cymbal picture, and I get a lower level of piano spill (which is also better sounding) into the bargain. The new overhead seems to blend a little better with the kick-drum spot mic's tone too, although I have to revisit their polarity/phase relationship to get the best out of it.

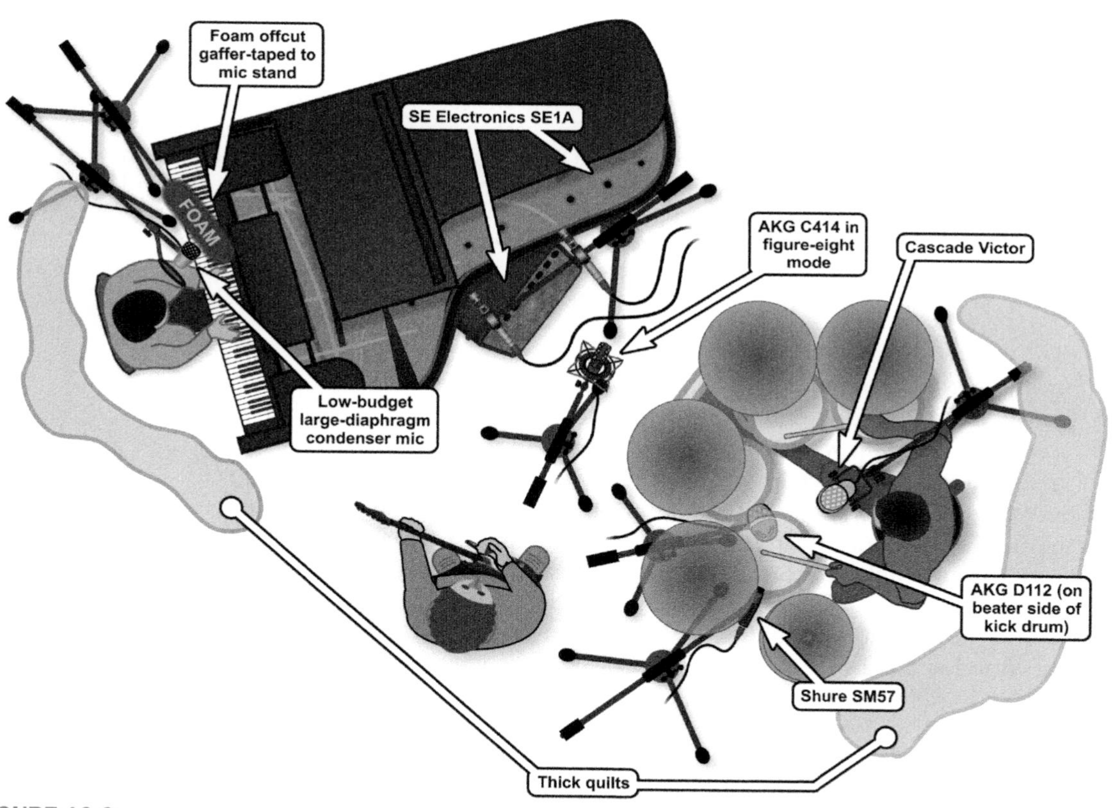

FIGURE 10.9
A diagram illustrating this chapter's second case-study setup, as described in Section 10.3.2.

Now the Meld proceeds more promisingly, with the high frequencies in the piano peer array's backdrop complementing the duller pickup from the drum peer array's ribbon mic, although I still feel the need to add a little high-end shelving boost to the drum-mic submix to get enough brightness overall. Some muddiness has crept into the drum sound because of its spill onto the piano mics too, and although some adjustments to the phase relationship between the piano and drum peer arrays improves this a bit, I shelve a couple of decibels off the ribbon mic below around 250 Hz as well. With the C414B switched to its figure-eight pickup mode, I'm able to get a great deal of isolation on the bass mic even at a distance of 9 inches or so from the cab, so it's not too difficult to pull off a reasonable bass sound. It also Melds fairly easily with the other instruments because it's not picking up too much spill and isn't too loud in the recording room.

Once the full mix is in front of me I zap a quick test recording and then call a break so that I can have a closer listen with the singer. Although the sound seems to be fairly well received, both of us feel that the kick-drum is sounding rather too much like a rock kick drum on account of the close D112 spot mic, and doesn't blend that well with the rest of the ensemble because it's not creating very much spill on other mics. I'm loath to move the D112 significantly further away, however, because I'm concerned that the unusual tone of its spill may damage the rest of the drum sound, so instead I mute the D112 temporarily and try picking up a slightly more holistic view of the kick drum using the rear sensitivity lobe of the bass mic's figure-eight polar pattern. This involves sliding the bass cabinet (with its mic) to angle it slightly toward the drums—the bass-versus-kick balance subsequently being adjusted by the C414's angle and by the distance between the kick drum and the bass mic. This immediately helps give the kick (and indeed the whole kit) a more well-balanced sound,

COULD YOU GO ROUND THOSE TOMS ONCE MORE FOR ME...?

The combined mic-setup/mixing scheme I outline in this chapter shouldn't require anyone in the ensemble to play on their own just for your benefit, so if the ensemble are happy just jamming or rehearsing for a while, it's perfectly possible to set up and mix the entire rig very inconspicuously, without bothering the performers very much at all. In fact, scheduling a rehearsal in the recording venue prior to the session proper is often a great way to get the performers comfortable with the recording layout, while giving you valuable extra setup time.

Nevertheless, there are occasions where you may want to ask the musicians to play certain things specifically for you, so that you can deal with technical matters more efficiently. If you sense that physical stamina may be a limiting factor when it comes to capturing enough takes (particularly when working with singers, percussionists, and wind/brass players), try not to have them performing for any longer than absolutely necessary—you don't want to cross the line between "warmed up" and "burnt out" before something's actually been recorded! Also, where certain instruments are played only sporadically (e.g., the toms in a drum kit), you'll generally conserve everyone's time and energy if you ask players to loop around a certain musical section where those instruments are featured.

and allows me to use the D112 at a much lower level in the mix just for a sprinkling of extra attack definition, although its phase relationship does naturally require careful rechecking at the Meld.

An additional worry of mine is that I might want to rebalance the snare brushwork a little at the mix, so I put up the SM57 as a spot mic to catch this—not ideal for high frequencies, but a dynamic's all I've got left! Finally, seeing as we didn't need to use the second quilt elsewhere, I sling it up above and behind the drummer to kill a bit more of the room reverb coming into his overhead mic. One quick burger pitstop later, and we're ready for the proper takes!

10.3.3 An Exploratory Campaign

As in Section 9.5, these two case studies aren't meant to provide you with some kind of "cut out and keep" ensemble-recording recipe. On the contrary: I hope I've shown how the nature of a miking setup is inextricably linked to the unique combination of musicians, equipment, and venue it's contending with. You can't just adopt some kind of cookie-cutter attitude—apart from anything, the restrictions of low-budget operation frequently demand considerable lateral thinking from small-studio engineers, so leftfield recording rigs (such as the ones mentioned above) often become eminently justifiable.

I also hope I've been able to communicate how exploratory this kind of session typically is. The way instruments, acoustics, mics, and processing interact on even fairly small peer-array sessions is so complicated that you really can't know for sure what's going to work until you try it, so you must always be willing (and able) to probe new options on the fly if you don't hit paydirt straight away. That said, there's no point spending ages seeking sonic perfection if that leaves insufficient time to get the necessary takes, so it's important to prioritize those aspects of the recorded sound that are of greatest value to the music you're serving. That way, if you feel forced to hit Record earlier than you'd wish (which is most of the time!), any remaining sonic compromises are relegated to less critical aspects of the production. In Section 10.3.2, for example, I was willing to compromise on the stereo width, HF clarity, and overall spill levels of the drum kit, because I wanted the piano to be wider, clearer, and more upfront than the drums in the mix.

"The way instruments, acoustics, mics, and processing interact on even fairly small peer-array sessions is so complicated that you really can't know for sure what's going to work until you try it, so you must always be willing (and able) to probe new options on the fly if you don't hit paydirt straight away."

I'm not wanting to promote defeatism, though, because I firmly believe you should always be striving for the best possible sound you can get on every session. It's just that learning about peer-array recording is a long-term campaign, rather than a single battle. The first sessions you do, you'll have to work out most of your mic placements from scratch, so you'll be firing on all cylinders just to get a usable sound before everyone dies of boredom. But as you rack up a bit of experience with different ensembles/venues, you'll begin to find you have more of a head start. For example, the piano and vocal setup of Section 10.3.1 saved me time when setting up for the ensemble in Section 10.3.2. Likewise, if I had to record the same jazz trio again, I might be more efficient by starting from the previous session's mic setup; or, alternatively, my attempts at trying to mix the first trio recording might suggest that, say, arranging the quilt over the bass mic to reduce cymbal spill might be a better option in future.

THE KICKSNARE MIC

Here's an interesting little trick for drum recording mentioned by Tom Syrowski:[26] "We also had what we call a kicksnare mic, which is an AKG C414 in figure-eight placed just underneath the snare, pointing at the kick-drum beater and the bottom of the snare drum. We move it around until kick and snare have roughly the same volume level." Gus Oberg[27] used a Telefunken 260 small-diaphragm omni in a very similar way while recording The Strokes: "I put the 260 on the beater side of the rim of the kick drum, it's like two inches from the top of the kick drum and two inches from the side of the snare." The advantages of this placement for me are that it keeps cymbal spill low and it provides an alternate tonal picture that supplements more traditional close-mic placements rather well, typically adding beater definition to the kick and tonal "beef" to the snare.

If you scratch the surface of any large studio's "standard" band setup, you'll find a whole history of similar session-by-session experiments which led to it, through which the engineers endeavored to get the best from the specific players, rooms, and equipment they had. "We figured out how to get good sounds in the room," recalls Robert Musso[28] of his work at Greenpoint Studio, for example. "At first it was a matter of trial and error, until we finally got to a point when it really came together. Then we stayed with what worked." One of the things you're paying for in a well-run professional studio is this accumulation of uniquely location-specific experience, which makes it possible for their staff to get beautifully refined peer-array recordings in a flash. Sadly, you can't really hope to compete with any such recording by copying its session layout, because so many of the nuanced engineering decisions that originally spawned that layout will be invalidated the moment you substitute its specific players/instruments/gear/acoustics with yours. In other words, there's no alternative but to build up your own recording experience, based on the musicians, kit, and rooms *you* have to work with, and develop your own "standard" procedures which fit those resources.

Even then, though, there's only a certain amount you can ever predict about how a setup will sound based on previous sessions, if only because of atmospheric conditions affecting the way sound travels through the air. "It can be a big change," observes Bruce Botnick.[29] "I've had it where the first two days [of a session] sounded amazing, and all of a sudden the third day is as dry as can be, and you have to either raise the mics or go in closer, or change them."

EDITING ENSEMBLE RECORDINGS

Although all-in-one-room ensemble recording is very popular in classical and acoustic music, many small-studio recordists shy away from it when working in more mainstream rock and pop styles, primarily on the basis that spill between the mics might make it difficult to correct mistakes or to tighten ensemble timing after the fact. This is a shame, because the simple ability to comp together slices from three or four different multitrack takes usually affords more than enough editing/fixing power for those purposes, in my experience, especially if there's any prominent percussion part (such as a drum kit) providing lots of easy edit points in every bar of the music. Furthermore, most commercial styles will usually have enough repetition in them that you can fix up missed snare hits or flubbed guitar chords even when there's only one take to work from, just by copying between repeated sections.

However, to get the best out of editing ensemble takes, there are a few things to keep in mind:

- The most effective edits will be those in which you cut across all the ensemble mics at once, so you shouldn't expect to be able to patch up subsets of the ensemble session's mic signals using tracks from a separate take. You might just get away with dropping in an individual part for a small snippet without too many odd-sounding side-effects, but I wouldn't recommend ever relying on that—do another take to get the performance you need while you're still tracking.

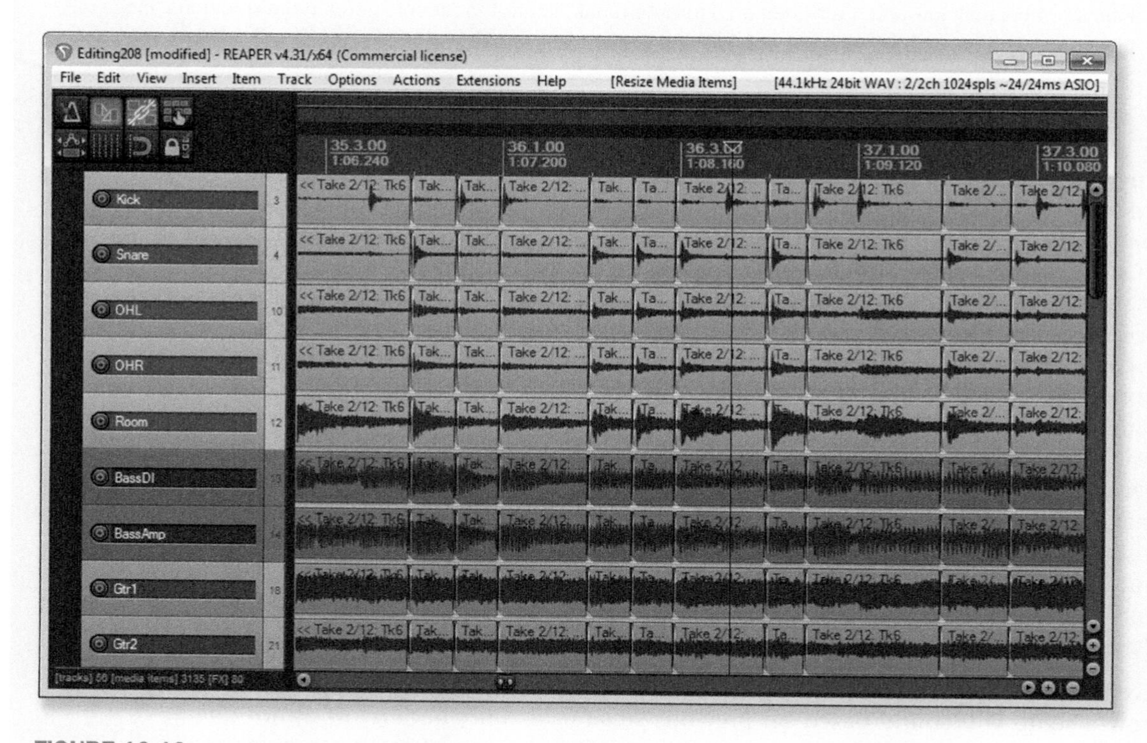

FIGURE 10.10
You can often get away with editing multitrack ensemble recordings in surprising detail without audible problems—just so long as you're careful to maintain the phase-relationships between the tracks.

- You should be careful not to change the time-alignment between the different tracks in the multitrack recording, otherwise you may mess with internal phase-relationships, potentially rubbishing your entire ensemble timbre. "I'll tell you," says Wyn Davis,[30] "as soon as you start moving things around and start taking things out of their phase-relationship to each other, it's over!" Most editing systems will provide the option to time-lock a set of tracks together for editing purposes, so enable that feature if it's available, to avoid slip-ups.
- Should you record lead vocals and improvised solo parts as part of your full-ensemble take, or separately as a later overdub? If such lines vary much between performances, you may find it almost impossible to make finer edits between takes, or indeed between repeated sections of the same take. With a lot of amateur ensembles, it's often pragmatic to postpone such parts to the overdubbing stage to make best use of limited session time—something we'll discuss further next chapter.

Make sure you keep good clear notes of which takes are which, and where to find each one within your recording system, because it saves all manner of confusion come editing time. Within this context, most engineers who regularly work with ensembles develop their own shorthand for certain common session occurrences. For example, I normally use "T3" for "take three," "FS" for "false start," "M" for "master," "X" for "discard," and "TUN"/"TIM" to flag tuning/timing concerns. It's also sensible to write down something about the purpose of each take, so that you know what to listen for when comparing takes post-session—particularly if that happens to be several weeks later!

Occasionally you'll come into contact with bands who are determined to avoid all studio trickery and present the listening public with nothing but complete, unedited takes. As defensible as such indier-than-thou posturing might be philosophically, I don't think it actually gives the best results in most small-studio situations. The first problem is that it tends to expand the number of takes required to achieve a suitable quality level, especially where several band members have equal voting rights—to the point where the stamina of the players begins to become a serious limiting factor. But that's not the only reason, as Steven Epstein[31] recalls from the age of direct-to-vinyl orchestral recording: "When you got to the last two minutes of a direct-to-disc recording, the artist would tend to give a very safe performance because they didn't want to make any mistakes and have to start all over again...Having the option of editing allows the artist to be much more free with the performance."

10.4 COPING WITH LARGE-SCALE SESSIONS

10.4.1 Delegating the Workload

Already in Chapter 9 the logistics involved with all those mics and signals may have begun to feel fairly taxing, but peer-array recording is usually even more so, such that most engineers will struggle to do a good job of it single-handed—especially if their brief extends beyond sonics into wider production matters. "Honestly, I don't know how some engineer-producers can handle it on their own," remarks Steve Bush.[32] "Why do you think that, on most pro sessions, there's generally a producer, and engineer, and an assistant or two?" Given that employing staff will be out of the budget of small-studio engineers, one of the best ways you can improve your recorded results on

"Given that employing staff will be out of the budget of small-studio engineers, one of the best ways you can improve your recorded results on peer-array sessions is to find ways of delegating some of the work to willing volunteers."

peer-array sessions is to find ways of delegating some of the work to willing volunteers—whether those are members of the ensemble (who should have a clear interest in helping you make them sound good), students in need of recording experience, or other guerrilla recordists helping you out on a favors-for-favors basis.

The simplest thing other people can do is usually loading/unloading gear if you're working on location, although I usually try to avoid getting musicians involved with packing/unpacking mic stands and cables unless I've been able to demonstrate how I'd prefer it done—misguided assistance in this respect can easily lose you more time than it saves! Nominating someone to keep track of the session schedule and organize any breaks/refreshments often works well, especially if you (like me) only remember to take a break when you're beginning to hallucinate from malnutrition...

On the production side, there are lots of tasks that are eminently delegable. For example, it's a godsend having someone who reminds everyone to tune up before takes and who listens specifically for sour tuning during playbacks. Likewise, a "Timing Tsar" might issue pre-take tempo reminders and scout for timing concerns post-recording, and a conductor or musical director might take charge of spotting and remedying internal balance issues via adjustments

FIGURE 10.11
One of the most useful things an assistant or willing volunteer can do for you while setting up an ensemble recording is to put on headphones and move microphones on your behalf, while you listen and direct their actions from the control room using talkback.

to the ensemble's musical performance. A designated mic-checker/mover can be another fantastically useful person, because they'll save you a lot of legwork running back and forth between the mics and the mixer. If you've managed to set up a separate control room for your ensemble session, then ideally it's nice to provide this helper with a pair of headphones to direct their activities, but don't underestimate what can be achieved just by agreeing a few bits of sign-language in advance. I like to find someone with a little studio experience for this role if possible, but I don't let it put me off if I can't—even a technophobe will usually save me significant effort here.

Where you have someone much more studio-savvy available, then you'll get the best assistance from them if you've got a decent setup list they can refer to. This should show at least what mics you're planning to use for which instruments, and which stand, cable, and recorder input each mic should be connected to, but there's lots of other useful information you might also provide on it: how any on-body mic switches should be set, for instance, or whether a pop shield will be required. An annotated venue floorplan can also be invaluable—it doesn't need to look like an architect's blueprint to show all the instruments and mics, as well as the positioning of any furniture, gobos, headphone amps, and music stands. Labeling your recording system's inputs and monitor faders with masking tape to show the mic signals they're responsible for saves a lot of aggro as well, by preventing people working at cross-purposes.

Once everything's plugged in, it makes for quite a good division of labor if you ask your fellow engineer to handle foldback and talkback arrangements for you, seeing as these can get quite complicated on larger sessions. This is one situation where DAW-based recording engineers can really benefit from having a multi-output audio interface and a separate analog mixer, because this makes it possible for two people to work in tandem: the recordist building the main control-room monitor mix within the DAW mixer, while sending per-instrument submixes to the channel line inputs of the analog console; and the guest engineer sitting at the analog desk creating any necessary foldback mixes for the musicians via its auxiliary sends, as well as being responsible for talkback functions.

THE LONESOME LINE-CHECKER

If I've no choice but to line-check a large ensemble-recording setup on my own, here's what I do. First I take a cheap phantom-powered test-signal generator and plug it into the end of each mic cable. This gives me a consistent sinewave signal that I can use to check routing and which will also highlight any audio clicks or distortion pretty clearly when auditioned. When I know that the post-mic input chains are all working properly, I connect up the mics, switch on a portable radio in the middle of the ensemble, and quickly Solo each of the mic signals to check that none of the mics is malfunctioning. Because I know how loud the radio is, and where it is in the room, I can normally use it to set up approximate preamp gain settings too.

10.4.2 Mic Stands and Cabling for Ensemble Work

So far I've been assuming that you're using general-purpose mic stands, of the type that can lift your mics around 7 feet (2.2 m) into the air. As you start working seriously with ensembles, however, you may have to begin investing in more specialized products. Shorter mic stands specifically designed for addressing things like kick drums and small amplifiers are a useful first step (and not too pricey), because they allow you to position mics more freely when multimiking within a restricted space. Where floor area really is at a premium, you can also get little clips which will attach mics directly to drums and speaker cabinets, although I'm not too fond of those myself, because they often make it more difficult to reposition mics and generally promote rather extreme close-miking.

Elevating your mics well above head height is typically a more expensive proposition, with dedicated double-height mic stands typically retailing at more than twice the price. One budget workaround here is to investigate cheaper lighting stands. If you fix a length of metal curtain-rod tubing to the cross-bar of one of these, you can hang mics from it using screw-on clips such as K&M's 238 Microphone Holders. In fact, because some lighting stands have multiple arm attachments, one stand base may even be able to accommodate large multimic dominant arrays such as omni "curtain," Decca Tree, or phased-array rigs. Whatever design of large-scale stand you're using, though, take extra care to avoid it overbalancing, because the consequences hardly bear thinking about on a big ensemble date. "All the cables have to be nicely dressed," says Armin Steiner,[33] "because anyone who is stepping around on a break or something can cause you irreparable damage if everything isn't tied down properly." Add weight to the stand bases where possible, and consider counterweighting the stand's boom arm wherever a longer horizontal reach is required.

FIGURE 10.12
An affordable lighting stand repurposed as an extended-height multi-mic stand—the mics are fixed to the main frame with curtain-rail tubing and screw-on microphone mounts.

FIGURE 10.13
A multicore cable such as this one (which has a stage box built into the cable reel) makes it a lot easier to set up for larger-scale sessions.

Speaking of counterweights, remember that the point of them is to reduce strain on the clutch mechanism which holds the boom arm in position, so try to make sure the arm actually balances on its own before tightening the clutch. Also, do endeavor to set up any counterweight well above head height, otherwise you're just begging for someone to come along and brain themselves on that big lump of metal!

When you've got lots of mic stands in a small space, there are a couple of other general pitfalls to watch out for. Firstly, try not to place the foot of one stand on top of the foot of another. Not only does that potentially destabilize the upper stand, but it also means its position will be disturbed if you need to move the one underneath. By the same token, try not to place any mic stand's base over a cable, so you limit the damage if anyone does trip over your audio wiring.

Where you're using just a handful of mics within about a 30-feet (10 m) range of your recording system, ordinary mic cables will usually do the job fine, even though it may mean you have to chain mic cables from time to time. However, once you're using more than about eight mics, or you need to set up your multitrack recorder significantly further away from the ensemble, using lots of individual cables becomes rather ungainly, and it's usually wise to invest in some thick "multicore" cables, each of which can carry multiple audio signals. These are sometimes called "snakes," and are usually terminated at one end with short "tails" or "flying connectors" for plugging into the inputs of your recording system. The other end of the snake feeds into a heavy-duty "stage box" covered with audio sockets, and there are labels on the tails and stage box to show how their connections correspond. The idea is that you use your regular mic cables to get from the microphones to the stage box, and then the snake carries those signals all the way back to the recording system without any extra fuss. Where the total cable length ends up being more than about 100 feet (30 m), the cables may begin to become responsible for unwelcome amounts of added noise, especially if you're recording quieter acoustic sources. Some professional engineers tackle this by preamplifying their mics in the live room, so that they can send them through the snake at line level, but I wouldn't get involved with that unless you really hear that you have to, because it makes managing levels a bit of hassle if you're engineering single-handed.

DROP ARMS

There are some situations where large-scale microphone stands are used for ordinary-height sources. Vocals are common recipients of this treatment, because it allows the stand's base to be moved out of harm's way, providing more room for the singer to move, as well as allowing freer positioning of their music stand and/or any instrument they're accompanying themselves with. Drum overheads may also benefit from larger stands too, leaving more floor space around the kit clear for spot-mic stands. However, in both these cases it can prove awkward to avoid putting the stand's counterweight at head height, so high-end mic stands frequently feature an extra little attachment at the end of the boom arm, called a "drop arm," which simply lowers the mic by roughly a foot (30 cm), allowing the boom and its counterweight to be elevated further out of harm's way. You rarely see drop arms in use in small studios, because cheaper mic stands don't include them as standard, but if the idea appeals to you, then check out K&M's $30 (£20) Mini Boom Arm, because this will fit to the end of any general-purpose mic stand's boom to give a very similar result.

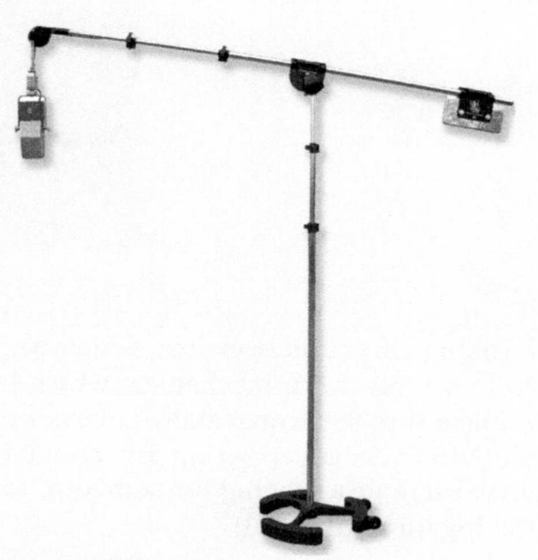

FIGURE 10.14
This high-spec Latch Lake MicKing 2200 stand has a built-in drop arm, but you can also retrofit drop arms to more affordable stands.

10.4.3 Equipment Redundancy

The more musicians you're recording at once, the more of a bummer it's likely to be if some technical gremlin renders the recording system inoperable. For this reason, you should always try to plan some equipment redundancy into your recording rig on ensemble sessions. The easiest way to do this is to go through each piece of equipment in your rig and ask yourself what you'd do if it went up in smoke. "You absolutely have to assume that whatever can go wrong most likely will," says Frank Filipetti.[34] Clearly there will be few small-studio recordists who can afford to bring along direct replacements for every bit of kit they're using on a session, but there are some sensible precautions you can take that shouldn't break the bank:

- Have at least one more of each type of cable/adaptor than you need. Cable faults are the most frequent problem you're likely to encounter on any location session. With mic cables, I'd personally plan to bring 50% more of those, so you can also use them as extensions if necessary.

- Take a couple more mics than strictly necessary, even if they're not all that suitable. If one of your first-call mics gives up on you during the session, it's usually better to replace it with an inappropriate mic than to have none at all, especially within a peer-array recording situation. This also gives you some flexibility if the ensemble line-up changes on the day of the recording—a dispiritingly common occurrence on low-budget sessions.

"Take a couple more mics than strictly necessary, even if they're not all that suitable. This gives you some flexibility if the ensemble line-up changes on the day of the recording—a dispiritingly common occurrence on low-budget sessions."

- Always have a spare pair of headphones, a spare headphone output, and a long headphone extension cable with you, whether you're planning to use headphone foldback or not. It's amazing how often this will come in useful: for bailing you out when your main monitoring system fails; for directing an assistant who's moving mics; for protecting your ears while you rearrange drum overheads; or for providing an unforeseen additional foldback feed.
- Have some kind of feasible backup plan in case your main recording system dies. This might simply be an agreement with a friendly co-conspirator that you'll come to each other's rescue if either of you finds yourselves in such dire straights. Alternatively, you might continue to maintain an outdated recording system as a fallback after upgrading, or else invest in a secondhand multitracker or even one of those ubiquitous portable stereo recorders—recording direct to stereo may be your idea of a complete nightmare, but would you prefer no recording at all?
- Whatever data storage space you *think* you need for your session, make sure you've got twice as much available. You never want to have to stop recording because you've run out of space. In a similar vein, work out how you're planning to back up your work as you go, in case any of your storage media fail mid-session.

Planning for the unexpected may also involve working out what you'd do in the event of a power cut. Backup power systems that are capable of keeping a whole recording rig running seamlessly through a mains outage will be well out of budget for most small-studio denizens, but I wouldn't worry about those much anyway—most ensembles will need power to perform anyway, if only for lights. However, if you're recording digitally, it's still worth considering whether a power cut might lose you unsaved data. This is where laptop computer systems are great for location recording, because they'll switch over to battery power automatically if the mains supply vanishes, giving enough time to save whatever you're working on. With other computers and digital recorders, however, I'd recommend investing in a small Uninterruptible Power Supply (UPS). This is effectively a backup battery that switches in if the mains power fails, giving you a few minutes of extra running time to save in-progress project files and shut everything down correctly. To protect just one or two recorders/computers, you should be able to get hold of a UPS with a suitable power rating for under $150 (£100).

FIGURE 10.15
The more difficult or expensive a session is to reschedule, the more it makes sense to protect your recordings against unforeseen power outages—for example with an affordable uninterruptible power supply such as this one.

SAGGING PHANTOM POWER

If you're running lots of microphones simultaneously on an ensemble session, their combined phantom-power draw can overstress the hardware of a multi-input preamp, mixer, or audio interface within your recording system. This is something that's much more likely to happen in cheaper equipment, especially in audio interfaces which draw their own power from a computer via a data cable (rather than directly from the mains). If the load on the phantom power supply causes its voltage level to drop by more than a few volts below the nominal 48 V standard, you'll find that your mics will lose headroom—i.e., they'll distort more easily on loud sounds.

To check for sagging phantom-power levels on any multi-input device, plug in all the mics/DIs you're going to use and engage phantom powering for those mics that need it. Then find a spare mic input (unplug one of the mics if necessary) and measure the voltage between Pin 1 (earth) and either Pin 2 or Pin 3 using a regular voltmeter—it should read above 44 V to be within internationally agreed standards. If you had to unplug a mic to make the measurement, it's worth repeating this test using a different mic's socket, just in case the mic you unplugged first of all was particularly power-hungry! (Also, be careful not to short-circuit the phantom-power supply by connecting unbalanced outputs to phantom-powered inputs, as that can also cause the phantom power supply's voltage to droop in less well-designed equipment.)

10.5 SESSION PACING AND PLAYBACKS

10.5.1 Workarounds for Limited Setup Time

In previous chapters, I've repeatedly stressed the importance of working at high speed so that you're ready to capture moments of magic whenever they happen. It's no different when working with ensembles. "I think [engineers are] very unaware of how important it is for the band to come into the studio and not have to wait around," says Phil Ramone.[35] "If the band feels like playing in the first five minutes, there's your record." "You just want to get everything set up so everybody can play," adds Nigel Godrich.[36] "My engineering may not be the most refined of anybody's, but I have learned that that's not what it's about. It's not about getting the most hi-fi sounding thing; what makes a good record good is the vibe, and how everything just falls together."

Where you're recording single instruments (or ensembles via the dominant-array approach), copious preparation can usually equip you to hit Record the moment the musicians arrive, if necessary, while still ending up with a usable recording—even when you're working with a group of performers for the first time in an unfamiliar venue. Sure, there'll always be risk involved in doing this, but it'll be a comparatively calculated risk. With peer-array recording, on the other hand, unfortunate spill pickup on just one of your peer arrays may cause ferocious problems at the mix, and it's very difficult to predict such things in advance before you've heard what your specific ensemble sounds like through your specific mics in your specific room. So although it's still your duty to press Record within the first few moments of a peer-array session should the occasion demand, I'd also highly recommend doing everything within your power to avoid getting into that situation in the first place!

> "Try to get some of the musicians into the studio in advance of the others, so that you get a head-start on sorting out their peer arrays. An added bonus of bringing in the performers piecemeal is that you get more opportunity to talk with each of them individually and get constructive communication going before you need to make use of it in earnest."

A primary tactic here is to get some of the musicians into the studio in advance of the others, so that you get a head-start on sorting out their peer arrays before everyone needs to start playing together. An added bonus of bringing in the performers piecemeal like this is that you get more opportunity to talk with each of them individually, so you can get constructive communication going before you need to make use of it in earnest. This isn't nearly as easy to pull off when a coachload of people all rock up at the same time! Another good trick is to start the session with some kind of warm-up jam or rehearsal (or by deliberately recording some piece of music inessential to the production as a whole), on the understanding that it's expressly for the purpose of letting the performers settle in and allowing you to get sounds. "[If] I've decided which song I'd like to record first," says Richard Dodd,[37] "I'll actually have them warm up and get my sounds on another song, so that when we start the song we're going to record, I'm ready."

MINIMAL DRUM MIKING

Some small-studio engineers labor under the misapprehension that you can only capture commercial-grade drum sounds via elaborate multimiking, and therefore avoid recording a drum kit in ensemble situations simply because they don't feel they have enough hardware. In fact, lots of commercial drum recordings have been carried out with very few mics indeed. Many pre-multitrack-era big-band recordings have a single overhead mic on the drums, for instance, as did Cliff Richard's rock-and-roll classic "Move It," according to Malcolm Addey.[38] Two-mic drum setups were fairly common in the '60s and '70s, when few people would bat an eyelid at a mono drum sound. For instance, Andy Johns[39] used just two Beyerdynamic M160 ribbon mics for Led Zeppelin's oft-imitated "When The Levee Breaks"; both Steve Albini[40] and Eddie Kramer[41] have talked about creating full mono kit sounds with one overhead and a kick mic; and Bones Howe[42] used an overhead-plus-snare/hat setup for many artists, including Ornette Coleman and Tom Waits.

There are some classic three-mic setups too: For instance, Robbie Adams[43] used a single overhead mic with kick and snare spots for U2's *Achtung Baby*; Bruce Botnick[44] used a very similar miking scheme for The Doors, except that his snare mic was underneath the drum; and Al Stone[45] used a three-mic technique for Jamiroquai's break-through album *Traveling Without Moving* ("One mic a few feet in front of the bass drum, one over the drummer's shoulder covering the hi-hat and snare, and another to the side to cover the cymbals," he explains."). Even if you insist on stereo, there are plenty of well-known engineers who've recorded kits by supplementing a main pair with only one or two spot mics—for example, both Tony Platt[46] and Mike Hedges[47] recall Glyn Johns working with just stereo overheads and a kick-drum mic, and Jacquire King[48] used a similar setup with Kings Of Leon. Steve Churchyard[49] and Peter Henderson[50] have also talked about Geoff Emerick using a four-mic setup with spots for just kick and snare.

It also stands to reason that peer-array methods will work best on longer sessions where more than one piece of music is to be recorded, because this gives you the opportunity to continue finessing the setup on a rolling basis. My favorite situation, in fact, is to get a sound for the ensemble's first piece of music at a fairly leisurely pace on day one, ending up with a couple of preliminary takes before supper time. Then I reassess my initial rough mix later that evening and make any setup tweaks which seem necessary with hindsight. The following day the musicians can then return to the session fresh and rested, and work at whatever pace suits them, because the lion's share of the technical legwork is already complete.

Unfortunately, such utopian scenarios are rarely feasible, and sometimes you've simply got to grit your teeth and cut corners to hit the session schedule. "How can you get great sounds without hearing the band?" asks David Thoener.[51] "The answer: you wing it!" You might, for example, mic a little closer and use a bit more absorption overall to reduce interaction between the peer arrays, making the Meld stages a little less critical (and time consuming). Another shortcut would be to concentrate on getting the Focus of each array sorted out first, and then skipping directly to a full-ensemble Melding free-for-all so you can highlight the most obvious spill-related issues. Both of these tricks will likely cause you headaches at mixdown, but at the same time they should hopefully minimize the chances of capturing signals that prove to be truly unmixable. "If I'm going into the session, and it's essential that it runs fast," says Frank Filipetti,[52] "I'll go with what I consider the safest and not necessarily what I consider the best."

But what if something still sounds hopeless during the first take? "Don't just stand there and wait until the take is over!" asserts Phil Ramone.[53] "I've got a take with me adjusting mics in the middle of the take...and me moving a whole guitar rig over, and it's in the record! It was the best take!" And if the first take isn't the keeper, then you won't have wasted the opportunity to better prepare for the next one. In fact, Ed Cherney[54] actually takes the view that leaving sonic refinements until after the first take can actually be beneficial, because at that point there's actually something concrete to discuss: "It's real at that point... I've been in places where you mess around a lot before you play any music, and the session doesn't move forward."

10.5.2 The Power of Playbacks

As I said in Section 10.2, you can only properly evaluate the quality of a peer-array recording by creating a basic mix of it during the tracking session. But this isn't the only reason for spending as much time on your control-room monitor mix as you can. When you've got a lot of sessions under your belt, it's easy to forget just how exhilarating it is for musicians to hear themselves sounding like something out of their own record collections, so if you can hit them with that after their first take, they'll almost always respond by delivering a second take that utterly smokes it, setting up a classic virtuous cycle. "The best thing you can do for a musician is to get them excited," says Mark Howard:[55] "As soon as you win their confidence with a great sound, and they know they don't have to worry about anything technical, you are going to get great performances out of them. They are just going to nail it." But even with more seasoned players who are more likely to deliver maximum expression from the get-go, it's still worth giving them a playback sound that's within spitting distance of the final release, because they'll respond to any audible imbalances by moderating their playing—in other words, they'll help mix the record for you!

"Even with more seasoned players, it's still worth giving them a playback sound that's within spitting distance of the final release, because they'll respond to any audible imbalances by moderating their playing—in other words, they'll help mix the record for you!"

And let's not forget how much the first playback reflects upon you. "You must wow your band," says Eric "Mixerman" Sarafin.[56] "A poor job on tones right from the beginning of your session can be unrecoverable. Job number one on the first day is to solidify the band's trust in you." In this light, having speakers in the control room for playback is usually a massive advantage, because everyone can then listen and respond to the playback communally, and you can also push up the listening level if you think that'll help the general vibe. However, take care that you don't make a rod for your own back by shepherding listeners into the wrong areas of the control room. For example, sitting a guitarist closer to the right speaker when he's panned left in the balance is a bit daft. Likewise I'd think twice about sitting that guitarist close to a wall where boundary effect will tend to overemphasize the level of the bass part, or to put the bass player somewhere where room resonances are murdering the low-end response.

One specific thing I like to do is vacate the sweetspot during playbacks, and go stand somewhere at the back of the room. Not only does this put the musicians into the best listening location, but also positions me where I can watch everyone's immediate reactions to the music at once. Another tip I picked up from Chuck Ainlay[57] was to angle the speakers more toward the back wall of the control room for playbacks, so that you improve their high-frequency coverage across the floor area, a stunt which somehow seems to make people feel more comfortable moving around the room while listening too.

RIDING FADERS TO TAPE

When you're working with ensembles, one way to improve your monitor mixes is to compensate for any unwanted balance changes by massaging recording levels on the fly (or "riding the faders to tape"). This was a day-to-day requirement back in the days of direct-to-mono/stereo recording, but the idea rarely seems to occur to small-studio engineers these days. "If the trumpets change to mutes, or something like that," says Leslie Ann Jones,[58] "I try to bring them up while recording. If I don't, every time we play the tune back, I have to make a level ride for people to hear it the way the arrangement is supposed to be heard." Joe Chiccarelli[59] says something similar: "I definitely record things so you can leave the faders at one level and there's your record. If you know this part needs to be subtler in the verses, you record it a little lower...You don't have to go, 'Oh yeah, I forgot the bridge is coming up, I've got to crank that guitar way down.' All the stuff that takes you out of the picture for a second." This is one situation where hardware faders really come into their own, because you can ride several at once during a take.

FIGURE 10.16

Bruce Botnick,[60] Jeff Powell,[61] and Elliott Scheiner[62] also mention riding their input levels as a matter of course, the latter explaining another benefit he feels it brings to a production. "I watch guys come in, throw on a compressor, throw on an EQ, and then never touch the fader again. Everything is predetermined. For me, part of it is that I feel, as an engineer, you are as much a part of the performance, especially when you're doing a live band, as any member of the band. And you're interpreting what they're doing, and your hand rides dictate that. It's not a constant compression where everything is safe, everything is slammed, and everything is at a certain level. A lot of times, the dynamics are totally missing when people do that."

But don't just abandon your monitor mix after the first playback—keep working on it throughout the session whenever you have a moment spare. On the most basic level, solo each mic signal from time to time to scout for subtle equipment faults before they have the chance to spoil dozens of takes. Experimenting with your mix settings will also frequently alert you to weaknesses in your recording setup that you overlooked during the initial setup melee. But on a more general level, improving the control-room monitor mix on a rolling basis helps build on any good first impression you've already created with the musicians, as well as consolidating your own decisions about how you ultimately want the finished production to sound. "If, upon completing your last overdub, you have no earthly idea how you want your tracks balanced," says Eric "Mixerman" Sarafin,[63] "then you have been completely ignoring how the track is supposed to make you move and feel...Therefore, you should be toying with your balances each and every time you touch the track."

Finally, I always like to bounce down my monitor mixes at the end of every tracking session, because they can be a tremendously useful reference-point at the mixdown stage. "The first initial thought is always the best," explains Rodney Jerkins.[64] "It's the raw feeling, that first, 'Whoa—why did I have that triangle so loud?' But there was a reason for it." Daniel Lanois[65] has made whole albums from such work-in-progress mixes: "On that day, there will be a certain unspoken knowledge of that piece of music. The people in the room are operating on instinct, and so your blend is not cerebral."

CUT TO THE CHASE

- Compared with the dominant-array method, peer arrays allow greater scope for reshaping the timbres and balances within an ensemble. However, the downside is peer-array sessions are usually more complicated and time consuming to set up. Although many small-studio users have misgivings about using peer-array recording where corrective edits may be required, in practice these concerns are often unfounded.
- Planning a peer-array session involves bridging the gap between the sound that's likely to be happening in the recording room and your vision for the production's final mix sonics. The closer you can bring the two, the less demanding the job becomes.
- Quieter instruments will be drier and more upfront than louder instruments in a peer-array recording unless you take deliberate spill-reduction measures. Spill may also need careful handling if you're planning on heavily processing any mic signals or if stereo-positioning disagreements arise between different peer arrays. Attempting to cut spill by spacing instruments further apart is often counterproductive.
- Each peer array can be anything from a single mic to a full "dominant array plus spot mics" setup. Part of your engineering task is to optimize each peer array's Focus (what it's ostensibly pointing at) and Backdrop (everything else it's picking up), but with the needs of the Focus taking precedence if conflicts arise.
- The only reliable way to evaluate a peer-array setup is to build a mix of it while tracking. Start by mixing those arrays which pick up the most spill and which are subject to mic/instrument-placement restrictions. For each array you add to the mix, consider its Ghost (the spill from its Focus instrument/section across the whole mix so far) and Meld (any interactions with other peer arrays already present in the mix). Whenever you make adjustments to the Focus or Backdrop of any individual peer array, always recheck its Meld.

- Although it's possible to compensate for the sonic shortcomings of a peer array's Ghost by tweaking its Focus, this may reduce your balancing flexibility.
- When dealing with comb-filtering problems at the Meld, polarity/phase adjustments and mic/instrument repositioning tend to be more effective remedies than equalization. Time-alignment methods are less useful for Melding purposes, even though they remain fair game within individual peer arrays.
- Room mics are tricky to handle within peer-array systems unless your ensemble is well balanced acoustically, so artificial reverb is frequently necessary to address depth-perspective concerns.
- Minimize the impact of peer-array setup technicalities for the musicians by: dry-running the recording layout during rehearsals; involving volunteers to split the workload; staggering the arrival of the performers on the first day of the session; starting the session with some kind of warm-up number; and recording multiple pieces of music over a longer continuous session.
- Talkback speakers can be very useful for ensemble work, not least because they allow easy slating of takes.
- Don't underestimate the importance of a good playback mix, because it has the potential to improve an ensemble's performance tremendously—and will also naturally reflect on your abilities as an engineer!
- If you're regularly recording ensembles, consider investing in some smaller/larger mic stands. Take extra care to make sure your stands are stable, staying mindful of where cables and counterweights are positioned. Backup power and a certain degree of equipment redundancy are also advisable.

Assignment: Ensemble Recording with Peer Arrays

Record several different ensembles using the peer-array approach. You're free to choose any size of ensemble, but do try to find at least one group with more than four players. You may use whatever recording equipment you wish, but be sure to research and plan the session well in advance, doing your best to make engineering technicalities as unobtrusive for the musicians as possible. Be sure to capture more than one take of each piece of music you record, so that you can experiment with the possibilities of full-multitrack editing after the session.

WEB RESOURCES

On this book's companion website you'll find a selection of resources to support this chapter, including:

- audio demonstrations of a variety of peer-array setup issues;
- real-world peer-array recordings in multitrack format, with accompanying notes about their session setup and mixability;
- further reading on phantom-power problems and backup powering;
- links to useful mic-stand products and various related widgets.

http://www.cambridge-mt.com/rs-ch10.htm

CHAPTER 11

Going Freestyle

In order to introduce new studio techniques in a progressive manner throughout this course, all the chapter assignments so far have imposed artificial limitations on what and how you recorded. The time has finally come to remove those restrictions completely. You should now have enough experience to go totally freestyle with your recordings, building whole productions from scratch using your own unique blend of the techniques in this book. With this in mind, I'd like to delve into a few of the broader issues involved with piloting a full project under your own steam.

11.1 MULTIPLE SESSIONS AND MULTIPLE ROOMS

Freestyle recording of entire productions means making the following choices that we've side-stepped up until now:

- Should the recordings be carried out over more than one session?
- How many musicians should play together on each session?
- Should any ensemble instruments be acoustically isolated from others?

The practical challenges of working on a budget are so great that small-studio engineers all too often fall back on the "safe option" of building full arrangements out of single-instrument overdubs, despite the fact that group performances are normally so much vibier for the musicians. Tony Hoffer[1] also points out that ensemble recording is much more efficient in terms of session time, because making sonic decisions becomes more straightforward in context. "I think the best way to shape a drum sound is to have your guitar sound and you're sort of shaping all the sounds together. You're going for a guitar sound that's probably going to be the final sound, you're going for the bass sound that's going to be the final sound. And that way I'm able to do [complete albums] in three or four weeks...I'd rather work fast, make quick decisions, and not waste time going back and forth, because you can get a lot more done in a shorter period of time."

A good way of fighting Excessive Overdubbing Syndrome is to start your planning by asking yourself how you'd go about capturing the entire arrangement as a one-room full-ensemble recording. If it transpires that this "all in one go" approach would be impractical in your situation for any reason, then adapt your plans as little as possible to enable a workable setup. To illustrate this concept, let's return to the imaginary five-piece band introduced at the start of Chapter 10: a singer, an acoustic guitar, an electric guitar, an electric bass, and a drum kit. In principle, such a line-up is eminently recordable within a single room, albeit probably using peer-array methods, given that the vocal and acoustic guitar are likely to be outgunned acoustically by the other instruments. So what factors might prompt me to deviate from this one-stop shop?

THE RIGHT KEY AND THE RIGHT TEMPO

A couple of decisions that are frequently taken rather too lightly in small-studio productions are the key and tempo the music is going to be performed in. A poor choice of key can be particularly counterproductive where a lead vocalist is involved. "If they can't get the bloody notes," says Gus Dudgeon,[2] "you spend ages trying to drop in over and over again—it's just hell." But it's not just a case of whether the line's possible, says Phil Thornalley[3]—it's also a question of the tone you're after: "Some singers, especially pop singers, may have soul and a sound, but also a limited effective range, say an octave or less. It's not that they can't sing higher or lower; it's more that they have an area where their voice becomes 3D, and outside of that range they have no character or expression. So avoid that at your peril."

Tommy LiPuma[4] is equally emphatic about finding the right tempo for each piece of music. "Many years ago, I read something that [band leader] Harry James once said; that you can lose a record with the wrong tempo. That stuck with me, because he's absolutely right. You can lose a hit by having the wrong tempo, so I'm very, very conscious of that." It can, however, easily become a point of contention in the studio, so try to agree tempos with all the players well in advance if you can—if only so that they rehearse at the same speed they'll be recording at. Where this isn't possible, Jack Endino[5] has a nice little trick up his sleeve: making a note of the actual tempos used at a series of rehearsals/shows. "When they get in the studio it's a good argument-settling device. People freak out and they play everything at half speed in the studio, and you can say, 'Wait a minute! Here's how fast you played it the last five times I saw you.' That really saves a lot of hassle."

11.1.1 Personnel and Equipment

The most obvious restriction in most popular styles is that there will frequently be many more parts in a recorded arrangement than there are players available. If our five-piece band wanted to double-track both guitars and add tambourine and backing-vocal parts, you'd obviously have to plan for capturing those overdubs after the initial full-band takes. Even for those extra parts, though, I'd still suggest retaining the "everything at once" mindset if you can, by recording group overdubs rather than laying down each as a separate pass by default.

For those on a budget, the kit list may also be a deciding factor. So if you had, say, only three condenser mics for the five-piece band recording, you might allocate those as a vocal mic and a drum-kit stereo pair, overdubbing the acoustic guitar later to avoid having to record it on a dynamic. Or perhaps your studio setup has inadequate facilities for acoustic absorption/isolation and you have a quiet vocalist, in which case overdubbing that might prevent other instruments drowning under the room reverb from the vocal mic's Backdrop. This kind of reasoning can be a bit of a slippery slope for small-studio engineers, though, because it's extremely tempting to retreat to the argument that you couldn't *possibly* do justice to the sound of such-and-such an instrument without using all of your microphones, preamps, and acoustic treatment for that alone. Unassailable as this standpoint can seem from an engineering perspective, it should still take second place to capturing a performance that really sets your pants on fire.

11.1.2 A Question of Timing

Recording ensembles can occasionally be counterproductive on a performance level, though, if certain members of the group require more takes than others to adequately nail their part. This may be because some players simply aren't up to the standard of the others, but it could also be that some musical parts are much more challenging—as will frequently be the case with lead-vocal or hook/solo lines, especially in genres where detailed comping of these is par for the course. "If the acoustic guitar is going to be the lead solo instrument, I'll put them in a booth to get the necessary isolation," comments Al Schmitt.[6] "However, if he's part of the rhythm section, I'll keep him in the room with the other guys." There are also issues of stamina to take into account, because no matter how easy a loud drum, trumpet, or vocal part may be technically, there's a limit to how long any player can hammer the volume before their strength gives out.

"Any player you suspect might frustrate the others by slowing down the communal work rate, or who might just run out of steam too early, should ideally have their instrument acoustically isolated from the core group, allowing you to edit that instrument's part freely between takes and/or patch it up with punch-ins."

So what you need to do is identify those ensemble players you reckon will be able to deliver the goods when their instruments are all in the same room. Any player you suspect might frustrate the others by slowing down the communal work rate, or who might just run out of steam too early, should then ideally have their instrument acoustically isolated from the core group, allowing you to edit that instrument's part freely between takes and/or patch it up with additional single-instrument punch-ins once you've finished the main ensemble takes. Postponing the recording of such troublesome instruments to the overdubbing stage can seem attractive where isolation and foldback facilities are scarce, but this option will, of course, reduce the degree of live human interaction within the ensemble, so I'd personally advise against it unless it's truly unavoidable.

Although few small studios have dedicated isolation booths available, you can frequently bodge together a reasonably operational alternative with a little lateral thinking. Electric instruments afford the greatest flexibility, because you can simply DI them during the ensemble session (perhaps in conjunction with a digital amplifier emulation to give an appropriate foldback sound) so that they're not picked up by any live-room mics, and then reamp them at leisure afterwards. But even when using a real amp you can usually move a few bits of spare furniture to box off the speaker cabinet pretty effectively, or move it into a suitable cupboard or hallway. David Barbe[7] recalls another ingenious work-around: "Andy Le Master had a studio in Toccoa called The Furry Vault…a small cinder-block shed behind his parents' house. It was so small that the guitar tracks would be done in a car or van. They'd back the van up to the door of the place and put the amps out there and the mics in the van." Keeping the player in the live room then requires only a long instrument cable and some head-phone foldback. In a similar vein, using a digital keyboard to emulate acoustic piano, organ, or harp for the purposes of an ensemble recording (rather than recording the real thing) can buy you additional post-recording flexibility, albeit usually with some trade-off in terms of the naturalness of the recorded sound.

If you're lucky enough to be working in a separate control room, then another scheme that can work well is bringing some quieter instrument in there with you. While this may cramp your style a little while setting up that instrument's sonics, it shouldn't make building the sound for the core ensemble instru-ments any harder, and the opportunity to use the control-room loudspeakers for headphone-free foldback monitoring (see Section 5.3) in this configuration can pay dividends with many musicians, especially vocalists.

WHEN SHOULD I RECORD THE VOCALS?

One of the things that really defines someone's production style in commercial genres is how soon they choose to record lead vocals. The most common approach is to deal with them after most of the other parts have been captured, so that the singer can respond to a backing track that's pretty much in its final form. But that's not the only school of thought. "During the preproduction period we do a lot of experimentation with rhythms, with textures, and especially with the vocals," says Pierre Marchand,[8] for instance. "Most of those early vocals are retained. We still try to get better vocals later on, but it seems that when there's little on tape, the vocal is more focused. If you record a vocal to a finished backing track it often doesn't work. Moreover, this way of working means that everybody in the band plays to the vocal, which helps to keep *them* focused."

"The singing has to set the level for the energy of the track," adds Steve Bush.[9] "If it were practical, you'd almost rather do the vocals first, because the thing which will be heard and focused on is the singing. If you start with the drums, bass, and one guitar, and you're obsessed with those parts sounding as good as they can, it's easy to forget the fact that it's not going to be very prominent in the end. The tone of the vocal makes every track sound a particular way, so [I] try to put a guide vocal down as soon as [I] start work, because you have to tailor the music to fit what the singer's doing." John Leckie[10] and Tony Brown[11] both also insist that singers perform guide vocals with maximum enthusiasm and emotional commitment. "If they give it their all, it's like a kick in the ass for the band," says Brown, "and the tracking session can turn into a lovefest." (In addition to any psychological benefits, though, Clarke Schleicher[12] considers it an invaluable opportunity to audition potential vocal mics within the context of a real performance: "When the singer is out there consciously doing a shootout, that's not a real test.")

Although most commercial lead vocals are the result of multiple takes and comping, you certainly shouldn't assume that guide vocals won't end up on the finished record. John Leckie,[13] for instance, estimates that 50% of all his final mixed vocals have been guides; Darryl Swann[14] says a similar proportion of Macy Gray's performances on the Grammy-winning *On How Life Is* were from their original nocturnal song-writing sessions; and Taylor Swift's original "demo" lead vocals for her album *Speak Now* made it all the way to the final mixes (according to Nathan Chapman[15]), as did Adele's vocals on her single "Rolling In The Deep" (according to Paul Epworth[16]). Jimmy Jam[17] also favors the earliest takes, even when there is no full-band ensemble session involved: "Nine times out of ten, the scratch vocals are better than the real thing, because the artist doesn't have the pressure that 'this is it!' rolling around in his or her mind.... You can catch gold while an artist is in the process of learning a song and playing around with addictive new melodies."

FIGURE 11.1
Don't assume lead vocals have to be recorded last—all these hits, for instance, feature lead vocals recorded at the demo stage, while the song was still being written.

11.1.3 Postponing Decisions

In Section 7.4 I extolled the virtues of making creative and sonic commitments from the earliest stages of a production, but there's only so much you can do on this front if you happen to be working with artists who start recording before they've actually finished writing the music! Don't get me wrong here—I'm not frowning on such practices (many fantastic albums have been made like that, after all), but they do make it difficult for an engineer to record one-room ensemble takes where all the instruments get into all the mics to some extent. With spill all over the place, it becomes impossible to completely mute individual instruments, so any replacement you put in later may conflict with the Ghost of its predecessor—something which rarely ends happily for Danish princes in particular.

The fear of such complications seems to drive some small-studio engineers to extremes, such that they'll only record multiple musicians simultaneously where separate recording rooms are available for each instrument. Honestly, though, I think such a hard-line stance is rarely necessary in practice, and it actively discriminates against recordists who don't have space or time on their side. A more pragmatic stance is to target any available isolation methods where they'll be most use, but not to lose sight of the advantages of having people playing together. "I don't care whether it's classical or rock or country," declares Eddie Kramer,[18] "you've got to capture that performance, and the hell with the bloody leakage!"

Returning to our five-piece band example, let's say I wanted to hedge my bets against song-writing changes, but only had a two-room setup to work with—in other words, a live room and a control room, but no iso booths at all. Here's what I might do:

- Bring the vocalist into the control room and mic them up there, using the control-room loudspeakers for the singer's foldback. If I got a truly inspired lead-vocal take, I could record a cancellation pass of loudspeaker spill (see Section 5.3.2) to improve my flexibility at mixdown; if I didn't, I could either punch in on the vocal track directly after I'd finished recording the full-band takes, or replace it completely later on without affecting the sound of my ensemble recording.
- Record a DI signal from the electric guitarist (using a digital amp modeler for foldback purposes), with the intention of reamping the part properly later if the performance was worth keeping.
- Use ensemble-miking techniques to record the drums, bass, and acoustic guitar together in the live room. I'd intercept some of the acoustic-guitar spill before it reached the drum overheads, using a couple of quilts draped over high-backed chairs (which shouldn't interfere with any sight-lines), and then I'd put a reflective board or three within this padded enclosure to provide the guitarist with improved acoustic monitoring. In addition to miking up the bass cab, I'd also take a DI signal from the instrument, and I'd record drum samples after each song's takes. I'd now likely be able to replace the acoustic guitar part without worrying about conflicts with its Ghost, and I could also solo the bass line if I wished.

- Provide vocal and electric-guitar foldback to the live-room musicians via headphones. I could probably get away with using just a single cue mix, but the electric guitarist would probably appreciate an independent "more me" mix if that were available. Between takes, the live-room players could converse directly, while hearing the singer in their headphones; and the singer could hear the rest of the players through the ensemble-rig mics.

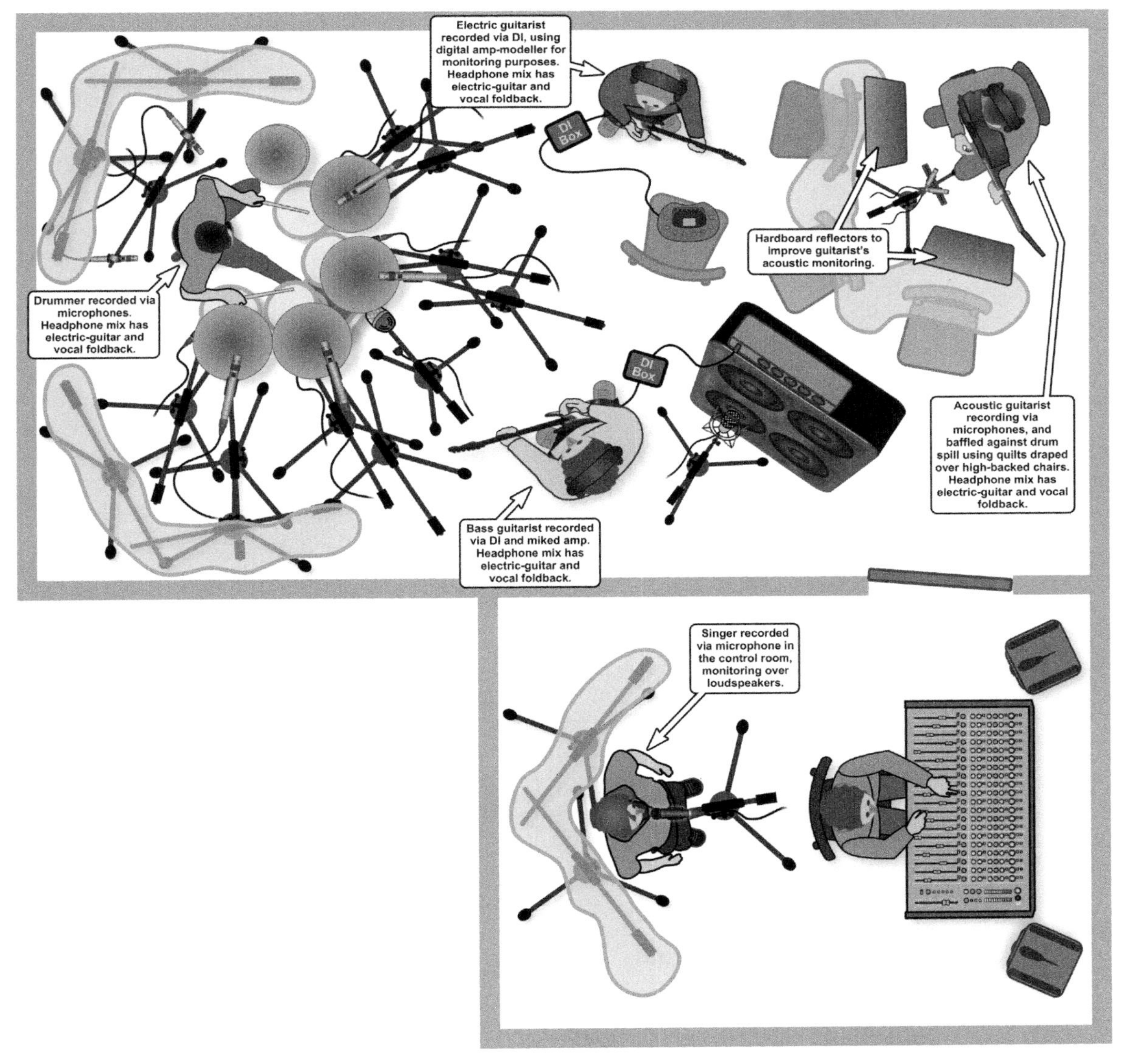

FIGURE 11.2
Here's an example of how you might go about tracking Chapter 10's five-piece band within a simple two-room setup, while still retaining masses of flexibility in terms of changing the song structure and arrangement post-recording.

This configuration is hardly very complex, and only gives total isolation for two of the parts, but it nonetheless affords tremendous rewriting flexibility while sacrificing very little of the full-band performance experience as far as the musicians are concerned. If the vocalist and engineer worked on headphones instead, the options would expand even further: I'd be free to edit the live-room rhythm tracks to tighten timing and rejig the bass line without making a nonsense of the vocal part; and I could comp the final vocal from pieces of all the full-band takes without affecting the sound of any other instrument. In fact, just about the only thing I'd really struggle to do in this scenario would be soloing the drums in the arrangement—although I might actually even get away with that for brief moments by substituting samples in place of a few of the live-performed hits!

In different circumstances, of course, this setup might not be effective at all, so it's important to question your studio-layout and isolation scheme for each new piece of music, or wherever the instrumentation or recording venue changes. Andrija Tokic[19] illustrates this mentality beautifully when talking about recording Alabama Shakes in his home studio: "I mostly tracked them on my ground floor…It was important to make sure that the core of the song was going down in one piece at the same time, [but] just like anything, you have to be flexible…On some songs, we put [lead singer] Brittany in a room right across from the room where everybody else was and then, if the drums weren't too loud, we'd leave the doors open so she could at least see the bass player and the bass player could see the drummer. There's a hallway between those two rooms, and we could either put the amps in the hallway or put them in the rooms on my basement floor…It would totally depend on what had to be where for the sound. Sometimes you want the bass to rattle the snare real hard, so you've got to have it right there."

11.2 IMPROVING YOUR EFFICIENCY

You're more likely to engineer complete productions efficiently if you maintain a clear overview of how much has already been recorded, and how much remains to be completed. The "to record" list I mentioned back in Section 2.3.1 can be easily expanded into tabular form for this purpose, with a row for each piece of music you're recording and columns to tick when each part of the arrangement has been successfully written, programmed, recorded, and/or edited. Rodaidh McDonald[20] used just such a tactic when working with The XX, for instance: "[We] had a chart on the wall with every track on it. We had boxes for 'bass done' and 'guitar done' and vocals…so we could visually see how far we were getting."

Where session time is limited, creating some kind of written schedule will usually help too, and can provide a means of managing the mood and momentum of the production process. "Falling behind on a project," comments Eric "Mixerman" Sarafin,[21] "can be a useful session tool, as it tends to ratchet up pressure. It can also bring down morale, particularly if you get too far behind.

Getting ahead relieves pressure, and allows everyone to relax. Between the push of falling behind and the pull of getting ahead, you can manipulate the forward push of your session...While you don't have total control over how things go on a session, you do have considerable influence. You should certainly have some idea of which tracks will be a breeze and which a struggle... If you're ahead, work on something that you believe will require time. If you're behind, knock out something easy." Gareth Jones[22] feels that each session should have clear boundaries too: "I try to schedule things so the sessions are not open ended and so everyone knows that we start at twelve noon and work until eight or nine, and in that time we're working and creating. That means that we don't waste time in the studio."

THE BENEFITS OF TAG-TEAM PRODUCTION

If you regularly find yourself having to both engineer and produce sessions, then there's a lot to be said for setting up a regular partnership with someone else in a similar position. "It works brilliantly," enthuses Steve Bush[23] about his fruitful collaboration with fellow producer Marshall Bird. "Marsh and I can both cover the engineering side and it's great to have someone other than the artist to bounce ideas off. How many times have you been doing something, and you just want to ask someone what they think? That extra opinion makes all the difference." L.A. Reid[24] concurs: "You don't have to second-guess yourself when working with someone whom you trust. As my partner, Babyface, says: 'By working as part of a team, one always has the benefit of a second opinion.'" Working in partnership also makes it easier to avoid hold-ups to the creative flow of a session. "If one of us is experiencing a creative block for some reason," says Jimmy Jam,[25] "the other one picks up the slack." Steve Bush[26] agrees: "Any problems can be dealt with by one of us while the other one carries on working with the band."

Gareth Jones[27] also has a clever approach to scheduling ensemble band recordings, tracking everything in the first half of the studio time, and then reevaluating and retracking those takes as necessary in the second half. "[Once] everyone is settled into the studio, I find that a band can do their best work and can usually do a good take in ten minutes," he says. "Bands love it, and I love it too, because we realize that what we've recorded is not so precious. That second half is when they can really start to deliver. We end up keeping some of the stuff from the first round of tracking, but a lot of it just gets done again."

Another means of increasing productivity is to cut out unnecessary setup time, and one of the easiest ways to do this is by consolidating recording projects into a smaller number of longer sessions. With ensemble recordings in particular, the heavy-duty technical setup involved at the outset will usually only delay the start of the first few takes, and once the session hits its stride you should find yourself in the enviable position of being able to hit Record at the drop of a hat without compromising on sonics. It also gives you the ability to hop back and forth between different pieces of music to keep players from getting bored or frustrated, as Butch Vig[28] did when recording Nirvana's *Nevermind*: "If we didn't get something right away, we'd just move on to something else and go back for it later. I think that's why the band always sounds

FIGURE 11.3
If you schedule electric-guitar overdubs to follow on directly from a drumkit-tracking session, you can try out all sorts of multimiking strategies in a flash, simply by reusing the drum mics. It's also very quick to try out new mics, because you can just plug them into the already line-checked drum-mic recording chains.

fresh—they kept their spontaneity while recording." Even where you're building an arrangement part by part via discrete overdubs, there are still significant time economies to be had by scheduling several overdubs for the same session. For example, if you leave a dozen mics active in the live room following a drum kit recording, you only need to move a few stands to carry out mic shootouts for other instruments or indulge in complicated multimiking rigs. Robert Carranza[29] also stresses the speed advantages of this approach, describing how he reused mics from his initial ensemble-miking setup while overdubbing for Jack Johnson's *Sleep Through The Static*: "The mic we used [for each overdub] was usually based on what was nearest and easiest to use. When you work like that, it creates a nice flow...Even if the sound wasn't right, at least the idea was there."

Leaving mics permanently set up during/between sessions is something that many high-profile producers endorse, particularly Daniel Lanois[30], who has often mentioned leaving instruments and their microphones set up as "stations" in the studio, their express purpose being so that musicians never have to wait around to start recording. Nathan Chapman[31] applied the same kind of principle in his own project studio while writing and recording with Taylor Swift: "It's just she and I. I do the engineering myself. It's not hard to do... I have all my different recording chains plugged in and ready to go. I don't ever have to reach up and turn a knob... I have three or four microphones that I just swing into place when I need them and swing back when I don't."

Another application of pre-configured recording hardware is to allow a choice of different room-miking options while overdubbing, a tactic used by Jacquire King[32] and Joe Chiccarelli for their work with Kings Of Leon and The White Stripes respectively. "[The White Stripes] work extremely quickly," recalls Chiccarelli,[33] "and expected me to be ready to record at any time, so I had [six to ten] room ambience microphones set up in the studio. When they began to play a song, I could quickly push up the faders and choose which microphones best suited the songs." Even if you don't require this kind of elaborate multimiking, though, it's excellent practice to have at least a vocal mic primed at all times when recording any group featuring a lead singer, irrespective of what instrument you're actually planning to record at any given moment. Many producers stress this point, especially in pop and hip-hop genres. "I always have an open mic in the studio," says Drake's engineer Noah "40" Shebib,[34] for instance. "When he says he wants to sing or rap or do an overdub, I just hit record."

Something that can seriously send a project off the rails is failing to acknowledge quickly enough that a mistake has been made. It's often better to head straight back to the drawing board on a recording that doesn't feel right, rather than endlessly dickering with details. "If you decide you have made a mistake with it and must rerecord it, that's only a day out of the schedule", says Kevin Killen.[35] "You haven't spent three months trying to make it pop or groove a little better by adding one more thing."

BEYOND THE STUDIO

Within these pages I've concentrated on engineering techniques for studio recording—whether that studio happens to be your bedroom, a purpose-built live room, or a location-recording venue. However, once you get outside such comparatively controlled situations into environments like open-air concerts, live musicals/opera, or audio-plus-video events, the game-plan can change completely. The need to keep mics and stands out of the way of audience members and camera shots can restrict your options tremendously, and you may also have to plan for performers moving around the stage. The sonic contributions of applause and any sound-reinforcement system need to be carefully considered too, and the stakes are frequently higher overall in terms of making sure that equipment is robust and/or easily replaced if a punter spills beer on it or the weather quite literally rains on your parade. Although such scenarios fall outside the scope of this book, the understanding of mic technique you've gained here should provide a solid foundation from which to grapple with the additional demands of live-performance and multimedia recording dates, and I've included some further reading suggestions in this chapter's web resources if you're interested in finding out more about these kinds of applications.

11.3 THE ANALOG MINDSET

Although small studios have now almost universally adopted computer-based recording, there are many producers who feel that the move away from analog recorders has had a negative impact on music-making—irrespective of any sonic concerns. Part of it is a question of session pacing. "One of the blessings about analog was that it was a slower process," explains Ken Caillat.[36] "While I'm rewinding I have to fill in the time, so it's like... 'Hey, Lindsey [Buckingham of Fleetwood Mac], on that second verse you started doing this syncopated thing...it was really cool.' He'd say, 'Really, could you play that for me?' I'd stop and play him back the part, and he'd say, 'Wow...I didn't realize that.' So I'd say, 'Do that some more.' You have all this talking. But in [a DAW] it's just, 'Boom, boom, boom. Start again.' Stop. Just stop. Let me talk to them. I want to bring talking and creativity back into it." Robert Carranza[37] felt the same when working on tape for Jack Johnson's *Sleep Through The Static*: "These are little moments of quiet in the recording process, and they made us pay more attention. The pauses forced me to refocus and get my thoughts together, and pay attention to what was happening next."

Carranza also cites the lack of a screen as a benefit: "[With DAWs] you start listening with your eyes. Working with tape forced us to sit back and really listen to takes and to what was good or not in terms of feel, and be a little bit more forgiving of little mistakes that happen." Jacquire King[38] concurs: "I have found that when I'm recording to tape, the musicians' performances are more focused and inspired....They're not looking at a screen, so it's all about using your ears. Making records with your eyes, as is a danger with [a DAW], is really a poor approach."

Philippe Zdar[39] also feels that the nondestructive nature of digital recording removes an important element of risk from the session dynamic: "The problem with [using a DAW] is that it's easy to forget to take risks. But you have to take risks...I hate the comfort and the safety net that digital provides." The whiff of danger in the air concentrates the minds of the musicians, feels Mike Vernon:[40] "For me, analogue makes you focus more on what you're doing. You have to actually give the performance of your life to be able to get the result. When you work digitally, you have so much more scope, and if you make a mistake, it can be corrected at the drop of a mouse."

"Although analog recording is rarely practical for most small-studio engineers, there's a lot to be said for retaining an analog attitude."

Nigel Godrich[41] expresses probably the most common complaint of all about computer-based recording, however: "When we first got [into] it, I was trying to do this and that with it, and ended up sitting in front of the thing for two days. And everybody gets pissed off, and you lose the feeling that you're excited about something. The trick is to do something fast enough and then keep going so you can't get bored with the thing you've just done...The best times have all been really fast, with everyone I've worked with."

So although analog recording is rarely practical for most small-studio engineers, there's a lot to be said for retaining an analog attitude. "[Don't] allow yourself to be tied to the ways in which the software tries to get you to work," advises Mike Poole.[42] "There will be always be younger people coming along that will...adapt DAWs to work for them, rather than allow the computerized, linear way of looking at and organizing music to become a stumbling block. They will just hit Record and Play and will be able to work beyond what the medium expects them to do." This is another reason why I think stand-alone digital multitrackers can still be tremendously useful for small-studio tracking sessions, even when your main system is computer based, because they don't offer as many distracting options during the earliest stages of the creative process.

But it's not just the recording medium that affects the workflow, because the ergonomics of other analog studio hardware can also affect your workflow. " When I'm standing at [an analog] desk," continues Mike Poole,[43] "my hands will reach intuitively for the EQ, and without even thinking about it I may at the same time do a fader move with my other hand...instantly and purely on feel. By contrast, with the digital devices, even with big controllers, I find I need to think in a very linear fashion. It's very difficult to ride and tweak things at the same time." This is partly why, in my opinion, small computer-based studios frequently benefit from having a small analog console for tracking purposes, especially as it makes it easier to set up true zero-latency monitoring (as discussed in Part 1) and to ride input levels while recording (see Section 10.5.2).

PREPARING MULTITRACKS FOR MIXDOWN

Although most small-studio engineers tend to mix their own recordings these days, if you're submitting your multitracks to a third party for mixdown, there are a few things you can do to help them make the best job of it. The biggest advice I can give is to finish making as many decisions as you can about the arrangement. "My goal," says Glen Ballard,[44] for instance, "is always to give something to a mixer where he can push up the faders and it's reasonably close to the architecture that I want in terms of the arrangement. If you give them that to start with, it frees them up to really focus on detailing the sounds and making everything sound great. I try not to add to their burden the idea that they've got to arrange the song as well." Mix engineer Randy Staub[45] reinforces Ballard's message: "I always tell people that you have to create the sound of the record that you want right from the beginning, so that when the mixer first pushes up the fader, the record is there. Everything sounds the way it should, the arrangements are good, the emotion in there, the balance is there, and as a mixer your job is to take it to another level. You pump it up and you make it sound larger than life. This rather than trying to create something that wasn't there to begin with."

Staub[46] also highlights another vital point: "It's important to send clear, clean sessions. We don't need playlists, we don't need stuff hidden. When I open a session, everything that the artist and producer want in there should come up immediately, so I can see it straight away." It also makes sense to bounce down edited tracks to prevent things being moved inadvertently during the mixing process—as David R Ferguson[47] did when preparing his Johnny Cash multitracks for mixdown, for example, and which Fraser T. Smith[48] likes to do as a matter of course. You can also save a mix engineer time by printing the returns of any favorite effects as audio so that they don't have to be recreated from scratch, and you might also submix multiple layered parts (such as backing vocals) to a single stereo file for similar reasons. "I am not going to mix two hundred tracks on the desk for you," remarks Michael Brauer,[49] "so the best thing for you to do is to give me the stereo blends that you like…I'm happy to mix sixteen tracks. The simpler the better. I don't want to have to fight my way through a mix."

If you're sending raw digital audio files, rather than a DAW project file, try to name the tracks sensibly and make sure they all start at exactly the same time so that they line up properly in the mixing system. I'd also suggest accompanying them with your most recent rough mix of the material and a document containing as much as possible of the following information:

- the tempo and key of the music;
- the sample-rate and bit-depth you recorded at;
- full details of any commercial releases that you'll be referencing the final mix against;
- your own contact details so that the mix engineer can easily make further inquiries if necessary.

11.4 USING COMMERCIAL STUDIOS

While it's perfectly possible to create release-quality productions without ever setting foot inside a commercial studio, that's no reason to disregard the idea of paying for time at a professional facility where budget allows. A studio with a great selection of instruments and microphones will make it easier to get decent raw sounds, for example, and the experience of the in-house staff can save you hours of experimentation with studio layout on ensemble-recording sessions (as mentioned in Section 10.3.3). In addition, the ability to record and monitor in purpose-designed acoustic environments is another massive advantage, and the availability of separate isolation booths and well-specified foldback facilities makes it much easier to maintain an ensemble vibe in situations where you require maximum mixdown flexibility.

> "Although any self-respecting engineer will endeavor to adapt their working methods to their client's preferences, it tends to save time if this doesn't demand a complete 'U'-turn relative to their own natural recording style."

If you do decide to go down the pro-studio route, however, it's vital that you find a facility that really suits your needs, so check over the contenders carefully and ask plenty of pointed questions before booking. Get a list of the instruments and mics, and confirm that they'll actually be available for your session date—the last thing you want is to be greeted by "Oh, those U87s are out with the location rig today" when you've planned your session around them! Can you leave mics and instruments set up overnight without them being disturbed? Is there enough redundancy in the studio setup, and are there technicians on site capable of fixing common faults so that work can continue without major disruption? "I like to work fast and get good momentum and it can be a bit frustrating if you're constantly coming up against technical worries and mechanical issues," comments Tony Hoffer.[50] "I learned early on that you've got to find the studios that have good maintenance."

If an engineer is included with the rate, find out how long they've been working for the studio, listen to some of their previous work, and preferably try to have a quick chat with them before booking. Ask them how they'd suggest recording your project in their environment, and ask yourself whether their instincts chime with yours—particularly if you're planning on capturing a full ensemble. Although any self-respecting engineer will endeavor to adapt their working methods to their client's preferences, it tends to save time if this doesn't demand a complete "U"-turn relative to their own natural recording style. If only an assistant is provided, then ask to send them a mic list and setup diagram in advance of the session, so that they can prepare the studio for you. That way you can hopefully start serious work the moment you walk through the door.

And it should probably go without saying by now that you and the musicians should be as well prepared as you can be before the session, in order to make the most of your expenditure once the meter's running. People more familiar with project-studio workflow are frequently caught off-guard by how swiftly a well-managed commercial room can operate. Don't say I didn't warn you!

CUT TO THE CHASE

- When planning the engineering for a complete production, it's a good idea to start by considering how you'd record the whole arrangement at once in a single room. Although this is rarely feasible in practice for most small-studio productions, sticking close to that kind of setup will usually improve the humanity and musicality of your end product. Scarcity of personnel and equipment may prevent the "all in one go" recording approach, but an equally good reason to deviate from it is if you need to match the stamina and work rates of different performers, or if the musical material has yet to be finalized at the time of recording.

- Clear written work/time plans will help you make more efficient progress, and you should do your best to manipulate the session scheduling to reduce total setup time over the lifetime of the project. Fewer, longer sessions are sensible in this respect, as is any workflow that allows you to leave equipment constantly primed for instant use.
- Tracking lead vocals early on in the production process can help you build a more mixable arrangement. Just don't settle for anything less than a compelling performance, whether or not you consider these early vocals "final," because their attitude will powerfully influence the way the whole project develops.
- Owners of small studios shouldn't reject out of hand the idea of booking commercial studio time, especially when recording ensembles. However, you need to ask the right questions when deciding which studio to book, as well as making sure that everyone involved with the session is well prepared for the speedy workflow of well-run commercial facilities.

Assignment

What are you looking at me for? You're the boss now—you call the shots!

WEB RESOURCES

On this book's companion website you'll find a selection of resources to support this chapter, including:

- links to a range of articles describing real-world ensemble recordings in multi-room scenarios, complete with audio demonstration files;
- suggested further reading about recording live concerts and multimedia events.

http://www.cambridge-mt.com/rs-ch11.htm

CHAPTER 12

Conclusion

Most of the best-known recording engineers learnt their craft by progressing through the ranks of some professional studio or other, carefully observing the work of their elders and betters. These days, however, the network of high-level recording facilities that supported this informal but effective apprenticeship system is a shadow of its former self, and only a lucky few are able to acquire a broad base of recording skills in that way. "With the studio culture disappearing, nobody is learning any more how to record properly," says Randy Staub.[1] "People buy studio equipment, put it in their house, and immediately think they're a recording engineer. But a lot of the stuff that I get in to mix has been recorded horrendously." This book has been my own response to this trend, and I hope it will help provide self-starting engineers with something akin to a "virtual apprenticeship"—a solid foundation of all-purpose recording technique combined with real-world insights from a wide range of top-level professionals.

The good news for small-studio operators, though, is that if you can master the fundamentals I've laid out here, there's really no reason why a slim budget should prevent you from putting together high-quality productions. "I've mixed stuff that's been done on really cheap equipment that sounds gorgeous," says Dave Jerden,[2] for instance, "and [other] stuff that's been done on all the proper high-end gear, and it sounds like crap. It's all in how you record it." Low-budget scenarios can give you a creative edge, in fact, says John Cuniberti:[3] "The project studio is where you can outdo the guys who are strapped into $2000-a-day rooms filled with expensive microphones they feel they *have* to use. It's very difficult to go into a big room with big money and high expectations and be experimental."

I've also touched on many broader issues of music production in these pages, in recognition of the fact that small-studio operators frequently need to fill the role of producer on low-budget projects. Inevitably, though, I've only been able to graze the surface of that enormous subject in a book about audio engineering. The best teacher there is experience, though, so up-and-coming engineers should seek out good experience wherever they can find it, whether it pays the bills or not. "Every great producer has had to go out on a limb and do projects where they weren't getting paid," says Justin Niebank.[4] "You should view every record you make as an investment in your future. That's really the bottom line." Even dreadful session experiences can yield positive experience, says Andy Bradfield:[5] "The way to view those kinds of sessions is that they're teaching you what *not* to do."

The main thing is that you should never stop learning in this business. Here's Al Schmitt,[6] who probably has more right than any of us to rest on his laurels: "When I go home at night, I'll think…'What did I learn today that I didn't know?' I consciously do that…You have to keep learning." "Never think you know everything," adds Toby Wright.[7] "The minute you do, you don't!" And it's not just a question of studio skills either, because John Simon[8] advises continually widening your musical tastes beyond your stylistic comfort zone, while Joe Chiccarelli[9] extols the virtues of a broader knowledge of art, literature, and psychology as well.

And, of course, if you get the chance to learn directly from the top names, then grasp that opportunity with both hands! It always amazes me when walking around the big music-industry trade fairs how few people make the effort to track down and ask a question of their favorite engineers and producers. In my experience, these people all acknowledge the gift they received from their own mentors, and can be extremely gracious and helpful when approached courteously.

Finally, remember that the way you work will influence the music itself—or, as John Hampton[10] puts it: "Fun sticks to the tape. If you're not having fun, that sticks to the tape too."

APPENDIX 1
Music Studios and the Recording Process: An Overview

Throughout most of this book I make the assumption that you have a basic overview of what the music-production process entails. However, because I can't *rely* on that, I wanted to lay out exactly what background knowledge I'm expecting of the reader, and in the most condensed form I can. (Health Warning: Anyone with a strong allergic reaction to sweeping generalizations should top up their medication now!)

I've put important pieces of technical jargon in bold-face to make them stand out. It's a fact of life that every engineer applies studio terminology slightly differently, so I hope that clarifying my own usage will help minimize confusion.

SOUND IN DIFFERENT FORMS

Sound is a series of pressure waves moving through the air at the speed of sound (roughly 343 meters/second). Any vibrating object can create these pressure waves, and when they reach your eardrums, those vibrate in sympathy, triggering the hearing sensation. Transducers such as pickups and microphones can convert air-pressure waves, or the vibrations which triggered them, into an electrical signal, representing their movements over time as voltage fluctuations—often displayed for inspection purposes as a waveform of voltage level (vertical axis) against time (horizontal axis). Other transducers can convert this electrical signal back into air-pressure waves for listening or **monitoring** purposes (via loudspeakers or headphones) or into some other form for storage/replay (e.g., the variable groove depths on a vinyl LP or the variable flux levels on a magnetic tape).

Once sounds have been converted to electrical signals, it becomes possible to process and combine them using all manner of electronic circuitry. In addition, the voltage variations of an electrical signal can also be represented as a stream of numbers in digital form, whereupon digital signal-processing (DSP) techniques can be applied to them. Transferring any signal between the **analog domain** (of electrical signals, vinyl grooves, magnetic flux, physical vibrations, and pressure waves) and the **digital domain** requires either an analog-to-digital converter (**ADC**) or a digital-to-analog converter (**DAC**). The fidelity of analog-to-digital conversion is primarily determined by two statistics: the frequency with which the analog signal's voltage level is measured (the **sample rate** or **sampling frequency**) and the resolution (or **bit depth**) of each measurement (or **sample**), expressed in terms of the length of the binary number required to store it.

SINEWAVES AND AUDIO FREQUENCIES

If an electrical signal's waveform looks chaotic, what you hear will usually feature noisy sounds, whereas repeating waveform patterns are heard as pitched events such as musical notes or overtones. However, before I say any more about complex sounds, it's useful first to understand the simplest repeating sound wave: a **sinewave tone**. Its pitch is determined by the number of times it repeats per second, referred to as its **frequency** and measured in Hertz (Hz). Roughly speaking, the human ear can detect sinewave tones across a 20 Hz–20 kHz frequency range (the **audible frequency spectrum**). Low-frequency tones are perceived as low pitched, while high-frequency tones are perceived as high pitched. Although a sinewave tone isn't exactly thrilling to listen to on its own, it turns out that all the more interesting musical sounds can actually be broken down into a collection of different sinewave tones. The mixture of different sinewave components within any given complex sound determines its timbre.

One way to examine these sinewave components is to use a **spectrum analyzer**, a real-time display of a sound's energy distribution across the audible frequency spectrum. On a spectrum analyzer, a simple sinewave tone shows up as a narrow peak, while real-world signals create a complex undulating plot. Narrow peaks in a complex spectrum-analyzer display indicate pitched components within the signal, while the distribution of energy across the frequency display determines the timbre of the sound—subjectively darker sounds are richer in low frequencies, whereas bright sounds tend to be strong on high frequencies. Although a single sinewave tone will be perceived as a pitched note, almost all real-world musical notes are actually made up of a **harmonic series** of related sinewaves. The most low-frequency of these, the **fundamental**, determines the perceived pitch, while a series of **harmonics** at multiples of the fundamental's frequency determine the note's timbre according to their relative levels.

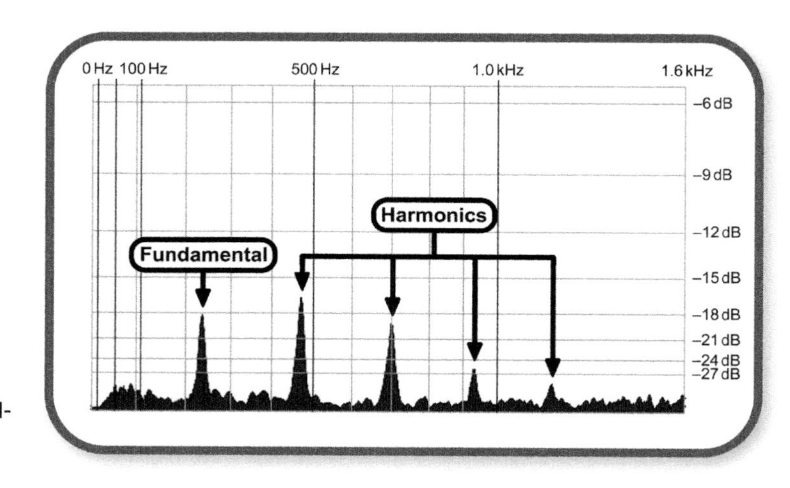

FIGURE A.1
On this spectrum-analyzer display you can clearly see the fundamental and first four harmonics of a single pitched note projecting out of the full-frequency background noise.

LOGARITHMIC SCALES FOR LEVEL & PITCH

Human hearing perceives both level and pitch in a roughly logarithmic way—in other words, we compare levels and pitches in terms of ratios. For example, the perceived pitch interval between a note with its fundamental at 100 Hz and a note with its fundamental at 200 Hz is the same as that between notes with fundamentals at 200 Hz and 400 Hz. Similarly, when dealing with sound in electrical form, the perceived volume difference on playback between signals peaking at 100 mV and 200 mV is roughly similar to that between signals peaking at 200 mV and 400 mV. In recognition of this, both pitch and signal level measurements are frequently made using a logarithmic scale. In the case of pitch, this is done in terms of traditional musical intervals: e.g., 200 Hz is an octave above 100 Hz, 200 Hz is an octave below 400 Hz, and so on. In the case of signal levels this is done using **decibels** (dB): e.g., 200 mV is 6 dB higher in level than 100 mV, 200 mV is 6 dB lower in level than 400 mV—or, to express it in another commonly used form, a +6 dB level change ("+6 dB gain") takes 100 mV to 200 mV, whereas a −6 dB level change ("−6 dB gain") takes 400 mV to 200 mV.

On their own, decibel values can only be used to indicate *changes* in signal level, which is why they are often used to label audio "gain controls" (such as faders) that are expressly designed for this purpose. However, it's important to remember that decibels (just like musical intervals) are always relative. In other words, it's meaningless to say that a signal level is "4.75 dB," much as it's nonsensical to say that any isolated note is a major sixth, because the question is "4.75 dB larger than what?" or "a major sixth above what?" Therefore, if you want to state absolute level values in terms of decibels, you need to express them relative to an agreed **reference level**, indicating this using a suffix. Common reference levels used for studio purposes include dBu and dBV (for electrical signals) and dBFS (for digital signals), but those are by no means the only ones out there.

FREQUENCY RESPONSE

Any studio device will alter the nature of sound passing through it in some way, however small, and the nature of any such effect on a signal's frequency balance is commonly expressed in terms of a **frequency response** graph, which shows the gain applied by the device across the frequency range. A device that left the frequency balance completely unchanged would show a straight horizontal frequency-response plot at the 0 dB level. However, real-world equipment deviates somewhat from this ideal **flat response**—indeed, some devices deliberately warp their frequency-response curve for creative purposes.

THE MULTITRACK RECORDING PROCESS

Modern studio production revolves around the concept of **multitrack recording**, whereby you can capture different electrical signals on different recorder **tracks**, retaining the flexibility to process and blend them independently afterwards. Furthermore, multitrack recorders also allow you to **overdub** new signals to additional tracks while listening back to (**monitoring**) any tracks that have already been recorded, which enables complicated musical arrangements to be built up one instrument at a time if required.

An equally important cornerstone of the production process in many styles is the use of **synthesizers** (which generate audio signals electronically) and **samplers** (which can creatively manipulate selected sections of prerecorded audio). These can sometimes mimic the performances of live musicians, but more importantly provide the opportunity to design sounds that reach beyond the realms of the natural. Typically these devices are digitally controlled using **MIDI** (Musical Instrument Digital Interface) messages, which can be programmed/recorded in multitrack form, edited, and replayed using a **MIDI sequencer**.

Although the production workflow in different musical styles can contrast radically, many people in the studio industry find it useful to discuss the progress of any given project in terms of a series of notional "stages." Everyone has a slightly different view of what constitutes each stage, what they're called exactly, and where the boundaries are between them, but roughly speaking they work out as follows:

- **Preproduction & Programming**. The music is written and arranged. Fundamental synth/sampler parts may be programmed at this stage and musicians rehearsed in preparation for recording sessions.
- **Recording** (or **Tracking**). The instruments and vocals required for the arrangement are recorded, either all at once or piecemeal via overdubbing. Audio editing and corrective processing may also be applied during this stage to refine the recorded tracks into their final form, in particular when pasting together (**comping**) the best sections of several recorded **takes** to create a single **master take**. MIDI-driven synth and sampler parts are **bounced down** to the multitrack recorder by recording their audio outputs.
- **Mixing** (or **Mixdown**). All the recorded tracks are balanced and processed to create a commercial-sounding stereo mix.
- **Mastering**. The mixdown file is further processed to adapt it to different release formats.

The most common professional studio setup for the recording stage involves two separate rooms, acoustically isolated from each other: a **live room** where musicians perform with their instruments; and a **control room** containing the bulk of the recording equipment, where the recording engineer can make judgments about sound quality without the direct sound from the performers interfering with what he's hearing from his monitoring loudspeakers (or **monitors**). Where several performers are playing together in the live room, each with their own microphone, every mic will not only pick up the sound of the instrument/voice it's pointing at, but will also pick up some of the sound from the other instruments in the room—something variously referred to as **spill**, **leakage**, **bleed**, or **crosstalk**, depending on who you speak to! In some studio setups, additional sound-proofed rooms (**isolation booths**) are provided to get around this and improve the **separation** of the signals.

AUDIO SIGNALS AND MIXERS

A typical multitrack recording session can easily involve hundreds of different audio signals. Every audio source (microphones, pickups, synths, samplers) needs routing to its own track of the multitrack recorder, often through a **recording chain** of studio equipment designed to prepare it for capture. Each playback signal from the recorder will pass through its own **monitoring chain**, being blended with all the other tracks so that you can evaluate your work in progress via loudspeakers or headphones. Additional **cue/foldback mixes** may be required to provide personalized monitoring signals for each different performer during a recording session. Further mixes might also feed external (or **outboard**) effects processors, the outputs of which must be **returned** to the main mix so you can hear their results.

The way studios marshal all these signals is by using **mixers** (aka **mixing desks**, **boards**, or **consoles**). At its most basic, a mixer accepts a number of incoming signals, blends them together in some way, and outputs the resulting blended signal. Within the mixer's architecture, each input signal passes through its own independent signal-processing path (or **channel**), which is furnished with a set of controls (the **channel strip**) for adjusting the level and sound character of that signal in the mixed output. In the simplest of mixers, each channel strip may have nothing more than a fader to adjust its relative level for a single output mix, but most real-world designs have many other features besides this:

- If the **main/master mix** output is stereo (which it usually will be) then each mono channel will have a **pan control** (or **pan pot**) which adjusts the relative levels sent from that channel to the left and right sides of the main mix. If the mixer provides dedicated stereo channels, these may have a **balance control** instead, which sets the relative level of the stereo input signal's left and right signal streams.
- An independent **monitor mix** or **control-room mix** may be available for your studio loudspeakers. Although this will usually receive the master mix signal by default, you can typically also feed it with any subset of the input signals for closer scrutiny by activating per-channel **solo** buttons.
- In addition to the faders that set each input signal's level in the main mix, there may be controls for creating further **auxiliary mixes** too—perhaps labeled as **cue sends** (for the purposes of foldback) and **effects sends** (for feeding external effects processors).
- There may be buttons on each channel strip that allow you to disconnect that channel from the main mix, routing it instead to a separate **group** or **subgroup** channel with its own independent output. This provides a convenient means of routing different input signals to different tracks of the multitrack recorder and of **submixing** several input signals together onto a single recorder track.
- Audio metering may be built in, visually displaying the signal levels for various channels as well as for the group, monitor, and master mix signals.

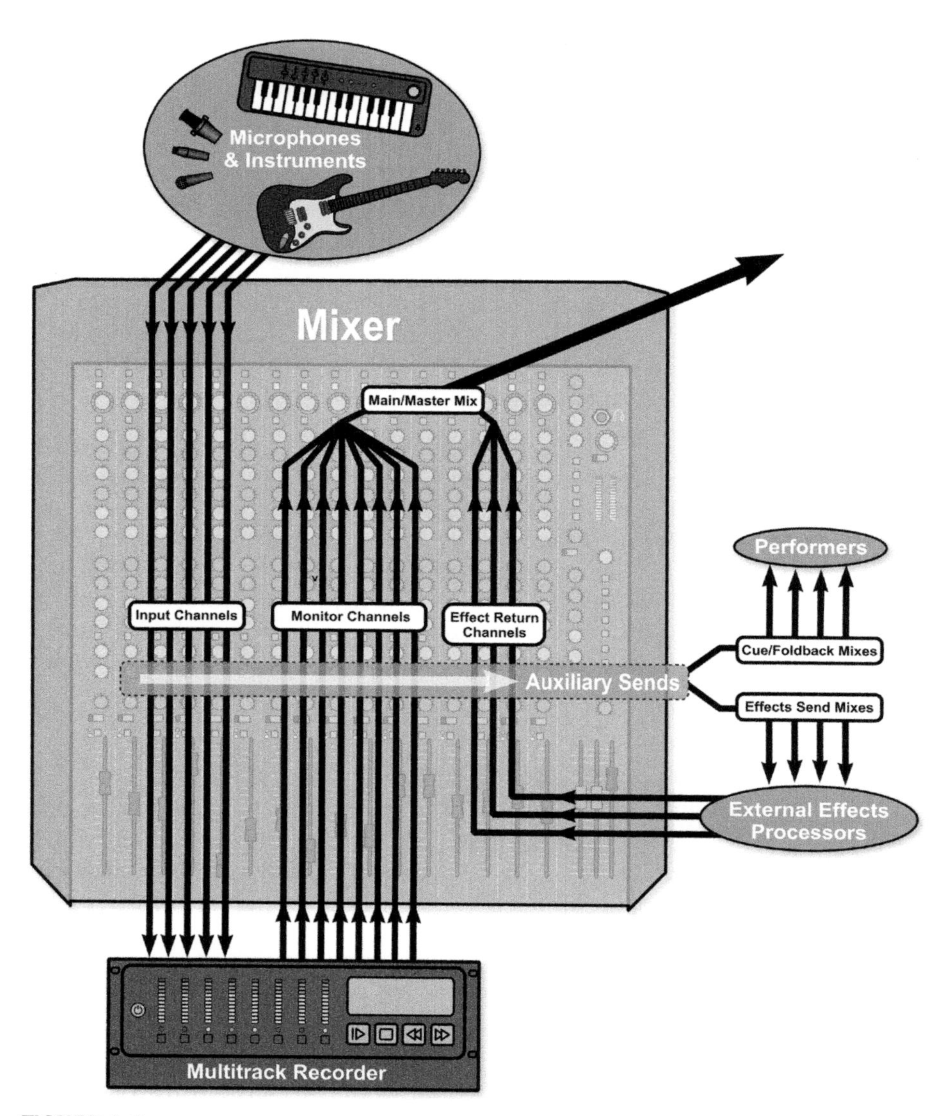

FIGURE A.2
Some of the basic functions of a mixer in the recording studio. Signals from microphones and instruments are relayed through the mixer's input channels to a multitrack recorder. Playback signals from the multitrack recorder are passed through the mixer's monitor channels on their way to the stereo main/master mix. Auxiliary send facilities on each channel allow independent mixes to be sent to performers in the studio (cue/foldback mixes) and to external effects processors (effects send mixes), the latter returning their signals to the main/master mix.

Mixer channels that are conveying signals to the multitrack recorder for capture are often referred to as **input channels**, whereas those which blend together the multitrack recorder's monitor outputs and send them to your loudspeakers/headphones are frequently called **monitor channels**. Some mixers just have a bunch of channels with identical functionality, and leave it up to you to decide which to use as input and monitor channels, while others have dedicated sections of input and monitor channels whose channel-strip facilities are specifically tailored for their respective tasks. Another design, the **in-line mixer**, combines the controls of both an input channel and a monitor channel within the same channel strip. This is popular in large-scale studio setups because it creates a physically more compact control layout, provides ergonomic benefits for advanced users, and allows the two channels to share some processing resources. (Plus there's the added benefit that it confuses the hell out of the uninitiated, which is always gratifying…)

Another specialized mixer, called a **monitor controller**, has evolved to cater for studios where several different playback devices and/or loudspeaker systems are available. It typically provides switches to select between the different audio sources and speaker rigs, as well as a master volume control for whichever speaker system is currently active.

SIGNAL PROCESSING

Beyond simply blending and routing signals, multitrack production invariably involves processing them as well. In some cases this may comprise nothing more than "preamplifying" the signal to a suitable level for recording purposes, but there are several other processes that are frequently applied as well:

- Spectral Shaping: Audio **filters** and **equalizers** may be used to adjust the levels of different frequencies relative to each other.
- **Dynamics**: Tools such as **compressors**, **limiters**, **expanders**, and **gates** allow the engineer to control the level-contour of a signal over time in a semi-automatic manner.
- **Modulation** Effects: A family of processes which introduce cyclic variations into the signal. Includes effects such as **chorusing**, **flanging**, **phasing**, **vibrato**, and **tremolo**.
- Delay-Based Effects: Another group of processes which involve overlaying one or more echoes onto the signal. Where these effects become complex, they can begin to artificially simulate the reverberation characteristics of natural acoustic spaces.

In some cases, such processing may be **inserted** into the signal path directly—rather than being fed from an independent effects send and then returned to the mix (a **send–return** configuration).

REAL-WORLD STUDIO SETUPS: SOMETHING OLD, SOMETHING NEW

Although every recording studio needs to route, record, process, and mix audio signals, every engineer's rig ends up being slightly different, either by virtue of the equipment chosen, or because of the way the gear is hooked up. One defining feature of many systems is the extent to which digital technology is used. While there are still some people who uphold the analog-only studio tradition of the 1970s, the reliability, features, and pricing of DSP processing and data storage have increasingly drawn small studios toward hybrid systems. Standalone digital recorders and effects processors began this trend within otherwise analog systems, but the advent of comparatively affordable digital mixers and "studio in a box" digital **multitrackers** during the 1990s eventually allowed project studios to operate almost entirely in the digital domain, converting all analog signals to digital data at the earliest possible opportunity and then transferring that data between different digital studio processors losslessly. These days, however, the physical hardware units of early digital studios have largely been superseded by Digital Audio Workstation (**DAW**) software, which allows a single general-purpose computer to emulate all their routing, recording, processing, and mixing functions at once, connecting to the analog world where necessary via an **audio interface**: a collection of audio input/output (**I/O**) sockets, ADCs, and DACs.

A similar trajectory can be observed with synths and samplers. Although early devices were all-analog designs, microprocessors quickly made inroads during the 1980s as the MIDI standard took hold. The low data bandwidth of MIDI messages and the plummeting price of personal computing meant that computer-based MIDI sequencing was already the norm 20 years ago, but in more recent years the synths and samplers themselves have increasingly migrated into that world too, in the form of software **virtual instruments**. As a result, most modern DAW systems integrate MIDI sequencing and synthesis/sampling facilities alongside their audio recording and processing capabilities, making it possible for productions to be constructed almost entirely within a software environment. In practice, however, most small studios occupy a middle ground between the all-analog and all-digital extremes, combining old and new, analog and digital, hardware and software—depending on production priorities, space/budget restrictions, and personal preferences.

SETTING UP A SMALL STEREO MONITORING SYSTEM

When it comes to audio engineering, the equipment you use for monitoring purposes is vital—the better you can hear what you're doing, the faster you'll learn how to improve your sonics. Loudspeakers are usually preferable over headphones, because the latter struggle to reproduce low frequencies faithfully and create an unnaturally wide "inside your head" stereo sensation. That said, if you've got less than $750 (£500) to spend, my view is that top-of-the-range studio headphones will give you better results for the money than any combination of speakers and acoustic treatment in that price range. Whatever monitoring system you use, it will only become useful to you once you know how your favorite productions sound through it.

In general, it's best to choose full-range speakers designed specifically for studio use, rather than general-purpose hi-fi models which tend to flatter sounds unduly. The speakers themselves should be firmly mounted (preferably on solid stands) and carefully positioned according to the manufacturer's instructions—setting up across the shorter dimension of a rectangular room will usually give the better sound. To present a good stereo image, the two speakers and the listening "**sweetspot**" should form an equilateral triangle, with the tweeter and woofer of each speaker vertically aligned and both tweeters angled toward the listener's ears. Speaker systems with built-in amplification ("active" or "powered" speakers) are not only convenient, but also offer sonic advantages because of the way the amplifier(s) can be matched to the specific speaker drivers.

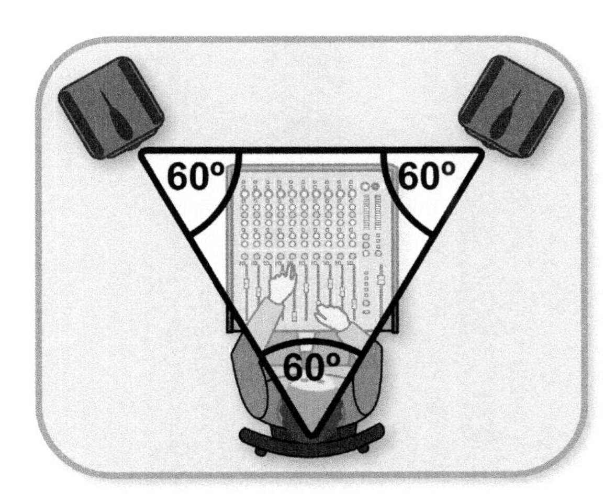

FIGURE A.3
For decent stereo imaging on loudspeakers, the listener's ears and the speakers should be positioned in an equilateral triangle formation, as shown here.

Although most small-studio speaker systems are **nearfield** models which are designed to operate within about 6 feet (2 m) of the listener, room acoustics can still have an enormous impact on their tone and fidelity. As such, you should probably spend roughly the same amount of money on acoustic treatment as on the speaker system itself if you're going to get the best bang for your buck. High-frequency reflections and room resonances can be absorbed very effectively with judiciously applied acoustic foam, but taming low-end problems requires more expensive and bulky acoustics measures.

WEB RESOURCES

On this book's companion website you'll find a selection of resources to support this overview, including:

- an interactive version of this chapter peppered with dozens of links to more detailed further reading and some supporting audio/video materials;
- links to some well-maintained audio glossaries in case you need to look up any technical terms;
- examples of entry-level recording systems suitable for use with this book, in case you don't already have access to any studio gear.

http://www.cambridge-mt.com/rs-app.htm

Who's Who: Selected Discography

Here's a list of all the producers I've cited in this book, along with some of the most influential recordings they've been involved with.

ABBISS, JIM

Arctic Monkeys: *Whatever People Say I Am, That's What I'm Not*; Ladytron: *Witching Hour*; Sneaker Pimps: *Becoming X*; Editors: *The Back Room*; Kasabian: *Kasabian, Empire*; Adele: *19, 21*; KT Tunstall: *Tiger Suit*; The Kooks: *Junk Of The Heart*; Stereophonics: *Keep Calm and Carry On*.

ADAMS, ROBBIE

U2: *Achtung Baby, Zooropa*; Smashing Pumpkins: *Adore*; Naked: *Naked*; Don Piper: *A Don Piper Situation*; Patty Griffin: *Impossible Dreams*.

ADDEY, MALCOLM

Judy Garland: *The London Studio Recordings 1957–1964*; Cliff Richard: *Cliff/The Young Ones, Summer Holiday*; The Beatles: *Sgt. Pepper's Lonely Hearts Club Band*; Kool & The Gang: *Kool & The Gang*.

AFANASIEFF, WALTER

Mariah Carey: *Without You, Never Forget You, Anytime You Need A Friend, All I Want For Christmas Is You*; Ricky Martin: *Ricky Martin*; Celine Dion: *My Heart Will Go On*; Destiny's Child: *Survivor*.

AGNELLO, JOHN

Sonic Youth: *Rather Ripped*; Dinosaur Jr.: *Beyond*; Kurt Vile: *Smoke Ring For My Halo, Wakin On a Pretty Daze*; Jennifer O'Connor: *Here With Me*; The Kills: *No Wow*.

AHERN, BRIAN

Neil Young: *Live At The Riverboat 1969*; Anne Murray: *What About Me, This Way Is My Way, Snowbird*; Emmylou Harris: *Pieces Of The Sky, Elite Hotel, Blue Kentucky Girl, Evangeline*.

AINLAY, CHUCK

Trisha Yearwood: *Where Your Road Leads, Thinkin' About You, Everybody Knows, How Do I Live, Real Live Woman*; George Strait: *Somewhere Down In Texas, Pure Country, Blue Clear*

Sky, Carrying Your Love With Me; Vince Gill: *High Lonesome Sound, The Key*; Dixie Chicks: *Wide Open Spaces*; Mark Knopfler: *Sailing To Philadelphia, The Ragpicker's Dream, Wag The Dog, Metroland*; Patty Loveless: *If My Heart Had Windows, Honky Tonk Angel.*

ALBINI, STEVE

The Pixies: *Surfer Rosa*; Nirvana: *In Utero*; Bush: *Razorblade Suitcase*; PJ Harvey: *Rid Of Me*; Jimmy Page & Robert Plant: *Walking Into Clarksdale*; Manic Street Preachers: *Journal For Plague Lovers*; The Wedding Present: *El Rey*; The Breeders: *Mountain Battles*; The Stooges: *The Weirdness*; Cheap Trick: *Rockford.*

ALLEN, BEN

Bombay Bicycle Club: *So Long, See You Tomorrow, A Different Kind Of Fix*; Lenka: *Two*; Christina Aguilera: *Back To Basics.*

ASTLEY, JOHN

The Who: *Who Are You*; Eric Clapton: *Just One Night.*

AUSTIN, DAN

Cherry Ghost: *A Thirst For Romance*; Doves: *Kingdom of Rust*; Arcane Roots: *Blood & Chemistry*; Airship: *Stuck In This Ocean.*

AVRON, NEAL

Fall Out Boy: *From Under The Cork Trees, Infinity On High, Folie A Deux*; Weezer: *Make Believe*; Everclear: *So Much For The Afterglow*; Linkin Park: *Minutes To Midnight*; The Wallflowers: *Bringing Down The Horse*; Lifehouse: *No Name Face*; Yellowcard: *Ocean Avenue.*

BAKER, ROY THOMAS

Queen: *Queen, Queen II, Jazz*; Ozzy Ozbourne: *No Rest For The Wicked*; The Smashing Pumpkins: *Zeitgeist, American Gothic*; The Darkness: *One Way Ticket To Hell And Back*; Chris de Burgh: *Spark To A Flame*; Mötley Crüe: *Too Fast For Love.*

BALLARD, GLEN

Wilson Phillips: *Wilson Phillips*; Alanis Morissette: *Jagged Little Pill, Supposed Former Infatuation Junkie*; Michael Jackson: *Thriller, Bad, Dangerous*; POD: *Testify*; Anastacia: *Anastacia*; Goo Goo Dolls: *Let Love In*; Christina Aguilera: *Stripped*; No Doubt: *Return Of Saturn*; Aerosmith: *Nine Lives*; Paula Abdul: *Forever Your Girl.*

BARBE, DAVID

Deerhunters: *Halcyon Digest*; Patterson Hood: *Heat Lightning Rumbles In The Distance*; Dennis Ellsworth: *Dusk Dreams*; Drive-By Truckers: *Go-Go Boots, The Big To-Do, Brighter Than Creation's Dark*; New Madrid: *Yard Boat*; Dead Confederate: *In The Marrow.*

BARBIERO, MICHAEL

Cutting Crew: *Broadcast*; Peter Frampton: *Fingerprints*; Blues Traveler: *Run-around*; Ziggy Marley: *Fallen Is Babylon*; Whitney Houston: *Whitney*.

BARRESI, JOE

Queens Of The Stone Age: *Queens Of The Stone Age, Lullabies To Paralyze*; Tool: *10000 Days*; The Melvins: *Stag, Honky*; Hole: *Celebrity Skin*; Limp Bizkit: *Chocolate Starfish & The Hotdog Flavored Water*; The Lost Prophets: *Start Something*; Skunk Anansie: *Stoosh*; Bad Religion: *The Empire Strikes First, New Maps Of Hell, The Dissent Of Man*.

BHASKER, JEFF

Kanye West: *808s & Heartbreak, My Beautiful Dark Twisted Fantasy*; Jay-Z: *The Blueprint 3*; Alicia Keys: *The Element Of Freedom*; Mary J. Blige: *Stronger With Each Tear*; Snoop Dogg: *Doggumentary*; Beyoncé: *4*; Lana Del Rey: *Born To Die*; Fun: *Some Nights*; Pink: *The Truth About Love*; Eminem: *The Marshall Mathers LP 2*; Dido: *Girl Who Got Away*; Taylor Swift: *Red*; The Rolling Stones: *GRRR!*; Bruno Mars: *Unorthodox Jukebox*.

BISHOP, MICHAEL

Engineer and producer for many Grammy-winning Telarc releases, including *Copland: The Music Of America, Vaughan Williams: A Sea Symphony*, and *Elgar/Britten: Enigma Variations, The Young Person's Guide To The Orchestra, and Four Sea Interludes*, all of which won Best Engineered Classical Album Grammies.

BLAKE, TCHAD

Sheryl Crow: *The Globe Sessions*; Bonnie Raitt: *Souls Alike*; Phish: *Undermind*; Suzanne Vega: *Nine Objects Of Desire, 99.9°F, Beauty And Crime*; Crowded House: *Farewell To The World*; Paul Simon: *Surprise*; Peter Gabriel: *Ovo, Long Walk Home, Up*; The Dandy Warhols: *Come Down, Odditorium*; Neil Finn: *One Nil, Try Whistling This*; Pearl Jam: *Binaural*.

BOTNIK, BRUCE

The Doors: *L.A. Woman*; Eddie Money: *Eddie Money, Life For The Taking*; The Beat: *The Beat, The Kids Are The Same*; Love: *Forever Changes*.

BOTTRILL, BILL

The Traveling Wilburys: *The Traveling Wilburys Volume 1*; Tom Petty: *Full Moon Fever*; Madonna: *Like A Prayer*; The Jacksons: *Victory*; Michael Jackson: *Bad, Dangerous*; Sheryl Crow: *Tuesday Night Music Club, Sheryl Crow*; Shelby Lynne: *I Am Shelby Lynne*; Elton John: *Songs From The West Coast*.

BOYS, JERRY

REM: *Fable Of The Reconstruction*; Everything But The Girl: *Amplified Heart*; The Cat Empire: *Two Shoes*; Steeleye Span: *Parcel Of Rogues*; Buena Vista Social Club: *Buena Vista Social Club*.

BRADFIELD, ANDY

Robbie Robertson: *Contact From The Underworld Of Redboy*; Rufus Wainwright: *Want One, Want Two, Release The Stars*; Josh Groban: *Awake*; Alanis Morissette: *Flavors of Entanglement*; Spice Girls: *Spice*.

BRAUER, MICHAEL

Coldplay: *Parachutes, X&Y, Viva La Vida*; Athlete: *Tourist*; The Fray: *The Fray*; Evans Blue: *The Pursuit Begins When This Portrayal of Life Ends*; Paolo Nutini: *These Streets*; John Mayer: *Continuum*; James Morrison: *Undiscovered*; The Kooks: *Inside In/Inside Out*; My Morning Jacket: *Evil Urges*; Travis: *The Boy With No Name*; Doves: *Kingdom Of Rust*; KT Tunstall: *Drastic Fantastic*; Idlewild: *Warnings/Promises*; Fountains Of Wayne: *Traffic And Weather*; Aimee Mann: *Lost In Space*; Bob Dylan: *Lovesick*.

BRIGHT, MARK

Carrie Underwood: *Some Hearts, Blown Away*; Scotty McCreery: *Clear As Day*; Rascal Flatts: *Rascal Flatts, I Melt, Feels Like Today*.

BRION, JON

Fiona Apple: *When The Pawn, Extraordinary Machine*; Dido: *Safe Trip Home*; Sky Ferreira: *Ghost*; Keane: *Perfect Symmetry*; Kanye West: *Late Registration, Graduation*; Rufus Wainwright: *Rufus Wainwright*; Aimee Mann: *I'm With Stupid, Whatever, Bachelor No. 2*.

BROWN, PHILL

Dido: *Life For Rent*; Paper Aeroplanes: *Little Letters*; Robert Plant: *Nine Lives*; Talk Talk: *Spirit Of Eden, Laughing Stock*; Throwing Muses: *Bright Yellow Gun, University*; Paul Carrack: *Nightbird*; Amazing Blondel: *Mulgrave Street*.

BROWN, TONY

George Strait: *It Just Comes Natural, Troubadour, Here For A Good Time*; Vince Gill: *I Still Believe In You, Let's Make Sure We Kiss Goodbye*; Steve Earle: *Guitar Town*; Patty Loveless: *Honky Tonk Angel*; Lyle Lovett: *Lyle Lovett And His Large Band*; Trisha Yearwood: *Love Songs*; Reba McEntire: *Duets*; Lucinda Williams: *Car Wheels On A Gravel Road*.

BULLOCK, BOB

Shania Twain: *Come On Over, Up!*; Crazy Horse: *Crazy Moon*; Hank Williams Jr.: *Five-O*; Kenny Rogers: *Love Will Turn Around*; Patty Loveless: *Honky Tonk Angel, When Fallen Angels Fly*; George Strait: *Holding My Own, Livin' It Up, Beyond The Blue Neon*; Nancy Griffith: *Little Love Affairs*; Reba McEntire: *What Am I Gonna Do About You, My Kind Of Country*; Waylon Jennings: *Will The Wolf Survive*.

BURNETT, T-BONE

Elvis Costello: *King Of America, Out Of Our Idiot, Spike*; Roy Orbison: *In Dreams, A Black And White Night Live, Mystery Girl, King of Hearts*; The Wallflowers: *Bringing Down The Horse*; Diana Krall: *Glad Rag Doll*.

BUSH, BILLY

Garbage: *Not Your Kind Of People, Absolute Garbage, Beautiful Garbage, Bleed Like Me, Version 2.0*; The Boxer Rebellion: *Promises*; Muse: "Neutron Star Collision"; Snow Patrol: "In The End"; The Naked And Famous: *In Rolling Waves, Passive Me, Aggressive You.*

BUSH, STEVE

Stereophonics: *Word Gets Around, Performance & Cocktails, Just Enough Education To Perform*; Corinne Bailey Rae: *Corinne Bailey Rae.*

CAILLAT, KEN

Fleetwood Mac: *Rumours, Tusk, Mirage*; Colbie Caillat: *Coco, Breakthrough*; Christine McVie: *In The Meantime*; Lionel Richie: *Dancing On The Ceiling.*

CARRANZA, ROBERT

Salvador Santana: *Keyboard City*; Zee Avi: *Zee Avi*; Jack Johnson: *Sleep Through The Static, Jack Johnson En Concert*; The Mars Volta: *The Bedlam In Goliath.*

CHANDLER, STEVE

James King: *The Bluegrass Storyteller, Three Chords And The Truth*; Dry Branch Fire Squad: *Echoes Of The Mountains*; JD Crowe: *Lefty's Old Guitar, Old Friends Get Together*; Mike Scott: *Home Sweet Home, Take Me Lord And Use Me*; Gene Watson: *A Taste For The Truth*; The Happy Goodman Family: *The Final Stand, Set Your Sails*; Hazel Dickens: *By The Sweat Of My Brow*; Keith Whitley: *Sad Songs And Waltzes.*

CHAPMAN, NATHAN

Taylor Swift: *Taylor Swift, Faceless, Speak Now, Red*; The Band Perry: *The Band Perry*; Point Of Grace: *No Changin' Us, Home For The Holidays*; Martina McBride: *Shine*; Jewel: *Sweet And Wild.*

CHERNEY, ED

The Rolling Stones: *Stripped, Bridges To Babylon, No Security*; Bonnie Raitt: *Nick Of Time, Luck Of The Draw, Longing In Their Hearts, Road Tested*; Bob Dylan: *Under The Red Sky, Unplugged.*

CHICCARELLI, JOE

White Stripes: *Icky Thump*; Frank Zappa: *Sheik Yerbouti, Joe's Garage, Tinseltown Rebellion*; My Morning Jacket: *Evil Urges*; The Shins: *Wincing The Night Away*; Counting Crows: *The Desert Life.*

CHURCHYARD, STEVE

The Pretenders: *Learning To Crawl*; Counting Crows: *Recovering The Satellites*; Celine Dion: *Falling Into You*; Ricky Martin: *Vuelve, Almas Del Silencio, Sound Loaded*; Shakira: *Laundry Service*; The Stranglers: *La Folie, Feline*; Big Country: *Wonderland*; Bryan Ferry: *Boys & Girls*; INXS: *Listen Like Thieves*; Kelly Clarkson: *Thankful.*

CLARKE, ALEX

Tom Jones/Mousse T: "Sex Bomb."

CLEARMOUNTAIN, BOB

The Pretenders: *Get Close*; Bryan Adams: *Into The Fire, Reckless, Cuts Like A Knife, So Far So Good, 18 Till I Die, Room Service, 11*; Bruce Springsteen: *Born In The USA*; The Rolling Stones: *Tattoo You*; Bon Jovi: *These Days, Bounce, Crush*; Roxy Music: *Avalon*; David Bowie: *Let's Dance*; INXS: *Kick, Full Moon, Dirty Hearts, Welcome To Wherever You Are*; The Corrs: *Talk On Corners, Forgiven Not Forgotten, Unplugged*; Robbie Williams: *Intensive Care*; Simple Minds: *Once Upon A Time, Black & White 050505, Graffiti Soul*; Sheryl Crow: *Wildflower*; Aimee Mann: *Whatever, Bachelor No. 2*; Rufus Wainwright: *Rufus Wainwright*.

CLIMIE, SIMON

Eric Clapton: *Pilgrim, Back Home*; BB King & Eric Clapton: *Riding With The King*; Michael McDonald: *Motown I, Motown II*; Lara Fabian: *TLFM*.

CLINK, MIKE

Guns N' Roses: *Appetite For Destruction, GNR Lies, Use Your Illusion, The Spaghetti Incident*; Survivor: *Eye Of The Tiger*; Megadeth: *Rust In Peace*; Mötley Crüe: *New Tattoo*; Whitesnake: *Whitesnake*.

COAD, JEZ

Simple Minds: *Black & White 050505, Graffiti Soul, Stranger*; Andrew Strong: *Gypsy's Kiss*; The Surfing Brides: *Sparky's Dinner*.

COSARO, JASON

Duran Duran: *The Biggest And The Best*; Billy Squier: *Icon;* Madonna: *Like A Virgin*; Skylar May: *Married To The Moon*; Soundgarden: *Phantasm, A-Sides, Alive In The Superunknown, Superunknown*; Soulfly: *Soulfly*; Cheap Trick: *Sex, America, Cheap Trick*; Duran Duran: *Thankyou*; Cyndi Lauper: *Seven Deadly Sins And Then Some*; Deep Purple: *The Battle Rages On*; Robert Palmer: *Riptide, Addictions Vol. 1 & 2*; The Ramones: *Brain Drain*; Iggy Pop: *Instinct*; Motörhead: *Rock 'n' Roll*; Steve Winwood: *Back In The High Life*; Cyndi Lauper: *Tru Colors*; Jeff Beck: *Flash*; Madonna: *Like A Virgin*; Chic: *Believer, Take It Off*; Paul Simon: *Hearts And Bones*.

COSTEY, RICH

Muse: *Absolution, Black Holes & Revelations*; Interpol: *Our Love To Admire*; Polyphonic Spree: *Together We're Heavy*; Franz Ferdinand: *You Could Have It So Much Better*; The Mars Volta: *Frances The Mute, Deloused In The Comatorium*; Audioslave: *Audioslave*; Weezer: *Make Believe*; Bloc Party: *Silent Alarm*; Doves: *Some Cities*; My Chemical Romance: *Three Cheers For Sweet Revenge*; Three Days Grace: *Three Days Grace*; Jimmy Eat World: *Futures*; POD: *Payable On Death*; Rage Against The Machine: *Renegades*; Fiona Apple: *When The Pawn*; Jurassic 5: *Quality Control*; Nine Inch Nails: *With Teeth*.

CUNIBERTI, JOHN

Joe Satriani: *Joe Satriani, Time Machine, Cryin', The Extremist, Flying In A Blue Dream;* Jack Miller: *Dead Lock Rock;* Michael Manring: *Thonk;* Lynch Mob: *Sacred Grove;* Dead Kennedys: *Plastic Surgery Disasters.*

DAVIDGE, NEIL

Massive Attack: *Mezzanine, 100th Window, Heligoland;* UNKLE: *Rabbit In Your Headlights, War Stories;* Sinead O'Connor: *She Who Dwells...;* Soundtrack: *Halo 4, Luther.*

DAVIES, RHETT

Genesis: *Selling England By The Pound;* Dire Straits: *Dire Straits;* Bryan Ferry: *Another Time, Another Place, Boys and Girls, Dylanesque, Olympia;* Brian Eno: *Taking Tiger Mountain, Another Green World.*

DAVIS, WYN

Foreigner: *Feels Like The First Time;* Paul Chesne: *White Man's Curse;* Van Morrison: *Moondance;* Dio: *Last In The Line, Angry Machines, Magica, Master Of The Moon;* Ritchie Valens: *La Bamba 87;* Dokken: *Breaking The Chains, Dysfunctional, Erase The Slate, Long Way Home, Hell To Pay, Lightning Strikes Again;* Great White: *Shot In The Dark, Icon.*

DAWSON, SIMON

The Stone Roses: *Second Coming;* John Squire: *Time Changes Everything, Marshall's House;* New Model Army: *Anthology;* T'Pau: *Red;* XTC: *Apple Venus Vol. 1, Wasp Star;* The Darkness: *One Way Ticket To Hell And Back;* Simple Minds: *Graffiti Soul;* New Model Army: *Strange Brotherhood.*

DODD, RICHARD

Uriah Heep: *Raging Silence, Time of Revelation;* Clannad: *Harry's Game, Magical Ring, Pastpresent, Anam;* Leo Sayer: *Silverbird;* Rosemary Clooney: *Nice To Be Around;* Tina Charles: *I Love To Love, Heart & Soul;* George Harrison: *Cloud Nine;* The Traveling Wilburys: *The Traveling Wilburys Vol. 1 & 3;* Freddie Mercury: *Barcelona;* Jeff Lynne: *Armchair Theatre;* Ringo Starr: *Time Takes Time, Weight Of The World;* Roy Orbison: *King Of Hearts;* Joe Cocker: *Night Calls;* Hank Marvin: *Heartbeat;* Tom Petty: *Wildflowers;* Steve Earle: *I Feel Alright;* Keith Urban: *The Ranch;* Green Day: *Nimrod.*

DOLLAR, JONNY

Neneh Cerry: *Raw Like Sushi, Homebrew, Man;* Gabrielle: *Rise;* Eliza Dolittle: *Eliza Dolittle.*

DORFSMAN, NEIL

Dire Straits: *Brothers In Arms, Love Over Gold;* Bruce Hornsby: *Scenes From The South Side;* Paul McCartney: *Flowers In The Dirt;* Bruce Springsteen: *Tracks, The River;* Sting: *Nothing Like The Sun, Brand New Day;* Bjork: *Medulla.*

DOUGLAS, JACK

Aerosmith: *Get Your Wings, Toys In The Attic, Rocks, Draw The Line, Rock In A Hard Place, Honkin' On Bobo*; John Lennon: *Imagine, Double Fantasy*; Cheap Trick: *Cheap Trick, At The Budokan, Standing On The Edge.*

DUDGEON, GUS

Elton John: *Elton John, Madman Across The Water, Honky Chateau, Goodbye Yellow Brick Road, Caribou, Captain Fantastic and the Brown Dirt Cowboy*; David Bowie: *Space Oddity*; Original Soundtrack: *Tommy.*

EASTER, MITCH

REM: *Murmur, Reckoning*; Suzanne Vega: *Solitude Standing*; Ride: *Tarantula*; Pavement: *Brighten The Corners*; Wilco: *Summerteeth*; Velvet Crush: *Melody Freaks, Timeless Melodies, Heavy Changes, Hold Me Up*; Motocaster: *Acid Rock, Stay Loaded*; Dinosaur Jr.: *Whatever's Cool With Me.*

EMERICK, GEOFF

The Beatles: *Revolver, Sgt. Pepper's Lonely Hearts Club Band, Magical Mystery Tour, Abbey Road*; Paul McCartney: *Band On The Run, Flaming Pie*; Elvis Costello: *Imperial Bedroom, All This Useless Beauty*; Badfinger: *No Dice*; Robin Trower: *Bridge Of Sighs.*

ENDINO, JACK

Soundgarden: *Screaming Life*; Mudhoney: *Superfuzz Bigmuff, Mudhoney, My Brother The Cow*; Screaming Trees: *Buzz Factory*; Tad: *God's Balls*; Nirvana: *Bleach, Incesticide*; Afghan Whigs: *Up In It*; Gas Huffner: *Janitors Of Tomorrow*; Supersuckers: *The Smoke Of Hell*; Seven Year Bitch: *Viva Zapata*; Murder City Devils: *Empty Bottles, Broken Hearts*; Black Halo: *The Violent Years*; Zen Guerilla: *Shadows On The Sun*; Therapy: *Shameless.*

EPSTEIN, STEVEN

Produced records for many jazz and classical artists, such as Wynton Marsalis, Joshua Bell, Yo-Yo Ma, The Three Tenors, Isaac Stern, Juilliard String Quartet, Murray Perahia, Zubin Mehta, Emanuel Ax, Midori, Claudio Abbado, Bobby McFerrin, Radu Lupu. Four times winner of the Grammy Award for Classical Producer Of The Year.

EPWORTH, PAUL

George Martin: *In My Life*; Tom Jones: *Reload*; The Futureheads: *The Futureheads*; Bloc Party: *Two More Years, Banquet, Tulips, Silent Alarm*; Maximo Park: *Going Missing*; The Rakes: *Retreat, Capture/Release, 22 Grand Job*; The Rapture: *Pieces Of The People We Love*; Kate Nash: *Made Of Bricks*; Sam Sparro: *Sam Sparro*; Friendly Fires: *Friendly Fires, Pala*: Adele: *21*, "Skyfall"; Florence & The Machine: *Lungs, Ceremonials*; Cee Lo Green: *The Lady Killer*; Wonderland: *Wonderland*; Bruno Mars: *Unorthodox Jukebox*; Paul McCartney: *New*; Primal Scream: *Beautiful Future.*

ERINGA, DAVE

Manic Street Preachers: *Generation Terrorists, Gold Against The Soul, The Holy Bible, Everything Must Go, This Is My Truth Tell Me Yours, Send Away The Tigers, Journal For Plague Lovers, Rewind The Film*; Soup Dragons: *Divine Thing*; Tom Jones: *Reload*; Toploader: *Onka's Big Moka, At The Magic Hotel*; Idlewild: *100 Broken Windows, The Remote Part, Warnings/Promises, Make Another World, Post Electric Blues*; Ash: *Intergalactic Sonic 7s*; Kylie Minogue: *Confide In Me, Impossible Princess*; Suede: *A New Morning*; 3 Colours Red: *If You Ain't Got A Weapon You'll Never Get A Say*; Ocean Colour Scene: *A Hyperactive Workout For The Flying Squad*.

FAULKNER, TONY

Produced records for many classical artists, such as The Tallis Scholars, Allegri String Quartet, Murray Perahia, London Symphony Orchestra, Royal Philharmonic Orchestra, Joanna McGregor, Vladimir Ashkenazy, Evgeny Kissin, Emma Kirkby, Trevor Pinnock, Endellion String Quartet, Placido Domingo, BBC Symphony Orchestra, Colin Davis. Recorded the chart-topping Elektra-Nonesuch production of Gorecki: Symphony No. 3 with Dawn Upshaw.

FERGUSON, DAVID R.

Johnny Cash: *We The People, Water From The Wells Of Home, The Mystery Of Life, American I, II, III, IV & V*; Marty Stuart: *Marty Stuart, Let There Be Country*; John Prine: *Souvenirs, In Spite Of Ourselves*; Mac Wiseman: *Standard Songs For Average People*; Charley Pride: *Pride & Joy, Choices*; Del McCoury Band: *Family Circle*; Ronnie McCoury: *Heartbreak Town, Little Mo McCoury*.

FILIPETTI, FRANK

Foreigner: *Agent Provocateur, Inside Information*; Hole: *Celebrity Skin*; Korn: *Untouchables, Here To Stay*; Carly Simon: *Hello Big Man, Spoiled Girl, Coming Around Again, Film Noir, Bedroom Tapes*; Barbra Streisand: *Higher Ground*; James Taylor: *Hourglass, That's Why I'm Here*; Kiss: *Lick It Up*; Bangles: *Everything*; Survivor: *Too Hot To Sleep*.

FLOOD

Editors: *In This Light And On This Evening*; 30 Seconds To Mars: *This Is War*; Goldfrapp: *Seventh Tree*; PJ Harvey: *White Chalk, Is This Desire, To Bring You My Love*; The Killers: *Sam's Town*; U2: *The Joshua Tree, Achtung Baby, Pop, Zooropa, How To Dismantle An Atomic Bomb*; Smashing Pumpkins: *Mellon Collie And The Infinite Sadness, Machina/Machines Of God, Adore*; Nine Inch Nails: *Downward Spiral, Pretty Hate Machine*; Depeche Mode: *Violator, Pop Will Eat Itself, Songs Of Faith And Devotion*; Nick Cave And The Bad Seeds: *From Her To Eternity, The Firstborn Is Dead, Kicking Against The Pricks, Your Funeral My Trial, Tender Prey, The Good Son*; Erasure: *Wonderland, The Circus*.

FORD, JAMES

Haim: *Days Are Gone*; Arctic Monkeys: *Suck It And See, Favourite Worst Nightmare, Humbug, AM*; Florence & The Machine: *Lungs, Ceremonials*; Beth Ditto: *Beth Ditto*;

Klaxons: *Myths Of The Near Future*; Mystery Jets: *Making Dens*; Simian Mobile Disco: *Delicacies, Temporary Pleasure, Attack Decay Sustain Release*; Peaches: *I Feel Cream*.

FRASER, MIKE

AC/DC: *Razor's Edge, Ballbreaker, Stiff Upper Lip, Black Ice, Plug Me In, No Bull*; Aerosmith: *Permanent Vacation, Pump*; Metallica: *Live Shit*; Joe Satriani: *Crystal Planet, Engines Of Creation, Is There Love In Space, Black Swans And Wormhole Wizards*; Van Halen: *Balance*; Slipknot: *Iowa*; Poison: *Flesh And Blood*; Mötley Crüe: *Supersonic And Demonic Relics*; Krokus: *Blitz*; Franz Ferdinand: *Tonight*; Enter Shikari: *A Flash Flood Of Colour*; Dio: *Strange Highways*; Bryan Adams: *Reckless*.

FRITZ, OZ

Deadline: *Dissident*; The Last Poets: *Holy Terror, Time Has Come*; Buckethead: *Day Of The Robot, Monsters & Robots*; Tom Waits: *Mule Variations, Blood Money, Alice, Orphans*; Bill Laswell: *Imaginary Cuba*; Pain Killer: *Buried Secrets, Execution Ground*.

GATICA, HUMBERTO

Michael Jackson: *Bad*; Celine Dion: *Celine Dion, D'eux, Falling Into You, Let's Talk About Love, A New Day Has Come, One Heart*; Michael Bublé: *Call Me Irresponsible, Crazy Love*; Ricky Martin: *Vuelve*; Cher: *Believe*; Julio Iglesias: *Non Stop, Un Hombre Solo, Crazy, La Carretera, Tango*; Barbra Streisand: *The Mirror Has Two Faces, Higher Ground*.

GODRICH, NIGEL

Radiohead: *The Bends, OK Computer, Kid A, Amnesiac, Hail To The Thief, In Rainbows*; Thom Yorke: *The Eraser*; Natalie Imbruglia: *Left Of The Middle*; Beck: *Mutations, Sea Change, Guero, The Information*; The Divine Comedy: *Regeneration, Absent Friends*; Travis: *The Man Who, The Invisible Band, The Boy With No Name*; Paul McCartney: *Chaos And Creation In The Backyard*; Air: *Talkie Walkie, Pocket Symphony*; Pavement: *Terror Twilight*; REM: *Up*.

GRAYDON, JAY

Al Jarreau: *Jarreau, Breakin' Away, This Time, High Crime*; Earth Wind & Fire: *I Am*; George Benson: "Turn Your Love Around"; Steeley Dan: *Aja, Citizen Steeley Dan*; Dionne Warwick: *Love At First Sight, Friends In Love*; Barry Manilow: *Even Now*; Cher: *I'd Rather Believe In You, Take Me Home*; Barbara Streisand: *Lazy Afternoon, Streisand Superman, Songbird, Wet*.

HAGUE, STEPHEN

Peter Gabriel: *Up*; A-ha: *Lifelines*; Siouxie & The Banshees: *Superstition*; Public Image Ltd: *9*; Orchestral Manoeuvres In The Dark: *The Pacific Age, Crush*; Pet Shop Boys: *Please*; Melanie C: *I Want Candy*; Dubstar: *Disgraceful, Goodbye*; Erasure: *The Innocents*; Robbie Robertson: *Storyville*.

HAMPTON, JOHN

The Replacements: *Pleased To Meet Me, All For Nothing/Nothing For All*; Todd Rundgren: *Nearly Human*; Gin Blossoms: *New Miserable Experience, Congratulations I'm Sorry*; Afghan Whigs: *Gentlemen*; BB King: *Blues Summit*; Rhett Akins: *A Thousand Memories, Somebody New*; Mudhoney: *Tomorrow Hit Today*; Travis Tritt: *The Rockin' Side, The Lovin' Side, T-r-o-u-b-l-e, It's All About To Change*; Ocean Six: *Somewhere Between Day And Night*; Todd Agnew: *Better Questions, Do You See What I See, Reflection Of Something, Grace Like Rain.*

HARRIS, STEPHEN

Dave Matthews Band: *Busted Stuff, Some Devil*; Ben Kweller: *Sha Sha*; Santana: *Supernatural*; Aqualung: *Still Life*; Kula Shaker: *K*; The Kaiser Chiefs: *Employment*; The Automatic: *Not Accepted Anywhere, This Is A Fix*; Blue Merle: *Burning In The Sun*; U2: *All That You Can't Leave Behind.*

HEDGES, MIKE

Orchestra Manoeuvres In The Dark: *Crush, The Pacific Age*; Blur: *Park Life*; Erasure: *The Innocents*; Siouxie And The Banshees: *Superstition, Peepshow*; Pet Shop Boys: *Please, Very, Disco, Disco 2*; The Cure: *Three Imaginary Boys, Seventeen Seconds*; The Undertones: *The Sin Of Pride*; Marc Almond: *Vermin In Ermine, Stories Of Johnny, Mother Fist And Her Five Daughters*; The Las: *The Las*; The Beautiful South: *Choke*; McAlmont & Butler: *Yes, The Sound Of*; Manic Street Preachers: *Everything Must Go, This Is My Truth Tell Me Yours, Know Your Enemy*; Travis: *The Man Who, The Boy With No Name*; U2: *All That You Can't Leave Behind*; Dido: *Life For Rent*; Alfie Bow: *Storyteller.*

HENDERSON, PETER

Jeff Beck: *Wired*; Paul McCartney: *Wings At The Speed Of Sound, London Town, Flowers In The Dirt*; Split Enz: *DizRhythmia*; Frank Zappa: *Sheikh Yerbouti*; Supertramp: *Breakfast In America, Famous Last Words*; Rush: *Grace Under Pressure.*

HILLIER, BEN

U2: *Pop*; Lighthouse Family: *Postcards From Heaven*; Blur: *13, Think Tank*; Suede: *Head Music*; The Wannadies: *Yeah*; Elbow: *Powder Blue, Asleep In The Back, Fugitive Motel, Cast Of Thousands*; Echobelly: *People Are Expensive*; Erasure: *Moon & The Sky*; Tom McRae: *Just Like Blood*; Doves: *Some Cities*; Depeche Mode: *Playing The Angel, Delta Machine*; Natalie Imbruglia: *Come To Life.*

HOBBS, PHILIP

Long-time producer for Linn Records, recording artists such as The Dunedin Consort, Scottish Chamber Orchestra, Swedish Radio Symphony Orchestra, The Tallis Scholars, Nicola Benedetti, Gottlieb Wallisch, James Gilchrist, Capella Nova, Royal Liverpool Philharmonic, Valery Gergiev, Charles Mackerras, Phantasm, Alfie Bow, and The Brodsky Quartet.

HOFFER, TONY

Beck: *Midnite Vultures, Guero*; Supergrass: *Life On Other Planets*; Suede: *A New Morning*; The Thrills: *So Much For The City, Teenager*; Phoenix: *Alphabetical*; Idlewild: *Warnings/Promises*; The Kooks; *Inside In/Inside Out, Konk, Junk Of The Heart*; Belle & Sebastian: *The Life Pursuit, Write About Love*; The Fratellis: *Costello Music*; Goldfrapp: *Seventh Tree*; Ladyhawke: *Ladyhawke*; Thirty Seconds To Mars: *Love, Lust, Faith, And Dreams*; Turin Brakes: *Ether Songs*.

HOMME, JOSH

Member of Queens Of The Stone Age, Kyuss, and Eagles Of Death Metal. Collaborations/productions include: Arctic Monkey: *AM*, Humbug; Biffy Clyro: *Only Revolutions*; Primal Scream: *Beautiful Future*; UNKLE: *War Stories*; Peaches: *Impeach My Bush*; The Hives: *Lex Hives*; The Strokes: *You Only Live Once*; Foo Fighters: *In Your Honor*.

HORN, TREVOR

Buggles: *Adventures In Modern Recording*; Art Of Noise: *Who's Afraid Of The Art Of Noise, The Seduction Of Claude Debussy*; Yes: *Drama, 90125, Big Generator*; ABC: *Lexicon Of Love*; Dollar: *The Dollar Album*; Frankie Goes To Hollywood: *Welcome To The Pleasuredome, Liverpool*; Band Aid: "Do They Know It's Christmas"; Grace Jones: *Slave To The Rhythm*; Pet Shop Boys: *Introspective*; Simple Minds: *Street Fighting Years*; Paul McCartney: *Flowers In The Dirt*; Seal: *Seal* (1991), *Seal* (1994), *Human Being, Seal IV, Soul 2*; Marc Almond: *Tenement Symphony*; LeAnn Rimes: *I Need You*; Tatu: *200kmh In The Wrong Lane, Dangerous And Moving*; John Legend: *Evolver*; Robbie Williams: *Reality Killed The Video Star*.

HOSKULDS, S. HUSKY

Norah Jones: *Come Away With Me*; Solomon Burke: *Don't Give Up On Me*; Tom Waits: *Blood Money*; Sheryl Crow: *The Globe Sessions*; Turin Brakes: *Ether Songs*; The Wallflowers: *Breach*; Vanessa Paradis: *Bliss*; The Gipsy Kings: *Roots*; Aimee Mann: *Bachelor No. 2*; Soulwax: *Much Against*; Joe Henry: *Scar*.

HOWARD, MARK

Bob Dylan: *Time Out Of Mind, Oh Mercy*; Tom Waits: *Real Gone*; Lucinda Williams: *World Without Tears*; Emmylou Harris: *Wrecking Ball*; U2: *All That You Can't Leave Behind*; REM: *Automatic For The People*; Avril Lavigne: *Let Go*; Natalie Imbruglia: *White Lilies Island*; Red Hot Chili Peppers: *Californication*; Iggy Pop: *Avenue B, American Caesar*; Willie Nelson: *Teatro*; Marianne Faithfull: *Vagabond Ways*; Chris Whitley: *Living With The Law*.

HOWE, "BONES"

The Turtles: "It Ain't Me Babe"; The Association: "Windy," "One Less Bell To Answer"; The Mamas & The Papas: "California Dreamin'," "Monday, Monday"; 5th Dimension: "Go Where You Wanna Go," "Up, Up, and Away," "Aquarius"; Ornette Coleman: *The Shape Of Jazz to Come, Change Of The Century*; Mel Torme: *Comin' Home Baby*;

The Crickets: *California Sun, Out Of Limits*; Tom Waits: *The Heart Of Saturday Night, Nighthawks At The Diner, Small Change, Foreign Affairs, Blue Valentine, Heartattack And Vine*; Jerry Lee Lewis: *Rocket 88*.

HUDSON, JOHN

Gary Glitter: *Glitter, Gary Glitter, Rock 'n' Roll*; Alvin Stardust: *The Untouchable*; Lou Reed: *Metal Machine Music*; Bucks Fizz: *Bucks Fizz, Are You Ready, Hand Cut, I Hear Talk*; Captain Sensible: *Power Of Love*; Tina Turner: *Private Dancer, Break Every Rule, Simply The Best*; Ultravox: *Lament*; A-ha: *Stay On These Roads*; The Moody Blues: *Keys Of The Kingdom, Time Traveller*; Tanita Tikaram: *Lovers In The City*; Pulp: *Disco 2000*; Elvis Costello: *All This Useless Beauty*; Wynonna Judd: *What The World Needs Now Is Love*; Toots & The Maytals: *True Love*.

JAM, JIMMY

Janet Jackson: *Control, Janet, Rhythm Nation 1814, All For You, 20 YO, The Velvet Rope*; The Human League: *Crash*; Mariah Carey: *Rainbow, Glitter, Charmbracelet*; Beverley Knight: *100%*; Michael Jackson: *HIStory, Blood On The Dance Floor*; Jordan Knight: *Jordan Knight*; Boyz II Men: "I Will Get There," *II*; TLC: *Fanmail*; Mary J Blige: *Share My World*; George Michael: *Faith*; Johnny Gill: *Johnny Gill*; Karyn White: *Ritual Of Love*; Patti LaBelle: *Gems*.

JEAN, WYCLEF

The Fugees: *Blunted On Reality, The Score*; Wyclef Jean: *The Carnival, The Ecleftic, Masquerade, The Preacher's Son*; Santana: *Supernatural*; Destiny's Child: *Destiny's Child*; Pras: *Ghetto Supastar; We Lose Or Draw*; Mya: *Fear Of Flying*; Michael Jackson: *Blood On The Dance Floor*; Shakira: *Oral Fixation Vol. 2, She Wolf*.

JERDEN, DAVE

Talking Heads: *Remain In Light*; Brian Eno & David Byrne: *My Life In The Bush Of Ghosts*; Frank Zappa: *The Man From Utopia*; Herbie Hancock: *Future Shock*; Red Hot Chili Peppers: *Red Hot Chili Peppers, Mother's Milk*; The Rolling Stones: *Dirty Work*; Jane's Addiction: *Nothing's Shocking, Ritual De Lo Habitual*; Alice In Chains: *Facelift, Sap, Dirt*; Spinal Tap: *Break Like The Wind*; Anthrax: *Sound Of White Noise*; Dig: *Dig*; The Offspring: *Ixnay On The Hombre, Americana*; Pitchshifter: *Deviant*; Rust: *Bar Chord Ritual*.

JERKINS, RODNEY

Brandy & Monica: "The Boy Is Mine"; Mary J. Blige: *Share My World, The Breakthrough*; Whitney Houston: *My Love Is Your Love*; Jennifer Lopez: *On The 6*; Destiny's Child: *The Writing's On The Wall, Destiny Fulfilled*; Spice Girls: *Forever*; Toni Braxton: *The Heat*; Michael Jackson: *Invincible*; Britney Spears: *Britney*; Brandy: *Full Moon, Never Say Never*; Monica: *All Eyez On Me*; TLC: *3D*; Beyoncé: *B'Day*; Mariah Carey: *The Emancipation Of Mimi, The Art Of Letting Go*; Janet Jackson: *Discipline*; Pussycat Dolls: *Doll Domination*; Lady Gaga: *The Fame Monster*; Kelly Clarkson: *Stronger*; Nelly Furtado: *The Spirit Indestructible*; Justin Bieber: *Believe*.

JOHNS, ANDY

Led Zeppelin: *Led Zeppelin III, Led Zeppelin IV, Houses Of The Holy, Physical Graffiti*; The Rolling Stones: *Sticky Fingers, Exile On Main Street, Goat's Head Soup*; Van Halen: *For Unlawful Carnal Knowledge*; Television: *Marquee Moon*; Jethro Tull: *Stand Up*.

JOHNSTON, BOB

Patti Page: *Hush, Hush, Sweet Charlotte*; Bob Dylan: *Highway 61 Revisited, Blonde On Blonde*; Simon & Garfunkel: *Sounds Of Silence, Parsley Sage Rosemary & Thyme*; Johnny Cash: *At Folsom Prison, I Walk The Line*; Leonard Cohen: *Songs From A Room, Songs Of Love And Hate*; Lindisfarne: *Fog On The Tyne*; New Riders Of The Purple Sage: *Oh, What A Mighty Time*; John Mayall: *Bottom Line*; Pete Seeger: *Link In The Chain*.

JONES, GARETH

Erasure: *Wild!, Erasure, Cowboy, Other People's Songs, Light At The End Of The World*; Depeche Mode: *Construction Time Again, Some Great Reward, Exciter*; Wire: *The Ideal Copy; A Bell Is A Cup... Until It Is Struck*.

JONES, LESLIE ANN

Rosemary Clooney: *Sentimental Journey, Brazil, Mothers & Daughters, White Christmas*; Herbie Hancock: *Mr. Hands, Monster, Light Me Up*; Michael Feinstein: *Michael & George, With The Israel Philharmonic*; Bobby McFerrin: *Bobby McFerrin, Spontaneous Inventions*; Carlos Santana: *The Swing Of Delight*; Wayne Shorter: *High Life*; Dianne Reeves: *Goodnight And Good Luck*; Marcus Miller: *Marcus Miller, The Sun Don't Lie*; Manhattan Transfer: *The Offbeat Of Avenues*; BB King & Diane Schuur: *Heart To Heart*.

KELLY, JON

Tori Amos: *Little Earthquakes*; Kate Bush: *The Kick Inside, Lionheart*; Beautiful South: *Quench, Carry On Up The Charts, Blue Is The Colour*; Heather Nova: *Siren*; Paul McCartney: *Tug Of War, Pipes Of Peace*; Chris Rea: *Auberge, Road To Hell Vol. 1, Chris Rea, Dancing With Strangers*; The Damned: *Phantasmagoria, Anything*; Deacon Blue: *Raintown, Fellow Hoodlums*.

KILLEN, KEVIN

Peter Gabriel: *So*; U2: *War, The Unforgettable Fire, Rattle and Hum*; Elvis Costello: *Spike, The Juliet Letters, Kojak Variety, North, Mighty Like A Rose, Cruel Smile*; Shakira: *Oral Fixation Vol. One, Oral Fixation Vol. Two*; Tori Amos: *Under The Pink*; Bryan Ferry: *Boys And Girls*; Kate Bush: *The Sensual World, This Woman's Work*; Paula Cole: *This Fire, Courage*.

KING, JACQUIRE

Kings Of Leon: *Aha Shake Heartbreak, Only By The Night, Come Around Sundown*; Norah Jones: *The Fall*; Tom Waits: *Blood Money, Mule Variations, Alice*; Orphans: *Brawlers, Bawlers And Bastards*; Modest Mouse: *Good News For People Who Like Bad News*.

KITCHINGHAM, GERRY

A-Ha: *Scoundel Days, Stay On These Roads*; Pulp: *Disco 2000*; Cliff Richard: *Small Corners, Always Guaranteed, Stronger, Together With Cliff Richard*; The Moody Blues: *Time Traveller, Keys Of The Kingdom.*

KOLOTKIN, GLEN

Santana: *Moonflower, Abraxas, Supernatural*; Joan Jett & The Blackhearts: *I Love Rock 'N' Roll, Flashback*; Pete Seeger: *Pete*; Janis Joplin: *Pearl*; Barbara Streisand: *Stoney End, Release Me*; The Jimi Hendrix Experience: *Electric Ladyland*; Moby Grape: *Grape Jam, Wow, Moby Grape '69.*

KOSTEN, DAVID

Bat For Lashes: *The Haunted Man, Two Suns, Fur And Gold*; Everything Everything: *Man Alive*; Guillemots: *Walk The River.*

KRAMER, EDDIE

Jimi Hendrix: *Are You Experienced? Axis: Bold As Love, Electric Ladyland, Band of Gypsys*; Led Zeppelin: *Led Zeppelin II, Led Zeppelin III, How The West Was Won, Houses Of The Holy, Physical Graffiti, Coda*; Traffic: *Traffic*; The Nice: *Nice, Five Bridges*; Peter Frampton: *Frampton Comes Alive!*

KURLANDER, JOHN

Badfinger: *No Dice*; Toto: *Toto IV*: Soundtracks: *Lord Of The Rings* (all three films), *Ice Age, 3:10 To Yuma, The Hurt Locker, Master And Commander*; Christopher Tin: *Calling All Dawns.*

LANOIS, DANIEL

U2: *The Joshua Tree, Achtung Baby, All That You Can't Leave Behind, No Line On The Horizon*; Peter Gabriel: *So, Us*; Bob Dylan: *Oh Mercy, Time Out Of Mind.*

LAUNAY, NICK

Nick Cave And The Bad Seeds: *Dig Lazarus Dig, Abattoir Blues, Nocturama*; Grinderman: *Grinderman, Grinderman 2*; Maximö Park: *Quicken The Heart*; Supergrass: *Diamond Hoo Ha*; Yeah Yeah Yeahs: *Is Is, It's Blitz*; Arcade Fire: *Neon Bible, The Suburbs*; Public Image Ltd: *The Flowers Of Romance*; Killing Joke: *Follow The Leaders, What's THIS For*; INXS: *The Swing*; Semisonic: *Feeling Strangely Fine.*

LECKIE, JOHN

Pink Floyd: *Meddle*; Radiohead: *The Bends*; Muse: *Showbiz, Origin Of Symmetry*; The Stone Roses: *The Stone Roses*; The Verve: *A Storm In Heaven*; Kula Shaker: *K*; My Morning Jacket: *Z*; The Coral: *Butterfly House*; Cast: *All Change, Mother Nature Calls*; Doves: *Kingdom Of Rust.*

LEVINE, LARRY

Beach Boys: *Pet Sounds*; Herb Alpert And The Tijuana Brass: "A Taste Of Honey"; Eddie Cochran: "Summertime Blues," "Twenty Flight Rock," "C'mon Everybody"; The Crystals: "He's A Rebel," "Da Doo Ron Ron"; Bob B Soxx & The Blue Jeans: "Zip-A-Dee-Doo-Dah"; Ronettes: "Be My Baby"; Righteous Brothers: "You've Lost That Lovin' Feelin'"; Ike & Tina Turner: "River Deep Mountain High"; Leonard Cohen: *Death Of A Ladies' Man*; The Ramones: *End Of The Century*.

LEVINE, STEVE

Culture Club: *Don't Mind If I Do, Kissing To Be Clever, Colour By Numbers, Waking Up With The House On Fire*; The Beach Boys: *The Beach Boys*; Gary Moore: *Corridors Of Power, Ballads & Blues*; The Honeyz: *Wonder No. 8*; Deniece Williams: *Water Under The Bridge*; 911: *There It Is*.

LIDDELL, FRANK

Miranda Lambert: *Crazy Ex-girlfriend, Revolution, Four The Record*; Lee Ann Womack: *I Hope You Can Dance, Something Worth Leaving Behind, The Reason For Romance*; Eli Young Band: *Life At Best*.

LILLYWHITE, STEVE

U2: *Boy, October, War, Achtung Baby, All That You Can't Leave Behind, How To Dismantle An Atomic Bomb, No Line On The Horizon*; The Pogues: *If I Should Fall From Grace With God*; Dave Matthews Band: *Under The Table And Dreaming, Crash, Before These Crowded Streets, Busted Stuff*; The Las: *The Las*; Phish: *Billy Breathes, Joy*; Jason Mraz: *Mr A-Z*; Crowded House: *Time On Earth*; Thirty Seconds To Mars: *This Is War*; XTC: *Drums & Wires, Black Sea*; Ultravox: *Ultravox, Ha Ha Ha*; Thompson Twins: *Set*; The Rolling Stones: *Dirty Work*; The Psychedelic Furs: *The Psychedelic Furs, Talk Talk Talk*; The Killers: *Battle Born*; Talking Heads: *Naked*; Simple Minds: *Sparkle In The Rain*; Peter Gabriel: *Peter Gabriel III*; Joan Armatrading: *Walk Under Ladders, The Key, Sleight Of Hand*; Kirsty MacColl: *Kite, Electric Landlady, Galore*; Beady Eye: *Still Speeding, Different Gear*.

LIPSON, STEVE

Annie Lennox: *Diva*; Frankie Goes To Hollywood: *Welcome To The Pleasuredome, Liverpool*; Paul McCartney: *Flowers In The Dirt*; Grace Jones: *Slave To The Rhythm*; Jordin Sparks: *Battlefield*; Boyzone: *Where We Belong, By Request*; Rachel Stevens: *Funky Dory*.

LIPUMA, TOMMY

Diana Krall: *When I Look Into Your Eyes, The Look Of Love, Live In Paris*; Paul McCartney: *Kisses On The Bottom*; Natalie Cole: *Unforgettable, Take A Look, Ask A Woman Who Knows, Unforgettable With Love*; George Benson: *Breezin', Weekend In LA*; Barbra Streisand: *The Way We Were*; Al Jarreau: *Glow, Look To The Rainbow*; Miles Davis: *TuTu*.

LORD-ALGE, CHRIS

James Brown: *Living In America*; Prince: *Batman* soundtrack; Joe Cocker: *Unchain My Heart*; Chaka Khan: *Destiny*; Green Day: *Nimrod, American Idiot, 21st Century Breakdown*; My Chemical Romance: *The Black Parade*; Stevie Nicks: *Trouble In Shangri-La*; POD: *Testify*; AFI: *Decemberunderground*; Darren Hayes: *Spin*; Creed: *Full Circle*; Sum 41: *Underclass Hero*; Switchfoot: *The Beautiful Letdown, Nothing Is Sound*; Slipknot: *Dead Memories, Sulfur*; Stone Temple Pilots: *Stone Temple Pilots*.

LUKE, DAVID

Earth, Wind & Fire: *Heritage*; MC Hammer: *Too Legit To Quit*; Bobby McFerrin: *Bang Zoom*; Johnny Nocturne Band: *Wild And Cool, Million Dollar Secret, Shake 'Em Up, Wailin' Daddy*; Dave Ellis: *In The Long Run*; The Delfonics: *Forever New*; Herbie Hancock: *Possibilities*; Ladysmith Black Mambazo: *Long Walk To Freedom*; The Spinners: *Down To Business*.

MACK, REINHOLD

Black Sabbath: *Dehumanizer*; Extreme: *Extreme*; Queen: *A Kind Of Magic, Flash Gordon, The Game, The Works*; Meat Loaf: *Bad Attitude*; Sparks: *Angst In My Pants, Whomp That Sucker*; Billy Squier: *Don't Say No*; David Coverdale: *White Snake*; Deep Purple: *Stormbringer, Come Taste The Band*; Rolling Stones: *It's Only Rock 'n' Roll*; Scorpions: *Fly To The Rainbow*; Electric Light Orchestra: *Balance Of Power, Time, Xanadu, Discovery, Out Of The Blue, A New World Record, Face The Music*; T.Rex: *Zinc Alloy And The Hidden Riders Of Tomorrow*.

MALOUF, BRIAN

Everclear: *Sparkle And Fade*; David Gray: *White Ladder*; Lit: *A Place In The Sun*; Madonna: *I'm Breathless*; Pearl Jam: *Ten*; Slaughter: *Stick It To Ya*.

MARCANTONIO, STEVE

Tim McGraw: *Not A Moment Too Soon*; Reba McEntire: *All The Woman I Am*; Lionel Richie: *Tuskegee*; Vince Gill: *Guitar Slinger, These Days, Let's Make Sure We Kiss Goodbye, The Key*; Kenny Chesney: *The Road And The Radio, Welcome To The Fishbowl, Just Who I Am*; LeAnn Rimes: *Family*; Brooks & Dunn: *If You See Her, Cowboy Town*; Rascal Flatts: *Still Feels Good*; Faith Hill: *Fireflies*; Keith Urban: *Keith Urban, The Fuse*; Kenny Rogers: *You Can't Make Old Friends*.

MARCHAND, PIERRE

Sarah McLachlan: *Solace, Fumbling Towards Ecstasy, Surfacing, Afterglow, Wintersong, Laws Of Illusion*; Rufus Wainwright: *Rufus Wainwright, Poses, All Days Are Nights*; Kate & Anna McGarrigle: *Matapedia, Heartbeats Accelerating*; Ron Sexsmith: *Ron Sexsmith*; The Devlins: *Waiting*; Stevie Nicks: *Trouble In Shagri-la*; Patty Larkin: *Strangers World*.

MASSENBURG, GEORGE

Little Feat: *Feats Don't Fail Me Now, Hoy-Hoy, Let It Roll, Shake Me Up*; Linda Ronstadt: *Mas Canciones, Frenesi, Feels Like Home, Dedicated To The One I Love, We Ran*; Lyle Lovett: *Joshua Judges Ruth*; Bonnie Raitt: *Nine Lives*; Toto: *The Seventh One*; Earth, Wind & Fire: *Gratitude, That's The Way Of The World, Spirit, I Am, All N All*; Journey: *Trial By Fire, When You Love A Woman*.

MASSY, SYLVIA

Johnny Cash: *Unchained*; Tool: *Opiate, Undertow*; Red Hot Chili Peppers: *Love Rollercoaster*; Machines Of Loving Grace: *Gilt*; Love And Rockets: *Sweet FA*; Powerman 5000: *Tonight The Stars Revolt*; System Of A Down: *System Of A Down*; Skunk Anansie: *Paranoid & Sunburnt*; Sevendust: *Scream 3*; Prince: *Diamonds & Pearls*; Lustra: *Lustra*; Deftones: *Escape From LA*; Cog: *The New Normal, Sharing Space*.

MAZER, ELLIOT

Linda Ronstadt: *Silk Purse*; Bob Dylan: *The Basement Tapes, Isle Of Wight*; The Band: *The Last Waltz*; Janis Joplin: *Farewell Song, Joplin In Concert*; Neil Young: *Lucky 13, Old Ways, Everybody's Rockin', Hawks & Doves, Decade, American Stars & Bars, Tonight's The Night, Time Fades Away, Journey Through The Past, Harvest*.

MCDONALD, RODAIDH

Gil Scott-Heron: *I'm New Here, We're New Here*; Adele: *21*; How To Dress Well: *Total Loss*; The XX: *XX, Coexist*; King Krule: *6 Feet Beneath The Moon*; Vampire Weekend: *Contra*.

MERCHANT, JOHN

Mika: *Life In Cartoon Motion*; Barbra Streisand: *Guilty Pleasures, Night Of My Life*; Cliff Richard: *I Cannot Give You My Love*; Lenny Kravitz: *Baptism*; Michael Bublé: *How Can You Mend A Broken Heart*; Bee Gees: *The Record, This Is Where I Come In, One Night Only, Still Waters, Storytellers*; Celine Dion: *Immortality*; Natural: *It's Only Natural*; Ronan Keating: *Lovers And Friends*.

MESSINA, JAY

Booker T & The MGs: *Melting Pot*; Aerosmith: *Get Your Wings, Toys In The Attic, Rocks, Draw The Line*; Supertramp: *Crisis What Crisis, Crime Of The Century, Some Things Never Change*; Kiss: *Destroyer, Dynasty, Unmasked*; Cheap Trick: *Cheap Trick*; George Benson: *In Your Eyes*; Lou Reed: *Berlin*; New York Dolls: *One Day It Will Please Us To Remember This*.

MEYERSON, ALAN

Soundtracks: *Despicable Me 1 & 2, Man Of Steel, Halo 4, Battleship, Kung Fu Panda 2, Pirates Of The Caribbean 1, 2, 3, & 4, Tron Legacy, Transformers 1 & 2, I Am Legend, Iron Man, Madagascar 1 & 2, The Dark Knight, The Dark Knight Rises, Taken, The Da Vinci Code, Black Hawk Down, The Bourne Identity, Gladiator, Shrek, Mission Impossible 2, Armageddon, Speed*.

MILLER, JACK

Duane Eddy: *Have Twangy Guitar Will Travel, The Twang's The Thang, Especially For You, Girls Girls Girls*; Anthony Newley: *Newly Recorded, Who Can I Turn To*; also produced records for Herman's Hermits, The Monkees, The Animals, and The Rolling Stones.

MINCIELI, ANN

Alicia Keys: *Songs In A Minor, Diary Of Alicia Keys, As I Am, The Element Of Freedom, Girl On Fire*; Emeli Sandé: *Our Version Of Events*; Drake: *Thank Me Later*; Whitney Houston: *I Look To You, Million Dollar Bill*; Robin Thicke: *Sex Therapy*; Keyshia Cole: *The Way It Is*; Usher: *Confessions*; Angie Stone: *Stone Love*; Dru Hill: *Dru World Order*; Yoko Ono: *Blueprint For A Sunrise.*

MOSS, WAYNE

Charlie McCoy: *Out On A Limb, Harpin' The Blues, The World Of*; Memphis Slim: *Goin' Back To Tennessee*; Barefoot Jerry: *Keys To The Country, You Can't Get Off With Your Shoes On, Watchin' TV, Southern Delight*; Tony & Terry: *Cross Country*; Mickey Newbury: *Heaven Help The Child, It Looks Like Rain, Live At Montezuma Hall*; John Hambrick: *Windmill In A Jet Filled Sky, Frisco Mabel Joy*; Leo Kottke: *Mudlark*; Area Code 615: *A Trip In The Country, Area Code 615*; Steve Miller Band: *Number 5*; Linda Ronstadt: *Silk Purse*; Eric Andersen: *Eric Andersen*; Wes Miller: *9 Out Of 10, Knee Deep In Grass.*

MURPHY, SHAWN

Soundtrack: *Braveheart, Dances With Wolves, E.T. The Extra-Terrestrial, Ghost, Jurassic Park, Men In Black, Pretty Woman, Saving Private Ryan, The Sixth Sense, Star Wars: Episode I—The Phantom Menace, Episode II— Attack Of The Clones, Titanic.*

MUSSO, ROBERT

Material: *Hallucination Engine, The Third Power, Memory Serves, Seven Souls, One Down*; Sly & Robbie: *Rhythm Killers, Language Barrier*; Tom Waits: *Rain Dogs*; Ginger Baker: *Horses & Trees*; Iggy Pop: *Instinct*; Bill Laswell: *Hear No Evil, Psychonavigation, Silent Recoil, Cymatic Scan*; The Ramones: *Brain Drain*; Al Green: *Don't Look Back, Your Heart's In Good Hands*; The Last Poets: *Holy Terror*; Bootsy Collins: *Blasters Of The Universe, The Funk Capital Of The World*; George Clinton: *Get Yo Ass In The Water And Swim Like Me*; Sonny Sharrock: *Into Another Light, Ask The Ages, Faith Moves, Seize The Rainbow.*

NEILL, MARK

Dan Auerbach: *Keep It Hid*; The Black Keys: *Brothers*; The Paladins: *The Paladins, Slippin' In, Pavoline No. 7*; Deke Dickerson: *Rhythm, Rhyme, & Truth, In 3 Dimensions, The Melody, My Name Is Deke*; Los Straitjackets: *Supersonic Guitars In 3D, Encylopaedia Of Sound*; The Computers: *Love Triangles And Hate Squares.*

NELSON, KEN

Paolo Nutini: *These Streets*; Coldplay: *Parachutes, X&Y, A Rush Of Blood To The Head*; Kings Of Convenience: *Quiet Is The New Loud*; The Charlatans: *Up At The Lake*; Embrace:

If You've Never Been; Snow Patrol: *When It's All Over We'll Still Have To Clear Up*; Badly Drawn Boy: *Hour Of The Bewilderbeast*; Feeder: *Pushing The Senses*; Gomez: *Bring It On, Liquid Skin*; Skin: *Flesh Wounds*.

NEWLAND, JAY

Norah Jones: *Come Away With Me, Feels Like Home*; Ayo: *Joyful, Gravity At Last, Ticket To The World*; Richie Havens: *Nobody Left To Crown*; Charlie Haden: *Land Of The Sun, Nocturne, Come Sunday*; Michael Brecker: *Wide Angles*; Herbie Hancock: *Directions In Music*; Rumer: *Seasons Of My Soul*; Nigel Kennedy: *Blue Note Sessions*; Etta James: *Blue Gardenia, The Heart Of A Woman, Time After Time, Mystery Lady, Love Songs, Burnin' Down The House*.

NICHOLS, ROGER

Steeley Dan: *Can't Buy A Thrill, Countdown To Ecstasy, Pretzel Logic, Gaucho, Aja, Two Against Nature, Everything Must Go*.

NIEBANK, JUSTIN

Marty Stuart: *Country Music*; Keith Urban: *Be Here, Fuse, Get Closer, Defying Gravity, Love Pain And The Whole Crazy Thing*; Patty Loveless: *On Your Way Home, Dreamin' My Dreams*; Vince Gill: *Next Big Thing, These Days*; Rascal Flatts: *Changed, Nothing Like This, Unstoppable, Still Feels Good, Me And My Gang*; George Strait: *George Strait, Always Never The Same*; Lady Antebellum: *Golden*; Kenny Chesney: *Life Of A Rock, Welcome To The Fishbowl, Hemingway's Whiskey, Lucky Old Sun*; Taylor Swift: *Red, Speak Now, Fearless*; Lionel Richie: *Tuskegee*.

NORRELL, CLIF

Indigo Girls: *Indigo Girls, Nomads Indians Saints*; REM: *Automatic For The People*; Inspiral Carpets: *I Want You, Uniform, Cool As, Devil Hopping*; Jeff Buckley: *Grace*; Rush: *Test For Echo*; Faith No More: *King For A Day Fool For A Lifetime*; Weezer: *Pinkerton*; Mansun: *Attack Of The Grey Lantern*; Echo & The Bunnymen: *Evergreen*; Babyshambles: *Back To The Bus*; Miley Cyrus: *Start All Over, Breakout*.

OBERG, GUS

The Strokes: *Angles, Comedown Machine*; The Postelles: *The Postelles*; Albert Hammond Jr.: *Como Te Llama*; Turbonegro: *Sexual Harassment*; Har Mar Superstar: *Bye Bye 17*.

OLSEN, KEITH

Fleetwood Mac: *Fleetwood Mac*; Foreigner: *Double Vision*; Scorpions: *Crazy World*; Whitesnake: *Whitesnake, Slide It In, Slip Of The Tongue*.

PACZOSA, GARY

Alison Krauss: *I Know Who Holds Tomorrow, Forget About It, Essential Alison Krauss, Lonely Runs Both Ways, New Favorite, So Long So Wrong*; Dolly Parton: *Those Were The Days, Little Sparrow, The Grass Is Blue, Treasures, Something Special, Slow Dancing With The Moon, Eagle When She Flies*; Harry Connick Jr.: *Every Man Should Know*; Steep Canyon

Rangers: *Nobody Knows You, Rare Bird Alert*; Yo-Yo Ma: *Songs Of Joy & Peace*; John Prine: *Fair & Square*.

PADGHAM, HUGH

Genesis: *Abacab, Genesis, Invisible Touch*; Phil Collins: *Face Value, Hello I Must Be Going, No Jacket Required, But Seriously, Both Sides*; The Police: *Ghost In The Machine, Synchronicity*; Sting: *Nothing Like The Sun, The Soul Cages, Ten Summoner's Tales, Mercury Falling*; Peter Gabriel: *Peter Gabriel*; XTC: *Black Sea*.

PARSONS, ALAN

The Beatles: *Abbey Road*; Pink Floyd: *Dark Side Of The Moon*; Al Stewart: *Year Of The Cat*; Paul McCartney: *Wild Life, Red Rose Symphony*; The Hollies: *Hollies*, "He Ain't Heavy, He's My Brother"; Ambrosia: *Somewhere I've Never Travelled*; Alan Parsons Project: *Tales Of Mystery And Imagination, Pyramid, Eve, The Turn Of A Friendly Card, Eye In The Sky*.

PETTIBONE, SHEP

Madonna: *Celebration, You Can Dance, I'm Breathless, Erotica, Like A Prayer, Bedtime Stories, Something To Remember*; Duran Duran: *Big Thing*; Taylor Dayne: *Soul Dancing*; Pet Shop Boys: *Actually*; Gary Barlow: *Love Won't Wait*.

PLATT, TONY

Bob Marley: *Catch A Fire, Burnin'*; Toots & The Maytals: *Funky Kingston*; Aswad: *Aswad*; AC/DC: *Highway To Hell, Back In Black*; Foreigner: *4*; Boomtown Rats: *The Fine Art Of Surfacing*; Anathema: *Eternity*.

POOLE, MIKE

Robert Plant: *Band Of Joy*; Martina McBride: *Emotion, Evolution, Wild Angels, White Christmas, The Way That I Am*; Patty Griffin: *American Kid, Downtown Church*; Rick Trevino: *In My Dreams*; Keith Urban: *The Ranch*; Little Big Town: *Little Big Town*; Travis Tritt: *The Lovin' Side, The Rockin' Side, It's All About To Change, Country Club*; Ricky Skaggs: *Kentucky Thunder, Comin' Home To Stay*.

PORTER, JOHN

The Smiths: *The Smiths, Meat Is Murder, The Queen Is Dead*; Billy Bragg: *Talking With The Taxman About Poetry, Preaching To The Converted*; Buddy Guy: *Damn Right I've Got The Blues, Feels Like Rain*; Taj Mahal: *Phantom Blues*; Keb Mo: *Keb Mo, Just Like You, Keep It Simple, Suitcase*; Maria Muldaur: *Meet Me At Midnite*; John Mayall: *Blues For The Lost Days*; John Lee Hooker: *Smile Like Yours*; BB King: *Blues On The Bayou, Makin' Love Is Good For You*; RL Burnside: *Wish I Was In Heaven Sitting Down*; Jimmy Smith: *Dot Com Blues*; Lucky Peterson: *Double Dealin'*; Los Lonely Boys: *Los Lonely Boys*.

POWELL, JEFF

Big Star: *In Space*; Stevie Ray Vaughan & Double Trouble: *Live At Carnegie Hall*; Afghan Whigs: *Black Love, Honky's Ladder, Gentlemen, 1965, Somethin' Hot*; Primal Scream: *Give Out But Don't Give Up*; The Bottle Rockets: *Zoysia*; Centro-Matic: *Fort Recovery*.

PRICE, BILL

The Sex Pistols: *Never Mind The Bollocks*; The Clash: *The Clash, Give 'Em Enough Rope, London Calling, Sandinista!*; The Pretenders: *Pretenders, Pretenders II*; Elton John: *Too Low For Zero*; Pete Townshend: *Empty Glass*; The Jesus & Mary Chain: *Darklands*; Babyshambles: *Down In Albion*.

PRICE, STUART

Madonna: *American Life, Confessions On A Dance Floor*; New Order: *Waiting For The Siren's Call*; Seal: *System*; The Killers: *Sawdust, Day & Age, Battle Born*; Keane: *Perfect Symmetry*; Scissor Sisters: *Night Work, Magic Hour*; Kylie Minogue: *Aphrodite*; Brandon Flowers: *Flamingo*; Take That: *Progress, Progressed*; Duffy: *Endlessly*; Pet Shop Boys: *Electric*.

PUIG, JACK JOSEPH

Snow Patrol: *Eyes Open*; Goo Goo Dolls: *Let Love In*; Black Eyed Peas: *Monkey Business*; Fergie: *The Dutchess;* Mary J. Blige: *The Breakthrough*; Pussy Cat Dolls: *PCD*; Stereophonics: *You've Got To Go There To Come Back*; Sheryl Crow: *C'mon C'mon*; The Rolling Stones: *Forty Licks, A Bigger Bang, Biggest Mistake*; Greenday: *Warning*; No Doubt: *Return Of Saturn*; Hole: *Celebrity Skin*; Weezer: *Pinkerton*; Jellyfish: *Spilt Milk, Bellybutton*.

PUTNAM, NORBERT

New Riders Of The Purple Sage: *Wasted Tasters, The Adventures Of Panama Red*; Flying Burrito Brothers: *Flying Again*; Dan Fogelberg: *Love Songs, Windows And Walls, The Innocent Age, Phoenix, Nether Lands, Home Free*; Jimmy Buffett: *Boats, Beaches, Bars, and Ballads, Somewhere Over China, Coconut Telegraph, Volcano, Son Of A Son Of A Sailor, Changes In Latitudes, Changes In Attitudes*; Joan Baez: *The Contemporary Ballad Book, The Night They Drove Old Dixie Down, Where Are You Now My Son*.

RAMONE, PHIL

Paul Simon: *There Goes Rhymin' Simon, Still Crazy After All These Years*; Bob Dylan: *Blood On The Tracks*; Sinead O'Connor: *Am I Not Your Girl?*; Billy Joel: *52nd Street, Glass Houses, The Nylon Curtain, The Bridge*.

RAPHAEL, GORDON

The Strokes: *Is This It, Room On Fire, First Impressions Of Earth*; Regina Spektor: *Soviet Kitsch*; The Plastics: *Shark*.

REID, L.A.

Kanye West: *My Beautiful Dark Twisted Fantasy*; Rihanna: *Rated R, Loud*; Justin Bieber: *My World, My World 2.0*; Jay-Z: *Kingdom Come, American Gangster*; Mariah Carey: *The Emancipation Of Mimi, E=MC²*; Boyz II Men: *I'll Make Love To You, Cooleyhighharmony, II, Full Circle*; Toni Braxton: *Toni Braxton, Secrets, You're Makin' Me High, Unbreak My Heart, I Don't Want To, Heat, Spanish Guitar, Snowflakes, More Than A Woman*; TLC: *3D, On The TLC Tip, CrazySexyCool, FanMail*; Avril Lavigne: *Let Go*; Whitney Houston: *Love Whitney,*

Just Whitney, I'm Your Baby Tonight, Queen Of The Night; Usher: *Confessions, Usher, Many Ways, My Way, Nice & Slow*; Outkast: *Outkast, Speakerboxx/The Love Below*.

RICHARDSON, GARTH

You Me At Six: *Sinners Never Sleep*; Skunk Anansie: *Stoosh*; Red Hot Chili Peppers: *Mother's Milk*; Rage Against The Machine: *Rage Against The Machine*; The Jesus Lizard: *Shot*; Biffy Clyro: *Puzzle, Only Revolutions, Opposites*; Melvins: *Stoner Witch, Stag*.

ROCK, BOB

Bon Jovi: *New Jersey, Slippery When Wet, Keep The Faith*; Survivor: *Survivor*; The Cult: *Sonic Temple, The Cult, Beyond Good And Evil*; Mötley Crüe: *Dr. Feelgood, Decade Of Decadence, Mötley Crüe*; Metallica: *Metallica* (The Black Album), *Load, ReLoad, Garage Inc., S&M, St. Anger*; Bryan Adams: *On A Day Like Today*; Lost Prophets: *Liberation Transmission*; Michael Bublé: *Call Me Irresponsible, Crazy Love, To Be Loved*; Offspring: *Rise And Fall Rage And Grace, Days Go By*; 311: *Uplifter, Universal Pulse*.

ROSSE, ERIC

Tori Amos: *Little Earthquakes, Under The Pink*; Lisa Marie Presley: *To Whom It May Concern*; Anna Nalick: *Wreck Of The Day*; Sara Bareilles: *Little Voice*; Nerina Pallot: *Fires*.

SANDS, DENNIS

Soundtrack: *American Beauty, Back To The Future 1 & 2, Erin Brokovich, Forrest Gump, Independence Day, Who Framed Roger Rabbit, Contact, Cast Away, The Polar Express, Men In Black 2 & 3, Terminator Salvation, Terminator 3, Bolt, Spiderman 1 & 2, The Bourne Supremacy, Finding Nemo, Corpse Bride, The Green Mile, Pleasantville, Good Will Hunting, The Shawshank Redemption, Edward Scissorhands*.

SARAFIN, ERIC (MIXERMAN)

Pharcyde: *Bizarre Ride To The Pharcyde, Instrumentals*; Brand New Heavies: *Brother Sister*; Ben Harper: *Fight For Your Mind, The Will To Live, Burn To Shine*; David Cassidy: *Old Trick New Dog*; Hilary Duff: *Hilary Duff, Most Wanted*; Barenaked Ladies: *Everything To Everyone*; Terence Trent D'Arby: *Wildcard*; Kathy Troccoli: *Love And Mercy, Sounds Of Heaven*.

SARDINA, RAFA

Luis Miguel: *Vivo, Romances, Todos Los Ramances, Amarte Es Un Placer, Mis Romances, Complices, Mexico En La Piel, Grandes Exitos, Mis Boleros Favoritos, 33*; Alejandro Sanz: *No Es Lo Mismo, El Tren De Los Momentos Tu No Tienes Alma*; Lady Gaga: *Born This Way*; Stevie Wonder: *A Time To Love*; Dru Hill: *Enter The Dru*.

SCHEINER, ELLIOT

Steely Dan: *Aja, Gaucho, Two Against Nature*; Donald Fagan: *Nightfly*; Billy Joel: *Songs In The Attic*; Fleetwood Mac: *The Dance*; Roy Orbison: *Black And White Night*; John Fogerty: *Premonition*; Van Morrison: *Moondance*; Bruce Hornsby & The Range: *The Way It Is*.

SCHILLING, ERIC

Gloria Estefan: *Into The Light*; Ricky Martin: *Sound Loaded*; Julio Iglesias: *Quelque Chose de France*; Jon Secada: *Secada*; Bacilos: *Caraluna, Sin Vergüenza*.

SCHLEICHER, CLARKE

Martina McBride: *The Time Has Come, Evolution, Emotion, Martina*; Big & Rich: *Comin' To Your City, Between Raising Hell And Amazing Grace*; Dixie Chicks: *Wide Open Spaces, Fly*; Pam Tillis: *Thunder and Roses*; Sara Evans: *Born To Fly, Restless*; Mark Chesnutt: *Savin' The Honky-Tonk*; Taylor Swift: *Taylor Swift*.

SCHMITT, AL

George Benson: *Breezin'*; Steely Dan: *Aja, FM (No Static At All)*; Toto: *Toto IV*; Natalie Cole: *Unforgettable*; Diana Krall: *When I Look In Your Eyes, The Look Of Love*; Ray Charles: *Genius Loves Company*; Jefferson Airplane: *After Bathing At Baxter's, Crown Of Creation, Volunteers*; Luis Miguel: *Amarte Es Un Placer*.

SCOTT, JIM

Sting: *Bring On The Night, Dream Of The Blue Turtles*; Santana: *Freedom, Blues For Salvador, Supernatural*; Lonesome Romeos: *Lonesome Romeos*; Roy Orbison: *King Of Hearts*; Tom Petty & The Heartbreakers: *Wildflowers, Playback, The Last DJ, She's The One*; Slayer: *Divine Intervention*; Ride: *Carnival Of Light*; Johnny Cash: *American Recordings*; Screaming Trees: *Dust*; The Rolling Stones: *Bridges To Babylon*; Reef: *Glow*; Natalie Merchant; *Ophelia*; Lucinda Williams: *Car Wheels On A Gravel Road*; Wilco: *Being There, Summerteeth*; Red Hot Chili Peppers: *Californication, By The Way*; Counting Crows: *This Desert Life, Films About Ghosts*; Rage Against The Machine: *Renegades*; Barenaked Ladies: *Maroon*; Foo Fighters: *One By One*; System Of A Down: *Lonely Day*; Matchbox 20: *Exile On Mainstream, More Than You Think You Are*; Dixie Chicks: *Taking The Long Way*.

SCOTT, TOBY

Bruce Springsteen: *High Hopes, Wrecking Ball, Working On A Dream, Magic, Devils & Dust, The Rising, The Ghost Of Tom Joad, Human Touch, Lucky Town, Tunnel Of Love, Born In The USA, The River, We Shall Overcome*; Bob Dylan: *Shot Of Love*; Manhattan Transfer: *Pastiche*; Booker T & The MGs: *Universal Language*; The Replacements: *All For Nothing/ Nothing For All*.

SENGPIEL, EBERHARD

Produced records for classical artists such as Gidon Kremer, Sviatoslav Richter, Daniel Barenboim, Alban Berg Quartet, Mstislav Rostropovich, London Symphony Orchestra, András Schiff, Nikolaus Harnoncourt, Il Giardino Armonico, Kurt Masur, Kent Nagano, Hélène Grimaud, The Brodsky Quartet, Maxim Vengerov, Chicago Symphony Orchestra, New York Philharmonic, Concerto Koeln, Radu Lupu, and Berlin Philharmonic. He has won Grammy awards for his recordings of Richard Strauss's Wind Concertos and Richard Wagner's *Tannhäuser*.

SHEBIB, NOAH "40"

Drake: *So Far Gone, Thank Me Later, Take Care, Nothing Was The Same*; Trey Songz: *Ready, Passion Pain And Pleasure*; Lil Wayne: *No Ceilings, I'm Not A Human Being*; Alicia Keys: *The Element Of Freedom*; Jamie Foxx: *Best Night Of My Life*; Usher: *Looking 4 Myself*; Nas: *Life Is Good*; DJ Khaled: *Suffering From Success, We The Best Forever*; Beyoncé: *Beyoncé*.

SHIPLEY, MIKE

Nickelback: *Dark Horse*; Faith Hill: *Breathe*; Def Leppard: *High 'n' Dry, Pyromania, Hysteria, Adrenalize*; The Cars: *Heartbeat City*; Shania Twain: *The Woman In Me, Come On Over, Up*; The Corrs: *In Blue*; Nickelback: *Dark Horse*; Maroon 5: *Hands All Over*.

SHOEMAKER, TRINA

Sheryl Crow: *Sheryl Crow, The Globe Sessions, C'mon C'mon*; Queens Of The Stone Age: *R*; Steven Curtis Chapman: *All Things New*; Dixie Chicks: *Home*.

SIMON, JOHN

Leonard Cohen: *Songs Of*; Simon & Garfunkel: *Bookends*; Blood Sweat And Tears: *Child Is Father To The Man*; Gordon Lightfoot: *Did She Mention My Name*; Big Brother And The Holding Company: *Cheap Thrills*; The Band: *Music From Big Pink, The Band, The Last Waltz*; John Hartford: *Morning Glory*; Steve Forbert: *Jackrabbit Slim*; Gil Evans: *Priestess*; Mama Cass: *Dream A Little Dream Of Me*; Michael Franks: *Tiger In The Rain*.

SMITH, ANDY

Paul Simon: *So Beautiful Or So What, Surprise, You're The One, Songs From The Caperman*; Carl Perkins: *Go Cat Go*; Mariah Carey: *Daydream, Merry Christmas*; The Waterboys: *Dream Harder*; Frank Sinatra: *Duets I & II*; The Gaddabouts: *Look Out Now, The Gaddabouts*; Edie Brickell: *Volcano, Picture Perfect Morning*.

SMITH, DON

The Rolling Stones: *Voodoo Lounge*; Ry Cooder: *Chavez Ravine, My Name Is Buddy*; Stevie Nicks: *Rock A Little, Trouble In Shangri-La*; The Tragically Hip: *Up To Here, Road Apples*; Tom Petty: *Long After Dark, Southern Accents, Full Moon Fever, The Last DJ*; Roy Orbison: *Mystery Girl*; Eurythmics: *Be Yourself Tonight*.

SMITH, FRASER T.

Craig David: *Born To Do It, Slicker Than Your Average, Trust Me*; Rachel Stevens: *Come And Get It*; Tinchy Stryder: *Catch 22, Third Strike*; Taio Cruz; *Rokstarr*; Cheryl Cole: *Three Words*; Ellie Goulding: *Lights*; Kano: *Home Sweet Home, London Town*; Beyoncé: *B'Day*; James Morrison: *Songs For You Truths For Me*; N-Dubz: *Uncle B*; Jennifer Hudson: *Jennifer Hudson*; Pixie Lott: *Turn It Up*; Chipmunk: *I Am Chipmunk*.

STAUB, RANDY

Nickelback: *Silver Side Up, The Long Road, Here And Now, All The Right Reasons, Dark Horse*; Alice In Chains: *The Devil Put Dinosaurs Here, Black Gives Way To Blue*; Metallica: *Some Kind Of Monster, Metallica* (The Black Album), *Reload, Load;* Our Lady Peace: *Healthy In Paranoid Times, Gravity*; Mötley Crüe: *Mötley Crüe, Decade Of Decadence, Dr. Feelgood;* U2: *Rattle & Hum*; David Bowie: *Black Tie White Noise*: The Cult: *The Cult, Beyond Good And Evil*; POD: *Satellite*; Avril Lavigne: *My Happy Ending*; Simple Plan: *Still Not Getting Any*.

STAVROU, MIKE

Siouxie And The Banshees: *Join Hands, The Scream;* T.Rex: *Dandy In The Underworld*; Cat Stevens: *Back To Earth*; Pretenders: *Pretenders, Pirate Radio, I'll Stand By You*; Paul McCartney: *Tug Of War*; John Williams: *Spanish Guitar Music*; Joan Baez: *Blowin' Away*; Steeleye Span: *All Around My Hat.*

STEINER, ARMIN

Bread: *Baby I'm A Want You, Guitar Man, Manna, On The Waters*; Neil Diamond: *Moods, Stones*; Gladys Knight: *Miss Gladys Knight*; Barbra Streisand: *Songbird, Streisand Superman, What Matters Most*; Soundtrack: *Home Alone, Presumed Innocent, The Matrix, The Matrix Reloaded, The Matrix Revolutions, Jurassic Park 3.*

STONE, AL

Jamiroquai: *Return Of The Space Cowboy, Travelling Without Moving, Synkronized*; Daniel Bedingfield: *Gotta Get Through This*; Stereo MCs: *Connected*; Bjork: *Debut, Post*; Turin Brakes: *The Optimist*; Lamb: *Fear Of Fours*; Eagle Eye Cherry: *Sub Rosa;* Spice Girls: *Spice.*

STREET, STEPHEN

The Smiths: *Meat Is Murder, The Queen Is Dead, Strangeways Here We Come*; Morrissey: *Viva Hate, Bona Drag*; Blur: *Modern Life Is Rubbish, Parklife, The Great Escape, Blur*; Cranberries: *Everybody Else Is Doing It So Why Can't We, No Need To Argue, Wake Up And Smell The Coffee, Roses*; Kaiser Chiefs: *Employment, Yours Truly Angry Mob*; Babyshambles: *Shotter's Nation.*

STREICHER, RON

Produced recordings for: Bolshoi Theatre Orchestra, Philadelphia Orchestra, Los Angeles Philharmonic, The Orford String Quartet, Shanghai Symphony Orchestra.

SULLIVAN, STUART

Sublime: *Sublime*; Sublime with Rome: *Yours Truly*; Meat Puppets: *Too High To Die, Huevos*; Willie Nelson: *How Great Thou Art, Let's Face The Music And Dance*; Dead Milkmen: *Metaphysical Graffiti, Beelzebubba*; James Cotton: *Cotton Mouth Man, Giant, Baby Don't You Tear My Clothes, Mighty Long Time.*

SWANN, DARRYL

Macy Gray: *On How Life Is, The Id, The Trouble With Being Myself, The World Is Yours.*

SWEDIEN, BRUCE

Michael Jackson: *Off The Wall, Thriller, Bad, Dangerous*; Quincy Jones: *Back On The Block, Q's Juke Joint*; The Jacksons: *Victory*; George Benson: *Give Me The Night*; Jennifer Lopez: *This Is Me ... Then, Rebirth, Brave.*

SYROWSKI, TOM

Incubus: *If Not Now When, Light Grenades*; Pearl Jam: *Lightning Bolt, Backspacer*; Seether: *Holding Onto Strings Better Left To Fray*; The Fray: *Scars And Stories*; Avril Lavigne: *The Best Damn Thing*; Cheap Trick: *The Latest*; Brandon Flowers: *Flamingo.*

TARSIA, MIKE

Teddy Pendergrass: *You And I*; Patti LaBelle: *Timeless Journey, Flame, Gems, Burnin', This Christmas, Be Yourself*: Dru Hill: *Dru Hill*; Sister Sledge: *African Eyes*; The O'Jays: *Home For Christmas, Love You To Tears, Serious.*

THOENER, DAVID

AC/DC: *For Those About To Rock We Salute You*; Aerosmith: "I Don't Wanna Miss A Thing," *Get A Grip, Nine Lives*; Brooks & Dunn: *Red Dirt Road, Tight Rope*; John Mellencamp: *Dance Naked, Uh-Huh, John Mellencamp*; Willie Nelson: *The Great Divide, Rainbow Connection*; Matchbox 20: *Exile On Mainstream, Mad Season.*

THORNALLEY, PHIL

Natalie Imbruglia: *Left Of The Middle, White Lillies Island*; The Cure: *Pornography*; Johnny Hates Jazz: *Turn Back The Clock, Tall Stories*; Pixie Lott: *Young Foolish Happy, Turn It Up*; Eliza Doolittle: *Eliza Doolittle*; Melanie C: *This Time, Reason, Beautiful Intentions*; Ronan Keating: *Ronan Keating, Destination, Ronan*; Lulu: *Back On Track.*

TOKIC, ANDRIJA

Alabama Shakes: *Boys & Girls*; Josephine Foster: *Blood Rushing, I'm A Dreamer*; Promised Land Sound: *Promised Land Sound;* Cheap Time: *Exit Smiles.*

VALENTINE, ERIC

Slash: *Apocalyptic Love, Slash*; Aqualung: *Memory Man*; Lostprophets: *Start Something*; Good Charlotte: *The Young And The Hopeless*; Queens Of The Stone Age: *Songs For The Deaf*; Smash Mouth: *Smash Mouth, Astro Lounge, Fush Yu Mang*; Third Eye Blind: *Third Eye Blind*; Hot Hot Heat: *Happiness Ltd*; All-American Rejects: *When The World Comes Down.*

VERNON, MIKE

John Mayall & The Bluesbreakers: *With Eric Clapton, A Hard Road, Crusade, Bare Wires, Diary Of A Band 1 & 2*; John Mayall: *Blues Alone, Blues From Laurel Canyon, Thru The Years*; Eric Clapton: *Crossroads*; Fleetwood Mac: *Fleetwood Mac, Mr Wonderful, English Rose, The Pious Bird Of Good Omen, Black Magic Woman*; Peter Green: *The End Of The Game*; Savoy Brown: *Shake Down, Getting To The Point*; David Bowie: *David Bowie*.

VIG, BUTCH

Nirvana: *Nevermind*; Smashing Pumpkins: *Gish, Siamese Dream*; Sonic Youth: *Dirty, Experimental Jet Set Trash And No Star*; Garbage: *Garbage, Version 2.0, Beautiful Garbage, Bleed Like Me, Not Your Kind Of People*; Green Day: *20th Century Breakdown*; Foo Fighters: *Wasting Light*.

VISCONTI, TONY

T.Rex: *Electric Warrior, The Slider*; David Bowie: *Diamond Dogs, Young Americans, Heroes, Low, Scary Monsters, Heathen, Reality*; Iggy Pop: *The Idiot*; The Moody Blues: *The Other Side of Life, Sur La Mer*; Thin Lizzy: *Bad Reputation, Live And Dangerous, Black Rose*.

VORNDICK, BIL

Charlie Haden: *Rambling Boy*; Lynn Anderson: *The Bluegrass Sessions*; Rhonda Vincent: *One Step Ahead*; Ralph Stanley II: *Stanley Blues, Clinch Mountain Sweethearts*; Bela Fleck: *Left Of Cool; Tales From The Acoustic Planet 1 & 2, UFO TOFU, Flight Of The Cosmic Hippo, Sinister Minister*; Doc Watson: *On Praying Ground, My Dear Old Southern Home*; Alison Krauss: *I've Got That Old Feeling*; Mark O'Connor: *The New Nashville Cats, Heroes*; Nashville Bluegrass Band: *Waitin' For The Hard Times To Go, Home Of The Blues*; Jerry Douglas: *Slide Rule, Glide, Great Dobro Sessions, Plant Early, Changing Channels, Under The Wire*.

WAGENER, MICHAEL

Skid Row: *Slave To The Grind, Skid Row*; Metallica: *Master Of Puppets*; Megadeth: *So Far So Good So What*; Ozzy Ozbourne: *No More Tears*; Extreme: *Pornograffiti*; Dokken: *Breaking The Chains, Tooth And Nail, Under Lock And Key*; Alice Cooper: *Constrictor, Raise Your Fist And Yell*; Great White: *Great White*.

WAS, DON

Carly Simon: *Spoilt Girl*; Bonnie Raitt: *Nick Of Time, Luck Of The Draw, Road Tested*; The B52s: *Cosmic Thing, Good Stuff*; Bob Dylan: *Under The Red Sky*; Elton John: *To Be Continued*; Willie Nelson: *Across The Borderline, Countryman*; Jackson Browne: *I'm Alive*; The Rolling Stones: *Stripped, Voodoo Lounge, Bridges To Babylon, A Bigger Bang*; Roy Orbison: *King Of Hearts*; Barenaked Ladies: *Maroon*; Hootie And The Blowfish: *Hootie And The Blowfish*; Stone Temple Pilots: *Stone Temple Pilots*; Ziggy Marley: *Spirit Of Music, Wild and Free*.

WESTON, BOB

Sebadoh: *Bakesale, Bubble & Scrape*; Superchunk: *Come Pick Me Up*; June Of 44: *Anahata, Four Great Points, Tropics And Meridians*; Polvo: *Shapes, Exploded Drawing, Today's Active Lifestyles*; Coctails: *Coctails*; Archers Of Loaf: *Vee Vee, White Trash Heroes, All The Nations Airports*.

WINSTANLEY, ALAN

Madness: *One Step Beyond, Absolutely, 7, The Rise & Fall, Keep Moving, The Liberty Of Norton Folgate, Mad Not Mad*; Stranglers: *The Raven*; The Teardrop Explodes: *Kilimanjaro*; Dexys Midnight Runners: *Too-Rye-Ay*; Elvis Costello & The Attractions: *Punch The Clock, Goodbye Cruel World*; Hothouse Flowers: *People, Horne*: Bush: *Sixteen Stone, The Science Of Things*; A-ha: *Lifelines*.

WORLEY, PAUL

Martina McBride: *The Time Has Come, The Way That I Am, Evolution, Emotion, Martina*; Pam Tillis: *Put Yourself In My Place, Homeward Looking Angel, Thunder And Roses*; The Dixie Chicks: *Wide Open Spaces, Fly*; Big & Rich: *Horse Of A Different Color, Comin' To Your City*.

WRIGHT, TOBY

Slayer: *Divine Intervention, Soundtrack To The Apocalypse*; Alice In Chains: *Jar Of Flies, Alice In Chains, Unplugged*; Korn: *Follow The Leader*; Metallica: *And Justice For All*; Mötley Crüe: *Girls Girls Girls*; Soulfly: *Primitive*.

YAKUS, SHELLY

Tom Petty: *Let Me Up, Southern Accents, Long After Dark, Hard Promises, Damn The Torpedoes*; Blue Öyster Cult: *Spectres, Agents Of Fortune*; Dire Straits: *Making Movies*; Don Henley: *The End Of The Innocence*; Van Morrison: *Moondance*; Alice Cooper: *Billion Dollar Babies, School's Out*; The Band: *Music From The Big Pink*; Stevie Nicks: *Rock A Little, The Wild Heart*; Lou Reed: *Berlin*; Patti Smith: *Dream Of Life, Easter*; Suzanne Vega: *Solitude Standing*.

YOUNG GURU

Jay-Z: *The Blueprint 3, American Gangster, Kingdom Come, Collision Course, The Blueprint*; Beyoncé: *B'Day*; Ludacris: *Theater Of The Mind*; Ghostface Killah; *Ghostdeini The Great*; Method Man: *Blackout 2*; Mariah Carey: *Charmbracelet*.

ZDAR, PHILIPPE

Cassius: *I Love Techno, The Rawkers, 15 Again, Au Reve, 1999*; Phoenix: *Wolfgang Amadeus Phoenix, United*; The Rapture: *In The Grace Of Your Love*; Two Door Cinema Club: *Tourist History*; MC Solaar: *Paradisiaque*; Drake: *Nothing Was The Same*.

ZOOK, JOE

Modest Mouse: *We Were Dead Before The Ship Even Sank*; Sheryl Crow: *C'mon C'mon*; Courtney Love: *America's Sweetheart*; One Republic: *Dreaming Out Loud*; Rancid: *Life Won't Wait*; Brooke Fraser: *Albertine*.

APPENDIX 3

Quote References

CHAPTER 1

1 Mix With The Masters (2012). Al Schmitt seminar publicity video. http://www.youtube.com/watch?v=LaEjRx9Ix2Q, April.

2 Massey, H. (2009). *Behind The Glass: Top Record Producers Tell How They Craft The Hits* (Vol. II). Backbeat Books.

3 Massey, H. (2000). *Behind The Glass: Top Record Producers Tell How They Craft The Hits* (Vol. I). Miller Freeman Books.

CHAPTER 2

1 Massey, H. (2009). *Behind The Glass: Top Record Producers Tell How They Craft The Hits* (Vol. II). Backbeat Books.

2 Massey, H. (2009). *Behind The Glass: Top Record Producers Tell How They Craft The Hits* (Vol. II). Backbeat Books.

3 Robbins, J. (1999). Interview with John Agnello. *Tape Op*, Fall.

4 Hatschek, K. (2006) *The Golden Moment: Recording Secrets From The Pros*. Backbeat Books.

5 Fritz, O. (2000). Visual Images In The Recording Studio. *Tape Op*, September/October. And Crews, E. (2010). Interview with Oz Fritz. *Tape Op*, January/February.

6 Buskin, R. (2005). Classic Tracks: Eddie Kramer. Jimi Hendrix Experience: "All Along The Watchtower." *Sound On Sound*, November.

7 Baccigaluppi, J. (2013). Steve Lillywhite Will Steer Your Boat Safely To Port. *Tape Op*, January/February.

8 Hatschek, K. (2006). *The Golden Moment: Recording Secrets From The Pros*. Backbeat Books.

9 Weiss, P. (2013). Interview with Jeff Powell. *Tape Op*, May/June.

10 Savona, A. (2005). *Console Confessions: The Great Producers In Their Own Words*. Backbeat Books.

11 Daley, D. (2004). Interview with Wyclef Jean. *Sound On Sound*, July.

12 Crane, L. (2012). Jacquire King: Finding The Truth. *Tape Op*, March/April.

13 Massey, H. (2009). *Behind The Glass: Top Record Producers Tell How They Craft The Hits* (Vol. II). Backbeat Books.

14 Crane, L. (2012). Jacquire King: Finding The Truth. *Tape Op*, March/April.

15 Massey, H. (2009). *Behind The Glass: Top Record Producers Tell How They Craft The Hits* (Vol. II). Backbeat Books.

16 Tingen, P. (2010). Inside Track: Mike Poole. Robert Plant: "Angel Dance." *Sound On Sound*, December.

17 Tingen, P. (2006). Roy Thomas Baker: Producing The Darkness' *One Way Ticket To Hell…And Back*. *Sound On Sound*, February.

18 Inglis, S. (2003). Rhett Davies & Bob Clearmountain: Recording & remixing Roxy Music's *Avalon*. *Sound On Sound*, August.

19 Massey, H. (2009). *Behind The Glass: Top Record Producers Tell How They Craft The Hits* (Vol. II). Backbeat Books.

20 Buskin, R. (2009). Classic Tracks: Jack Douglas & Jay Messina. Aerosmith: "Walk This Way." *Sound On Sound*, August.

21 Buskin, R. (1994). Don Was & Don Smith: Recording The Rolling Stones. *Sound On Sound*, December.

22 Droney, M. (2002). Interview With Dennis Sands. *Mix*, September.

23 Frost, M. (2012). John Porter: From The Smiths To The Blues. *Sound On Sound*, September.

24 Doyle, T. (2012). David Kosten: Producing Bat For Lashes & Everything Everything. *Sound On Sound*, December.

25 Sarafin, E. ("Mixerman") (2012). *Zen And The Art Of Producing*. Hal Leonard.

26 Lockwood, D. (2001). Jay Graydon: Musician & Producer. *Sound On Sound*, December.

27 Massey, H. (2009). *Behind The Glass: Top Record Producers Tell How They Craft The Hits* (Vol. II). Backbeat Books.

28 Tingen, P. (2010). Josh Homme: Producing Arctic Monkeys. *Sound On Sound*, February.

29 Savona, A. (2005) *Console Confessions: The Great Producers In Their Own Words*. Backbeat Books.

CHAPTER 3

1 Tingen, P. (2009). Inside Track: Mike Fraser. AC/DC: *Black Ice*. *Sound On Sound*, January.

2 Clark, R. (2011). *Mixing, Recording, And Producing Techniques Of The Pros*. Cengage Learning PTR.

3 Tingen, P. (2010). Inside Track: David R. Ferguson. Johnny Cash: "Ain't No Grave." *Sound On Sound*, June.

4 Senior, M. (2009). Bruce Swedien: Recording Michael Jackson. *Sound On Sound*, November.

5 Tingen, P. (2009). Inside Track: Mike Fraser. AC/DC: *Black Ice*. *Sound On Sound*, January.

6 Massey, H. (2009). *Behind The Glass: Top Record Producers Tell How They Craft The Hits* (Vol. II). Backbeat Books.

7 Crane, L. (2013). Ken Caillat: Making History And Rumours. *Tape Op*, July/August.

8 Flint, T. (2001). Tony Platt: Rock Island Life. *Sound On Sound*, April.

9 Crane, L. & Smith, C. (1999). Interview with Joe Chiccarelli. *Tape Op*, Fall.

10 Daley, D. (2004). Mixing Engineers: The Next Generation. *Sound On Sound*, February.

11 Savona, A. (2005). *Console Confessions: The Great Producers In Their Own Words*. Backbeat Books.

12 Frost, M. (2009). Jez Coad. Simple Minds: Recording *Graffiti Soul*. *Sound On Sound*, July.

13 Bilerman, H. & Foot, J. (2010). Interview with Bob Johnston. *Tape Op*, November/December.

14 Massey, H. (2009). *Behind The Glass: Top Record Producers Tell How They Craft The Hits* (Vol. II). Backbeat Books.

15 Droney, M. (2002). Interview with Neil Dorfsman. *Mix*, April.

16 Droney, M. (1999) Interview with Brian Malouf. *Mix*, December.

CHAPTER 4

1 Massey, H. (2009). *Behind The Glass: Top Record Producers Tell How They Craft The Hits* (Vol. II). Backbeat Books.

2 Stavrou, M.P. (2003). *Mixing With Your Mind*. Flux Research.

3 Simons, D. (2011). Mark Neill: Recording The Black Keys At Muscle Shoals. *Sound On Sound*, August.

4 Massey, H. (2000). *Behind The Glass: Top Record Producers Tell How They Craft The Hits* (Vol. I). Miller Freeman Books.

5 Buskin, R. (2003). Classic Tracks: Bruce Botnick. The Doors: "Strange Days." *Sound On Sound*, December.

6 Massey, H. (2000). *Behind The Glass: Top Record Producers Tell How They Craft The Hits* (Vol. I). Miller Freeman Books.

7 Massey, H. (2000). *Behind The Glass: Top Record Producers Tell How They Craft The Hits* (Vol. I). Miller Freeman Books.

8 Massey, H. (2000). *Behind The Glass: Top Record Producers Tell How They Craft The Hits* (Vol. I). Miller Freeman Books.

9 Massey, H. (2000) *Behind The Glass: Top Record Producers Tell How They Craft The Hits* (Vol. I). Miller Freeman Books.

10 Massey, H. (2000). *Behind The Glass: Top Record Producers Tell How They Craft The Hits* (Vol. I). Miller Freeman Books.

11 Tingen, P. (2005). Steve Albini: Sound Engineer Extraordinaire. *Sound On Sound*, September.

12 Buskin, R. (2004). Classic Tracks: John Hudson. Tina Turner: "What's Love Got To Do With It?" *Sound On Sound*, May.

13 Lockwood, D. (2001). Jay Graydon: Musician & Producer. *Sound On Sound*, December.

14 Tingen, P. (2000). Pierre Marchand: Producing Sarah McLachlan. *Sound On Sound*, March.

15 Recording seminar at British Grove Studios, May 2010.

16 Buskin, R. (2004). Classic Tracks: John Hudson. Tina Turner: "What's Love Got To Do With It?" *Sound On Sound*, May.

17 Massey, H. (2009). *Behind The Glass: Top Record Producers Tell How They Craft The Hits* (Vol. II). Backbeat Books.

18 Massey, H. (2009). *Behind The Glass: Top Record Producers Tell How They Craft The Hits* (Vol. II). Backbeat Books.

19 Buskin, R. (2004) Classic Tracks: John Hudson. Tina Turner: 'What's Love Got To Do With It?'. *Sound On Sound*, May.

20 Massey, H. (2000). *Behind The Glass: Top Record Producers Tell How They Craft The Hits* (Vol. I). Miller Freeman Books.

21 Massey, H. (2000). *Behind The Glass: Top Record Producers Tell How They Craft The Hits* (Vol. I). Miller Freeman Books.

22 Massey, H. (2000). *Behind The Glass: Top Record Producers Tell How They Craft The Hits* (Vol. I). Miller Freeman Books.

23 Owsinski, B. (2009). *The Recording Engineer's Handbook*. Cengage Learning PTR.

24 Tingen, P. (2010). Phil Thornalley: From Rock Producer To Pop Songwriter. *Sound On Sound*, June.

25 Savona, A. (2005). *Console Confessions: The Great Producers In Their Own Words*. Backbeat Books.

26 Tingen, P. (2009). Inside Track: Young Guru. Jay Z: *The Blueprint 3*. *Sound On Sound*, December.

27 Senior, M. (2009). Bruce Swedien: Recording Michael Jackson. *Sound On Sound*, November.

28 Sillitoe, S. (1994). Stephen Street: Recording Blur & The Cranberries. *Sound On Sound*, July.

29 Senior, M. (2000). Jonny Dollar & Simon Palmskin: Recording Gabrielle's "Rise." *Sound On Sound*, July.

30 Buskin, R. (2006) Classic Tracks: Alan Winstanley. Madness: "Our House." *Sound On Sound*, June.

31 Massey, H. (2000). *Behind The Glass: Top Record Producers Tell How They Craft The Hits* (Vol. I). Miller Freeman Books.

32 Humberstone, N. (2003) Neil Davidge: Recording Massive Attack's *100th Window*. *Sound On Sound*, April.

33 Senior, M. (2000) Jonny Dollar & Simon Palmskin: Recording Gabrielle's "Rise." *Sound On Sound*, July.

34 Stavrou, M P. (2003). *Mixing With Your Mind*. Flux Research.

35 Tingen, P. (2005). Joe Barresi: Recording Queens Of The Stone Age. *Sound On Sound*, July.

36 Massey, H. (2000). *Behind The Glass: Top Record Producers Tell How They Craft The Hits* (Vol. I). Miller Freeman Books.

37 Buskin, R. (2003). Rich Costey: Recording Muse's *Absolution*. *Sound On Sound*, December.

38 Droney, M. (1999). Interview with Leslie Ann Jones. *Mix*, August.

39 Buskin, R. (2009). Classic Tracks: Jack Douglas & Jay Messina. Aerosmith: "Walk This Way." *Sound On Sound*, August.

40 Crane, L. & Smith, C. (1999). Interview with Joe Chiccarelli. *Tape Op*, Fall.

41 Sillitoe, S. (1999). Stephen Street: Producing Blur, Cranberries & Catatonia. *Sound On Sound*, August.

42 Inglis, S. (2009). Paul Epworth: Producing Almost Everyone. *Sound On Sound*, January.

43 Owsinski, B. (2010). *The Music Producer's Handbook*. Hal Leonard.

44 McKenzie, A. (2010). Interview with Ben Allen. *Tape Op*, March/April.

45 Massey, H. (2009). *Behind The Glass: Top Record Producers Tell How They Craft The Hits* (Vol. II). Backbeat Books.

46 Saxon, J. (2011). Dave Jerden: Really Wrong Productions, Handle With Care. *Tape Op*, November/December.

47 Senior, M. Author's personal experience, Spring 1995.

48 Owsinski, B. (2010). *The Music Producer's Handbook*. Hal Leonard.

49 Saxon, J. (2011). Dave Jerden: Really Wrong Productions, Handle With Care. *Tape Op*, November/December.

50 Massey, H. (2009). *Behind The Glass: Top Record Producers Tell How They Craft The Hits* (Vol. II). Backbeat Books.

51 Massey, H. (2000). *Behind The Glass: Top Record Producers Tell How They Craft The Hits* (Vol. I). Miller Freeman Books.

52 Massey, H. (2000). *Behind The Glass: Top Record Producers Tell How They Craft The Hits* (Vol. I). Miller Freeman Books.

53 Doyle, T. (2008). Stuart Price: Producing Seal & Madonna. *Sound On Sound*, February.

54 Massey, H. (2000). *Behind The Glass: Top Record Producers Tell How They Craft The Hits* (Vol. I). Miller Freeman Books.

55 Massey, H. (2000). *Behind The Glass: Top Record Producers Tell How They Craft The Hits* (Vol. I). Miller Freeman Books.

56 Flint, T. (2001). Tony Platt: Rock Island Life. *Sound On Sound*, April.

57 Massey, H. (2000). *Behind The Glass: Top Record Producers Tell How They Craft The Hits* (Vol. I). Miller Freeman Books.

58 Tingen, P. (2011). Inside Track: Mike Shipley. Alison Krauss: "Paper Airplane." *Sound On Sound*, July.

59 Savona, A. (2005). *Console Confessions: The Great Producers In Their Own Words*. Backbeat Books.
60 Tingen, P. (2012). Inside Track: Jeff Bhasker. Fun: "We Are Young." *Sound On Sound*, December.
61 Massey, H. (2000). *Behind The Glass: Top Record Producers Tell How They Craft The Hits* (Vol. I). Miller Freeman Books.
62 Inglis, S. (2002). Butch Vig, Duke Erikson, & Billy Bush: Recording Garbage. *Sound On Sound*, June.
63 Tingen, P. (2010). Phil Thornalley: From Rock Producer To Pop Songwriter. *Sound On Sound*, June.
64 Flint, T. (2001). Tony Platt: Rock Island Life. *Sound On Sound*, April.

CHAPTER 5

1 Hatschek, K. (2006). *The Golden Moment: Recording Secrets From The Pros*. Backbeat Books.
2 Massey, H. (2000). *Behind The Glass: Top Record Producers Tell How They Craft The Hits* (Vol. I). Miller Freeman Books.
3 Hatschek, K. (2006). *The Golden Moment: Recording Secrets From The Pros*. Backbeat Books.
4 Massey, H. (2000). *Behind The Glass: Top Record Producers Tell How They Craft The Hits* (Vol. I). Miller Freeman Books.
5 Senior, M. (2000). Mousse T: Recording Tom Jones' "Sex Bomb." *Sound On Sound*, August.
6 Buskin, R. (2004). Classic Tracks: Bill Price. The Sex Pistols: "Anarchy In The UK." *Sound On Sound*, September.
7 Tingen, P. (1994). Robbie Adams: Recording U2's *Achtung Baby* & *Zooropa*. *Sound On Sound*, March.
8 Tingen, P. (1997). Flood & Howie B: Producing U2's *Pop*. *Sound On Sound*, July.
9 Massey, H. (2009). *Behind The Glass: Top Record Producers Tell How They Craft The Hits* (Vol. II). Backbeat Books.
10 Inglis, S. (2001). Stephen Harris: The Waiting Game. *Sound On Sound, October.*
11 Buskin, R. (2005). Classic Tracks: The Stone Roses "Fool's Gold." *Sound On Sound*, February.
12 Sokal, R. (2002). GGGarth Richardson: A Sergeant Leading His Troop Into Battle. *Tape Op*, March.
13 Inglis, S. (1999). Dave Eringa: Recording The Manic Street Preachers' "If You Tolerate This Your Children Will Be Next." *Sound On Sound*, April.
14 Buskin, R. (2006). Classic Tracks: Bob Clearmountain. Bryan Adams: "Run To You." *Sound On Sound*, July.
15 Tingen, P. (1999). Jim Scott: Recording Red Hot Chili Peppers' *Californication*. *Sound On Sound*, December.
16 Massey, H. (2000). *Behind The Glass: Top Record Producers Tell How They Craft The Hits* (Vol. I). Miller Freeman Books.
17 Buskin, R. (2003). Rich Costey: Recording Muse's *Absolution*. *Sound On Sound*, December.
18 Tingen, P. (2007). Inside Track: Joe Chiccarelli. The White Stripes: *Icky Thump*. *Sound On Sound*, October.
19 Tingen, P. (2009). Inside Track: Mike Fraser. AC/DC: *Black Ice*. *Sound On Sound*, January.
20 Buskin, R. (2012). Classic Tracks: Narada Michael Walden & Michael Barbiero. Whitney Houston: "I Wanna Dance With Somebody." *Sound On Sound*, May.
21 Senior, M. (2009). Bruce Swedien: Recording Michael Jackson. *Sound On Sound*, November.

22 Buskin, R. (1996). Hugh Padgham: The Master Craftsman Behind Sting & Phil Collins. *Sound On Sound*, October.

23 Buskin, R. (2005). Classic Tracks: Stephen Hague. New Order: "True Faith." *Sound On Sound*, March.

24 Massey, H. (2009). *Behind The Glass: Top Record Producers Tell How They Craft The Hits* (Vol. II). Backbeat Books.

25 Tingen, P. (2005). Steve Albini: Sound Engineer Extraordinaire. *Sound On Sound*, September.

26 Sillitoe, S. (1999). Mike Hedges: Recording Travis' "Why Does It Always Rain On Me?" *Sound On Sound*, November.

27 Buskin, R. (2005). Classic Tracks: Stephen Hague. New Order: "True Faith." *Sound On Sound*, March.

28 Massey, H. (2009). *Behind The Glass: Top Record Producers Tell How They Craft The Hits* (Vol. II). Backbeat Books.

29 Hatschek, K. (2006). *The Golden Moment: Recording Secrets From The Pros*. Backbeat Books.

30 Senior, M. (2009). Bruce Swedien: Recording Michael Jackson. *Sound On Sound*, November.

31 Senior, M. (2000). Bird & Bush: Producing Stereophonics. *Sound On Sound*, March.

32 Inglis, S. (2009). Paul Epworth: Producing Almost Everyone. *Sound On Sound*, January.

33 Senior, M. (2009). Bruce Swedien: Recording Michael Jackson. *Sound On Sound*, November.

34 Hatschek, K. (2006). *The Golden Moment: Recording Secrets From The Pros*. Backbeat Books.

35 Clark, R. (2011). *Mixing, Recording, And Producing Techniques Of The Pros*. Cengage Learning PTR.

36 Massey, H. (2000). *Behind The Glass: Top Record Producers Tell How They Craft The Hits* (Vol. I). Miller Freeman Books.

37 Tingen, P. (1996). Steve Levine: Digital Domain. *Sound On Sound*, December.

38 Bruce, B. (1999). Al Stone: Recording Jamiroquai's "Supersonic." *Sound On Sound*, December.

39 Buskin, R. (2009). Classic Tracks: Jack Douglas & Jay Messina. Aerosmith: "Walk This Way." *Sound On Sound*, August.

CHAPTER 6

1 Droney, M. (1998). Interview with Al Schmitt. *Mix*, October.

2 Senior, M. (2009). Bruce Swedien: Recording Michael Jackson. *Sound On Sound*, November.

3 Owsinski, B. (2009). *The Recording Engineer's Handbook*. Cengage Learning PTR.

4 Jackson, M. (2013). Stuart Sullivan. Ready For It. *Tape Op*, March.

5 Massey, H. (2009). *Behind The Glass: Top Record Producers Tell How They Craft The Hits* (Vol. II). Backbeat Books.

6 Davies, B. (2009). Elliot Mazer, A Da Vinci In His Own Time. *Tape Op*, July.

7 Owsinski, B. (2009). *The Recording Engineer's Handbook*. Cengage Learning PTR.

8 Mix With The Masters (2012). Al Schmitt seminar publicity video. http://www.youtube.com/watch?v=LaEjRx9Ix2Q, April.

9 Tingen, P. (2008). Inside Track: Joe Zook. One Republic: "Stop & Stare." *Sound On Sound*, June.

10 Baccigaluppi, J. & Crane L. (2012). Who's Trevor Horn? *Tape Op*, May.

11 Flint, T. (2001). Tony Platt: Rock Island Life. *Sound On Sound*, April.

12 Tingen, P. (2000). Pierre Marchand: Producing Sarah McLachlan. *Sound On Sound*, March.

13 Daley, D. (2002). Nashville Recording: The American Way. *Sound On Sound*, October.

14 Multi-Platinum Production (2010). *The Magic Of A Nashville Tuned Guitar.* http://www.youtube.com/watch?v=YFjhTdmbSIg, December.

15 Massey, H. (2009). *Behind The Glass: Top Record Producers Tell How They Craft The Hits* (Vol. II). Backbeat Books.

16 Massey, H. (2009). *Behind The Glass: Top Record Producers Tell How They Craft The Hits* (Vol. II). Backbeat Books.

17 Hatschek, K. (2006). *The Golden Moment: Recording Secrets From The Pros*. Backbeat Books.

18 Crane, L. (2000). Interview With Bob Weston. *Tape Op*, July/August.

19 Massey, H. (2000). *Behind The Glass: Top Record Producers Tell How They Craft The Hits* (Vol. I). Miller Freeman Books.

20 Tingen, P. (2011). Inside Track: Phil Ramone & Andy Smith. Paul Simon: *So Beautiful So What*. *Sound On Sound*, October.

21 Savona, A. (2005). *Console Confessions: The Great Producers In Their Own Words*. Backbeat Books.

22 Owsinski, B. (2009). *The Recording Engineer's Handbook*. Cengage Learning PTR.

23 Massey, H. (2000). *Behind The Glass: Top Record Producers Tell How They Craft The Hits* (Vol. I). Miller Freeman Books.

24 Stavrou, M.P. (2003). *Mixing With Your Mind*. Flux Research.

25 Massey, H. (2009). *Behind The Glass: Top Record Producers Tell How They Craft The Hits* (Vol. II). Backbeat Books.

26 Lockwood, D. (2001). Jay Graydon: Musician & Producer. *Sound On Sound*, December.

27 Savona, A. (2005). *Console Confessions: The Great Producers In Their Own Words*. Backbeat Books.

28 Senior, M. (2009). Bruce Swedien: Recording Michael Jackson. *Sound On Sound*, November.

29 Borgerson, B. (2002). Shelly Yakus: Nuggets Of Wisdom. *Tape Op*, September.

30 Savona, A. (2005). *Console Confessions: The Great Producers In Their Own Words*. Backbeat Books.

31 Mix With The Masters (2012). Al Schmitt seminar publicity video. http://www.youtube.com/watch?v=LaEjRx9Ix2Q, April.

32 Droney, M. (2001). Interview with David Thoener. *Mix*, January.

33 Clark, R. (2011). *Mixing, Recording, And Producing Techniques Of The Pros*. Cengage Learning PTR.

34 Jackson, B. (2008). Recording Bluegrass Instruments. *Mix*, May.

35 Daley, D. (2002). Nashville Recording: The American Way. *Sound On Sound*, October.

36 Recording seminar at British Grove Studios, May 2010.

37 Stavrou, M.P. (2003) *Mixing With Your Mind*. Flux Research.

38 Clark, R. (2011). *Mixing, Recording, And Producing Techniques Of The Pros*. Cengage Learning PTR.

39 Tingen, P. (2005). Steve Albini: Sound Engineer Extraordinaire. *Sound On Sound*, September.

40 Massey, H. (2000). *Behind The Glass: Top Record Producers Tell How They Craft The Hits* (Vol. I). Miller Freeman Books.

41 Tingen, P. (1999). Jim Scott: Recording Red Hot Chili Peppers' *Californication*. *Sound On Sound*, December.

42 Clark, R. (2011). *Mixing, Recording, And Producing Techniques Of The Pros*. Cengage Learning PTR.

43 Crane, L. (2000). Jon Brion: Producer, Session Player, Film Scorer, Etc…*Tape Op*, July/August.

44 Massey, H. (2000). *Behind The Glass: Top Record Producers Tell How They Craft The Hits* (Vol. I). Miller Freeman Books.

45 Tingen, P. (2009). Inside Track: Mike Fraser. AC/DC: *Black Ice*. *Sound On Sound*, January.

46 Nichols, R. (2013). *The Roger Nichols Recording Method: A Primer For The 21st Century Audio Engineer.* Alfred Publishing.

47 Frost, M. (2012). John Porter: From The Smiths To The Blues. *Sound On Sound*, September.

48 Cunningham, M. (1997). John Leckie: True Brit. *Sound On Sound*, May.

49 Massey, H. (2000). *Behind The Glass: Top Record Producers Tell How They Craft The Hits* (Vol. I). Miller Freeman Books.

50 Massey, H. (2000). *Behind The Glass: Top Record Producers Tell How They Craft The Hits* (Vol. I). Miller Freeman Books.

51 Buskin, R. (2009). Classic Tracks: Mitch Easter. REM: "Radio Free Europe." *Sound On Sound*, November.

52 Droney, M. (1998). Interview with Al Schmitt. *Mix*, October.

53 Owsinski, B. (2009). *The Recording Engineer's Handbook*. Cengage Learning PTR.

54 Crane, L. (2000). Interview with Bob Weston. *Tape Op*, July/August.

55 Massey, H. (2009). *Behind The Glass: Top Record Producers Tell How They Craft The Hits* (Vol. II). Miller Freeman Books.

56 Cunningham, M. (1995). Roy Thomas Baker & Gary Langan: The Making Of Queen's "Bohemian Rhapsody." *Sound On Sound*, October.

57 Tingen, P. (2011). Inside Track: Tom Syrowski. Incubus: "Adolescents." *Sound On Sound*, November.

58 Savona, A. (2005). *Console Confessions: The Great Producers In Their Own Words*. Backbeat Books.

59 Royer Microphones *Royer Demo CD#1*. http://www.royerlabs.com/royerdemocd1.html.

60 Stavrou, M.P. (2003). *Mixing With Your Mind*. Flux Research.

61 Hatschek, K. (2006). *The Golden Moment: Recording Secrets From The Pros*. Backbeat Books.

62 Savona, A. (2005). *Console Confessions: The Great Producers In Their Own Words*. Backbeat Books.

63 Owsinski, B. (2009). *The Recording Engineer's Handbook*. Cengage Learning PTR.

64 Tingen, P. (2008). Inside Track: Robert Carranza. Jack Johnson: *Sleep Through The Static*. *Sound On Sound*, May.

65 Massey, H. (2009). *Behind The Glass: Top Record Producers Tell How They Craft The Hits* (Vol. II). Backbeat Books.

CHAPTER 7

1 Massey, H. (2000). *Behind The Glass: Top Record Producers Tell How They Craft The Hits* (Vol. I). Miller Freeman Books.

2 Daley, D. (2002). Nashville Recording: The American Way. *Sound On Sound*, October.

3 Massey, H. (2000). *Behind The Glass: Top Record Producers Tell How They Craft The Hits* (Vol. I). Miller Freeman Books. And Stevenson, P. (2004). Andy Johns archival interview. *Tape Op*, January.

4 Tingen, P. (2006). Roy Thomas Baker: Producing The Darkness' *One Way Ticket to Hell…And Back*. *Sound On Sound*, February.

5 Inglis, S. (2000). Ken Nelson: Recording Coldplay's *Parachutes*. *Sound On Sound*, October.

6 Cunningham, M. (1997). John Leckie: True Brit. *Sound On Sound*, May.

7 Hatschek, K. (2006). *The Golden Moment: Recording Secrets From The Pros*. Backbeat Books.

8 Buskin, R. (2005). Classic Tracks: Stephen Street. The Smiths: "The Queen Is Dead." *Sound On Sound*, January.

9 Tingen, P. (2005). Steve Albini: Sound Engineer Extraordinaire. *Sound On Sound*, September. And B. Owsinski (2009). *The Recording Engineer's Handbook*. Cengage Learning PTR.

10 Tingen, P. (1999). Jim Scott: Recording Red Hot Chili Peppers' *Californication*. *Sound On Sound*, December.

11 Owsinski, B. (2009). *The Recording Engineer's Handbook*. Cengage Learning PTR.

12 Buskin, R. (2004). Classic Tracks: John Hudson. Tina Turner: "What's Love Got To Do With It?" *Sound On Sound*, May.

13 Tingen, P. (2006). Jim Abbiss: Producing Kasabian & Arctic Monkeys. *Sound On Sound*, September.

14 Tingen, P. (2005). Joe Barresi: Recording Queens Of The Stone Age. *Sound On Sound*, July.

15 Mayes-Wright, C. (2007). Jerry Boys & Livingston Studios. *Sound On Sound*, September.

16 Tingen, P. (2009). Inside Track: Mike Fraser. AC/DC: *Black Ice*. *Sound On Sound*, January.

17 Doyle, T. (2009). Dan Austin & Jez Williams: Producing Doves' *Kingdom Of Rust*. *Sound On Sound*, July.

18 Tingen, P. (2010). Inside Track: Mike Poole. Robert Plant: "Angel Dance." *Sound On Sound*, December.

19 Sarafin, E. ("Mixerman") (2008). The Womb Forums post. http://www.thewombforums.com/showpost.php?p=122671&postcount=15, April.

20 Wagener, M. (2002) Gearslutz forums post. http://www.gearslutz.com/board/1461-post11.html, June.

21 Lewisohn, M (1988). *The Beatles Recording Sessions*. Harmony Books.

22 Jackson, B. (2008). Recording Bluegrass Instruments. *Mix*, May.

23 Jackson, B. (2008). Recording Bluegrass Instruments. *Mix*, May.

24 Tingen, P. (2010). Inside Track: Eric Valentine. Recording Slash. *Sound On Sound*, July.

25 Massey, H. (2000). *Behind The Glass: Top Record Producers Tell How They Craft The Hits* (Vol. I). Miller Freeman Books.

26 Daley, D. (2002). Nashville Recording: The American Way. *Sound On Sound*, October.

27 Massey, H. (2000). *Behind The Glass: Top Record Producers Tell How They Craft The Hits* (Vol. I). Miller Freeman Books.

28 Massey, H. (2000). *Behind The Glass: Top Record Producers Tell How They Craft The Hits* (Vol. I). Miller Freeman Books.

29 Jackson, B. (2003). Recording Piano. *Mix*, November.

30 Massey, H. (2000). *Behind The Glass: Top Record Producers Tell How They Craft The Hits* (Vol. I). Miller Freeman Books.

31 Buskin, R. (1994). Don Was & Don Smith: Recording The Rolling Stones. *Sound On Sound*, December.

32 Daley, D. (2005). Toby Wright: Recording Hard Rock. *Sound On Sound*, December.

33 Massey, H. (2000). *Behind The Glass: Top Record Producers Tell How They Craft The Hits* (Vol. I). Miller Freeman Books.

34 Daley, D. (2002). Nashville Recording: The American Way. *Sound On Sound*, October.

35 Massey, H. (2000). *Behind The Glass: Top Record Producers Tell How They Craft The Hits* (Vol. I). Miller Freeman Books.

36 Tingen, P. (1994). Robbie Adams: Recording U2's *Achtung Baby* & *Zooropa*. *Sound On Sound*, March.

37 Tingen, P. (2005). Steve Albini: Sound Engineer Extraordinaire. *Sound On Sound*, September.

38 Jackson, B. (2008). Recording Bluegrass Instruments. *Mix*, May.

39 Jackson, B. (2008). Recording Bluegrass Instruments. *Mix*, May.

40 Buskin, R. (2005). Classic Tracks; Glyn Johns & Jon Astley. The Who: "Who Are You?" *Sound On Sound*, May.

41 Buskin, R. (2004). Classic Tracks: Bill Price. The Sex Pistols: "Anarchy In The UK." *Sound On Sound*, September.

42 Buskin, R. (2006). Classic Tracks: Alan Winstanley. Madness: "Our House." *Sound On Sound*, June.

43 Massey, H. (2000). *Behind The Glass: Top Record Producers Tell How They Craft The Hits* (Vol. I). Miller Freeman Books.

44 Massey, H. (2000). *Behind The Glass: Top Record Producers Tell How They Craft The Hits* (Vol. I). Miller Freeman Books.

45 Droney, M. (2003). Interview with Elliott Scheiner. *Mix*, January.

46 Buskin, R. (2003). David Bowie & Toni Visconti: Recording *Reality*. *Sound On Sound*, October.

47 Massey, H. (2000). *Behind The Glass: Top Record Producers Tell How They Craft The Hits* (Vol. I). Miller Freeman Books.

48 Massey, H. (2000). *Behind The Glass: Top Record Producers Tell How They Craft The Hits* (Vol. I). Miller Freeman Books.

49 Owsinski, B. (2009) *The Recording Engineer's Handbook*. Cengage Learning PTR.

50 Massey, H. (2000). *Behind The Glass: Top Record Producers Tell How They Craft The Hits* (Vol. I). Miller Freeman Books.

51 Buskin, R. (2004). Classic Tracks: Jon Kelly. Kate Bush: "Wuthering Heights." *Sound On Sound*, June.

52 Tingen, P. (1994). Robbie Adams: Recording U2's *Achtung Baby* & *Zooropa*. *Sound On Sound*, March.

53 Daley, D. (2004). Bob Bullock: Recording Shania Twain's *Up! Sound On Sound*, August.

54 Buskin, R. (2010). Classic Tracks: Bob Clearmountain & Toby Scott. Bruce Springsteen: "Born In The USA." *Sound On Sound*, March.

55 Senior, M. (2003). Husky S. Hoskulds: Recording Solomon Burke's *Don't Give Up On Me*. *Sound On Sound*, July.

56 Massey, H. (2000). *Behind The Glass: Top Record Producers Tell How They Craft The Hits* (Vol. I). Miller Freeman Books.

57 Jackson, M. (2013). Stuart Sullivan. Ready For It. *Tape Op*, March.

58 Bell, M. (1995). Simon Dawson: Recording The Stone Roses' *Second Coming*. *Sound On Sound*, May.

59 Greeves, D. (2003). Ben Hillier: Recording Blur, Tom McRae & Elbow. *Sound On Sound*, July.

60 Buskin, R. (2006). Classic Tracks: Bob Clearmountain. Bryan Adams: "Run To You." *Sound On Sound*, July.

61 Clark, R. (2011) *Mixing, Recording, And Producing Techniques Of The Pros*. Cengage Learning PTR.

62 Buskin, R. (2007). Classic Tracks: Jason Corsaro. Madonna "Like A Virgin." *Sound On Sound*, September.

63 Tingen, P. (2005). Steve Albini: Sound Engineer Extraordinaire. *Sound On Sound*, September.

64 Stevenson, P. (2004). Andy Johns Archival Interview. *Tape Op*, January.

65 Buskin, R. (2000). Glen Kolotkin: Recording Santana. *Sound On Sound*, June.

66 Droney, M. (1998). Interview With Dave Jerden. *Mix*, December.

67 Tingen, P. (2010). Inside Track: Eric Valentine. Recording Slash. *Sound On Sound*, July.

68 Massey, H. (2000). *Behind The Glass: Top Record Producers Tell How They Craft The Hits* (Vol. I). Miller Freeman Books.

69 Inglis, S. (2000). Ken Nelson: Recording Coldplay's *Parachutes*. *Sound On Sound*, October.

70 Frost, M. (2009). Jez Coad. Simple Minds: Recording *Graffiti Soul*. *Sound On Sound*, July.

71 Doyle, T. (2009). Dan Austin & Jez Williams: Producing Doves' *Kingdom Of Rust*. *Sound On Sound*, July.

72 Senior, M. (2009). Bruce Swedien: Recording Michael Jackson. *Sound On Sound*, November.

73 Buskin, R. (2007). Classic Tracks: Gareth Jones. Depeche Mode: "People Are People." *Sound On Sound*, February.

74 Tingen, P. (1994). Daniel Lanois: For The Beauty Of Wynona, Vintage Gear. *Sound On Sound*, September.

75 Tingen, P. (2008). Inside Track: Eric Rosse. Sara Bareilles: "Love Song." *Sound On Sound*, September.

76 Tingen, P. (2011). Daniel Lanois & Mark Howard: Recording Neil Young's *Le Noise*. *Sound On Sound*, February.

77 Massey, H. (2000). *Behind The Glass: Top Record Producers Tell How They Craft The Hits* (Vol. I). Miller Freeman Books.

78 Frost, M. (2012). John Porter: From The Smiths To The Blues. *Sound On Sound*, September.

CHAPTER 8

1 Stavrou, M.P. (2003). *Mixing With Your Mind*. Flux Research.

2 Murray, S. (2000). Interview with Tchad Blake. *Tape Op*, March/April.

3 Owsinski, B. (2009). *The Recording Engineer's Handbook*. Cengage Learning PTR.

4 Jackson, B. (2008). Recording Bluegrass Instruments. *Mix*, May.

5 Tingen, P. (2011). Inside Track: Mike Shipley. Alison Krauss: "Paper Airplane." *Sound On Sound*, July.

6 Jackson, B. (2008). Recording Bluegrass Instruments. *Mix*, May. And Daley, D. (2010). Nashville Guitars: Recording Today's Country Guitar Sounds. *Sound On Sound*, December.

7 Flint, T. (2001). Tony Platt: Rock Island Life. *Sound On Sound*, April.

8 Jackson, B. (2008). Recording Bluegrass Instruments. *Mix*, May.

CHAPTER 9

1 Vdovin, M. (2009). Leslie Ann Jones: Anticipating And Succeeding. *Tape Op*, November.

2 Crane, L. (1999). Interview With Jack Endino. *Tape Op*, Summer.

3 Clark, R. (2011). *Mixing, Recording, And Producing Techniques Of The Pros*. Cengage Learning PTR.

4 Buskin, R. (2002). Gordon Raphael: Producing The Strokes. *Sound On Sound*, April.

5 Massey, H. (2009). *Behind The Glass: Top Record Producers Tell How They Craft The Hits* (Vol. II). Backbeat Books.

6 Inglis, S. (2013). Philip Hobbs & Linn Records: Recording The Dunedin Consort. *Sound On Sound*, April.

7 Owsinski, B. (2009). *The Recording Engineer's Handbook*. Cengage Learning PTR.

8 Sengpiel, E. *Decca Tree Recording mit Neumann-Druckempfängern M50*. http://www.sengpielaudio.com?DeccaTreeRecordingM50.pdf.

9 Streicher, R. *The Decca Tree—It's Not Just For Stereo Any More.* http://www.wesdooley.com/pdf/Surround_Sound_Decca_Tree-urtext.pdf.

10 Royer Microphones. *Alan Meyerson Recording The Score For* Pirates Of The Caribbean: Dead Man's Chest. http://www.royerlabs.com/pirates.html.

11 Streicher, R. *The Decca Tree—It's Not Just For Stereo Any More.* http://www.wesdooley.com/pdf/Surround_Sound_Decca_Tree-urtext.pdf.

12 Massey, H. (2009). *Behind The Glass: Top Record Producers Tell How They Craft The Hits* (Vol. II). Backbeat Books.

13 Faulkner, T. (2009). Gearslutz forums post. http://www.gearslutz.com/board/3981037-post29.html, March.

14 Owsinski, B. (2009). *The Recording Engineer's Handbook.* Cengage Learning PTR.

15 Droney, M. (1994). Interview With Shawn Murphy. *Mix*, February.

16 Stavrou, M.P. (2003). *Mixing With Your Mind.* Flux Research.

17 Buskin, R. (1997). Butch Vig: Nevermind The Garbage. *Sound On Sound*, March.

18 Swedien, B. (2009). *Make Mine Music.* Hal Leonard.

19 Massey, H. (2009). *Behind The Glass: Top Record Producers Tell How They Craft The Hits* (Vol. II). Backbeat Books.

20 Droney, M. (1994). Interview with Shawn Murphy. *Mix*, February.

21 Crane, L. (2011). Reinhold Mack: The Invisible Man. *Tape Op*, January.

22 Owsinski, B. (2009). *The Recording Engineer's Handbook.* Cengage Learning PTR.

23 Faulkner, T. (2009). Gearslutz forums post. http://www.gearslutz.com/board/3981037-post29.html, March.

24 Daley, D. (2004). Bones Howe & Tom Waits: The Odd Couple? *Sound On Sound*, February.

25 Crane, L. (2011). Reinhold Mack: The Invisible Man. *Tape Op*, January.

26 Crane, L. (2011). Jack Miller: Arizona's National Treasure. *Tape Op*, November.

27 Clark, R. (2011). *Mixing, Recording, And Producing Techniques Of The Pros.* Cengage Learning PTR.

28 Owsinski, B. (2009). *The Recording Engineer's Handbook.* Cengage Learning PTR.

29 Clark, R. (2011). *Mixing, Recording, And Producing Techniques Of The Pros.* Cengage Learning PTR.

30 Clark, R. (2011). *Mixing, Recording, And Producing Techniques Of The Pros.* Cengage Learning PTR.

CHAPTER 10

1 Senior, M. (2003). S. Husky Hoskulds: Recording Solomon Burke's *Don't Give Up On Me. Sound On Sound*, July.

2 Owsinski, B. (2009). *The Recording Engineer's Handbook.* Cengage Learning PTR.

3 Crane, L. & Eckman, C. (1999). Sharing Food And Conversation With Phill Brown. *Tape Op*, Spring.

4 Owsinski, B. (2009). *The Recording Engineer's Handbook.* Cengage Learning PTR.

5 Daley, D. (2004). Bones Howe & Tom Waits: The Odd Couple? *Sound On Sound*, February.

6 Fiegel, E. (2011). Wayne Moss: Cinderella Sound, Nashville. *Sound On Sound*, October.

7 Droney, M. (1998). Interview with Al Schmitt. *Mix*, October.

8 Droney, M. (2001). Interview with Armin Steiner. *Mix*, May.

9 Droney, M. (1998). Interview with Al Schmitt. *Mix*, October.

10 Massey, H. (2000). *Behind The Glass: Top Record Producers Tell How They Craft The Hits* (Vol. I). Miller Freeman Books.

11 Doyle, T. (2008). Nick Launay: Producing Nick Cave, Kate Bush…And Snoop Dogg. *Sound On Sound*, July.

12 Massey, H. (2000). *Behind The Glass: Top Record Producers Tell How They Craft The Hits* (Vol. I). Miller Freeman Books.

13 Stavrou, M.P. (2003). *Mixing With Your Mind*. Flux Research.

14 Hatschek, K. (2006). *The Golden Moment: Recording Secrets From The Pros*. Backbeat Books.

15 Buskin, R. (2004). Classic Tracks: Bill Price. The Sex Pistols: "Anarchy In The UK." *Sound On Sound*, September.

16 Buskin, R. (2005). Classic Tracks: Steve Churchyard. The Pretenders: "Back On The Chain Gang." *Sound On Sound*, September.

17 Saxon, J. (2011). Dave Jerden: Really Wrong Productions, Handle With Care. *Tape Op*, November/December.

18 Sokal, R. (2000). The Db Of Dave Bottrill. *Tape Op*, September/October.

19 Massey, H. (2009). *Behind The Glass: Top Record Producers Tell How They Craft The Hits* (Vol. II). Backbeat Books. And Tingen, P. (2010). Paul Worley Producing Lady Antebellum. *Sound On Sound*, September.

20 Owsinski, B. (2010). *The Music Producer's Handbook*. Hal Leonard.

21 Owsinski, B. (2010). *The Music Producer's Handbook*. Hal Leonard.

22 Massey, H. (2009). *Behind The Glass: Top Record Producers Tell How They Craft The Hits* (Vol. II). Backbeat Books.

23 Tingen, P. (2008). Inside Track: Neal Avron. Fall Out Boy: "This Ain't A Scene, It's An Arms Race." *Sound On Sound*, January.

24 Owsinski, B. (2010). *The Music Producer's Handbook*. Hal Leonard.

25 Doyle, T. (2011). James Ford: Producing Arctic Monkeys. *Sound On Sound*, July.

26 Tingen, P. (2011). Inside Track: Tom Syrowski. Incubus: "Adolescents." *Sound On Sound*, November.

27 Tingen, P. (2011) Inside Track: Gus Oberg. The Strokes: *Angles*. *Sound On Sound*, June.

28 Wetherbee, P. (2013). Robert Musso: Working With Laswell And More. *Tape Op*, January.

29 Owsinski, B. (2009). *The Recording Engineer's Handbook*. Cengage Learning PTR.

30 Owsinski, B. (2009). *The Recording Engineer's Handbook*. Cengage Learning PTR.

31 Massey, H. (2009). *Behind The Glass: Top Record Producers Tell How They Craft The Hits* (Vol. II). Backbeat Books.

32 Senior, M. (2000). Bird & Bush: Producing Stereophonics. *Sound On Sound*, March.

33 Droney, M. (2001). Interview with Armin Steiner. *Mix*, May.

34 Droney, M. (2001). Interview with Frank Filipetti. *Mix*, March.

35 Savona, A. (2005). *Console Confessions: The Great Producers In Their Own Words*. Backbeat Books.

36 Robinson, A. (1997). Interview With Nigel Godrich. *The Mix*, August.

37 Hatschek, K. (2006). *The Golden Moment: Recording Secrets From The Pros*. Backbeat Books.

38 Buskin, R. (2003). Classic Tracks: Malcolm Addey. Cliff Richard: "Move It." *Sound On Sound*, November.

39 Stevenson, P. (2004). Andy Johns Archival Interview. *Tape Op*, January.

40 Tingen, P. (2005). Steve Albini: Sound Engineer Extraordinaire. *Sound On Sound*, September.

41 Massey, H. (2000). *Behind The Glass: Top Record Producers Tell How They Craft The Hits* (Vol. I). Miller Freeman Books.

42 Daley, D. (2004). Bones Howe & Tom Waits: The Odd Couple? *Sound On Sound*, February.

43 Tingen, P. (1994). Robbie Adams: Recording U2's *Achtung Baby* & *Zooropa*. *Sound On Sound*, March.

44 Buskin, R. (2003). Classic Tracks: Bruce Botnick. The Doors: "Strange Days." *Sound On Sound*, December. And Baccigaluppi, J., Crane L., Hiller, J. (2009) Bruce Botnick: A Perception Of The Doors And Much More. *Tape Op*, November.

45 Bruce, B. (1999). Al Stone: Recording Jamiroquai's "Supersonic." *Sound On Sound*, December.

46 Flint, T. (2001). Tony Platt: Rock Island Life. *Sound On Sound*, April.

47 Sillitoe, S. (1999). Mike Hedges: Recording Travis' "Why Does It Always Rain On Me?" *Sound On Sound*, November.

48 Crane, L. (2012). Interview with Jacquire King. *Tape Op*, March/April.

49 Massey, H. (2000). *Behind The Glass: Top Record Producers Tell How They Craft The Hits* (Vol. I). Miller Freeman Books.

50 Buskin, R. (2005). Classic Tracks: Peter Henderson. Supertramp: "Logical Song." *Sound On Sound*, July.

51 Droney, M. (2001). Interview with David Thoener. *Mix*, January.

52 Owsinski, B. (2009). *The Recording Engineer's Handbook*. Cengage Learning PTR.

53 Massey, H. (2000). *Behind The Glass: Top Record Producers Tell How They Craft The Hits* (Vol. I). Miller Freeman Books.

54 Owsinski, B. (2009). *The Recording Engineer's Handbook*. Cengage Learning PTR.

55 Tingen, P. (2011). Daniel Lanois & Mark Howard: Recording Neil Young's *Le Noise*. *Sound On Sound*, February.

56 Sarafin, E. ("Mixerman") (2012). *Zen And The Art Of Producing*. Hal Leonard.

57 Recording seminar at British Grove Studios, May 2010.

58 Droney, M. (1999). Interview with Leslie Ann Jones. *Mix*, August.

59 Crane, L. & Smith, C. (1999). Interview with Joe Chiccarelli. *Tape Op*, Fall.

60 Baccigaluppi, J., Crane L., & Hiller, J. (2009). Bruce Botnick: A Perception Of The Doors And Much More. *Tape Op*, November.

61 Clark, R. (2011). *Mixing, Recording, And Producing Techniques Of The Pros*. Cengage Learning PTR.

62 Droney, M. (2003). Interview with Elliott Scheiner. *Mix*, January.

63 Sarafin, E. ("Mixerman") (2012). *Zen And The Art Of Producing*. Hal Leonard.

64 Massey, H. (2009). *Behind The Glass: Top Record Producers Tell How They Craft The Hits* (Vol. II). Backbeat Books.

65 Massey, H. (2009). *Behind The Glass: Top Record Producers Tell How They Craft The Hits* (Vol. II). Backbeat Books.

CHAPTER 11

1 Doyle, T. (2008). Tony Hoffer: Producing Belle & Sebastian, The Fratellis & The Kooks. *Sound On Sound*, May.

2 Massey, H. (2000). *Behind The Glass: Top Record Producers Tell How They Craft The Hits* (Vol. II). Miller Freeman Books.

3 Tingen, P. (2010). Phil Thornalley: From Rock Producer To Pop Songwriter. *Sound On Sound*, June.

4 Buskin, R. (2012). Inside Track: Tommy LiPuma. Paul McCartney: *Kisses On The Bottom. Sound On Sound*, May.

5 Crane, L. (1999). Interview with Jack Endino. *Tape Op*, Summer.

6 Hatschek, K. (2006). *The Golden Moment: Recording Secrets From The Pros*. Backbeat Books.

7 Owings, H. (1999). Interview with David Barbe. *Tape Op*, Fall.

8 Tingen, P. (2000). Pierre Marchand: Producing Sarah McLachlan. *Sound On Sound*, March.

9 Senior, M. (2000). Bird & Bush: Producing Stereophonics. *Sound On Sound*, March.

10 Massey, H. (2000). *Behind The Glass: Top Record Producers Tell How They Craft The Hits* (Vol. I). Miller Freeman Books.

11 Massey, H. (2009). *Behind The Glass: Top Record Producers Tell How They Craft The Hits* (Vol. II). Backbeat Books.

12 Massey, H. (2009). *Behind The Glass: Top Record Producers Tell How They Craft The Hits* (Vol. II). Backbeat Books.

13 Massey, H. (2000). *Behind The Glass: Top Record Producers Tell How They Craft The Hits* (Vol. I). Miller Freeman Books.

14 Massey, H. (2009). *Behind The Glass: Top Record Producers Tell How They Craft The Hits* (Vol. II). Backbeat Books.

15 Tingen, P. (2011). Inside Track: Nathan Chapman & Justin Niebank. Taylor Swift: *Speak Now. Sound On Sound*, February.

16 Tingen, P. (2011). Inside Track: Tom Elmhirst. Adele: "Rolling In The Deep." *Sound On Sound*, September.

17 Savona, A. (2005). *Console Confessions: The Great Producers In Their Own Words*. Backbeat Books.

18 Owsinski, B. (2009). *The Recording Engineer's Handbook*. Cengage Learning PTR.

19 Frost, M. (2012). Andrija Tokic: Recording Alabama Shakes' *Boys & Girls. Sound On Sound*, July.

20 Frost, M. (2011). Rodaidh McDonald: Recording The XX. *Sound On Sound*, June.

21 Sarafin, E. ("Mixerman") (2012). *Zen And The Art Of Producing*. Hal Leonard.

22 Owsinski, B. (2010). *The Music Producer's Handbook*. Hal Leonard.

23 Senior, M. (2000). Bird & Bush: Producing Stereophonics. *Sound On Sound*, March.

24 Savona, A. (2005). *Console Confessions: The Great Producers In Their Own Words*. Backbeat Books.

25 Savona, A. (2005). *Console Confessions: The Great Producers In Their Own Words*. Backbeat Books.

26 Senior, M. (2000). Bird & Bush: Producing Stereophonics. *Sound On Sound*, March.

27 Owsinski, B. (2010). *The Music Producer's Handbook*. Hal Leonard.

28 Savona, A. (2005). *Console Confessions: The Great Producers In Their Own Words*. Backbeat Books.

29 Tingen, P. (2008). Inside Track: Robert Carranza. Jack Johnson: *Sleep Through The Static. Sound On Sound*, May.

30 Massey, H. (2009). *Behind The Glass: Top Record Producers Tell How They Craft The Hits* (Vol. II). Backbeat Books.

31 Tingen, P. (2011). Inside Track: Nathan Chapman & Justin Niebank. Taylor Swift: *Speak Now. Sound On Sound*, February.

32 Tingen, P. (2008). Inside Track: Jacquire King. Kings Of Leon: "Sex On Fire." *Sound On Sound*, December.

33 Tingen, P. (2007). Inside Track: Joe Chiccarelli. The White Stripes: *Icky Thump. Sound On Sound*, October.

34 Tingen, P. (2012). Inside Track: Noah "40" Shebib. Drake: "Headlines." *Sound On Sound*, March.

35 Savona, A. (2005). *Console Confessions: The Great Producers In Their Own Words*. Backbeat Books.

36 Crane, L. (2013). Ken Caillat: Making History And Rumours. *Tape Op*, July/August.

37 Tingen, P. (2008). Inside Track: Robert Carranza. Jack Johnson: *Sleep Through The Static*. *Sound On Sound*, May.

38 Tingen, P. (2008). Inside Track: Jacquire King. Kings Of Leon: "Sex On Fire." *Sound On Sound*, December.

39 Tingen, P. (2011). Inside Track: Philippe Zdar. Beastie Boys: "Make Some Noise." *Sound On Sound*, August.

40 Frost, M. (2010). Mike Vernon: Producing British Blues. *Sound On Sound*, December.

41 Robinson, A. (1997). Interview with Nigel Godrich. *The Mix*, August.

42 Tingen, P. (2010) Inside Track: Mike Poole. Robert Plant: "Angel Dance." *Sound On Sound*, December.

43 Tingen, P. (2010). Inside Track: Mike Poole. Robert Plant: "Angel Dance." *Sound On Sound*, December.

44 Senior, M. (2003). Glen Ballard. Songwriter & Producer: Alanis Morissette. *Sound On Sound*, March.

45 Tingen, P. (2012). Inside Track: Randy Staub. Evanescence: "What You Want." *Sound On Sound*, January.

46 Tingen, P. (2012). Inside Track: Randy Staub. Evanescence: "What You Want." *Sound On Sound*, January.

47 Tingen, P. (2010). Inside Track: David R. Ferguson. Johnny Cash: "Ain't No Grave." *Sound On Sound*, June.

48 Tingen, P. (2009). Inside Track: Fraser T Smith. Tinchy Stryder: "Number One." *Sound On Sound*, November.

49 Tingen, P. (2008). Inside Track: Michael Brauer. Coldplay: "Viva La Vida." *Sound On Sound*, November.

50 Doyle, T. (2008). Tony Hoffer: Producing Belle & Sebastian, The Fratellis & The Kooks. *Sound On Sound*, May.

CHAPTER 12

1 Tingen, P. (2012). Inside Track: Randy Staub: 'What You Want.' *Sound On Sound*, January.

2 Droney, M. (1998). Interview With Dave Jerden. *Mix*, December.

3 Hatschek, K. (2006). *The Golden Moment: Recording Secrets From The Pros*. Backbeat Books.

4 Massey, H. (2009). *Behind The Glass: Top Record Producers Tell How They Craft The Hits* (Vol. II). Backbeat Books.

5 Massey, H. (2009). *Behind The Glass: Top Record Producers Tell How They Craft The Hits* (Vol. II). Backbeat Books.

6 Droney, M. (1998). Interview with Al Schmitt. *Mix*, October.

7 Massey, H. (2009). *Behind The Glass: Top Record Producers Tell How They Craft The Hits* (Vol. II). Backbeat Books.

8 Massey, H. (2009). *Behind The Glass: Top Record Producers Tell How They Craft The Hits* (Vol. II). Backbeat Books.

9 Massey, H. (2009). *Behind The Glass: Top Record Producers Tell How They Craft The Hits* (Vol. II). Backbeat Books.

10 Faulhaber, H.D. (2012). John Hampton: You Call This Work? *Tape Op*, November.

APPENDIX 4

Picture Credits

CHAPTER 3

Figure 3.1: Image courtesy of Radial Engineering®.
Figure 3.2: Images courtesy of Palmer® and L.R. Baggs®.
Figure 3.6: Image courtesy of Palmer®.

CHAPTER 4

Figure 4.1: Image courtesy of AKG®.
Figure 4.9: Microphone specifications courtesy of the Microphone Data archive (www.microphone-data.com).

CHAPTER 5

Figure 5.2: Image courtesy of Neumann®. Microphone specifications courtesy of the Microphone Data archive (www.microphone-data.com).
Figure 5.3: Images courtesy of Shure®, Beyerdynamic®, and Electrovoice®.
Figure 5.4: Images courtesy of Sontronics®, Superlux®, and Cloud®.
Figure 5.7: Image courtesy of DAP Audio®.

CHAPTER 6

Figure 6.3: Image courtesy of Stewis Media (www.stewismedia.com).
Figure 6.4: Image courtesy of *Sound On Sound*® magazine (www.soundonsound.com).
Figure 6.5: Image courtesy of Stewis Media (www.stewismedia.com).
Figure 6.6: Image courtesy of *Sound On Sound*® magazine (www.soundonsound.com).
Figure 6.7: Image courtesy of Stewis Media (www.stewismedia.com).
Figure 6.9: Image courtesy of *Sound On Sound*® magazine (www.soundonsound.com).
Figure 6.10: Ibid.
Figure 6.12: Ibid.
Figure 6.13: Image courtesy of Cornford®.
Figure 6.19: Image courtesy of Pearl®.
Figure 6.20: Images courtesy of Superlux®, AKG®, and Samson®.
Figure 6.23: Images courtesy of Sennheiser®, Shure®, and AKG®.

CHAPTER 7

Figure 7.3: Central image courtesy of *Sound On Sound*® magazine (www.soundonsound.com).
Figure 7.6: Right-hand image courtesy of *Sound On Sound*® magazine (www.soundonsound .com).
Figure 7.8: Image courtesy of Little Labs®.
Figure 7.10: Image courtesy of Radial Engineering®.

CHAPTER 8

Figure 8.2: Images courtesy of Studio Projects®, Rode®, and Royer®.

CHAPTER 9

Figure 9.8: Image by John McBride, courtesy of Linn Records (www.linnrecords.com).
Figure 9.11: Image by Jason Williams; courtesy of Neil Rogers at Half Ton Studios, Cambridge, UK (www.halftonstudios.co.uk).
Figure 9.12: Images courtesy of AKG®, Shure®, SE Electronics®, and Cascade®. Microphone specifications courtesy of the Microphone Data archive (www.microphone-data.com).

CHAPTER 10

Figure 10.2: Image courtesy of *Sound On Sound*® magazine (www.soundonsound.com).
Figure 10.14: Image courtesy of Latch Lake®.
Figure 10.16: Image courtesy of Stewis Media (www.stewismedia.com).

Index of Names

Subject Index